BIOLOGY of the NMDA RECEPTOR

FRONTIERS IN NEUROSCIENCE

Series Editors
Sidney A. Simon, Ph.D.
Miguel A.L. Nicolelis, M.D., Ph.D.

Published Titles

Apoptosis in Neurobiology
Yusuf A. Hannun, M.D., Professor of Biomedical Research and Chairman/Department
of Biochemistry and Molecular Biology, Medical University of South Carolina,
Charleston, South Carolina
Rose-Mary Boustany, M.D., tenured Associate Professor of Pediatrics and Neurobiology,
Duke University Medical Center, Durham, North Carolina

Methods of Behavioral Analysis in Neuroscience
Jerry J. Buccafusco, Ph.D., Alzheimer's Research Center, Professor of Pharmacology
and Toxicology, Professor of Psychiatry and Health Behavior,
Medical College of Georgia, Augusta, Georgia

Neural Prostheses for Restoration of Sensory and Motor Function
John K. Chapin, Ph.D., Professor of Physiology and Pharmacology, State University
of New York Health Science Center, Brooklyn, New York
Karen A. Moxon, Ph.D., Assistant Professor/School of Biomedical Engineering, Science,
and Health Systems, Drexel University, Philadelphia, Pennsylvania

Computational Neuroscience: Realistic Modeling for Experimentalists
Eric DeSchutter, M.D., Ph.D., Professor/Department of Medicine, University of Antwerp,
Antwerp, Belgium

Methods in Pain Research
Lawrence Kruger, Ph.D., Professor of Neurobiology (Emeritus), UCLA School of Medicine
and Brain Research Institute, Los Angeles, California

Motor Neurobiology of the Spinal Cord
Timothy C. Cope, Ph.D., Professor of Physiology, Wright State University, Dayton, Ohio

Nicotinic Receptors in the Nervous System
Edward D. Levin, Ph.D., Associate Professor/Department of Psychiatry and Pharmacology
and Molecular Cancer Biology and Department of Psychiatry and Behavioral Sciences,
Duke University School of Medicine, Durham, North Carolina

Methods in Genomic Neuroscience
Helmin R. Chin, Ph.D., Genetics Research Branch, NIMH, NIH, Bethesda, Maryland
Steven O. Moldin, Ph.D., University of Southern California, Washington, D.C.

Methods in Chemosensory Research
Sidney A. Simon, Ph.D., Professor of Neurobiology, Biomedical Engineering,
and Anesthesiology, Duke University, Durham, North Carolina
Miguel A.L. Nicolelis, M.D., Ph.D., Professor of Neurobiology and Biomedical Engineering,
Duke University, Durham, North Carolina

The Somatosensory System: Deciphering the Brain's Own Body Image
Randall J. Nelson, Ph.D., Professor of Anatomy and Neurobiology,
    University of Tennessee Health Sciences Center, Memphis, Tennessee

The Superior Colliculus: New Approaches for Studying Sensorimotor Integration
William C. Hall, Ph.D., Department of Neuroscience, Duke University,
    Durham, North Carolina
Adonis Moschovakis, Ph.D., Department of Basic Sciences, University of Crete,
    Heraklion, Greece

New Concepts in Cerebral Ischemia
Rick C. S. Lin, Ph.D., Professor of Anatomy, University of Mississippi Medical Center,
    Jackson, Mississippi

DNA Arrays: Technologies and Experimental Strategies
Elena Grigorenko, Ph.D., Technology Development Group, Millennium Pharmaceuticals,
    Cambridge, Massachusetts

Methods for Alcohol-Related Neuroscience Research
Yuan Liu, Ph.D., National Institute of Neurological Disorders and Stroke,
    National Institutes of Health, Bethesda, Maryland
David M. Lovinger, Ph.D., Laboratory of Integrative Neuroscience, NIAAA,
    Nashville, Tennessee

***In Vivo* Optical Imaging of Brain Function**
Ron Frostig, Ph.D., Associate Professor/Department of Psychobiology,
    University of California, Irvine, California

Primate Audition: Behavior and Neurobiology
Asif A. Ghazanfar, Ph.D., Princeton University, Princeton, New Jersey

Methods in Drug Abuse Research: Cellular and Circuit Level Analyses
Dr. Barry D. Waterhouse, Ph.D., MCP-Hahnemann University, Philadelphia, Pennsylvania

Functional and Neural Mechanisms of Interval Timing
Warren H. Meck, Ph.D., Professor of Psychology, Duke University, Durham, North Carolina

Biomedical Imaging in Experimental Neuroscience
Nick Van Bruggen, Ph.D., Department of Neuroscience Genentech, Inc.
Timothy P.L. Roberts, Ph.D., Associate Professor, University of Toronto, Canada

The Primate Visual System
John H. Kaas, Department of Psychology, Vanderbilt University
Christine Collins, Department of Psychology, Vanderbilt University, Nashville, Tennessee

Neurosteroid Effects in the Central Nervous System
Sheryl S. Smith, Ph.D., Department of Physiology, SUNY Health Science Center,
    Brooklyn, New York

Modern Neurosurgery: Clinical Translation of Neuroscience Advances
Dennis A. Turner, Department of Surgery, Division of Neurosurgery,
    Duke University Medical Center, Durham, North Carolina

Sleep: Circuits and Functions
Pierre-Hervé Luoou, Université Claude Bernard Lyon, France

Methods in Insect Sensory Neuroscience
Thomas A. Christensen, Arizona Research Laboratories, Division of Neurobiology, University of Arizona, Tuscon, Arizona

Motor Cortex in Voluntary Movements
Alexa Riehle, INCM-CNRS, Marseille, France
Eilon Vaadia, The Hebrew University, Jerusalem, Israel

Neural Plasticity in Adult Somatic Sensory-Motor Systems
Ford F. Ebner, Vanderbilt University, Nashville, Tennessee

Advances in Vagal Afferent Neurobiology
Bradley J. Undem, Johns Hopkins Asthma Center, Baltimore, Maryland
Daniel Weinreich, University of Maryland, Baltimore, Maryland

The Dynamic Synapse: Molecular Methods in Ionotropic Receptor Biology
Josef T. Kittler, University College, London, England
Stephen J. Moss, University College, London, England

Animal Models of Cognitive Impairment
Edward D. Levin, Duke University Medical Center, Durham, North Carolina
Jerry J. Buccafusco, Medical College of Georgia, Augusta, Georgia

The Role of the Nucleus of the Solitary Tract in Gustatory Processing
Robert M. Bradley, University of Michigan, Ann Arbor, Michigan

Brain Aging: Models, Methods, and Mechanisms
David R. Riddle, Wake Forest University, Winston-Salem, North Carolina

Neural Plasticity and Memory: From Genes to Brain Imaging
Frederico Bermudez-Rattoni, National University of Mexico, Mexico City, Mexico

Serotonin Receptors in Neurobiology
Amitabha Chattopadhyay, Center for Cellular and Molecular Biology, Hyderabad, India

Methods for Neural Ensemble Recordings, Second Edition
Miguel A.L. Nicolelis, M.D., Ph.D., Professor of Neurobiology and Biomedical Engineering, Duke University Medical Center, Durham, North Carolina

Biology of the NMDA Receptor
Antonius M. VanDongen, Duke University Medical Center, Durham, North Carolina

BIOLOGY of the NMDA RECEPTOR

Edited by

Antonius M. VanDongen
Duke University Medical Center
North Carolina

CRC Press
Taylor & Francis Group
Boca Raton London New York

CRC Press is an imprint of the
Taylor & Francis Group, an **informa** business

CRC Press
Taylor & Francis Group
6000 Broken Sound Parkway NW, Suite 300
Boca Raton, FL 33487-2742

First issued in paperback 2019

© 2009 by Taylor & Francis Group, LLC
CRC Press is an imprint of Taylor & Francis Group, an Informa business

No claim to original U.S. Government works

ISBN-13: 978-1-4200-4414-0 (hbk)
ISBN-13: 978-0-367-38652-8 (hbk)

Library of Congress Cataloging-in-Publication Data

Biology of the NMDA receptor / editor, Antonius M. VanDongen.
    p. ; cm. -- (Frontiers in neuroscience)
  Includes bibliographical references and index.
  ISBN 978-1-4200-4414-0 (alk. paper)
    1. Methyl aspartate--Receptors. I. VanDongen, Antonius M. II. Title. III. Series.
    [DNLM: 1. Receptors, N-Methyl-D-Aspartate. WL 102.8 B6528 2008]

QP364.7.B564 2008
612.8--dc22
    2008007233

Visit the Taylor & Francis Web site at
http://www.taylorandfrancis.com

and the CRC Press Web site at
http://www.crcpress.com

Contents

Series Preface

Our goal in creating the Frontiers in Neuroscience Series is to present the insights of experts on emerging fields and theoretical concepts that are, or will be, in the vanguard of neuroscience. Books in the series cover genetics, ion channels, apoptosis, electrodes, neural ensemble recordings in behaving animals, and even robotics. The series also covers new and exciting multidisciplinary areas of brain research, such as computational neuroscience and neuroengineering, and describes breakthroughs in classical fields like behavioral neuroscience. We hope every neuroscientist will use these books to become acquainted with new ideas and frontiers in brain research. These books can be given to graduate students and postdoctoral fellows when they are looking for guidance to start a new line of research.

Each book is edited by an expert and consists of chapters written by the leaders in a particular field. Books are richly illustrated and contain comprehensive bibliographies. Chapters provide substantial background material relevant to the particular subject. We hope that as the volumes become available, the efforts of the publisher, the book editors, and individual authors will contribute to the further development of brain research. The extent to which we achieve this goal will be determined by the utility of these books.

Sidney A. Simon, Ph.D.
Miguel A.L. Nicolelis, M.D., Ph.D.
Series Editors

Preface

The subject of this book is the NMDA receptor (NMDAR), a glutamate-gated cation channel that plays myriad roles in the biology and pathophysiology of higher organisms, from fruit flies to humans. The NMDAR is critical for setting up the correct neuronal wiring diagram during brain development, by preventing the elimination of properly functioning synapses[1] and neurons.[2] Starting at birth, NMDARs are involved in generating rhythms, repetitive patterns of burst firing, that organisms use for very basic processes, including breathing and locomotion. The most widely studied aspect of NMDAR function is, however, the role it plays in supporting activity-dependent synaptic plasticity underlying processes of learning and memory. In addition, NMDAR function is important for many higher cognitive brain functions including fear, anxiety, attention, mood, and cognition.

NMDARs belong to the class of ionotropic glutamate receptors (iGluRs),[3] ion channel receptors activated by the excitatory amino acid L-glutamate.[4] They are pharmacologically distinguished from other glutamate receptors (AMPA, kainite, and delta receptors) by their sensitivity to the specific synthetic agonist N-methyl-D-aspartate (NMDA), discovered in the early 1960s by Curtis and Watkins[5,6] before L-glutamate was recognized as a *bona fide* neurotransmitter in the mammalian nervous system.

Development of radioligand binding assays and specific antagonists during the 1970s led to the idea that several distinct excitatory amino acid receptors existed in the mammalian brain,[7] one of which was specifically activated by NMDA. Cloning of the NMDAR NR1 subunit in 1991[8] and subsequent identification of four genes encoding different NR2 subunits[9] confirmed the existence of separate but related gene families for NMDA and non-NMDA glutamate receptors, and heralded an era of detailed molecular and cellular investigations into the many roles of NMDARs in brain (dys)function.

Important lessons about the various roles of NMDARs in brain function have been learned from recombinant mouse models. The importance of NMDARs for processes involving rhythm generation in the central nervous system is illustrated clearly by the phenotype of the NR1 knockout mouse, which dies hours after birth due to an inability to breathe or suckle.[10] Apparently, NMDARs were employed for rhythm generation very early in evolution. Experiments in *Caenorhabditis elegans* worms have shown that the NMR-1 subunit, a homologue of the mammalian NR1 subunit, is required for slow NMDA-activated currents in neurons that regulate reversal frequency.[11]

Despite the obvious importance of rhythms for the sustenance of life, this aspect of NMDAR function remains poorly studied. Recombinant mice studies have confirmed the importance of NMDAR function for memory formation and consolidation. Overexpression of the immature NR2B subunit, which is highly expressed in developing animals following insertion of an extra copy of the GRIN2B gene in a transgenic mouse, has been reported to enhance learning and memory performance in both adult[12] and aged animals.[13] The role of NMDAR subtypes in specific forms of memory is currently under investigation using mice in which the expression levels of NMDAR genes are altered in specific brain regions and during defined periods.[14–17] Novel functions of NMDARs in memory maintenance are still being discovered. Mice in which NR1 expression is controlled by an inducible, reversible, and region-specific knockout provided evidence for a role of NMDAR reactivation in long-term memory consolidation.[18]

This book covers many aspects of the biology of NMDARs: their role in controlling structure and function of synapses and neurons during early development; how overstimulation of NMDARs results in excitotoxicity and contributes to several progressive brain disorders, including Huntington's disease; the newly discovered and intriguing interactions of NMDARs and dopamine receptors that mediate reward in the central nervous system; the role of NMDARs in alcohol dependence and the promise of NMDAR-based therapeutics for treating alcoholism; how functional expression is controlled at the level of gene transcription by several families of transcription factors; how NMDAR activation regulates local synaptic protein synthesis required for long-term changes in synaptic strength; the modulation of NMDAR function by signaling cascades cumulating in activation of protein kinases and phosphatases; the importance of cellular mechanisms underlying trafficking and targeting of NMDAR protein for many of its physiological functions; how NMDAR-mediated calcium signaling in dendritic spines controls synaptic efficacy and spine morphology; the roles NMDARs play in different temporal phases of memory formation in *Drosophila*; the extracellular modulation of NMDARs by polyamines, subunit-specific inhibitors, zinc ions, and pH, and the structural bases for their effects; a detailed description of NMDAR pharmacology, structure–activity relationships of agonists and antagonists, and roads to therapeutic drug design; the physiological roles played by NMDARs and their molecular structures; NMDAR activation mechanisms and the therapeutic potential of allosteric modulators; and the novel role of presynaptically localized NMDARs in controlling synaptic plasticity.

An average of 350 papers have appeared annually on the subject of NMDARs since 1994, totaling approximately 5800 research reports by the end of 2007. Although it is impossible to deal with all aspects of NMDAR biology in a single book this size, we have attempted to cover a wide variety of topics and levels of description, from human disease and brain plasticity to gene promoters and X-ray protein structure, with emphasis on cellular and molecular mechanisms. It is hoped that bringing together all these vantage points in a single volume will encourage cross-fertilization among the different disciplines, resulting in a deeper understanding of the hierarchy of processes affected by NMDAR activation and deregulation, from synaptic strengthening to regulating higher cognitive processes. A more complete understanding of all aspects of NMDAR biology may also result in the development of successful therapeutic approaches targeting the NMDAR for the many acute and chronic brain disorders in which the receptor is deregulated.

Antonius M.J. VanDongen

REFERENCES

1. Cline, H.T. and Constantine-Paton, M., NMDAR antagonists disrupt the retinotectal topographic map, *Neuron*, 3, 413, 1989.
2. Ikonomidou, C. et al., Blockade of NMDARs and apoptotic neurodegeneration in the developing brain, *Science*, 283, 70, 1999.
3. Dingledine, R. et al., The glutamate receptor ion channels., *Pharmacol. Rev.*, 51, 7, 1999.
4. Hayashi, T., The effect of sodium glutamate on the nervous system, *Keio J. Med.*, 3, 183, 1954.
5. Curtis, D.R. and Watkins, J.C., Analogues of glutamic and gamma-amino-n-butyric acids having potent actions on mammalian neurones, *Nature*, 191, 1010, 1961.
6. Curtis, D.R. and Watkins, J.C., Acidic amino acids with strong excitatory actions on mammalian neurones, *J. Physiol.*, 166, 1, 1963.
7. McLennan, H., Hicks, T.P., and Hall, J.G., Receptors for the excitatory amino acids, *Adv. Biochem. Psychopharmacol.*, 29, 213, 1981.
8. Moriyoshi, K. et al., Molecular cloning and characterization of the rat NMDAR, *Nature*, 354, 31, 1991.
9. Monyer, H. et al., Heteromeric NMDARs: Molecular and functional distinction of subtypes, *Science*, 256, 1217, 1992.
10. Forrest, D. et al., Targeted disruption of NMDAR-1 gene abolishes NMDA response and results in neonatal death, *Neuron*, 13, 325, 1994.
11. Brockie, P.J. et al., The *C. elegans* glutamate receptor subunit NMR-1 is required for slow NMDA-activated currents that regulate reversal frequency during locomotion, *Neuron*, 31, 617, 2001.
12. Tang, Y.P. et al., Genetic enhancement of learning and memory in mice, *Nature*, 401, 63, 1999.
13. Cao, X. et al., Maintenance of superior learning and memory function in NR2B transgenic mice during ageing, *Eur. J. Neurosci.*, 25, 1815, 2007.
14. Cui, Z. et al., Inducible and reversible NR1 knockout reveals crucial role of the NMDAR in preserving remote memories in the brain, *Neuron*, 41, 781, 2004.
15. Rondi-Reig, L. et al., Impaired sequential egocentric and allocentric memories in forebrain-specific-NMDAR knock-out mice during a new task dissociating strategies of navigation, *J. Neurosci.*, 26, 4071, 2006.
16. McHugh, T.J. et al., Dentate gyrus NMDARs mediate rapid pattern separation in the hippocampal network, *Science*, 317, 94, 2007.
17. Niewoehner, B. et al., Impaired spatial working memory but spared spatial reference memory following functional loss of NMDARs in the dentate gyrus, *Eur. J. Neurosci.*, 25, 837, 2007.
18. Wang, H., Hu, Y., and Tsien, J.Z., Molecular and systems mechanisms of memory consolidation and storage, *Prog. Neurobiol.*, 79, 123, 2006.

About the Editor

Antonius VanDongen is an associate professor in the Department of Pharmacology of Duke University and is currently located at the Duke–NUS Graduate Medical School in Singapore. His current research interest is molecular and cellular aspects of NMDAR function.

Contributors

Rana A. Al-Hallaq
Laboratory of Neurochemistry
NIDCD
National Institutes of Health
Bethesda, Maryland, USA

Véronique M. André
Mental Retardation Research Center
David Geffen School of Medicine
University of California
Los Angeles, California, USA

Guang Bai
Department of Biomedical Sciences
Program in Neuroscience
University of Maryland Dental School
Baltimore, Maryland, USA

Marie L. Blanke
Department of Pharmacology
Duke University Medical Center
Durham, North Carolina, USA

Brenda L. Bloodgood
Department of Neurobiology
Harvard Medical School
Boston, Massachusetts, USA

Carlos Cepeda
Mental Retardation Research Center
David Geffen School of Medicine
University of California
Los Angeles, California, USA

Ann-Shyn Chiang
Institute of Biotechnology
Department of Life Science
National Tsing Hua University
Hsinchu, Taiwan

Hollis T. Cline
Watson School of Biological Sciences
and
Cold Spring Harbor Laboratory
Cold Spring Harbor, New York, USA

Yina Dong
Program in Neurosciences and Mental
  Health
The Hospital for Sick Children
and
Department of Physiology
University of Toronto
Toronto, Ontario, Canada

Ian C. Duguid
Wolfson Institute for Biomedical
  Research
University College London
London, England, U.K.

Rebecca C. Ewald
Watson School of Biological Sciences
and
Cold Spring Harbor Laboratory
Cold Spring Harbor, New York, USA

Herman B. Fernandes
Department of Psychiatry
Brain Research Centre
University of British Columbia
Vancouver, British Columbia, Canada

Charles A. Hoeffer
Center for Neural Science
New York University
New York, New York, USA

Peter W. Hoffman
Department of Biology
College of Notre Dame of Maryland
Baltimore, Maryland, USA

David E. Jane
Department of Physiology and
  Pharmacology
MRC Center for Synaptic Plasticity
University of Bristol
Bristol, England, U.K.

Emily L. Jocoy
Mental Retardation Research Center
David Geffen School of Medicine
University of California
Los Angeles, California, USA

Lorraine V. Kalia
Program in Neurosciences and Mental
  Health
The Hospital for Sick Children
and
Department of Physiology
University of Toronto
Toronto, Ontario, Canada

Eric Klann
Center for Neural Science
New York University
New York, New York, USA

Michael S. Levine
Mental Retardation Research Center
David Geffen School of Medicine
University of California
Los Angeles, California, USA

Xue Jun Liu
Program in Neurosciences and Mental
  Health
The Hospital for Sick Children
and
Department of Physiology
University of Toronto
Toronto, Ontario, Canada

Daniel T. Monaghan
Department of Pharmacology and
  Experimental Neuroscience
University of Nebraska Medical Center
Omaha, Nebraska, USA

Ronald S. Petralia
Laboratory of Neurochemistry
NIDCD
National Institutes of Health
Bethesda, Maryland, USA

Graham Pitcher
Program in Neurosciences and Mental
  Health
The Hospital for Sick Children
and
Department of Physiology
University of Toronto
Toronto, Ontario, Canada

Lynn A. Raymond
Department of Psychiatry
Brain Research Centre
University of British Columbia
Vancouver, British Columbia, Canada

Dorit Ron
The Gallo Research Center
Department of Neurology
University of California
Emeryville, California, USA

Bernardo L. Sabatini
Department of Neurobiology
Harvard Medical School
Boston, Massachusetts, USA

Michael W. Salter
Program in Neurosciences and Mental
  Health
The Hospital for Sick Children
and
Department of Physiology
University of Toronto
Toronto, Ontario, Canada

Trevor G. Smart
Department of Pharmacology
University College London
London, England, U.K.

Antonius M.J. VanDongen
Duke University Medical Center
and
Duke–National University of
  Singapore
Graduate Medical School
Singapore, Singapore

Jun Wang
The Gallo Research Center
Department of Neurology
University of California
Emeryville, California, USA

Robert J. Wenthold
Laboratory of Neurochemistry
NIDCD
National Institutes of Health
Bethesda, Maryland, USA

Keith Williams
Department of Physiology and
  Pharmacology
Downstate Medical Center
State University of New York
Brooklyn, New York, USA

Shouzhen Xia
Cold Spring Harbor Laboratory
Cold Spring Harbor, New York, USA

1 NMDA Receptors and Brain Development

Rebecca C. Ewald and Hollis T. Cline

CONTENTS

1.1 INTRODUCTION

More than 20 years have passed since the discovery that NMDAR-mediated synaptic activity is important for the functional development of sensory brain circuits[1] by specifically regulating the refinement of sensory topographic maps.[2] Since then we have learned in great detail how this receptor shapes the development of neurons and neuronal circuits. The diversity of NMDAR subunits and their unique developmental and spatial expression patterns add levels of complexity to the role of the NMDARs in development that are only beginning to be understood. In this chapter, we will examine NMDARs and their role in brain development from a circuit perspective, on the single neuron level and on the synaptic level.

1

1.2 DEVELOPMENTAL EXPRESSION PATTERNS OF NMDA RECEPTORS

All members of the NMDAR family are characterized by distinct but overlapping regional and developmental expression patterns, indicating that individual populations of neurons express age-dependent idiosyncratic signatures of NMDARs. This potentially suggests particular roles for NMDAR subtypes during development and in different brain areas.[3] Individual NMDAR subunits and their splice isoforms have been well characterized and found to have unique biophysical properties and structural domains for differential interactions with protein partners.[3-5] However, we are only starting to understand whether and how these differences impact brain development.

1.2.1 NR1

In situ hybridization studies[6,7] with pan- and splice variant-specific probes of rat brains at different pre- and postnatal ages show that in general NR1 expression begins as early as E14, peaks around the third postnatal week, and then declines slightly to adult levels. This pattern is followed approximately by the individual isoforms as well. Regional expression of the splice variants does not seem to change significantly during development once the expression patterns are established around birth. The NR1 isoforms without the N-terminal N1 exon (NR1-a) are expressed homogeneously throughout the brain, while those containing the N1 cassette (NR1-b) are more restricted to specific areas such as the neonatal caudate, sensorimotor cortex, and thalamus. Similarly, the NR1-2 variant is widely and more or less homogeneously expressed throughout the brain, while the NR1-1 and NR1-4 isoforms form almost complementary patterns, with NR1-1 restricted to more rostral parts (cortex, hippocampus, caudate) and NR1-4 to more caudal parts (thalamus, cerebellum, colliculi) of the central nervous system (CNS). NR1-3 expression is the weakest: it is barely detectable at birth and restricted postnatally to very low levels in cortex and hippocampus.

Overall, the approximate abundance pattern is NR1-2 > NR1-1 > NR1-4 >> NR1-3. Studies of the developmental expression patterns of NR1 isoforms in other species are less exhaustive or focus on a narrower window of development.[8,9] Interestingly, species-specific preferences seem to exist for particular NR1 splice variants, for example in *Xenopus* and fish, in which subsets of splice variants are expressed.[10-13] The functional significance of the differential expression patterns of NR1 splice variants remains to be determined. The alternatively spliced cassettes primarily modulate the biophysical and trafficking properties of the NR1 subunit. If and how this impacts the development of neurons and circuits is still unexplored.

1.2.2 NR2

1.2.2.1 Histochemical Expression Patterns of NR2 Subunits

NR2 subunits, in contrast to the NR1 splice variants, are characterized by large differences in their electrophysiological profiles. Since they also show very distinct expression patterns during development, it has been speculated that they may play unique roles in brain development.[3] *In situ* hybridization studies of rat brain slices[14] have shown that only NR2B and NR2D are expressed during prenatal development. Expression starts at low levels as early as E14 and becomes gradually more enriched as embryonic development continues. NR2B expression predominates in prenatal cortex, is high in the thalamus and spinal cord, and is expressed at lower levels in the colliculi, hippocampus, and hypothalamus.

Expression in these areas becomes stronger around birth and persists through P0, when it can also be detected in the cerebellum, where it is the predominant subunit during early postnatal development. NR2D, in contrast, is absent from the telencephalon but abundant in the diencephalon, mesencephalon, and spinal cord. At P0, these areas continue to show strong signals for NR2D mRNA, when the transcript can also be detected at very low levels in the cortex, hippocampus, and septum. Also at P0, NR2A and NR2C start to show expression at very low levels in the hippocampus and cerebellum, respectively. During the first postnatal week, the mRNA levels of all NMDAR subtypes increase and expression of NR2D transcripts peaks around P7 before it declines to very low levels in adults.

Overall, NR2B expression levels are highest among the NR2 subunits during this early postnatal stage. NR2B expression peaks in the hippocampus and cortex during the third postnatal week and then declines to moderate/low adult levels. NR2A expression continues to increase in the hippocampus and cortex, and eventually is expressed throughout the brain during the second postnatal week when it becomes dramatically upregulated. Levels of NR2A, like NR2B, peak in the third postnatal week before they decline to adult levels. NR2C expression is very low in the cerebellum and forebrain at P7 but becomes markedly increased in the cerebellum by P12, where its expression in granule cells peaks during the third postnatal week and where it continues to be expressed at high levels in the adult as the predominant subunit.

1.2.2.2 Electrophysiological Characterization of NR2 Subunits during Development

The developmental change of the NR2 subunits observed on a cellular level has also been documented electrophysiologically on a synaptic level, particularly for NR2A- and NR2B-containing receptors. The NR2A and NR2B subunits are characterized by an overlapping but developmentally distinct expression pattern and different electrophysiological properties. NR2A and NR2B show biophysical differences in their glutamate sensitivity and deactivation kinetics,[4,14] resulting in a fast decay time for NR2A-containing receptors and a three- to four-fold slower decay time for NR2B-containing receptors when expressed recombinantly in heterologous cells.[14,15] During development, NMDAR decay kinetics shorten due to early expression of NR2B and delayed expression of NR2A later in development. This shift in NMDAR decay kinetics has been observed in many model systems, from rats to frogs, and in different brain regions.[16–20] Furthermore, NR2B-containing receptors seem to be the predominant subunits in immature neurons and at nascent synapses, while NR2A-containing receptors are gradually added as neurons mature[20–23] to become the electrophysiologically predominant subunits.[24]

Beyond NR2A and NR2B, recordings from the cerebellum have shown a developmental increase in the NR2C subunit. The decay kinetics of the NMDA response shorten and the sensitivity to the NR2B-specific antagonist ifenprodil decreases, before the Mg^{2+}-sensitivity of the NMDA response decreases, suggesting the presence of NR2B-containing receptors at first, then the addition of synaptic NR2A-containing receptors and the subsequent incorporation of NR2C-containing receptors.[25] The physiological consequences of this developmental change in subunit composition, however, remain unclear.

1.2.2.3   Interactions with Downstream Signaling Partners

The developmental decrease in NMDAR decay kinetics may lead to a difference in synaptic Ca^{2+} influx[26] that may be functionally significant since Ca^{2+} is an important activator of Ca^{2+}-dependent signal transduction cascades that in turn may cause short- and long-term changes in cells.[27,28] Furthermore, NR2A and NR2B have been reported to interact differentially with binding partners in the postsynaptic density, such as CaMKII,[29–33] small GTPases,[34–36] and other postsynaptic density proteins like nNOS, Homer, β-catenin and CRMP2,[37] hinting at the potential activation of different downstream signaling pathways according to changes in NMDAR subunit composition during development.

1.2.2.4   Distinct Functional Roles of NR2 Subunits

Many attempts have been made to find contrasting roles for the NR2A and NR2B subunits but the functional significance of the developmental increase of the NR2A subunit to complement NR2B-containing receptors at the synapse remains unclear. For example, one proposal is that the shortened decay kinetics with increasing NR2A expression play a role in the regulation of critical period plasticity.[16,17,38] However this has not been supported experimentally, since NR2A knockout mice that do not experience shortenings of NMDAR decay kinetics still have normal critical periods in barrel cortex.[39] Pharmacological evidence for opposing roles of NR2A- and NR2B-containing receptors in synaptic plasticity suggested that NR2A is necessary for long-term potentiation (LTP) and NR2B for long-term depression (LTD).[40,41] However this interpretation was complicated by the finding that the NR2A antagonist NVP-AAM007 is less specific than assumed,[42] leaving the question of the actual functions of the different NMDARs unresolved.[43,44]

In addition, theses findings directly conflict with studies in transgenic animals; for example, LTP can still be induced in NR2A knockout mice[39,45] and mice with increased expression of NR2B show enhanced LTP.[46] Other electrophysiological data show overlapping roles of NR2A and NR2B in synaptic plasticity.[47–50] Kim et al. (2005) proposed that NR2A-containing receptors promote and NR2B-containing receptors inhibit synaptic insertion of the AMPAR subunit GluR1 and linked NR2B to the weakening of synaptic strength.[34] Although these results are consistent with the Liu and Massey papers,[40,41] another study suggests that NR2B mediates synaptic potentiation via the strong association of the receptor with CaMKII.[33] The question regarding the different functional roles of NR2A and NR2B, particularly during development, remains unresolved.

1.2.2.5 Dynamic Regulation of NR2A and NR2B Subunits at Synapses

One additional concept should be considered: the finding that the ratio of the NR2A and NR2B subunits is dynamically regulated in response to changes in synaptic activity.[51–55] This view originates from the theory of a "sliding" threshold of synaptic modification that depends on the previous level of activity at a synapse.[56] This may present interesting implications for the role of NR2A- and NR2B-containing receptors during development, as these subunits may be used to fine-tune NMDAR responses of developing neurons in their changing environment.[13]

1.2.2.6 Synaptic and Extrasynaptic NMDA Receptors during Development

Before recruitment to a synapse, NMDARs can cycle in and out of the plasma membranes of dendrites via exo- and endocytosis.[57] More than 65% of synaptic NMDARs can shuttle between synaptic and extrasynaptic sites.[58] While nascent synapses are populated predominantly by NR2B-containing receptors before supplementation with NR2A-containing receptors, extrasynaptic NMDARs are predominantly NR2B-containing receptors as shown by recordings from hippocampal, cortical, and cerebellar granule cell neurons.[20–23,59,60] What determines the synaptic or extrasynaptic localization of the NMDAR and the dynamic distribution of the receptors in and out of the plasma membrane remains unclear. Functionally, synaptic and extrasynaptic NMDARs may have opposing roles with regard to regulation of the transcription factor CREB, expression of the neurotrophic factor BDNF, and survival of the neuron.[61,62]

1.2.3 NR3

NR3A and NR3B are the most recently discovered members of the NMDAR family.[63–65] When coassembled with NR1, they form excitatory glycine receptors that are Ca^{2+}-impermeable, Mg^{2+}-insensitive, and unresponsive to glutamate or NMDA.[63] NR3A must be associated with NR1 for proper cell surface expression[66] and its expression as a triheteromer with NR1 and NR2 regulates NMDAR currents.[64,65,67] NR3 subunit expression is also developmentally and spatially regulated, In rats, NR3A is expressed prenatally at least as early as E15 in the spinal cord, medulla, pons, tegmentum, hypothalamus, and thalamus. Its expression level increases to include the entire brain stem, and postnatally through adulthood also includes the hippocampus, amygdale, and parts of cortex. Expression of NR3A peaks around P8 and then rapidly decreases to adult levels by P20.[64,65,68] NR3B expression in rodents, in contrast, is restricted to motoneurons of the brain stem and spinal cord.[63,69] Its expression levels peak postnatally around P14 and stay elevated through adulthood.[70] While both NR3 subunits are clearly subject to developmental regulation, their physiological role for brain development remains to be investigated.

1.3 NMDA RECEPTORS AND STRUCTURAL DEVELOPMENT OF NEURONS

The gross anatomy of brains of all transgenic mice lacking individual NMDAR subunits appears normal, indicating that NMDARs are not essential for brain development *per se*.[71–75] Nevertheless, NMDAR-mediated synaptic activity in adult animals has been shown to promote the integration of newly born neurons into the mature circuit.[76] NR1[-/-] and NR2B[-/-] animals die shortly after birth. Their lethality has been attributed to respiratory failure and impairment of the suckling response,[72,73] indicating a requirement for NMDAR function in circuit development. All other animals lacking NR subunits including NR2A/NR2C double knockout animals[75] survive until adulthood. Nevertheless, analysis of the fine structures of axonal and dendritic branches shows that NMDARs play a major role in brain development through the control of activity-dependent map formation on a circuit-wide scale and the structural development of neurons on a single cell level.

1.3.1 NMDA Receptors and Axon Arbor Elaboration

The first evidence that postsynaptic activity plays a role in axon arbor refinement came from experiments on retinotectal projection. Blockade of NMDARs with the NMDAR-specific drugs APV and MK-801 showed that NMDARs are important for the maintenance of eye-specific stripes of retinal ganglion cell axon arbors in the dually innervated optic tectum[2] and for the precise establishment of retinal topography.[77] Further experiments in the mammalian visual system have replicated these observations,[78] indicating that NMDARs play an evolutionarily conserved role in activity-dependent sensory map formation.

NMDAR activity has also been shown to be necessary for the initial segregation of ocular dominance bands.[79] Chronic blockade of NR1 translation with antisense technology prevents the development of orientation selectivity in ferret primary visual cortex.[80] These experiments indicate that postsynaptic NMDAR activation sends a retrograde signal that regulates the stability of presynaptic contacts and axon arbor structure.

Further evidence of impairment of axonal arbor establishment and refinement has been found in the whisker systems of NR1[-/-] and NR2B[-/-] transgenic mice.[73,81–84] Peripheral and central trigeminal ganglion cells innervate the whisker pads on the snouts of rodents and map the precise arrangement of the whiskers onto the brainstem trigeminal nuclei. The arrangement is then further mapped onto the thalamus and barrel cortex. These barrel-like structures fail to form in NR1 and NR2B knockout animals in the trigeminal nuclei, and as shown only for NR1, are also impaired in cortex.[73,81–84] Analysis of the axon arbor structures of single cells of NR1 knockout or knockdown mice shows an exuberant elaboration of the axonal arbor that prevents the development of segregated patches.[82]

During topographic map formation, axonal arbors initially overlap and are subsequently refined by selective elimination of branches from inappropriate areas.[79,85] NMDAR blockade increases the dynamics of the axonal arbor,[86] suggesting that axons require postsynaptic feedback mediated by NMDARs, possibly through coincident firing and retrograde messengers[87] in order to elaborate and refine their arbors properly and subsequently stabilize branches and synapses. Indeed, NMDAR activity is required for the selective elimination of axon branches,[79] suggesting that uncorrelated signaling through the NMDAR actively promotes branch destabilization analogous to heterosynaptic LTD, while correlated activity promotes selective branch stabilization analogous to LTP.[79]

1.3.2 DENDRITIC ARBOR ELABORATION AND NMDA RECEPTORS

In vivo time lapse images of optic tectal neurons from *Xenopus laevis* tadpoles, one of the few experimental systems that permits direct observation of neuronal development in intact animals, demonstrates that the dendritic arbor of a newly born neuron goes through distinct phases of growth before it reaches its mature structure. After extension of the axon, the dendritic arbor enters a highly dynamic and rapid phase of growth and elaboration, before its growth is slowed and the arbor stabilized.[88] Dendritic arbor growth is dependent on the interplay between synaptic activity, extracellular cues, and intracellular mechanisms.[89–91]

Blockade of NMDAR activity with APV reduces dendritic growth rate[92] and the dynamics of tectal cell dendrites *in vivo*.[86] Further *in vivo* evidence from the visual system shows that synaptic activity induced by visual stimulation promotes dendritic arbor growth and is mediated by NMDARs.[93] However, other studies reported increases in dendritic spine number and dendritic branch length and number after chronic APV-treatment *in vivo* and *in vitro*.[94,95]

One resolution of these apparent differences may relate to different developmental expression and activation of intracellular signaling pathways linking NMDAR to changes in the dendritic cytoskeleton. For instance, αCaMKII is expressed at low levels in rapidly growing young neurons, and increases in expression as neurons mature. Activation of αCaMKII, a downstream kinase activated by NMDAR signaling, leads to a "stop growing" signal and to the stabilization of the dendritic arbor in mature neurons.[96] Blocking NMDAR or αCaMKII in these neurons affects dendrite branch dynamics and growth rates, but has the opposite effect in young neurons that do not yet express much αCaMKII.

With respect to further aspects of intracellular signaling, changes in the cytoskeletal architecture important for NMDAR-dependent dendrite growth and branching are in part mediated by small RhoGTPases.[93,97,98] Rac and Cdc42 activities in particular promote dendritic arbor dynamics, while increased RhoA activity inhibits dendritic arbor development. Activation of Rac and inhibition of RhoA are dependent on synaptic activity mediated by NMDARs and AMPARs. In addition, considerable cross-regulation occurs between the small RhoGTPase signaling pathways that fine tunes dendritic arbor growth in response to NMDAR-mediated synaptic activity.[93,98]

NMDARs are also important for the refinement of the dendritic arbor. The inability to form barrel-like structures, as revealed by cytochrome oxidase staining, in the rodent whisker system of NR1 or NR2B knockout mice[73,81–84] may also be attributed to the improper clustering of dendrites during the formation of sensory topographic maps.[99] This was directly observed during analysis of the dendritic arbors of single neurons with reduced NR1 that lost the orientation of their dendritic arbors toward the barrel center.[82]

Central to dendritic arbor refinement is the decision where and when to make dendritic branches. *In vivo* imaging suggests that branch formation is promoted by NMDAR-mediated synaptic activity and involves the formation of dendritic filopodia-like structures that are subsequently retracted or stabilized and converted into a dendritic branch.[100–102] Dendritic filopodia are thin, highly motile, actin-based protrusions.[100,103–105] Dendritic branches are microtubule-based structures.[106,107] Synapse formation has been shown to stabilize dendritic filopodia and contribute to their conversion into dendritic branches.[100] Similarly, retraction of branches involves disassembly of synaptic structures.[100]

These observations support the synaptotropic hypothesis that synaptic activity guides dendritic growth.[108,109] Studies of dendritic spine morphology have shown that synaptic activity mediated through glutamate receptors stabilizes spines and synapses. NMDAR activity stabilizes the actin cytoskeleton[110] and increases the F-actin contents of spines.[111] In addition, the NMDAR-mediated synaptic recruitment of AMPA receptors stabilizes and maintains spines.[112,113] Finally, NMDAR activity can activate local protein synthesis[114] and the presence of local protein synthesis machinery potentially indicates stabilization of branches.[115] Because synapse stabilization is important for branch stabilization, these findings add further evidence that NMDAR activity is important for structural rearrangements of dendrites.

1.4 NMDA RECEPTORS AND ACTIVITY-DEPENDENT DEVELOPMENT OF GLUTAMATERGIC SYNAPSES

Glutamatergic synapse development is a fundamental part of proper brain development involving interactive signaling of the three main ionotropic receptors: $GABA_A$, NMDA, and AMPA. While many factors contribute to synapse development, synaptic activity is an important component of this process.[116] Early in development and at nascent synapses, glutamatergic synaptic transmission seems largely mediated by NMDARs.[21,117–121] Due to blockade of the NMDAR pore by Mg^{2+} ions and the initial lack of AMPARs, these synapses do not have significant ion conductance near the resting potential at hyperpolarized potentials and hence are known as "silent synapses."[121,122]

Early in development $GABA_A$-mediated synaptic transmission is excitatory rather than inhibitory due to delayed expression of the KCC2 transporter and a high concentration of internal Cl^-.[123,124] The depolarizing GABAergic conductance seems to facilitate the activation of NMDARs at nascent synapses by enabling the removal of the Mg^{2+} ion block of the NMDAR pore.[125,126] Such synergistic signaling permits further maturation of the glutamatergic synapse.[126]

NMDARs at nascent synapses are primarily composed of NR2B subunits[21] that are constitutively trafficked to synapses.[127] The subsequent delivery of NR2A-containing receptors, in contrast, is dependent on synaptic activity.[127] As described earlier, the addition of NR2A-containing NMDARs to synapses is a further indication of a maturing nervous system.[16–20] NMDAR-mediated synaptic transmission promotes the insertion of AMPARs—heterotetramers composed most commonly of GluR1 and GluR2 or GluR2 and GluR3 subunits.[113] Depending on the level of synaptic activity, NMDARs can promote the insertion of GluR1-containing receptors into synapses to result in synaptic potentiation[128–130] or lead to the removal of GluR1- and GluR2-containing receptors from synapses to lead to synaptic depression.[131–134]

During development, acquiring AMPAR-mediated synaptic transmission is a key step toward a mature glutamatergic synapse.[119] Interestingly, impairing early depolarizing GABAergic synaptic transmission by prematurely decreasing intracellular Cl⁻ levels blocks the developmental increase in AMPAR-mediated transmission.[126] Therefore depolarizing GABAergic transmission is important for the development of AMPAR-mediated transmission and may mediate this effect on synapse maturation by facilitating NMDAR activation. This observation underscores the interdependence of the GABAergic and glutamatergic signaling pathways in synaptic development.

1.5 CONCLUDING REMARKS

NMDAR activity is important at multiple levels of brain development. On the synaptic level, changes in NMDAR subunit composition and NMDAR-mediated synaptic activity clearly contribute to the maturation of glutamatergic synapses. Importantly, synapse formation and subsequent maturation within single neurons are processes that occur throughout the lifetime of an animal and contribute to synaptic plasticity during development and learning. Structurally, on a single neuron level, NMDAR activity is crucial in refining the axonal and dendritic arbors of a developing neuron. The resulting shape of the neuron defines its functional role within the neuronal circuit and underscores how NMDARs shape brain development. Future challenging work dissecting the significance of the NMDAR subunit diversity and its functional significance for brain development will yield exciting new insights into NMDAR biology.

REFERENCES

1. Kleinschmidt, A., Bear, M.F., and Singer, W., Blockade of "NMDA" receptors disrupts experience-dependent plasticity of kitten striate cortex, *Science*, 238, 355, 1987.
2. Cline, H.T., Debski, E.A., and Constantine-Paton, M., N-methyl-D-aspartate receptor antagonist desegregates eye-specific stripes, *Proc. Natl. Acad. Sci. USA*, 84, 4342, 1987.
3. Cull-Candy, S.G. and Leszkiewicz, D.N., Role of distinct NMDAR subtypes at central synapses, *Sci STKE*, 2004, re. 16, 2004.
4. Cull-Candy, S., Brickley, S., and Farrant, M., NMDAR subunits: diversity, development and disease, *Curr. Opin. Neurobiol.*, 11, 327, 2001.
5. Wenthold, R.J. et al., Trafficking of NMDARs, *Annu. Rev. Pharmacol. Toxicol.*, 43, 335, 2003.
6. Laurie, D.J. and Seeburg, P.H., Regional and developmental heterogeneity in splicing of the rat brain NMDAR1 mRNA, *J. Neurosci.*, 14, 318e, 1994.
7. Paupard, M.C., Friedman, L.K., and Zukin, R.S., Developmental regulation and cell-specific expression of N-methyl-D-aspartate receptor splice variants in rat hippocampus, *Neuroscience*, 79, 399, 1997.
8. Lee-Rivera, I., Zarain-Herzberg, A., and Lopez-Colome, A.M., Developmental expression of N-methyl-D-aspartate glutamate receptor 1 splice variants in the chick retina, *J. Neurosci. Res.*, 73, 369, 2003.
9. Magnusson, K.R., Bai, L., and Zhao, X., The effects of aging on different C-terminal splice forms of the zeta 1 (NR1) subunit of the N-methyl-D-aspartate receptor in mice, *Brain. Res. Mol. Brain Res.*, 135, 141, 2005.
10. Soloviev, M.M. et al., Functional expression of a recombinant unitary glutamate receptor from *Xenopus*, which contains N-methyl-D-aspartate (NMDA) and non-NMDAR subunits, *J. Biol. Chem.*, 271, 32572, 1996.

11. Cox, J.A., Kucenas, S., and Voigt, M.M., Molecular characterization and embryonic expression of the family of N-methyl-D-aspartate receptor subunit genes in the zebrafish, *Dev. Dyn.*, 234, 756, 2005.

12. Dunn, R.J., Bottai, D., and Maler, L., Molecular biology of the apteronotus NMDAR NR1 subunit, *J. Exp. Biol.*, 202, 1319, 1999.

13. Ewald, R.C. et al., Roles of NR2A and NR2B in the development of dendritic arbor morphology in vivo, *J. Neurosci.*, 23, 850, 2008.

14. Monyer, H. et al., Developmental and regional expression in the rat brain and functional properties of four NMDARs, *Neuron.*, 12, 529, 1994.

15. Vicini, S. et al., Functional and pharmacological differences between recombinant N-methyl-D-aspartate receptors, *J. Neurophysiol.*, 79, 555, 1998.

16. Carmignoto, G. and Vicini, S., Activity-dependent decrease in NMDAR responses during development of the visual cortex, *Science*, 258, 1007, 1992.

17. Roberts, E.B. and Ramoa, A.S., Enhanced NR2A subunit expression and decreased NMDAR decay time at the onset of ocular dominance plasticity in the ferret, *J. Neurophysiol.*, 81, 2587, 1999.

18. Shi, J., Aamodt, S.M., and Constantine-Paton, M., Temporal correlations between functional and molecular changes in NMDARs and GABA neurotransmission in the superior colliculus, *J. Neurosci.*, 17, 6264, 1997.

19. Hestrin, S., Developmental regulation of NMDAR-mediated synaptic currents at a central synapse, *Nature*, 357, 686, 1992.

20. Cline, H.T., Wu, G.Y., and Malinow, R., In vivo development of neuronal structure and function, *Cold Spring Harb. Symp. Quant. Biol.*, 61, 95, 1996.

21. Aizenman, C.D. and Cline, H.T., Enhanced visual activity *in vivo* forms nascent synapses in the developing retinotectal projection, *J. Neurophysiol.*, 97, 2949, 2007.

22. Tovar, K.R. and Westbrook, G.L., The incorporation of NMDARs with a distinct subunit composition at nascent hippocampal synapses *in vitro*, *J. Neurosci.*, 19, 4180, 1999.

23. Stocca, G. and Vicini, S., Increased contribution of NR2A subunit to synaptic NMDARs in developing rat cortical neurons, *J. Physiol.*, 507 (Pt 1), 13, 1998.

24. Kew, J.N. et al., Developmental changes in NMDAR glycine affinity and ifenprodil sensitivity reveal three distinct populations of NMDARs in individual rat cortical neurons, *J. Neurosci.*, 18, 1935, 1998.

25. Cathala, L., Misra, C., and Cull-Candy, S., Developmental profile of the changing properties of NMDARs at cerebellar mossy fiber-granule cell synapses, *J. Neurosci.*, 20, 5899, 2000.

26. Sobczyk, A., Scheuss, V., and Svoboda, K., NMDAR subunit-dependent $[Ca^{2+}]$ signaling in individual hippocampal dendritic spines, *J. Neurosci.*, 25, 6037, 2005.

27. Ghosh, A. and Greenberg, M.E., Calcium signaling in neurons: molecular mechanisms and cellular consequences, *Science*, 268, 239, 1995.

28. Kennedy, M.B. et al., Integration of biochemical signalling in spines, *Nat. Rev. Neurosci.*, 6, 423, 2005.

29. Strack, S., McNeill, R.B., and Colbran, R.J., Mechanism and regulation of calcium/calmodulin-dependent protein kinase II targeting to the NR2B subunit of the N-methyl-D-aspartate receptor, *J. Biol. Chem.*, 275, 23798, 2000.

30. Mayadevi, M. et al., Sequence determinants on the NR2A and NR2B subunits of NMDAR responsible for specificity of phosphorylation by CaMKII, *Biochim. Biophys. Acta.*, 1598, 40, 2002.

31. Bayer, K.U. et al., Interaction with the NMDAR locks CaMKII in an active conformation, *Nature*, 411, 801, 2001.

32. Leonard, A.S. et al., Calcium/calmodulin-dependent protein kinase II is associated with the N-methyl-D-aspartate receptor, *Proc. Natl. Acad. Sci. USA*, 96, 3239, 1999.

33. Barria, A. and Malinow, R., NMDAR subunit composition controls synaptic plasticity by regulating binding to CaMKII, *Neuron.*, 48, 289, 2005.
34. Kim, M.J. et al., Differential roles of NR2A- and NR2B-containing NMDARs in Ras-ERK signaling and AMPA receptor trafficking, *Neuron*, 46, 745, 2005.
35. Krapivinsky, G. et al., The NMDAR is coupled to the ERK pathway by a direct interaction between NR2B and RasGRF1, *Neuron*, 40, 775, 2003.
36. Schubert, V., Da Silva, J.S., and Dotti, C.G., Localized recruitment and activation of RhoA underlies dendritic spine morphology in a glutamate receptor-dependent manner, *J. Cell. Biol.*, 172, 453, 2006.
37. Al-Hallaq, R.A. et al., NMDA di-heteromeric receptor populations and associated proteins in rat hippocampus, *J. Neurosci.*, 27, 8334, 2007.
38. Flint, A.C. et al., NR2A subunit expression shortens NMDAR synaptic currents in developing neocortex, *J. Neurosci.*, 17, 2469, 1997.
39. Lu, H.C., Gonzalez, E., and Crair, M.C., Barrel cortex critical period plasticity is independent of changes in NMDAR subunit composition, *Neuron*, 32, 619, 2001.
40. Liu, L. et al., Role of NMDAR subtypes in governing the direction of hippocampal synaptic plasticity, *Science*, 304, 1021, 2004.
41. Massey, P.V. et al., Differential roles of NR2A and NR2B-containing NMDARs in cortical long-term potentiation and long-term depression, *J. Neurosci.*, 24, 7821, 2004.
42. Neyton, J. and Paoletti, P., Relating NMDAR function to receptor subunit composition: limitations of the pharmacological approach, *J. Neurosci.*, 26, 1331, 2006.
43. Morishita, W. et al., Activation of NR2B-containing NMDARs is not required for NMDAR-dependent long-term depression, *Neuropharmacology*, 52, 71, 2007.
44. Bartlett, T.E. et al., Differential roles of NR2A and NR2B-containing NMDARs in LTP and LTD in the CA1 region of two-week old rat hippocampus, *Neuropharmacology*, 52, 60, 2007.
45. Kiyama, Y. et al., Increased thresholds for long-term potentiation and contextual learning in mice lacking the NMDA-type glutamate receptor epsilon1 subunit, *J. Neurosci.*, 18, 6704, 1998.
46. Tang, Y.P. et al., Genetic enhancement of learning and memory in mice, *Nature*, 401, 63, 1999.
47. Kohr, G. et al., Intracellular domains of NMDAR subtypes are determinants for long-term potentiation induction, *J. Neurosci.*, 23, 10791, 2003.
48. Toyoda, H., Zhao, M.G., and Zhuo, M., Roles of NMDAR NR2A and NR2B subtypes for long-term depression in the anterior cingulate cortex, *Eur. J. Neurosci.*, 22, 485, 2005.
49. Weitlauf, C. et al., Activation of NR2A-containing NMDARs is not obligatory for NMDAR-dependent long-term potentiation, *J. Neurosci.*, 25, 8386, 2005.
50. Berberich, S. et al., Lack of NMDAR subtype selectivity for hippocampal long-term potentiation, *J. Neurosci.*, 25, 6907, 2005.
51. Quinlan, E.M. et al., Rapid, experience-dependent expression of synaptic NMDARs in visual cortex in vivo, *Nat. Neurosci.*, 2, 352, 1999.
52. Quinlan, E.M., Olstein, D.H., and Bear, M.F., Bidirectional, experience-dependent regulation of N-methyl-D-aspartate receptor subunit composition in the rat visual cortex during postnatal development, *Proc. Natl. Acad. Sci. USA*, 96, 12876, 1999.
53. Bear, M.F., Bidirectional synaptic plasticity: from theory to reality, *Philos. Trans. R Soc. Lond. B Biol. Sci.*, 358, 649, 2003.
54. Philpot, B.D. et al., Visual experience and deprivation bidirectionally modify the composition and function of NMDARs in visual cortex, *Neuron*, 29, 157, 2001.
55. Philpot, B.D., Cho, K.K., and Bear, M.F., Obligatory role of NR2A for metaplasticity in visual cortex, *Neuron*, 53, 495, 2007.

56. Bienenstock, E.L., Cooper, L.N., and Munro, P.W., Theory for the development of neuron selectivity: orientation specificity and binocular interaction in visual cortex, *J. Neurosci.*, 2, 32, 1982.

57. Washbourne, P. et al., Cycling of NMDARs during trafficking in neurons before synapse formation, *J. Neurosci.*, 24, 8253, 2004.

58. Tovar, K.R. and Westbrook, G.L., Mobile NMDARs at hippocampal synapses, *Neuron*, 34, 255–64, 2002.

59. Rumbaugh, G. and Vicini, S., Distinct synaptic and extrasynaptic NMDARs in developing cerebellar granule neurons, *J. Neurosci.*, 19, 10603, 1999.

60. Thomas, C.G., Miller, A.J., and Westbrook, G.L., Synaptic and extrasynaptic NMDAR NR2 subunits in cultured hippocampal neurons, *J. Neurophysiol.*, 95, 1727, 2006.

61. Vanhoutte, P. and Bading, H., Opposing roles of synaptic and extrasynaptic NMDARs in neuronal calcium signalling and BDNF gene regulation, *Curr. Opin. Neurobiol.*, 13, 366–71, 2003.

62. Hardingham, G.E., Fukunaga, Y., and Bading, H., Extrasynaptic NMDARs oppose synaptic NMDARs by triggering CREB shut-off and cell death pathways, *Nat. Neurosci.*, 5, 405, 2002.

63. Chatterton, J.E. et al., Excitatory glycine receptors containing the NR3 family of NMDAR subunits, *Nature*, 415, 793, 2002.

64. Ciabarra, A.M. et al., Cloning and characterization of chi-1: a developmentally regulated member of a novel class of the ionotropic glutamate receptor family, *J. Neurosci.*, 15, 6498, 1995.

65. Sucher, N.J. et al., Developmental and regional expression pattern of a novel NMDAR-like subunit (NMDAR-L) in the rodent brain, *J. Neurosci.*, 15, 6509, 1995.

66. Perez-Otano, I. et al., Assembly with the NR1 subunit is required for surface expression of NR3A-containing NMDARs, *J. Neurosci.*, 21, 1228, 2001.

67. Das, S. et al., Increased NMDA current and spine density in mice lacking the NMDAR subunit NR3A, *Nature*, 393, 377, 1998.

68. Wong, H.K. et al., Temporal and regional expression of NMDAR subunit NR3A in the mammalian brain, *J. Comp. Neurol.*, 450, 303, 2002.

69. Nishi, M. et al., Motoneuron-specific expression of NR3B, a novel NMDA-type glutamate receptor subunit that works in a dominant-negative manner, *J. Neurosci.*, 21, RC185, 2001.

70. Matsuda, K. et al., Cloning and characterization of a novel NMDAR subunit NR3B: a dominant subunit that reduces calcium permeability, *Brain Res. Mol. Brain Res.*, 100, 43, 2002.

71. Sakimura, K. et al., Reduced hippocampal LTP and spatial learning in mice lacking NMDAR epsilon 1 subunit, *Nature*, 373, 151, 1995.

72. Forrest, D. et al., Targeted disruption of NMDAR 1 gene abolishes NMDA response and results in neonatal death, *Neuron*, 13, 325, 1994.

73. Kutsuwada, T. et al., Impairment of suckling response, trigeminal neuronal pattern formation, and hippocampal LTD in NMDAR epsilon 2 subunit mutant mice, *Neuron*, 16, 333, 1996.

74. Ikeda, K. et al., Reduced spontaneous activity of mice defective in the epsilon 4 subunit of the NMDAR channel, *Brain Res. Mol. Brain Res.*, 33, 61, 1995.

75. Kadotani, H. et al., Motor discoordination results from combined gene disruption of the NMDAR NR2A and NR2C subunits, but not from single disruption of the NR2A or NR2C subunit, *J. Neurosci.*, 16, 7859, 1996.

76. Tashiro, A. et al., NMDA-receptor-mediated, cell-specific integration of new neurons in adult dentate gyrus, *Nature*, 442, 929, 2006.

77. Cline, H.T. and Constantine-Paton, M., NMDAR antagonists disrupt the retinotectal topographic map, *Neuron*, 3, 413, 1989.

78. Simon, D.K. et al., N-methyl-D-aspartate receptor antagonists disrupt the formation of a mammalian neural map, *Proc. Natl. Acad. Sci. USA*, 89, 10593, 1992.
79. Ruthazer, E.S., Akerman, C.J., and Cline, H.T., Control of axon branch dynamics by correlated activity in vivo, *Science*, 301, 66, 2003.
80. Ramoa, A.S. et al., Suppression of cortical NMDAR function prevents development of orientation selectivity in the primary visual cortex, *J. Neurosci.*, 21, 4299, 2001.
81. Li, Y. et al., Whisker-related neuronal patterns fail to develop in the trigeminal brainstem nuclei of NMDAR1 knockout mice, *Cell*, 76, 427, 1994.
82. Lee, L.J., Lo, F.S., and Erzurumlu, R.S., NMDAR-dependent regulation of axonal and dendritic branching, *J. Neurosci.*, 25, 2304, 2005.
83. Iwasato, T. et al., NMDAR-dependent refinement of somatotopic maps, *Neuron*, 19, 1201, 1997.
84. Iwasato, T. et al., Cortex-restricted disruption of NMDAR1 impairs neuronal patterns in the barrel cortex, *Nature*, 406, 726, 2000.
85. Rebsam, A., Seif, I., and Gaspar, P., Refinement of thalamocortical arbors and emergence of barrel domains in the primary somatosensory cortex: a study of normal and monoamine oxidase a knock-out mice, *J. Neurosci.*, 22, 8541, 2002.
86. Rajan, I., Witte, S., and Cline, H.T., NMDAR activity stabilizes presynaptic retinotectal axons and postsynaptic optic tectal cell dendrites in vivo, *J. Neurobiol.*, 38, 357, 1999.
87. Schmidt, J.T., Activity-driven sharpening of the retinotectal projection: the search for retrograde synaptic signaling pathways, *J. Neurobiol.*, 59, 114, 2004.
88. Wu, G.Y. et al., Dendritic dynamics in vivo change during neuronal maturation, *J. Neurosci.*, 19, 4472, 1999.
89. Hua, J.Y. and Smith, S.J., Neural activity and the dynamics of central nervous system development, *Nat. Neurosci.*, 7, 327, 2004.
90. Scott, E.K. and Luo, L., How do dendrites take their shape? *Nat. Neurosci.*, 4, 2001.
91. Cline, H.T., Dendritic arbor development and synaptogenesis, *Curr. Opin. Neurobiol.*, 11, 118, 2001.
92. Rajan, I. and Cline, H.T., Glutamate receptor activity is required for normal development of tectal cell dendrites *in vivo*, *J. Neurosci.*, 18, 7836, 1998.
93. Sin, W.C. et al., Dendrite growth increased by visual activity requires NMDAR and Rho GTPases, *Nature*, 419, 475, 2002.
94. Rocha, M. and Sur, M., Rapid acquisition of dendritic spines by visual thalamic neurons after blockade of N-methyl-D-aspartate receptors, *Proc. Natl. Acad. Sci. USA*, 92, 8026, 1995.
95. McAllister, A.K., Katz, L.C., and Lo, D.C., Neurotrophin regulation of cortical dendritic growth requires activity, *Neuron*, 17, 1057, 1996.
96. Wu, G.Y. and Cline, H.T., Stabilization of dendritic arbor structure in vivo by CaMKII, *Science*, 279, 1998.
97. Li, Z., Aizenman, C.D., and Cline, H.T., Regulation of rho GTPases by crosstalk and neuronal activity *in vivo*, *Neuron*, 33, 741, 2002.
98. Li, Z., Van Aelst, L., and Cline, H.T., Rho GTPases regulate distinct aspects of dendritic arbor growth in Xenopus central neurons *in vivo*, *Nat. Neurosci.*, 3, 217, 2000.
99. Chiaia, N.L., Bennett-Clarke, C.A., and Rhoades, R.W., Effects of cortical and thalamic lesions upon primary afferent terminations, distributions of projection neurons, and the cytochrome oxidase pattern in the trigeminal brainstem complex, *J. Comp. Neurol.*, 303, 600, 1991.
100. Niell, C.M., Meyer, M.P., and Smith, S.J., *In vivo* imaging of synapse formation on a growing dendritic arbor, *Nat. Neurosci.*, 7, 254, 2004.
101. Maletic-Savatic, M., Malinow, R., and Svoboda, K., Rapid dendritic morphogenesis in CA1 hippocampal dendrites induced by synaptic activity, *Science*, 283, 1923, 1999.

102. Dailey, M.E. and Smith, S.J., The dynamics of dendritic structure in developing hippocampal slices, *J. Neurosci.*, 16, 2983, 1996.
103. Portera-Cailliau, C., Pan, D.T., and Yuste, R., Activity-regulated dynamic behavior of early dendritic protrusions: evidence for different types of dendritic filopodia, *J. Neurosci.*, 23, 7129, 2003.
104. Fifkova, E. and Delay, R.J., Cytoplasmic actin in neuronal processes as a possible mediator of synaptic plasticity, *J. Cell. Biol.*, 95, 345, 1982.
105. Matus, A. et al., High actin concentrations in brain dendritic spines and postsynaptic densities, *Proc. Natl. Acad. Sci. USA*, 79, 7590, 1982.
106. Gray, E.G. et al., Synaptic organisation and neuron microtubule distribution, *Cell Tissue Res.*, 226, 579, 1982.
107. Westrum, L.E. et al., Synaptic development and microtubule organization, *Cell Tissue Res.*, 231, 93, 1983.
108. Vaughn, J.E., Barber, R.P., and Sims, T.J., Dendritic development and preferential growth into synaptogenic fields: a quantitative study of Golgi-impregnated spinal motor neurons, *Synapse*, 2, 69, 1988.
109. Vaughn, J.E., Fine structure of synaptogenesis in the vertebrate central nervous system, *Synapse*, 3, 255, 1989.
110. Star, E.N., Kwiatkowski, D.J., and Murthy, V.N., Rapid turnover of actin in dendritic spines and its regulation by activity, *Nat. Neurosci.*, 5, 239, 2002.
111. Fukazawa, Y. et al., Hippocampal LTP is accompanied by enhanced F-actin content within the dendritic spine that is essential for late LTP maintenance in vivo, *Neuron*, 38, 447, 2003.
112. Fischer, M. et al., Glutamate receptors regulate actin-based plasticity in dendritic spines, *Nat. Neurosci.*, 3, 887, 2000.
113. Malinow, R. and Malenka, R.C., AMPA receptor trafficking and synaptic plasticity, *Annu. Rev. Neurosci.*, 25, 103, 2002.
114. Huang, Y.S. et al., N-methyl-D-aspartate receptor signaling results in Aurora kinase-catalyzed CPEB phosphorylation and alpha CaMKII mRNA polyadenylation at synapses, *EMBO J.*, 21, 2139, 2002.
115. Bestman, J.E. and Cline, H.T., The RNA binding protein CPEB controls dendrite growth and neural circuit assembly *in vivo*, *J. Neurosci.*, 28, 850, 2008.
116. Garner, C.C., Waites, C.L., and Ziv, N.E., Synapse development: still looking for the forest, still lost in the trees, *Cell Tissue Res.*, 326, 249, 2006.
117. Zhu, J.J. et al., Postnatal synaptic potentiation: delivery of GluR4-containing AMPA receptors by spontaneous activity, *Nat. Neurosci.*, 3, 1098, 2000.
118. Durand, G.M., Kovalchuk, Y., and Konnerth, A., Long-term potentiation and functional synapse induction in developing hippocampus, *Nature*, 381, 71, 1996.
119. Wu, G., Malinow, R., and Cline, H.T., Maturation of a central glutamatergic synapse, *Science*, 274, 972, 1996.
120. Liao, D. and Malinow, R., Deficiency in induction but not expression of LTP in hippocampal slices from young rats, *Learn Mem.*, 3, 138, 1996.
121. Isaac, J.T. et al., Silent synapses during development of thalamocortical inputs, *Neuron*, 18, 269, 1997.
122. Liao, D., Hessler, N.A., and Malinow, R., Activation of postsynaptically silent synapses during pairing-induced LTP in CA1 region of hippocampal slice, *Nature*, 375, 400, 1995.
123. Akerman, C.J. and Cline, H.T., Refining the roles of GABAergic signaling during neural circuit formation, *Trends Neurosci.*, 30, 382, 2007.
124. Ben-Ari, Y. et al., GABAA, NMDA and AMPA receptors: a developmentally regulated "menage a trois," *Trends Neurosci.*, 20, 523, 1997.

125. Leinekugel, X. et al., Ca2+ oscillations mediated by the synergistic excitatory actions of GABA(A) and NMDARs in the neonatal hippocampus, *Neuron*, 18, 243, 1997.
126. Akerman, C.J. and Cline, H.T., Depolarizing GABAergic conductances regulate the balance of excitation to inhibition in the developing retinotectal circuit *in vivo*, *J. Neurosci.*, 26, 5117, 2006.
127. Barria, A. and Malinow, R., Subunit-specific NMDAR trafficking to synapses, *Neuron*, 35, 345, 2002.
128. Hayashi, Y. et al., Driving AMPA receptors into synapses by LTP and CaMKII: requirement for GluR1 and PDZ domain interaction, *Science*, 287, 2262, 2000.
129. Shi, S.H. et al., Rapid spine delivery and redistribution of AMPA receptors after synaptic NMDAR activation, *Science*, 284, 1811, 1999.
130. Shi, S. et al., Subunit-specific rules governing AMPA receptor trafficking to synapses in hippocampal pyramidal neurons, *Cell*, 105, 331, 2001.
131. Beattie, E.C. et al., Regulation of AMPA receptor endocytosis by a signaling mechanism shared with LTD, *Nat. Neurosci.*, 3, 1291, 2000.
132. Carroll, R.C. et al., Dynamin-dependent endocytosis of ionotropic glutamate receptors, *Proc. Natl. Acad. Sci. USA*, 96, 14112, 1999.
133. Carroll, R.C. et al., Rapid redistribution of glutamate receptors contributes to long-term depression in hippocampal cultures, *Nat. Neurosci.*, 2, 454, 1999.
134. Lee, S.H. et al., Clathrin adaptor AP2 and NSF interact with overlapping sites of GluR2 and play distinct roles in AMPA receptor trafficking and hippocampal LTD, *Neuron*, 36, 661, 2002.

2 NMDA Receptors and Huntington's Disease

Herman B. Fernandes and Lynn A. Raymond

CONTENTS

2.1 HUNTINGTON'S DISEASE

Huntington's disease (HD) is an inherited, progressive neurodegenerative disorder, with a prevalence of ~5 to 10 per 100,000 people.[1] This genetic, autosomal-dominant disease is caused by a mutation in exon 1 of the IT15 gene, resulting in the expansion of a CAG repeat[2] encoding a polyglutamine (polyQ) region near the N-terminus of the huntingtin protein. Normal individuals have 35 or fewer CAG repeats.[3] The presence of 36 or more CAG repeats leads to eventual development of the disease.[4] HD is one of nine currently identified neurodegenerative diseases resulting from the expansion of a CAG tract within the coding regions of nine different genes[5,6] that include the different spinocerebellar ataxias (SCAs) 1, 2, 3, 6, 7, and 17, spinal bulbar muscular atrophy (SBMA), and dentatorubral pallidoluysian atrophy (DRPLA).

2.1.1 CLINICAL FEATURES

The clinical presentation of HD includes a range of motor, cognitive, and mood changes. Disease onset generally occurs between the ages of 35 to 50, progressing over 15 to 20 years until death.[7] Motor symptoms include choreiform involuntary movements, postural imbalance, uncoordinated voluntary movements, and speech and swallowing difficulties, followed at later stages by akinesia and rigidity. Depression is a common symptom, and patients often display personality changes such as apathy and flashes of temper. Generally, emotional and cognitive changes precede the motor symptoms.[8,9] Cases of juvenile onset HD differ from adult onset in their clinical presentation, with symptoms including early bradykinesia, rigidity, dystonia, and often epileptic seizures.[1] In contrast to the adult onset form, juvenile onset HD patients often exhibit little or no chorea.[10]

Expanded CAG regions are relatively unstable, particularly when passed via the paternal germline, with expansions in CAG length occurring more often than reductions. This produces the phenomenon known as anticipation in which the CAG repeat number tends to increase in subsequent family generations.[11,12] The age of onset of symptoms correlates inversely with the length of the CAG expansion.[13–15] Most adult onset cases have CAG repeat lengths of 40 to 50, whereas juvenile onset cases have somewhat longer CAG expansions (>60).[10]

2.1.2 NEUROPATHOLOGY

The pattern of neurodegeneration in HD is particularly selective for the medium-sized spiny neurons (MSNs) of the striatum, which project to other areas of the basal ganglia, specifically the substantia nigra and globus pallidus,[16] and constitute approximately 95% of the neurons in the striatum.[17] In an extensive study characterizing postmortem HD brains, the highest degree of degeneration was found in the caudate and putamen nuclei comprising the neostriatum.[18] The authors used a five-point grading system (0 to 4) to describe striatal neuropathology in ascending order of severity, correlating closely with the extent of clinical disability. For example, in grade 0 HD brains, no gross neuropathological changes were seen, although 30 to 40% neuronal loss in the caudate was reported.[1] In grade 1 brains, only microscopic neuropathological changes (50% loss of neurons in the caudate nucleus) were evident. The more severe cases showed increasing neuronal loss within the caudate and other striatal regions. Grade 4 brains exhibited >95% neuronal loss in the caudate and extensive neuronal losses in the putamen, globus pallidus, and nucleus accumbens, accompanied by increasing numbers of astrocytes through grades 2 to 4.[18]

Striatal MSNs are the first population of neurons to die and generally show the greatest losses in numbers.[1] In the early stages of HD, the subpopulation of MSNs expressing enkephalin and projecting to the external segment of the globus pallidus (indirect basal ganglia pathway) die first, followed by the substance P-expressing MSNs[19] that project to the internal segment of the globus pallidus (direct basal ganglia pathway). Notably, the large aspiny cholinergic and nitric oxide synthase-containing interneurons are relatively spared.[20,21] In later stage HD, a number of brain regions display atrophy or a loss in cross-sectional area. The caudate and putamen show approximately 60% area loss and other brain regions (substantia nigra, globus pallidus, thalamus, hippocampus) display lesser atrophy of 20 to 30% in later stages.[22,23] Hence, the striatum undergoes the greatest extent of neuronal loss, and in the most severe cases (grade 4) this results in a significant decrease in neuronal density despite the concurrent regional atrophy. Losses of cortical neurons and volume also occur in more advanced cases of HD, particularly loss of the large pyramidal neurons in layers III, V, and VI that project directly to the striatum.[24,25]

2.1.3 HUNTINGTIN: DISTRIBUTION, CELLULAR ROLES, AND FUNCTION

The protein huntingtin (htt) is widely distributed throughout many tissues of the body[26] and throughout most brain regions.[27] No particular enrichment of the protein appears in the striatum.[28] Expansion of the CAG repeat region of the HD gene to produce polyglutamine-expanded, mutant htt (mhtt) does not appear to alter tissue distribution of the protein.[29] Htt is a 350 kDa cytosolic protein found in the soma and throughout the dendrites as well as in synaptic terminals.[29]

The function of htt is still unknown. It has been associated with membrane-bound organelles including mitochondria[30] and vesicular membranes[31,32] and is therefore believed to play a role in vesicular transport and endocytosis.[32] Htt can also be localized to the nucleus, and this presents implications for a possible role in gene transcription for both the normal and mutant forms.[33] Htt expression is required for normal development, as disruption of the endogenous htt protein results in embryonic lethality.[34–36] Interestingly, mhtt retains at least some functionality of the wild-type protein required for development; expression of mhtt can rescue the lethal disruption of endogenous htt expression.[36]

2.2 MODEL SYSTEMS FOR STUDY OF NMDA RECEPTOR FUNCTION IN HD

2.2.1 HETEROLOGOUS EXPRESSION SYSTEMS

The simplicity and degree of control possible in heterologous expression systems make them ideal starting points for examining the basis of mhtt-mediated toxicity. For example, the HEK293 cell line has been used extensively to study ionotropic glutamate receptor-mediated toxicity and function[37–41] and the influence of overexpression of both htt and mhtt on N-methyl-D-aspartate receptor (NMDAR) function and excitotoxicity.[42,43] Other lines such as COS and CHO cells also have been used to study NMDAR function and toxicity.[41,44,45] Notably, expression of mhtt along with NMDARs composed of NR1 and NR2B subunits enhanced NMDAR current amplitude,[42] similar to subsequent observations in acutely dissociated and cultured MSNs.[46–49]

2.2.2 MOUSE MODELS

A number of mouse models have been established as tools to examine the pre-symptomatic changes, pathogenic mechanisms, progression of disease pathology, and possible areas of therapeutic intervention in HD. Table 2.1 summarizes the relevant mouse models discussed here. A comprehensive summary of these and other models provided by the Hereditary Disease Foundation can be found at www. hdfoundation.org/PDF/hdmicetable.pdf. The most common models fall into three

TABLE 2.1
Selected HD Mouse Models

HD Model	YAC46, 72, 128	R6/1, R6/2	N171-82Q	tgHD100	CAG80	HdhQ92, Q111
Reference for HD mouse model	Hodgson et al., 1999; Slow et al., 2003	Mangiarini et al., 1996	Schilling et al., 1999	Laforet et al., 2001	Shelbourne et al., 1999	Wheeler et al., 2000
Promoter and transcript length	Human, full-length	Human, exon 1	Prion, N-terminal 171 amino acids	Rat neuron-specific enolase, N-terminal 3 kb	Murine, CAG knock-in	Murine, CAG knock-in
PolyQ length	46, 72, 128	120, 150	82	100	80	92, 111
NMDAR expression in striatum	Unchanged in YAC46, YAC72	NR1 unchanged or ↑; NR2A/B ↓	NR2A ↓ transiently	Unknown	Unknown	Unknown
MSN NMDAR currents	↑ in YAC72	↑	Unknown	↑	Unknown	Unknown
Striatal excitotoxicity	↑	↓ with time	↓ with time	Same as WT	Unknown	Unknown
NMDAR Ca2+ influx	↑ in YAC46, YAC72; recovery impaired in YAC128	↑	Unknown	↑	Unknown	Unknown
LTP/LTD	Altered	Altered	Unknown	Unknown	Altered	Unknown

categories: (1) transgenic mice expressing an N-terminal fragment of human htt containing an expanded CAG region (N-terminal fragment models); (2) transgenic mice expressing full-length human htt in addition to their own endogenous murine htt (full-length models); and (3) knock-in mouse models in which an expanded CAG repeat region is inserted into the endogenous Hdh gene.

The most common N-terminal fragment models are the R6/1 and R6/2 lines expressing exon 1 of the human HD gene containing 116 and 144 CAG repeats, respectively.[50] The R6/2 model shows the more aggressive disease phenotype. Generally both models are characterized by early death (approximately 12 mo for R6/1 and 4 mo for R6/2) preceded by early onset of motor symptoms (as early as 1 mo in R6/2) and overall brain atrophy (~20%)[51] apparent prior to neuronal loss.[50] Another well-characterized fragment model expresses an N-terminal fragment of 171 residues containing a CAG repeat length of 82. This model also dies prematurely after a lifespan of 2.5 to 11 mo.[52]

The R6/2 and N171-82Q models produce similar changes to striatal gene expression,[53] indicating common pathological alterations. Interestingly, both models also display striatal resistance to excitotoxins after HD-like symptoms develop,[54-56] but they also lack frank striatal neuronal loss, which is inconsistent with the notable striatal degeneration found in human postmortem HD brains. One key criticism of these models may be that expression of the full-length htt gene with all critical regulatory sequences is required to most accurately reproduce the human disease. However, the aggressive pathology and shortened lifespan make mhtt fragment models ideal for therapeutic testing in one sense because the beneficial effects of intervention will be more obvious than in a milder phenotype.

Knock-in HD mouse models generally exhibit very late onsets of motor symptoms (around 2 years of age), with a relatively mild disease phenotype and little to no neuropathology evident, aside from predominantly striatal aggregate formation, despite longer CAG repeat lengths ranging from 72 to 150.[57-61] One advantage of these models is the elimination of a key confounding factor in interpreting results from other HD models: htt (mutant or otherwise) overexpression. Also, their long lifespans make these models ideal for studying behavioral changes as a proxy for neuronal dysfunction prior to neuronal loss.

Full-length models such as the yeast artificial chromosome (YAC) mouse model[62,63] and the cytomegalovirus promoter model[64] recapitulate the pattern of selective striatal neuronal loss seen in human HD patients, making these models ideal for studying changes in neuronal function underlying selective neurodegeneration. These mouse models generally have a later onset (2 to 7 mo, depending on htt polyQ repeat length) of the HD motor phenotype, and have relatively normal lifespans.[62,63,65]

2.3 EXCITOTOXICITY HYPOTHESIS IN HD PATHOGENESIS

Several lines of evidence directly and indirectly support the role of excitotoxicity in HD, from alterations of NMDAR function to bioenergetic impairment. Following is an overview of evidence from human patients and *in vivo* and *in vitro* experiments using animal and cellular models in support of the excitotoxicity hypothesis.

2.3.1 Excitotoxicity Defined

Excitotoxicity in neurons is a toxic consequence of the actions of excitatory amino acids (EAAs), whether endogenous or exogenous (in the cases of some chemical models in animals or *in vitro*). Since glutamate is the major excitatory neurotransmitter, excitotoxicity in the central nervous system (CNS) is considered to result from glutamate exposure for prolonged periods or in excessive concentrations. This can lead to a number of pathological changes in neurons including ion influx, osmotic dysregulation, energy depletion, and biochemical changes, eventually causing cell death.[66,67]

Glutamate activates two classes of receptors in neurons: (1) metabotropic glutamate receptors (mGluRs) that exert their effects via coupling to G-proteins, and (2) ionotropic glutamate receptors (iGluRs), which upon binding of the appropriate ligands, allow passage of cations through a channel pore formed by the receptor subunits.[68] The most intensely studied iGluRs related to excitotoxicity are the NMDARs, a subclass of receptors exhibiting several features of relevance to neuronal death, i.e., relatively high permeability to Ca^{2+} and slow activation and deactivation kinetics.

2.3.2 NMDA Receptor Subunit Composition in Forebrain and Striatum

The combination of NR1 with different NR2 subunits alters NMDAR ion channel characteristics,[37,69–73] providing significant potential for functional diversity. NR2 subunit expression is both developmentally and spatially regulated.[70,74] In the adult forebrain, the main NR2 subunits expressed are NR2A and NR2B, indicating that most NMDARs in these regions are diheteromers composed of NR1/NR2A or NR1/NR2B or have a triheteromeric configuration of NR1/NR2A/NR2B.[75–77]

These subunit combinations produce NMDARs that are similar in certain channel properties such as permeability to Ca^{2+}, single channel conductance, and sensitivity to voltage-dependent Mg^{2+} block. However, NR2A and NR2B subunits have key differences in terms of the functional properties they convey to channels and how they are distributed on a subcellular level. NR2A and NR2B subunits have differential sensitivities to agonists and antagonists,[68,78–80] and channel gating properties are altered in a subunit-dependent fashion.[70,79] NR2A-containing receptors are generally expressed at synapses, and this subcellular expression pattern is considered a consequence of developmental regulation and synaptic maturation.[75–77,81,82] In contrast, NR2B-containing receptors appear to predominate at extrasynaptic sites.[76,81–83]

The specificity of this spatial distinction between NR2A and NR2B expression patterns may reflect differential roles in determining cell survival and cell death.[84–87] Recent studies suggest inclusion of the NR2B subunit in the NMDAR complex is sufficient to confer excitotoxic potential to NMDAR activation, regardless of subcellular localization, in relatively immature cortical cultures,[88] although NR2A-containing NMDARs showed similar excitotoxic potential in more mature cultured cortical neurons.[89] Interestingly, the striatum appears to express higher levels of NR2B relative to other NR2 subunits, compared to other regions of the brain; this pattern of expression is observed in several species including humans.[90–95]

2.3.3 EXCITOTOXIC HYPOTHESIS IN HD

Critical evidence to support a role for excitotoxicity in HD pathogenesis arose from studies of postmortem HD brains showing losses of striatal NMDAR binding sites.[96–98] These observations extended to brains from presymptomatic individuals, indicating that MSNs with high levels of NMDAR expression were at particular risk and losses occurred very early in disease progression, possibly contributing to symptom onset. These observations correlate with the selective loss of MSNs in HD patient brains.[16]

A survey of transgenic HD mouse models reveals less dramatic changes in NMDAR composition and subunit expression. In the R6/2 mouse model, immunostaining of striatal NR2A and NR2B were both reportedly decreased in symptomatic mice, although NR1 immunostaining was enhanced.[46] Single cell RT-PCR confirmed a decrease in NR2A mRNA even at presymptomatic stages, while NR2B was also reduced by 12 wk (fully symptomatic stage).[99] On the other hand, another study in the same model at 12 wk of age showed that striatal NR1 mRNA levels and NMDAR binding were similar to those of wild-type (WT) litter mates.[100] Together, there studies suggest a change in NMDAR subunit composition.

In another N-terminal fragment model (N171-82Q), no significant changes in striatal NR1, NR2A, or NR2B expression relative to controls were found.[56] In YAC mouse models, striatal NR1 and NR2B expression were similar in WT, YAC46, and YAC72 mice prior to onset of symptoms (2 mo).[95] Similarly, in the more aggressive YAC128 model in which striatal atrophy and neuronal loss (although only ~13% at 12 mo) are highly correlated with motor dysfunction,[63] no differences in striatal neuronal mRNA expression were found for NR1, NR2A, or NR2B subunits.[101] Protein levels in synaptic membrane fractions at symptomatic stages (12 mo) were similar to those of WT controls although expression in total striatal homogenates was not reported.[101] The lack of overt NMDAR loss in striata of transgenic HD mice models may reflect the failure of the models to reproduce the extensive neuronal losses observed in humans. Although further studies are required, it is likely that symptomatic HD is associated with a decrease in the overall number and/or a change in subunit composition of functional NMDARs within the striatum. This may contribute to a deficit in neuronal function that may precede neuronal death.[65]

2.3.4 EVIDENCE FROM CHEMICAL MODELS OF HD IN SUPPORT OF EXCITOTOXIC HYPOTHESIS

Several studies demonstrated that intrastriatal injection of NMDAR agonists results in the selective loss of MSNs while sparing interneurons, reproducing many behavioral and neuropathological characteristics of HD.[102–105] Additionally, striatal MSNs show increased sensitivity to NMDA-induced swelling (a correlate of current and toxicity) compared with large-sized striatal interneurons. Kainate produced similar swelling in both neuronal populations.[106] Moreover, NMDAR agonists are more effective than other GluR agonists for inducing striatal neuronal excitotoxicity.[107]

A different means of chemically inducing HD-like symptoms and neuropathology involves the use of inhibitors of mitochondrial complex II. Both malonate and 3-nitropropionic acid (3-NP) can be injected intrastriatally or systemically to rodents to produce selective degeneration of striatal MSNs[108–111] that may be blocked by NMDAR antagonists.[108,111]

While these models effectively replicate many aspects of advanced HD, they cannot be used to study early presymptomatic changes that may be critical to understanding disease pathogenesis and dysfunction prior to neuronal death. However, these models further support the excitotoxic involvement of NMDARs in HD pathogenesis, and in the case of mitochondrial complex II inhibitors, implicate mitochondrial dysfunction as a predisposing factor in HD excitotoxicity.

2.3.4.1 Alterations of NMDA Receptor Function in HD Models

Based on the probable involvement of NMDARs in HD pathology, a great deal of interest has focused on the question of whether NMDAR function in HD models is altered. Studies in intact HD animal models, *in vitro* preparations derived from HD model animals, and heterologous systems have been performed to answer this question. The most direct method for investigating HD-associated changes in NMDAR function is by assessing NMDAR-mediated currents with electrophysiological recordings. A number of studies have demonstrated enhancement of these currents in several transgenic HD mouse models.

2.3.4.2 NMDA Receptor Currents in HD Models

Enhancement of NMDAR current was observed in MSNs acutely dissociated from striata of presymptomatic and symptomatic R6/2 mice.[49] The same study also reported decreased sensitivity of NMDAR currents to Mg^{2+} block in a subpopulation of MSNs that exhibited significantly larger NMDAR currents, possibly reflecting an alteration of NMDAR subunit composition. This observation extended to MSNs from presymptomatic mice, indicating altered NMDAR function prior to apparent onset of symptoms. Additionally, NMDA-evoked current in striatal slices taken from both presymptomatic and symptomatic R6/2 mice were increased over WT. In the same mice, AMPA-evoked currents were actually smaller, indicating a process selective for enhancement of NMDAR function.[46]

Later studies of this model demonstrated that this enhancement of NMDAR current is specific for the striatum, as cortical NMDAR-mediated currents were not enhanced.[112] These enhanced striatal NMDAR currents also explain an earlier observation of increased swelling of MSNs in R6/2 striatal slices compared to slices from WT animals in response to exogenous NMDA application—an observation confirmed with MSNs in slices from a knock-in HD mouse model with a polyQ length of 94.[113] Similar observations were also documented in another N-terminal fragment model of HD (with a polyQ length of 100 contained within the N-terminal third of the human htt gene), showing significant enhancement of NMDAR peak currents and current density in MSNs from striatal slices.[114]

Studies of the full-length YAC72 HD mouse model largely confirmed findings in truncated fragment models. Acutely dissociated MSNs from YAC72 animals showed significantly increased NMDAR peak current densities relative to those in MSNs from WT animals at both presymptomatic (6 to 11 wk)[47] and symptomatic (>1 year)[46] ages. These results indicate that aberrant NMDAR activity is present prior to, and possibly plays a causative role in, behavioral changes associated with HD. Moreover, cultured MSNs taken from mice at birth also showed increased NMDAR peak current density for YAC72 compared with WT mice.[115]

Because these data were obtained using exogenously applied NMDA and stimulating presumably extrasynaptic NMDARs, other studies focused on possible changes in synaptic NMDAR properties produced by mhtt. Synaptic NMDAR currents recorded from MSNs in corticostriatal slices from YAC72 mice were also enhanced compared to those recorded from WT mice, and this enhancement was found to reflect a postsynaptic NMDAR-selective mechanism,[95,116] suggesting that mhtt preferentially modulates NMDAR function.

2.3.4.3 NR2B-Selective Hypothesis of Mhtt-Mediated Enhancement of NMDA Receptor Function

One of the first demonstrations of an effect of mhtt on NMDAR function arose from a study in which NMDARs composed of NR1/NR2A or NR1/NR2B were expressed in HEK293 cells in conjunction with full-length human htt containing 15 or 138 polyQ repeats.[42] Mhtt increased the responses of NMDARs composed of NR1/NR2B, whereas NR1/NR2A NMDARs were not differentially affected by the presence of WT or mutant htt.[42] These results were underscored by a later finding that mhtt selectively enhanced apoptotic cell death in HEK cells cotransfected with NR1/NR2B and not NR1/NR2A,[43] indicating a possible preferential modulation of NR2B-containing NMDARs by mhtt (the NR2B-selective hypothesis). Additional studies reported enhanced NMDAR-mediated toxicity for cells co-expressing NR1/NR2B, mhtt and PSD-95, which is dependent on tyrosine phosphorylation of NR2B.[117,118]

The possibility that the NR2B subunit may be necessary for selective modulation by mhtt is not surprising, considering that the adult striatum is enriched in NR2B (discussed above). Additional evidence for selective modulation of NR2B-containing NMDARs has been provided in studies of YAC HD mice. In acutely dissociated MSNs from juvenile mice and cultured MSNs from early postnatal YAC72 mice, more than 50% of the NMDAR current was mediated by NR1/NR2B NMDARs, whereas cultured cortical neurons with significantly lower expression of NR1/NR2B showed no increase in NMDAR current densities for YAC72 compared with WT mice.[47,115] Moreover, while NMDA-induced apoptosis was increased in YAC72 compared with WT MSNs, no differential effect was observed in cerebellar granule neurons (CGNs) that did not express NR2B under the culture conditions used.[47] The NR1/N2B selective antagonist ifenprodil (IFN) effectively eliminated apoptosis in both WT and YAC72 MSNs,[47] providing further evidence of the important role that this NMDAR subunit combination plays in NMDA-induced excitotoxic cell death in MSNs. An ifenprodil-sensitive enhancement of glutamate-induced apoptosis in YAC128 relative to WT MSNs was also reported.[119]

2.3.5 Relationship of Enhanced NMDA Receptor Activity, Excitotoxicity, and Mhtt PolyQ Length in MSNs

Because mhtt expression enhances NMDAR current in MSNs from a variety of HD mouse models along with excitotoxicity in full-length mhtt-expressing cells and neurons that also express high levels of NR1/NR2B, recent studies investigated the relationship of altered NMDAR signaling and mhtt polyQ length. Cultured YAC72 MSNs were more susceptible to NMDA-induced toxicity than WT MSNs, although no difference was observed between the two genotypes in response to AMPA application,[47] supporting observations of selective enhancement of NMDAR currents in corticostriatal slices[95,116] and correlating electrophysiological observations with neurotoxicity.[47] Additionally, the enhancement of apoptosis by mhtt in YAC46 and YAC72 MSNs was proportional to the length of the polyQ repeat,[47] and NMDAR current densities were significantly larger in YAC72 than YAC46 MSNs.[116]

Enhanced sensitivity to apoptosis in YAC128 compared with YAC72 MSNs was reported in response to subsaturating concentrations of NMDA supporting a further increase in excitotoxic susceptibility with larger polyQ expansions in mhtt, although apoptosis rates were equivalent when saturating concentrations of NMDA were used.[120] In contrast to the correlation between increasing NMDAR current and toxicity found with increasing polyQ length in YAC18, YAC46, and YAC72 MSNs,[47,115,120] MSNs from YAC128 mice exhibited NMDAR current densities similar to WT and YAC18, and significantly smaller than in YAC72 MSNs.[162] These findings led to the hypothesis that signaling events downstream of NMDAR activation contribute to enhanced excitotoxicity in mhtt-expressing MSNs, and that those downstream mechanisms are more important with extreme polyQ expansions.

2.4 DOWNSTREAM CONSEQUENCES OF NMDA RECEPTOR ACTIVATION IN CELLS EXPRESSING MHTT

Ca^{2+} is a central mediator of excitotoxic damage when it exceeds normal physiological concentrations. As Ca^{2+} homeostatic mechanisms are overwhelmed, a number of intracellular mechanisms are activated, leading to cellular damage and/or toxicity. These may include, among other signaling pathways, activation of Ca^{2+}-dependent enzymes such as proteases, phosphatases, and lipases. Other pathological events can ensue as a consequence of homeostatic mechanisms attempting to compensate for elevated Ca^{2+}, for example, when mitochondria buffer excitotoxic levels of Ca^{2+}, leading to loss of mitochondrial membrane potential, loss of adenosine 5′-triphosphate (ATP)-generating ability, free radical generation, and generalized mitochondrial dysfunction leading to neuronal death.[121,122] The presence of mhtt can further impair neuronal function at multiple points in this process downstream of NMDAR activity, exacerbating the deleterious effects that mhtt may exert at receptor level.

2.4.1 Effects of Mhtt on Ca^{2+} Homeostasis

Activation of the NMDAR channel under physiological conditions allows the conductance of cations, predominantly Na^+ and Ca^{2+} influx accompanied by K^+ efflux, with the Ca^{2+} component of the current representing ~10 to 18%.[123,124] Hence NMDARs are major routes for Ca^{2+} influx. A number of damaging events can ensue from excessive free cytosolic Ca^{2+}—a downstream consequence of NMDAR activation in MSNs expressing mhtt.[48,119,120] These may include inappropriate enzyme activation (i.e., calpains, calcineurin, other Ca^{2+}-regulated enzymes) and mitochondrial dysfunction (see discussion below) that in turn exerts a number of toxic consequences.

Evidence from several HD models indicates that regulation of intracellular Ca^{2+} is altered in several ways, although not always predictably. Resting Ca^{2+} levels are elevated in MSNs from R6/2 mice,[54] hippocampal neurons in symptomatic YAC72 mice,[62] and an immortalized cell line derived from striatal neurons from a knock-in HD mouse model.[125] Conversely, in studies of primary MSN cultures from YAC HD mice, resting Ca^{2+} levels were equivalent among WT, YAC46, and YAC72 MSNs[48] and WT, YAC18, and YAC128 MSNs.[119] Zeron et al.[48] found that stimulation of NMDARs in mhtt-expressing MSNs resulted in elevated Ca^{2+} levels relative to controls.

Similarly, Tang and colleagues[119] found increased cytosolic Ca^{2+} levels in YAC128 MSNs following repetitive glutamate stimulation. The increase was attributed in part to enhanced IP3 receptor (IP_3R)-mediated Ca^{2+} release from the ER, downstream of mGluR1/5 activation, as mhtt was found to sensitize the IP3R.[126] Hence, not only does mhtt appear to enhance NMDAR activity and consequently Ca^{2+} influx via NMDARs,[48] it also may enhance the probability of intracellular Ca^{2+} release.

Similar mhtt-enhanced NMDAR-mediated Ca^{2+} entries in MSNs were found in N-terminal fragment HD models including the R6/2[46] and tgHD100 mice.[114] These findings of altered Ca^{2+} regulation in YAC HD mice were observed in MSNs obtained from early postnatal mice,[48,119] implying that these changes are present at birth and may over time increase the risks of neuronal dysfunction and death.

NMDAR currents[47,48] and Ca^{2+} influx downstream of NMDAR activation[48] are both enhanced in YAC72 compared to WT MSNs, correlating with enhanced apoptosis levels in YAC72 MSNs compared to controls.[47,120] This difference in apoptosis was abolished by using the competitive NMDAR antagonist 2-amino-5-phosphonovalerate (APV) to reduce NMDAR-mediated Ca^{2+} influx in YAC72 MSNs down to a level equivalent to that seen in WT MSNs, indicating that enhanced neurotoxicity produced by NMDAR activation in YAC72 MSNs resulted from augmentation of NMDAR function by mhtt.[120] Similarly, NMDA-induced cytosolic Ca^{2+} and apoptosis are both enhanced in YAC46 compared with WT MSNs.[48] In YAC128 MSNs, cytosolic Ca^{2+} accumulated after repetitive glutamate application correlated with increased glutamate-induced neuronal death,[119] although acute NMDAR-evoked cytosolic Ca^{2+} was not significantly different from WT MSNs in spite of the enhanced NMDAR-mediated apoptosis.[120,162] Thus, NMDAR-mediated cytosolic Ca^{2+} increases and enhanced apoptosis correlate well for moderate mhtt polyQ expansions but other downstream pathways may play proportionately larger roles in extreme polyQ expansions.

2.4.2 MITOCHONDRIAL AND BIOENERGETIC IMPAIRMENT IN HD

Because of the influence of mhtt on both NMDAR function and subsequent Ca^{2+} entry and regulation, it follows that mitochondrial responses will also be affected.

Mitochondria have electrochemical gradients across their inner membranes established by the active extrusion of H^+ ions, in order to couple reentry of protons to oxidative phosphorylation in the generation of ATP. Mitochondria may also utilize this gradient to drive the uptake of Ca^{2+} from the cytosol when local concentrations exceed ~1 μM.[127] During this process, the mitochondrial membrane potential ($\Delta\Psi_m$) is dissipated; under normal conditions it is reestablished through continued activity of active H^+ transport, but excessive mitochondrial depolarization is associated with increased risk of apoptotic neuronal death.[128]

NMDAR activation in YAC46 and YAC72 MSNs has been shown to produce enhanced mitochondrial depolarization associated with increased cytosolic Ca^{2+} levels and apoptosis as noted earlier.[48,120] Furthermore, mitochondria isolated from HD patients and from YAC72 mice exhibited reduced resting $\Delta\Psi_m$ and depolarized to a greater extent when stressed.[128,130] Expression of mhtt reduces the ability of mitochondria to reestablish baseline $\Delta\Psi_m$.[131] These observations demonstrate that mhtt-mediated changes in NMDAR function can produce adverse consequences on neuronal health beyond altered activity of the receptor.

Ca^{2+} cycling between the mitochondria and cytosol occurs normally as a consequence of physiological activity, such as during synaptic activity. However, when excessive concentrations of Ca^{2+} are achieved, such as during excitotoxic stimuli, mitochondrial Ca^{2+} uptake leads to pathological activation of a conductance known as the mitochondrial permeability transition (mPT),[132-134] associated with apoptotic neuronal death processes.[135-137] The mPT is thought to be formed by the association of several mitochondrial membrane proteins including the voltage-dependent anion channel (VDAC), adenine nucleotide translocase (ANT), and cyclophilin D to create a pore that allows the movement of ions and small proteins (up to ~1.5 kDa) out of the mitochondria.[138-140]

Activation of the mPT short-circuits $\Delta\Psi_m$, preventing ATP generation and allowing the release of Ca^{2+} and apoptotic factors into the cytosol, including cytochrome c which leads to activation of caspase-9.[134,140-142] NMDA-induced mPT activity has been observed in murine MSNs,[143] and inhibitors of mPT formation have been shown to prevent NMDA-induced apoptosis in YAC HD MSN culture preparations, including YAC46 and YAC128.[48,119] Following prolonged NMDAR stimulation, the release of cytochrome c from mitochondria and activation of caspase-9 has been shown to occur in primary MSN cultures from YAC128 and YAC46 mice, respectively,[48,119] indicating induction of the mPT and intrinsic apoptotic pathways.

Although YAC128 MSNs are more sensitive than YAC18 MSNs to most toxic stimuli that increase intracellular Ca^{2+}, the largest differences were reported in response to NMDAR activation.[120] This finding is consistent with previous work suggesting that mitochondria preferentially buffer Ca^{2+} entering via NMDARs and that Ca^{2+} influx via these receptors has privileged access to mitochondria compared to the Ca^{2+} increases from other sources.[144,145] Thus, the route of Ca^{2+} entry may play a significant role in determining the extent of mhtt-induced enhanced toxicity.

Several observations indicated that mhtt may directly impair mitochondrial function. Studies of human HD brain tissue have noted decreased enzyme activity in a number of electron transport chain components, including complexes II, III and IV[146-149] and aconitase,[150] enzymes involved in mitochondrial ATP generation.

Indeed, defects in mitochondrial complex II have also been noted in MSNs expressing HttN171-82Q.[151] ATP-to-ADP ratios that reflect mitochondrial energy production decrease with increasing CAG repeat length in human lymphoblastoid cell lines and striatal neurons from a knock-in HD mouse model.[125] Other studies that used mitochondrial toxins[120,128] or conditions under which mitochondria are solely relied upon for energy production[131] indicate that mitochondrial function is impaired by expression of mhtt. Finally, studies using systemic inhibitors of mitochondrial complex II to mimic HD pathology in animal models in a manner sensitive to NMDAR antagonists[108,111] illustrate that mitochondrial dysfunction, alone or in conjunction with upstream alterations in NMDAR function and Ca^{2+} handling changes, may impact neuronal survival, especially of MSNs, in the context of HD.

2.4.3 IMPACTS OF NMDA RECEPTOR ALTERATIONS AND MITOCHONDRIAL DYSFUNCTION ON CELL DEATH

In summary, a variety of studies have indicated that NMDAR function and signaling and also mitochondrial function are altered in tissues from humans with HD and in mouse models of HD. Altered NMDAR activity and signaling result in enhanced calcium loads and mitochondrial stress. In turn, mitochondrial dysfunction leads to impaired ability to buffer NMDAR-mediated calcium loads and mitigate free radical damage. Resultant reduced ATP levels may impact ability to maintain resting membrane potential, leading to relief of magnesium block and further enhancement of NMDAR activity. In addition, a lower threshold for induction of the mPT[130,162] facilitates triggering of the apoptotic death program. Together, these changes increase vulnerability of striatal MSNs to excitotoxic cell death (Figure 2.1).

2.5 ALTERATIONS OF NMDA RECEPTOR-MEDIATED SYNAPTIC TRANSMISSION AND PLASTICITY IN HD

The use of animal models of HD (reviewed in Section 2.2) allows the detailed study of changes in synaptic function and neurotransmission that may be relevant to HD pathogenesis or progression, or underlie neuronal dysfunction prior to death. Studies of corticostriatal pathway function in this context revealed both pre- and post-synaptic changes associated with mhtt expression. Alterations of synaptic plasticity have been documented in several brain regions. The next section briefly reviews these changes focusing on those that directly impact or rely on NMDAR function.

2.5.1 CHANGES IN CORTICOSTRIATAL PATHWAY NEUROTRANSMISSION

Presynaptic changes in neurotransmitter release may presage synaptic dysfunction in HD. For example, while no changes in glutamate release probability are apparent in presymptomatic YAC72 mice compared with WT corticostriatal slices,[116] abnormally large spontaneous EPSCs in MSNs in corticostriatal slices that coincide with the onset of observable symptoms occur at higher frequencies in the R6/2 mouse model.[152] Interestingly, the frequency of spontaneous EPSCs decreased over time in R6/2 mice as the HD phenotype became more apparent[152] and paired-pulse

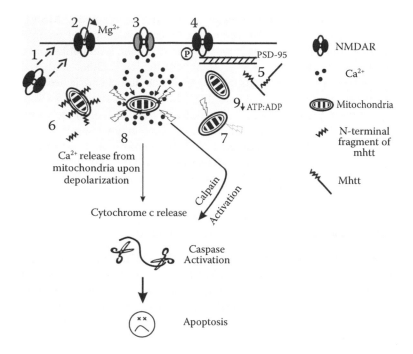

FIGURE 2.1 Overview of functional consequences of mhtt–NMDAR interaction. Mhtt influences MSN responses to NMDAR activation at several points including the NMDAR, cytosolic Ca^{2+} handling, and mitochondrial function. Reported effects include (1) increased rate of NMDAR forward trafficking leading to accumulation of NMDARs at plasma membranes; (2) decreased sensitivity of NMDARs to Mg^{2+} block; (3) possible changes in NMDAR subunit composition over time; (4) increased NMDAR function due to enhanced phosphorylation of serine and tyrosine residues; (5) decreased interaction between PSD-95 and mhtt that produces a net increase in NMDAR-mediated cell death; (6) embedding of mhtt fragments in mitochondrial membranes, associated with impaired mitochondrial function; (7) mitochondria in MSNs expressing mhtt show decreased resting membrane potentials and depolarize more readily (8) when challenged by increasing free Ca^{2+}; and (9) mitochondria in cells expressing mhtt have reduced ability to generate ATP via oxidative phosphorylation. These pathological events downstream of NMDAR activation can result in neurotoxicity via activation of the intrinsic apoptotic pathway where cytochrome c is released into the cytosol from injured mitochondria, activating caspases and resulting in apoptosis. Elevated levels of Ca^{2+} in the cytosol may also increase calpain activity, an event associated with neurotoxicity.

facilitation was reduced in both presymptomatic and symptomatic R6/2 mice.[153] Morphologically, symptomatic R6/2 mice had lower dendritic spine densities, reduced dendritic shaft diameters, and smaller dendritic fields—changes that would likely impair neurotransmission.[153]

One early postsynaptic change reported in presymptomatic YAC HD mice is selective enhancement of evoked synaptic NMDAR currents, as NMDAR- to AMPAR-mediated EPSC amplitude ratios were significantly increased in YAC72 mice in the absence of changes in presynaptic function.[114,116] Enhanced sensitivity of

MSNs in striatal slices to exogenous NMDA application has also been noted in other mouse models.[46,104,116] Hence, altered excitatory neurotransmission at corticostriatal synapses may play a role in neuronal dysfunction prior to neurodegeneration.

2.5.2 Synaptic Plasticity in HD Models

The cognitive changes that occur in HD patients relatively early in disease progression may, in part, reflect changes in NMDAR synaptic signalling associated with mhtt expression. A number of studies demonstrated alterations in activity-dependent synaptic function at a cellular level in HD models, implicating these changes as basic aberrations underlying higher order dysfunction.

A reduction in presynaptic function is a possible reason for impaired LTP induction noted in a knock-in HD mouse model.[154] Basal CA3-CA1 synaptic transmission in hippocampal slices is normal in these animals, despite a decrease in mean LTP magnitude.[57] Note that the mechanisms for LTP induction are present, but these synapses have increased thresholds of activity required for LTP expression, likely due to decreased release of glutamate during high frequency stimulation (HFS).[154]

Hodgson and colleagues[62] found that hippocampal LTP could not be induced in 10-month old YAC72 CA1 synapses despite observations of HFS-induced NMDAR-mediated hyperexcitability and enhanced short-term potentiation in hippocampal slices from the same animals at 6 months. In fact, HFS in hippocampal slices from 10-month old YAC72 CA1 synapses produced depression rather than potentiation.[62] These data further support a gradual mhtt-mediated impairment of LTP, possibly due to ongoing changes in postsynaptic NMDAR function.[62] Hippocampal LTD and LTP were also reported to be altered in R6 mouse models of HD. These changes have been ascribed to postsynaptic changes in NMDAR function occurring prior to an observable behavioral phenotype.[155,156] In addition to these examples of altered synaptic plasticity in the hippocampus in HD mouse models, changes in cortical plasticity in presymptomatic R6/1 HD transgenic mice have also been reported.[157–159]

Recent studies of the striatum—the most severely affected brain region in HD—revealed changes in corticostriatal plasticity. The administration of 3-NP to produce HD-like striatal lesions in rats produced MSN responses to HFS similar to those seen in symptomatic R6/2 mice; in both HD models, LTP could be induced, but the synapses were resistant to depotentiation of LTP by low frequency stimulation (LFS).[160] This alteration in depotentiation was sensitive to mAChR antagonists, suggesting that alterations in cholinergic neurotransmission rather than NMDAR-mediated activity are responsible.[160] Overall, the role of NMDARs in the corticostriatal synapse and the influence of altered NMDAR function and/or composition on synaptic plasticity in HD have only begun to be elucidated.

2.6 POSSIBLE MECHANISMS FOR MODULATION OF NMDA RECEPTOR FUNCTION BY MHTT

Mhtt has been demonstrated to alter NMDAR function in several ways. Peak NMDAR current density was increased in YAC72 MSNs relative to MSNs from WT animals, whereas no such potentiation was seen in cortical pyramidal neurons from the same animals.[115] In addition to a region-specific regulation of NMDAR currents, currents evoked by kainate application were not enhanced in YAC72 MSNs, indicating a receptor type-specific mechanism.[115] An observed increase in NMDAR: AMPAR-mediated EPSC amplitudes in MSNs in corticostriatal slices from YAC72 compared to WT mice indicates that the mechanisms underlying these alterations are not restricted to isolated MSNs expressed in primary cultures.[116]

Earlier research in HEK cells coexpressing NMDARs with mhtt did not reveal a specific effect on single channel function.[42] How was this selective enhancement of whole-cell NMDAR activity achieved? Fan and colleagues[115] demonstrated a shift of NMDARs from internal pools to plasma membranes and a significantly faster rate of NMDAR insertion to the surface in YAC72 MSNs than in WT MSNs in a manner selective for NMDAR subunits but not the AMPAR subunit GluR1. An enrichment of the C2′ cassette in NR1 that possibly accelerated the release of NMDARs from the ER[161] was also reported.[115] Hence mhtt expression produced a redistribution of NMDARs, resulting in increased surface NMDAR expression.

While no mhtt-induced shift of NMDAR subunit composition appeared in YAC mouse models of HD, a recent study in the R6/2 model of HD demonstrated a significant decrease in the percentage of MSNs expressing mRNA for NR2A at all ages including presymptomatic time points without significant changes in mRNA for NR1 and NR2B (at least until 12 weeks of age).[99] This early shift in NMDAR subunit composition coupled with increased sensitivity to NMDA and reduced sensitivity to block by Mg^{2+} may contribute to dysfunctional transmission at the corticostriatal synapse.[49]

Mhtt may also modulate NMDARs via intermediate interacting proteins. In a heterologous system, mhtt expression increased Src-mediated tyrosine phosphorylation of NMDARs, an effect enhanced by expression of PSD-95.[118] Other mhtt-mediated alterations of NMDAR phosphorylation were documented in N171-82Q mouse models of HD: down-regulation of PSD-95 expression and of the dopamine D1 receptor pathway that normally acts via protein kinase A activation to phosphorylate Ser897 of NR1 and increase NMDAR activity.[56] These events were interpreted as compensatory measures to reduce NMDAR-mediated excitotoxicity but could have the undesired effect of synaptic dysfunction.[56] Finally, while htt indirectly interacts with NMDARs via PSD-95, mhtt has a reduced ability to interact with PSD-95, increasing the vulnerability of neurons to glutamate-mediated excitotoxicity.[118] Thus, several possible mechanisms may allow mhtt to modulate NMDAR function at the receptor level and may contribute to excitotoxicity in HD (see Figure 2.1).

2.7 SUMMARY

Ample evidence suggests a critical role for NMDARs in initiating MSN dysfunction and death in HD. The expression of mhtt may produce changes in NMDAR composition, trafficking, and function. Additionally, it may affect neuronal responses to NMDAR activation, particularly those downstream of and dependent upon Ca^{2+} signalling and mitochondrial pathways (Figure 2.1). Evidence from cellular and animal models of HD suggests that these changes in turn can alter synaptic function and enhance the susceptibility of MSNs to NMDAR-mediated excitotoxicity in a polyQ length-dependent fashion. In summary, NMDARs may play a prominent role in the pathology of HD. The complexity of the interactions between mhtt and NMDARs presents a number of possible targets for effective and selective therapies.

REFERENCES

1. Vonsattel, J.P. and DiFiglia, M., Huntington disease, *J. Neuropathol. Exp. Neurol.*, 57, 369, 1998.
2. Huntington's Disease Collaborative Research Group, A novel gene containing a trinucleotide repeat that is expanded and unstable on Huntington's disease chromosomes, *Cell*, 72, 971, 1993.
3. Kremer, B. et al., A worldwide study of the Huntington's disease mutation: the sensitivity and specificity of measuring CAG repeats, *New. Engl. J. Med.*, 330, 1401, 1994.
4. Rubinsztein, D.C. et al., Phenotypic characterization of individuals with 30–40 CAG repeats in the Huntington disease (HD) gene reveals HD cases with 36 repeats and apparently normal elderly individuals with 36–39 repeats, *Am. J. Hum. Genet.*, 59, 16, 1996.
5. Tobin, A.J. and Signer, E.R., Huntington's disease: the challenge for cell biologists, *Trends Cell. Biol.*, 10, 531, 2000.
6. Ross, C.A., Polyglutamine pathogenesis: emergence of unifying mechanisms for Huntington's disease and related disorders, *Neuron*, 35, 819, 2002.
7. Hayden, M., *Huntington's Chorea*. 1981, Berlin: Springer-Verlag. 192.
8. Harper, P.S., *Huntington's Disease*, 2nd ed. Harper, P.S., Ed. 1996, London: W.B. Saunders. 438.
9. Paulsen, J.S. et al., Clinical markers of early disease in persons near onset of Huntington's disease, *Neurology*, 57, 658, 2001.
10. Brandt, J. et al., Trinucleotide repeat length and clinical progression in Huntington's disease, *Neurology*, 46, 527, 1996.
11. Myers, R.H. et al., Factors related to onset age of Huntington disease, *Am. J. Hum. Genet.*, 34, 481, 1982.
12. Duyao, M. et al., Trinucleotide repeat length instability and age of onset in Huntington's disease, *Nat. Genet.*, 4, 387, 1993.
13. Andrew, S.E. et al., The relationship between trinucleotide (CAG) repeat length and clinical features of Huntington's disease, *Nat. Genet.*, 4, 398, 1993.
14. Stine, O.C. et al., Correlation between the onset age of Huntington's disease and length of the trinucleotide repeat in IT-15, *Hum. Mol. Genet.*, 2, 1547, 1993.
15. Brinkman, R.R. et al., The likelihood of being affected with Huntington disease by a particular age, for a specific CAG size, *Am. J. Hum. Genet.*, 60, 1202, 1997.
16. Graveland, G.A., Williams, R.S., and DiFiglia, M., Evidence for degenerative and regenerative changes in neostriatal spiny neurons in Huntington's disease, *Science*, 227, 770, 1985.
17. Surmeier, D.J., Bargas, J., and Kitai, S.T., Voltage-clamp analysis of a transient potassium current in rat neostriatal neurons, *Brain Res.*, 473, 187, 1988.
18. Vonsattel, J.P. et al., Neuropathological classification of Huntington's disease, *J. Neuropathol. Exp. Neurol.*, 44, 559, 1985.
19. Richfield, E.K. et al., Preferential loss of preproenkephalin versus preprotachykinin neurons from the striatum of Huntington's disease patients, *Ann. Neurol.*, 38, 852, 1995.
20. Ferrante, R.J. et al., Selective sparing of a class of striatal neurons in Huntington's disease, *Science*, 230, 561, 1985.

21. Ferrante, R.J. et al., Morphologic and histochemical characteristics of a spared subset of striatal neurons in Huntington's disease, *J. Neuropathol. Exp. Neurol.*, 46, 12, 1987.

22. de la Monte, S.M., Vonsattel, J.P., and Richardson, E.P., Jr., Morphometric demonstration of atrophic changes in the cerebral cortex, white matter, and neostriatum in Huntington's disease, *J. Neuropathol. Exp. Neurol.*, 47, 516, 1988.

23. Cowan, C.M. and Raymond, L.A., Selective neuronal degeneration in Huntington's disease, *Curr. Top. Dev. Biol.*, 75, 25, 2006.

24. Cudkowicz, M. and Kowall, N.W., Degeneration of pyramidal projection neurons in Huntington's disease cortex, *Ann. Neurol.*, 27, 200, 1990.

25. Hedreen, J.C. et al., Neuronal loss in layers V and VI of cerebral cortex in Huntington's disease, *Neurosci. Lett.*, 133, 257, 1991.

26. Strong, T.V. et al., Widespread expression of the human and rat Huntington's disease gene in brain and nonneural tissues, *Nat. Genet.*, 5, 259, 1993.

27. Landwehrmeyer, G.B. et al., Huntington's disease gene: regional and cellular expression in brain of normal and affected individuals, *Ann. Neurol.*, 37, 218, 1995.

28. Sharp, A.H. et al., Widespread expression of Huntington's disease gene (IT15) protein product, *Neuron*, 14, 1065, 1995.

29. Aronin, N. et al., CAG expansion affects the expression of mutant huntingtin in the Huntington's disease brain, *Neuron*, 15, 1193, 1995.

30. Choo, Y.S. et al., Mutant huntingtin directly increases susceptibility of mitochondria to the calcium-induced permeability transition and cytochrome c release, *Hum. Mol. Genet.*, 13, 1407, 2004.

31. DiFiglia, M. et al., Huntingtin is a cytoplasmic protein associated with vesicles in human and rat brain neurons, *Neuron*, 14, 1075, 1995.

32. Velier, J. et al., Wild-type and mutant huntingtins function in vesicle trafficking in the secretory and endocytic pathways, *Exp. Neurol.*, 152, 34, 1998.

33. Kegel, K.B. et al., Huntingtin is present in the nucleus, interacts with the transcriptional corepressor C-terminal binding protein, and represses transcription, *J. Biol. Chem.*, 277, 7466, 2002.

34. Nasir, J. et al., Targeted disruption of the Huntington's disease gene results in embryonic lethality and behavioral and morphological changes in heterozygotes, *Cell*, 81, 811, 1995.

35. Zeitlin, S. et al., Increased apoptosis and early embryonic lethality in mice nullizygous for the Huntington's disease gene homologue, *Nat. Genet.*, 11, 155, 1995.

36. Leavitt, B.R. et al., Wild-type huntingtin reduces the cellular toxicity of mutant huntingtin in vivo, *Am. J. Hum. Genet.*, 68, 313, 2001.

37. Monyer, H. et al., Heteromeric NMDARs: molecular and functional distinction of subtypes, *Science*, 256, 1217, 1992.

38. Cik, M., Chazot, P.L., and Stephenson, F.A., Expression of NMDAR1-1a (N598Q)/NMDAR2A receptors results in decreased cell mortality, *Eur. J. Pharmacol.*, 266, R1, 1994.

39. Anegawa, N.J. et al., Transfection of N-methyl-D-aspartate receptors in a nonneuronal cell line leads to cell death, *J. Neurochem.*, 64, 2004, 1995.

40. Raymond, L.A. et al., Glutamate receptor ion channel properties predict vulnerability to cytotoxicity in a transfected nonneuronal cell line, *Mol. Cell. Neurosci.*, 7, 102, 1996.

41. Anegawa, N.J. et al., N-Methyl-D-aspartate receptor mediated toxicity in nonneuronal cell lines: characterization using fluorescent measures of cell viability and reactive oxygen species production, *Brain Res. Mol. Brain. Res.*, 77, 163, 2000.

42. Chen, N. et al., Subtype-specific enhancement of NMDAR currents by mutant huntingtin, *J. Neurochem.*, 72, 1890, 1999.

43. Zeron, M.M. et al., Mutant huntingtin enhances excitotoxic cell death, *Mol. Cell. Neurosci.*, 17, 41, 2001.
44. Boeckman, F.A. and Aizenman, E., Pharmacological properties of acquired excitotoxicity in Chinese hamster ovary cells transfected with N-methyl-D-aspartate receptor subunits, *J. Pharmacol. Exp. Ther.*, 279, 515, 1996.
45. Rameau, G.A. et al., Role of NMDAR functional domains in excitatory cell death, *Neuropharmacology*, 39, 2255, 2000.
46. Cepeda, C. et al., NMDAR function in mouse models of Huntington disease, *J. Neurosci. Res.*, 66, 525, 2001.
47. Zeron, M.M. et al., Increased sensitivity to N-methyl-D-aspartate receptor-mediated excitotoxicity in a mouse model of Huntington's disease, *Neuron*, 33, 849, 2002.
48. Zeron, M.M. et al., Potentiation of NMDAR-mediated excitotoxicity linked with intrinsic apoptotic pathway in YAC transgenic mouse model of Huntington's disease, *Mol. Cell. Neurosci.*, 25, 469, 2004.
49. Starling, A.J. et al., Alterations in N-methyl-D-aspartate receptor sensitivity and magnesium blockade occur early in development in the R6/2 mouse model of Huntington's disease, *J. Neurosci. Res.*, 82, 377, 2005.
50. Mangiarini, L. et al., Exon 1 of the HD gene with an expanded CAG repeat is sufficient to cause a progressive neurological phenotype in transgenic mice, *Cell*, 87, 493, 1996.
51. Davies, S.W. et al., From neuronal inclusions to neurodegeneration: neuropathological investigation of a transgenic mouse model of Huntington's disease, *Philos. Trans. R. Soc. Lond. B. Biol. Sci.*, 354, 981, 1999.
52. Schilling, G. et al., Intranuclear inclusions and neuritic aggregates in transgenic mice expressing a mutant N-terminal fragment of huntingtin, *Hum. Mol. Genet.*, 8, 397, 1999.
53. Luthi-Carter, R. et al., Decreased expression of striatal signaling genes in a mouse model of Huntington's disease, *Hum. Mol. Genet.*, 9, 1259, 2000.
54. Hansson, O. et al., Resistance to NMDA toxicity correlates with appearance of nuclear inclusions, behavioural deficits and changes in calcium homeostasis in mice transgenic for exon 1 of the Huntington gene, *Eur. J. Neurosci.*, 14, 1492, 2001.
55. MacGibbon, G.A. et al., Immediate-early gene response to methamphetamine, haloperidol, and quinolinic acid is not impaired in Huntington's disease transgenic mice, *J. Neurosci. Res.*, 67, 372, 2002.
56. Jarabek, B.R., Yasuda, R.P., and Wolfe, B.B., Regulation of proteins affecting NMDAR-induced excitotoxicity in a Huntington's mouse model, *Brain*, 127, 505, 2004.
57. Shelbourne, P.F. et al., A Huntington's disease CAG expansion at the murine Hdh locus is unstable and associated with behavioural abnormalities in mice, *Hum. Mol. Genet.*, 8, 763, 1999.
58. Wheeler, V.C. et al., Long glutamine tracts cause nuclear localization of a novel form of huntingtin in medium spiny striatal neurons in HdhQ92 and HdhQ111 knock-in mice, *Hum. Mol. Genet.*, 9, 503, 2000.
59. Lin, C.H. et al., Neurological abnormalities in a knock-in mouse model of Huntington's disease, *Hum. Mol. Genet.*, 10, 137, 2001.
60. Menalled, L.B. et al., Early motor dysfunction and striosomal distribution of huntingtin microaggregates in Huntington's disease knock-in mice, *J. Neurosci.*, 22, 8266, 2002.
61. Menalled, L.B. et al., Time course of early motor and neuropathological anomalies in a knock-in mouse model of Huntington's disease with 140 CAG repeats, *J. Comp. Neurol.*, 465, 11, 2003.
62. Hodgson, J.G. et al., A YAC mouse model for Huntington's disease with full-length mutant huntingtin, cytoplasmic toxicity, and selective striatal neurodegeneration, *Neuron*, 23, 181, 1999.

63. Slow, E.J. et al., Selective striatal neuronal loss in a YAC128 mouse model of Huntington disease, *Hum. Mol. Genet.*, 12, 1555, 2003.
64. Reddy, P.H. et al., Behavioural abnormalities and selective neuronal loss in HD transgenic mice expressing mutated full-length HD cDNA, *Nat. Genet.*, 20, 198, 1998.
65. Van Raamsdonk, J.M. et al., Cognitive dysfunction precedes neuropathology and motor abnormalities in the YAC128 mouse model of Huntington's disease, *J. Neurosci.*, 25, 4169, 2005.
66. Sattler, R. and Tymianski, M., Molecular mechanisms of glutamate receptor-mediated excitotoxic neuronal cell death, *Mol. Neurobiol.*, 24, 107, 2001.
67. Mattson, M.P., Excitotoxic and excitoprotective mechanisms: abundant targets for the prevention and treatment of neurodegenerative disorders, *Neuromolecular Med.*, 3, 65, 2003.
68. Dingledine, R. et al., The glutamate receptor ion channels, *Pharmacol. Rev.*, 51, 7, 1999.
69. Ishii, T. et al., Molecular characterization of the family of the N-methyl-D-aspartate receptor subunits, *J. Biol. Chem.*, 268, 2836, 1993.
70. Monyer, H. et al., Developmental and regional expression in the rat brain and functional properties of four NMDARs, *Neuron*, 12, 529, 1994.
71. Flint, A.C. et al., NR2A subunit expression shortens NMDAR synaptic currents in developing neocortex, *J. Neurosci.*, 17, 2469, 1997.
72. Chen, N., Luo, T., and Raymond, L.A., Subtype-dependence of NMDAR channel open probability, *J. Neurosci.*, 19, 6844, 1999.
73. Misra, C. et al., Slow deactivation kinetics of NMDARs containing NR1 and NR2D subunits in rat cerebellar Purkinje cells, *J. Physiol.*, 525 (Pt 2), 299, 2000.
74. Akazawa, C. et al., Differential expression of five N-methyl-D-aspartate receptor subunit mRNAs in the cerebellum of developing and adult rats, *J. Comp. Neurol.*, 347, 150, 1994.
75. Sheng, M. et al., Changing subunit composition of heteromeric NMDARs during development of rat cortex, *Nature*, 368, 144, 1994.
76. Li, J.H. et al. Developmental changes in localization of NMDAR subunits in primary cultures of cortical neurons, *Eur. J. Neurosci.*, 10, 1704, 1998.
77. Chapman, D.E., Keefe, K.A., and Wilcox, K.S., Evidence for functionally distinct synaptic NMDARs in ventromedial versus dorsolateral striatum, *J. Neurophysiol.*, 89, 69, 2003.
78. Buller, A.L. et al., The molecular basis of NMDAR subtypes: native receptor diversity is predicted by subunit composition, *J. Neurosci.*, 14, 5471, 1994.
79. Vicini, S. et al., Functional and pharmacological differences between recombinant N-methyl-D-aspartate receptors, *J. Neurophysiol.*, 79, 555, 1998.
80. Christie, J.M., Jane, D.E., and Monaghan, D.T., Native N-methyl-D-aspartate receptors containing NR2A and NR2B subunits have pharmacologically distinct competitive antagonist binding sites, *J. Pharmacol. Exp. Ther.*, 292, 1169, 2000.
81. Tovar, K.R. and Westbrook, G.L., The incorporation of NMDARs with a distinct subunit composition at nascent hippocampal synapses in vitro, *J. Neurosci.*, 19, 4180, 1999.
82. Barria, A. and Malinow, R., Subunit-specific NMDAR trafficking to synapses, *Neuron*, 35, 345, 2002.
83. Stocca, G. and Vicini, S., Increased contribution of NR2A subunit to synaptic NMDARs in developing rat cortical neurons, *J. Physiol.*, 507 (Pt 1), 13, 1998.
84. Hardingham, G.E., Fukunaga, Y., and Bading, H., Extrasynaptic NMDARs oppose synaptic NMDARs by triggering CREB shut-off and cell death pathways, *Nat. Neurosci.*, 5, 405, 2002.
85. Hardingham, G.E. and Bading, H., The yin and yang of NMDAR signalling, *Trends. Neurosci.*, 26, 81, 2003.

86. Liu, L. et al., Role of NMDAR subtypes in governing the direction of hippocampal synaptic plasticity, *Science*, 304, 1021, 2004.
87. Massey, P.V. et al., Differential roles of NR2A and NR2B-containing NMDARs in cortical long-term potentiation and long-term depression, *J. Neurosci.*, 24, 7821, 2004.
88. Liu, Y. et al., NMDAR subunits have differential roles in mediating excitotoxic neuronal death both *in vitro* and *in vivo*, *J. Neurosci.*, 27, 2846, 2007.
89. von Engelhardt, J. et al., Excitotoxicity in vitro by NR2A- and NR2B-containing NMDARs, *Neuropharmacology*, 53, 10, 2007.
90. Landwehrmeyer, G.B. et al., NMDAR subunit mRNA expression by projection neurons and interneurons in rat striatum, *J. Neurosci.*, 15, 5297, 1995.
91. Ghasemzadeh, M.B. et al., Multiplicity of glutamate receptor subunits in single striatal neurons: an RNA amplification study, *Mol. Pharmacol.*, 49, 852, 1996.
92. Rigby, M. et al., The messenger RNAs for the N-methyl-D-aspartate receptor subunits show region-specific expression of different subunit composition in the human brain, *Neuroscience*, 73, 429, 1996.
93. Kuppenbender, K.D. et al., Localization of alternatively spliced NMDAR1 glutamate receptor isoforms in rat striatal neurons, *J. Comp. Neurol.*, 415, 204, 1999.
94. Standaert, D.G. et al., Expression of NMDA glutamate receptor subunit mRNAs in neurochemically identified projection and interneurons in the striatum of the rat, *Brain Res. Mol. Brain Res.*, 64, 11, 1999.
95. Li, L. et al., Role of NR2B-type NMDARs in selective neurodegeneration in Huntington disease, *Neurobiol. Aging*, 24, 1113, 2003.
96. Young, A.B. et al., NMDAR losses in putamen from patients with Huntington's disease, *Science*, 241, 981, 1988.
97. Albin, R.L. et al., Striatal and nigral neuron subpopulations in rigid Huntington's disease: implications for the functional anatomy of chorea and rigidity-akinesia, *Ann. Neurol.*, 27, 357, 1990.
98. Albin, R.L. et al., Abnormalities of striatal projection neurons and N-methyl-D-aspartate receptors in presymptomatic Huntington's disease, *N. Engl. J. Med.*, 322, 1293, 1990.
99. Ali, N.J. and Levine, M.S., Changes in expression of N-methyl-D-aspartate receptor subunits occur early in the R6/2 mouse model of Huntington's disease, *Dev. Neurosci.*, 28, 230, 2006.
100. Cha, J.H. et al., Altered neurotransmitter receptor expression in transgenic mouse models of Huntington's disease, *Philos. Trans. R. Soc. Lond .B. Biol. Sci.*, 354, 981, 1999.
101. Benn, C.L. et al., Glutamate receptor abnormalities in the YAC128 transgenic mouse model of Huntington's disease, *Neuroscience*, 147, 354, 2007.
102. Beal, M.F. et al., Replication of the neurochemical characteristics of Huntington's disease by quinolinic acid, *Nature*, 321, 168, 1986.
103. Sanberg, P.R. et al., The quinolinic acid model of Huntington's disease: locomotor abnormalities, *Exp. Neurol.*, 105, 45, 1989.
104. Hantraye, P. et al., A primate model of Huntington's disease: behavioral and anatomical studies of unilateral excitotoxic lesions of the caudate-putamen in the baboon, *Exp. Neurol.*, 108, 91, 1990.
105. Beal, M.F. et al., Chronic quinolinic acid lesions in rats closely resemble Huntington's disease, *J. Neurosci.*, 11, 1649, 1991.
106. Cepeda, C. et al., Differential sensitivity of medium- and large-sized striatal neurons to NMDA but not kainate receptor activation in the rat, *Eur. J. Neurosci.*, 14, 1577, 2001.
107. DiFiglia, M., Excitotoxic injury of the neostriatum: a model for Huntington's disease, *Trends Neurosci.*, 13, 286, 1990.

108. Beal, M.F. et al., Neurochemical and histologic characterization of striatal excito-toxic lesions produced by the mitochondrial toxin 3-nitropropionic acid, *J. Neurosci.*, 13, 4181, 1993.
109. Greene, J.G. et al., Inhibition of succinate dehydrogenase by malonic acid produces an "excitotoxic" lesion in rat striatum, *J. Neurochem.*, 61, 1151, 1993.
110. Brouillet, E. et al., Chronic mitochondrial energy impairment produces selective striatal degeneration and abnormal choreiform movements in primates, *Proc. Natl. Acad. Sci. USA*, 92, 7105, 1995.
111. Bogdanov, M.B. et al., Increased vulnerability to 3-nitropropionic acid in an animal model of Huntington's disease, *J. Neurochem.*, 71, 2642, 1998.
112. Andre, V.M. et al., Altered cortical glutamate receptor function in the R6/2 model of Huntington's disease, *J. Neurophysiol.*, 95, 2108, 2006.
113. Levine, M.S. et al., Enhanced sensitivity to N-methyl-D-aspartate receptor activation in transgenic and knockin mouse models of Huntington's disease, *J. Neurosci. Res.*, 58, 515, 1999.
114. Laforet, G.A. et al., Changes in cortical and striatal neurons predict behavioral and electrophysiological abnormalities in a transgenic murine model of Huntington's disease, *J. Neurosci.*, 21, 9112, 2001.
115. Fan, M.M. et al., Altered NMDAR trafficking in a yeast artificial chromosome trans-genic mouse model of Huntington's disease, *J. Neurosci.*, 27, 3768, 2007.
116. Li, L. et al., Enhanced striatal NR2B-containing N-methyl-D-aspartate receptor-mediated synaptic currents in a mouse model of Huntington disease, *J. Neurophysiol.*, 92, 2738, 2004.
117. Sun, Y. et al., Polyglutamine-expanded huntingtin promotes sensitization of N-methyl-D-aspartate receptors via post-synaptic density 95, *J. Biol. Chem.*, 276, 24713, 2001.
118. Song, C. et al., Expression of polyglutamine-expanded huntingtin induces tyrosine phosphorylation of N-methyl-D-aspartate receptors, *J. Biol. Chem.*, 278, 33364, 2003.
119. Tang, T.S. et al., Disturbed Ca2+ signaling and apoptosis of medium spiny neurons in Huntington's disease, *Proc. Natl. Acad. Sci. USA*, 102, 2602, 2005.
120. Shehadeh, J. et al., Striatal neuronal apoptosis is preferentially enhanced by NMDAR activation in YAC transgenic mouse model of Huntington disease, *Neurobiol. Dis.*, 21, 392, 2006.
121. Khodorov, B., Glutamate-induced deregulation of calcium homeostasis and mito-chondrial dysfunction in mammalian central neurones, *Prog. Biophys. Mol. Biol.*, 86, 279, 2004.
122. Nicholls, D.G., Mitochondrial dysfunction and glutamate excitotoxicity studied in primary neuronal cultures, *Curr. Mol. Med.*, 4, 149, 2004.
123. Burnashev, N. et al., Fractional calcium currents through recombinant GluR channels of the NMDA, AMPA and kainate receptor subtypes, *J. Physiol.*, 485 (Pt 2), 403, 1995.
124. Schneggenburger, R., Simultaneous measurement of Ca^{2+} influx and reversal poten-tials in recombinant N-methyl-D-aspartate receptor channels, *Biophys. J.*, 70, 2165, 1996.
125. Seong, I.S. et al., HD CAG repeat implicates a dominant property of huntingtin in mitochondrial energy metabolism, *Hum. Mol. Genet.*, 14, 2871, 2005.
126. Tang, T.S. et al., Huntingtin and huntingtin-associated protein 1 influence neuronal calcium signaling mediated by inositol-(1,4,5) triphosphate receptor type 1, *Neuron*, 39, 227, 2003.
127. Nicholls, D.G. and Ward, M.W., Mitochondrial membrane potential and neuronal glutamate excitotoxicity: mortality and millivolts, *Trends. Neurosci.*, 23, 166, 2000.

128. Schinder, A.F. et al., Mitochondrial dysfunction is a primary event in glutamate neurotoxicity, *J. Neurosci.*, 16, 6125, 1996.
129. Sawa, A. et al., Increased apoptosis of Huntington disease lymphoblasts associated with repeat length-dependent mitochondrial depolarization, *Nat. Med.*, 5, 1194, 1999.
130. Panov, A.V. et al., Early mitochondrial calcium defects in Huntington's disease are a direct effect of polyglutamines, *Nat. Neurosci.*, 5, 731, 2002.
131. Oliveira, J.M. et al., Mitochondrial-dependent Ca^{2+} handling in Huntington's disease striatal cells: effect of histone deacetylase inhibitors, *J. Neurosci.*, 26, 11174, 2006.
132. White, R.J. and Reynolds, I.J., Mitochondrial depolarization in glutamate-stimulated neurons: an early signal specific to excitotoxin exposure, *J. Neurosci.*, 16, 5688, 1996.
133. Dubinsky, J.M. and Levi, Y., Calcium-induced activation of the mitochondrial permeability transition in hippocampal neurons, *J. Neurosci. Res.*, 53, 728, 1998.
134. Crompton, M., The mitochondrial permeability transition pore and its role in cell death, *Biochem. J.*, 341 (Pt 2), 233, 1999.
135. Marchetti, P. et al., Mitochondrial permeability transition is a central coordinating event of apoptosis, *J. Exp. Med.*, 184, 1155, 1996.
136. Nicholls, D.G. and Budd, S.L., Mitochondria and neuronal glutamate excitotoxicity, *Biochim. Biophys. Acta.*, 1366, 97, 1998.
137. Brustovetsky, N. et al., Calcium-induced cytochrome c release from CNS mitochondria is associated with the permeability transition and rupture of the outer membrane, *J. Neurochem.*, 80, 207, 2002.
138. Bernardi, P., Broekemeier, K.M., and Pfeiffer, D.R., Recent progress on regulation of the mitochondrial permeability transition pore; a cyclosporin-sensitive pore in the inner mitochondrial membrane, *J. Bioenerg. Biomembr.*, 26, 509, 1994.
139. Petit, P.X. et al., Mitochondria and programmed cell death: back to the future, *FEBS Lett.*, 396, 7, 1996.
140. Green, D.R. and Reed, J.C., Mitochondria and apoptosis, *Science*, 281, 1309, 1998.
141. Duchen, M.R., Mitochondria in health and disease: perspectives on a new mitochondrial biology, *Mol. Aspects Med.*, 25, 365, 2004.
142. Orrenius, S., Mitochondrial regulation of apoptotic cell death, *Toxicol. Lett.*, 149, 19, 2004.
143. Alano, C.C. et al., Mitochondrial permeability transition and calcium dynamics in striatal neurons upon intense NMDAR activation, *J. Neurochem.*, 80, 531, 2002.
144. Peng, T.I. and Greenamyre, J.T., Privileged access to mitochondria of calcium influx through N-methyl-D-aspartate receptors, *Mol. Pharmacol.*, 53, 974, 1998.
145. Sattler, R. et al., Distinct influx pathways, not calcium load, determine neuronal vulnerability to calcium neurotoxicity, *J. Neurochem.*, 71, 2349, 1998.
146. Brennan, W.A., Jr., Bird, E.D., and Aprille, J.R., Regional mitochondrial respiratory activity in Huntington's disease brain, *J. Neurochem.*, 44, 1948, 1985.
147. Beal, M.F., Aging, energy, and oxidative stress in neurodegenerative diseases, *Ann. Neurol.*, 38, 357, 1995.
148. Gu, M. et al., Mitochondrial defect in Huntington's disease caudate nucleus, *Ann. Neurol.*, 39, 385, 1996.
149. Browne, S.E. et al., Oxidative damage and metabolic dysfunction in Huntington's disease: selective vulnerability of the basal ganglia, *Ann. Neurol.*, 41, 646, 1997.
150. Tabrizi, S.J. et al., Biochemical abnormalities and excitotoxicity in Huntington's disease brain, *Ann. Neurol.*, 45, 25, 1999.
151. Benchoua, A. et al., Involvement of mitochondrial complex II defects in neuronal death produced by N-terminus fragment of mutated huntingtin, *Mol. Biol. Cell.*, 17, 1652, 2006.

152. Cepeda, C. et al., Transient and progressive electrophysiological alterations in the corticostriatal pathway in a mouse model of Huntington's disease, *J. Neurosci.*, 23, 961, 2003.

153. Klapstein, G.J. et al., Electrophysiological and morphological changes in striatal spiny neurons in R6/2 Huntington's disease transgenic mice, *J. Neurophysiol.*, 86, 2667, 2001.

154. Usdin, M.T. et al., Impaired synaptic plasticity in mice carrying the Huntington's disease mutation, *Hum. Mol. Genet.*, 8, 839, 1999.

155. Murphy, K.P. et al., Abnormal synaptic plasticity and impaired spatial cognition in mice transgenic for exon 1 of the human Huntington's disease mutation, *J. Neurosci.*, 20, 5115, 2000.

156. Milnerwood, A.J. et al., Early development of aberrant synaptic plasticity in a mouse model of Huntington's disease, *Hum. Mol. Genet.*, 15, 1690, 2006.

157. Cybulska-Klosowicz, A. et al., Impaired learning-dependent cortical plasticity in Huntington's disease transgenic mice, *Neurobiol. Dis.*, 17, 427, 2004.

158. Mazarakis, N.K. et al., Deficits in experience-dependent cortical plasticity and sensory-discrimination learning in presymptomatic Huntington's disease mice, *J. Neurosci.*, 25, 3059, 2005.

159. Cummings, D.M. et al., Aberrant cortical synaptic plasticity and dopaminergic dysfunction in a mouse model of Huntington's disease, *Hum. Mol. Genet.*, 15, 2856, 2006.

160. Picconi, B. et al., Plastic and behavioral abnormalities in experimental Huntington's disease: a crucial role for cholinergic interneurons, *Neurobiol. Dis.*, 22, 143, 2006.

161. Mu, Y. et al., Activity-dependent mRNA splicing controls ER export and synaptic delivery of NMDARs, *Neuron*, 40, 581, 2003.

162. Fernandes, H.B. et al., Mitochondrial sensitivity and altered calcium handling underlie enhanced NMDA-induced apoptosis in YAC128 model of Huntington's disease. *J. Neurosci.*, 27, 13614, 2007.

3 NMDA and Dopamine: *Diverse Mechanisms Applied to Interacting Receptor Systems*

Carlos Cepeda, Véronique M. André, Emily L. Jocoy, and Michael S. Levine

CONTENTS

3.1 INTRODUCTION

N-methyl-D-aspartate (NMDA) and dopamine (DA) receptors and their interactions control an incredible variety of functions in the intact brain and, when abnormal, these interactions underlie and contribute to numerous disease states. These receptor interactions are relevant in such diverse functions as motor control, cognition and memory, neurodegenerative disorders, schizophrenia, and addiction. It is thus not surprising that a wealth of information has been generated by the neuroscience community interested in the coordinated functions of NMDA and DA receptors. This chapter will describe the numerous mechanisms underlying DA–NMDA receptor interactions, particularly in the striatum, the main focus of our investigations.

DA modulation of spontaneous or glutamate-induced action potentials in the caudate nucleus has been known for some time.[1–3] Since the discoveries of different subtypes of glutamate and DA receptors, the number of potential interactions and their mechanisms has multiplied because the functions of glutamate and DA receptor subtypes are governed by multiple factors that tap into different types of signaling systems. Thus, the outcomes of interactions of these receptor families can be very diverse.

It has been 10 years since we published our first review summarizing known DA–NMDA receptor interactions and their mechanisms.[4] Since then, exciting findings have added new levels of complexity. For example, in addition to intracellular interactions via second messenger pathways, recent studies revealed the presence of physical interactions between NMDA and DA receptors at the membrane and cytoplasm levels. Furthermore, the generation of mice deficient of specific DA receptors or NMDAR subunits and mice expressing enhanced green fluorescent protein (EGFP) under the control of specific DA receptor subtype promoters has provided new tools for studying relationships of DA and NMDA receptors.

3.2 CLASSIFICATION AND MORPHOLOGICAL BASIS FOR INTERACTIONS AMONG GLUTAMATE AND DA RECEPTOR SUBTYPES

Glutamate receptors have been classified as ionotropic and metabotropic. Iono-tropic glutamate [α-amino-3-hydroxy-5-methyl-4-propionate (AMPA), kainate (KA), and NMDA] receptors are ligand-gated cation channels, whereas metabotropic glutamate receptors are coupled to various signal transduction systems.[5–7] NMDARs are unique in that their activation is governed by a strong voltage dependence due to receptor channel blockade by Mg^{2+} at hyperpolarized membrane potentials.[8] Mg^{2+} blockade gives NMDARs their characteristic negative slope conductance.

DA receptors also exhibit diversity. Five receptor subtypes have been cloned. They are classified into two main families: the D1 (D_1 and D_5 receptor subtypes) and the D2 families (D_2, D_3, and D_4 receptor subtypes).[9,10] All DA receptors are G protein-coupled and primarily alter the production of cAMP in cells when activated but also can affect other transduction systems. In this chapter, subscripted notations indicate DA receptor subtypes and nonsubscripted notations indicate the two DA receptor families.

The striatum is the main input structure of the basal ganglia. It is a central region where afferents from the cerebral cortex, thalamus, and substantia nigra converge and interact. Glutamate is released from cortical and, to a lesser extent, thalamic terminals.[11,12] DA is released from nigrostriatal terminals.[13] Because glutamate and DA inputs terminate on the same spines of striatal medium-sized spiny neurons (MSSNs), these sites offer the potential for physiological interactions between the glutamate and DA transmitter systems.[14] Morphological evidence demonstrates the presence of synaptic complexes formed by axospinous contacts in which the dendritic spine is the target of both an asymmetric (glutamatergic) bouton and a DA-positive symmetric synapse in striatal MSSNs.[15] This arrangement is also found in cortical pyramidal neurons[16] and provides a morphological basis for DA–glutamate receptor interactions at synapses. These interactions in the striatum support major sensory, motor, cognitive, and motivational functions.[17–21] In the cortex, they affect learning and memory[22] as well as normal and abnormal thought processes.[23]

DA receptors are also found presynaptically, where they can modulate neurotransmitter release. In the dorsal striatum, D_2 receptors are present on corticostriatal inputs and function to decrease glutamate release by presynaptic mechanisms.[24–27] Conversely, DA release can also be modulated by activation of glutamate receptors located on DA terminals.[28,29] Glutamate and DA receptor interactions are complex and their outcomes depend on a number of factors including receptor subtype, site of action (i.e., pre- or postsynaptic), timing of inputs, and concentration of neurotransmitter, to name only a few. For more exhaustive reviews see Cepeda and Levine[4] and Seamans and Yang.[30]

3.3 DA AND D1 RECEPTOR ENHANCEMENT OF NMDA RECEPTOR-MEDIATED RESPONSES

DA and D_1 receptor-mediated potentiation of NMDA responses was first described in human cortex and rodent striatum in the early 1990s.[31,32] Since then, with only a few notable exceptions,[33,34] this enhancement has been verified in these and other brain structures.[35–38] D_1 receptor potentiation of NMDA responses can lead to significant functional consequences. For example, potentiation of NMDAR-mediated responses can emphasize the most important input signals, but can also enhance glutamate activity, predisposing the system to excitotoxicity. In the striatum, activation of D_1 receptors is required for the induction of long-term potentiation (LTP),[39,40] suggesting further that activation of D1 receptors effectively amplifies cortical signals to the striatum.[41]

Although NMDA and D1 receptor interactions are clearly important, the nature and consequences of these interactions are complex and in some cases controversial or not fully elucidated. Multiple mechanisms underlie the interactions of D_1 and NMDARs and fall into two main categories: interactions through signal transduction systems and direct physical interactions (Table 3.1).

3.3.1 INTERACTIONS THROUGH SECOND MESSENGERS

D_1 receptor enhancement of NMDA responses can be mediated by a number of redundant and cooperative signaling cascades in the striatum.[4,30] The most prominent involve protein kinase A (PKA) and dopamine- and adenosine-3′,5′-monophosphate (cAMP)-regulated phosphoprotein of 32 kDa (DARPP-32),[37,42,43] phosphorylation of NMDAR NR1 subunits,[44] and activation of voltage-gated Ca^{2+} channels, particularly L-type channels.[45,46]

In other cerebral regions, different mechanisms may occur. For example, in the nucleus accumbens, NMDAR potentiation by phospholipase C-coupled

TABLE 3.1
NMDA–D1 and NMDA–D2 Receptor Interactions

Preparation	Region	Effect	Mechanism	References
		NMDA–D1 Receptor Interactions		
Brain slices, oocytes	Striatum	↑ NMDA responses	cAMP–PKA–DARPP-32	31, 32, 36, 43, 42
Dissociated cells	Cortex		Ca^{2+}	37, 44, 45, 46
Brain slices	N. Accumbens	$D_{1/5}$↑ NMDA responses	PKC, Ca^{2+}, PKA	38, 35, 112
Dissociated cells	Cortex			
Cell cultures	Striatum	↑ D_1 receptors in spines	Ca^{2+}-dependent	52
Organotypic cultures	Striatum	↑ D_1 receptors in spines	Allosteric change, diffusion trap	53
HEK293, cell cultures	Hippocampus	↓ NMDA currents ↓ Excitotoxicity	D_1-NR2A binding D_1-NR1 binding	49
HEK293, COS7, PSD	Striatum	Translocation of D_1-NR1 ↓ D_1 agonist-induced internalization	Oligomerization D_1-NR1	57
Synaptosomes from brain slices	Striatum	↑ NR1, NR2A, NR2B in synaptosomes	Fyn protein tyrosine kinase	55, 56
		NMDA-D2 Receptor Interactions		
Brain slices	Striatum	↓ NMDA responses	↓ cAMP, Ca^{2+}?	32
	Cortex	↓ NMDA responses	Activation of $GABA_A$ receptors	46
Brain slices	Cortex	D_4 ↓ NMDA responses	↓ PKA, CaM kinase II	69
Brain slices	Hippocampus	D_4 ↓ NMDA responses	PDGF β	70
	Cortex	$D_{2/3}$ ↓ NMDA responses	PDGF β	71

D_1-like receptors occurs via protein kinase C (PKC) activation.[35] Similarly, in cortical pyramidal neurons, intracellular application of the calmodulin Ca^{2+} chelator or inhibition of PKC activity significantly reduces the potentiation of NMDA currents, indicating that this interaction may be independent of PKA.[38]

3.3.2 PHYSICAL DA–NMDA RECEPTOR INTERACTIONS

In addition to modulation of NMDAR function through activation of signal transduction cascades,[47,48] recent studies have shown that physical interactions between these receptors allow cross-talk via receptor linkages. Direct physical interactions between the C-terminal tails of D_1 receptors and either the NR1 or NR2A NMDAR subunit have been demonstrated.[49] These protein–protein interactions are functionally relevant because D_1 receptor activation decreases NMDA currents when PKA and PKC activation are blocked.

Evidence indicates that the D_1 interaction with the NR2A subunit is involved in the inhibition of NMDAR-gated currents. The reduction of NMDA currents occurs via a decrease in the number of cell surface receptors.[49] The D_1 interaction with the NR1 subunit has been implicated in the attenuation of NMDAR-mediated excitotoxicity through a phosphatidylinositol 3-kinase (PI-3 kinase)-dependent pathway. The D_1–NR1 interaction also enables NMDAR activation to increase membrane insertions of D_1 receptors.[50]

The observation that physical receptor–receptor interactions reduce NMDA currents when second messenger pathways are blocked has been complicated by the demonstration that other mechanisms independent of D1 receptor activation may produce similar effects. A recent study revealed that one mechanism underlying reduction of NMDA currents is direct channel pore block of NMDARs by DA and several D1 receptor ligands.[51] Thus, without excluding the possibility that receptor–receptor interactions may lead to functional modulation, the inhibitory effects of DA or its agonists and antagonists require further examination since they may also occlude the channel.

3.4 RECIPROCAL D1–NMDA RECEPTOR INTERACTIONS

Activation of one type of receptor may alter the distribution of other types. In primary cultures of striatal neurons, activation of NMDARs increased the recruitment of D_1 but not D_2 receptors into the plasma membrane.[52] This translocation is abolished in the presence of an NMDAR antagonist or by removing Ca^{2+}. After NMDA treatment, a dramatic increase in the number of D_1 receptor-containing spines occurs.

The translocation of D_1 receptors to the plasma membrane was confirmed in subcellular fractionation experiments using slices of adult rat striatum. Furthermore, in striatal organotypic cultures from rat, application of NMDA caused an increase in D_1 receptor-positive spines.[53]

Surprisingly, under these conditions, this effect is independent of Ca^{2+} and also occurs in the presence of Mg^{2+}. Thus, in addition to the Ca^{2+}-dependent recruitment of D_1 receptors by activation of NMDARs seen in primary cultures, other NMDAR-dependent mechanisms may cause redistribution of D_1 receptors to spines. This is achieved by a diffusion trap mechanism in which subsets of D_1 receptors that typically move by lateral diffusion in the plasma membrane are trapped in the spines when NMDA binds to its receptor. Exposure to NMDA reduces the diffusion rate of D_1 receptors and allows the formation of D_1–NMDA heteroreceptor complexes.

This process may be explained by the allosteric theory of receptor activation.[54] After ligand binding, one conformation of the receptor is stabilized, shifting the equilibrium toward this state so that occupation of the binding site of the NMDAR favors a conformation that will bind to D_1 receptors and thus stabilize them in spines.

This mechanism is highly energy-efficient because it depends on D_1 receptor diffusion and NMDAR allosterism—not on activation of transduction systems and intracellular signaling.[53] One interesting caveat to these studies is that glutamate is the endogenous agonist for NMDARs and these experiments did not examine all the outcomes in the presence of glutamate rather than NMDA, bringing into question the natural relevance of some of these findings.

While activation of NMDARs induces changes in the distribution of D_1 receptors, the converse is also true. D_1 receptor activation produces an increase in NR1, NR2A, and NR2B proteins in the synaptosomal membrane fraction[55] that is dependent on Fyn protein tyrosine kinase but not DARPP-32.[56] Based on the partial overlap of NMDA and D_1 receptors in dendritic spines, protein–protein interactions may direct the trafficking of D_1 and NMDARs to the same subcellular domain.

The mechanism by which D_1 receptors are delivered to different spine domains was examined in co-immunoprecipitation studies.[57] In the striatal postsynaptic density (PSD), the D_1 receptor selectively complexes with the NR1 subunit of the NMDA channel through its C-terminal tail. The physical proximity between D_1 receptors and NR1 subunits can best be explained by the formation of constitutive protein dimers. Oligomerization with the NMDAR thus regulates D_1 receptor targeting to the plasma membrane. When the D_1 receptor and the NR1 subunit are coexpressed in HEK293 cells, the D_1 receptor is only partially targeted to the cell membrane, with most of the D_1 receptor staining retained in cytoplasmic structures where it is colocalized with NR1.

Coexpression of the D_1 receptor with both the NR1 and NR2B subunits relieves the cytoplasmic retention of the complex, allowing insertion of both the NR1 subunit and the D_1 receptor at the plasma membrane, where they are completely colocalized. These data suggest that D_1 and NMDARs are assembled as oligomeric units in the endoplasmic reticulum and transported to the cell surface as a preformed complex.[57] This implies that a direct protein–protein interaction with the NMDAR is one of the mechanisms directing the trafficking of D_1 receptors to specific subcellular compartments. This direct interaction may be crucial to recruit the D_1 receptor to the place where synaptic activity occurs and to keep it in close proximity to the NMDAR to allow rapid cAMP-PKA-DARPP-32-mediated potentiation of NMDA transmission.[57]

It is interesting that the current evidence indicates that most physical heteroreceptor interactions lead to mutual inhibitory effects. This idea seems to contrast with the well-known observation that D_1–NMDAR complexes play a role in enhancing synaptic plasticity and potentiating NMDA responses.

3.5 TOPOGRAPHIC AND TEMPORAL ASPECTS OF D1–NMDA OCCURS RECEPTOR INTERACTIONS

The ultimate outcome of D1–NMDAR interactions depends on a number of factors including temporal and topographic aspects (i.e., when and where the receptors are activated).[4,30] The outcome of activation of interacting receptors in the brain may depend on the temporal sequence of neurotransmitter release. For example, activation of D_1 receptors due to DA release caused by unexpected reward can prime particular corticostriatal synapses (including synapses in the nucleus accumbens) and recruit D_1–NMDAR complexes in a regulated manner.[57]

Reynolds et al.[58] measured responses to cortical afferents before and after intracranial self-stimulation of the substantia nigra that would release DA. Such stimulation of DA cells with behaviorally reinforcing parameters induces potentiation of glutamatergic corticostriatal synapses that is blocked by administration of a D1 receptor antagonist. Timing is an important requirement for this type of synaptic plasticity because DA release should occur before excitatory afferents are activated in order to induce potentiation.[59] It is tempting to speculate that if D_1 receptors are activated first, G protein- and Ca^{2+}-dependent oligomerization of D_1–NMDARs occurs, providing a regulated delivery of these complexes to plasma membranes, dendritic spines, or both. Massive DA release due to unexpected reward enhances the relevance of the stimulus by potentiating NMDA responses. This process is particularly important in MSSNs enriched with D_1 receptors.

DA concentration and the mode of release are also important. Phasic release may produce different effects from tonic release. MSSNs are constantly bombarded by cortical and thalamic inputs and tonic release of DA filters a sizable percentage of these glutamatergic inputs through D_2 receptors located on presynaptic terminals.[60] Higher local concentrations of DA occurring when it is phasically released are likely to activate D1 receptors and enhance selected corticostriatal synapses. For synaptic responses, studies in cortical pyramidal neurons revealed that the enhancement of NMDAR-mediated responses by DA follows an inverted U-shaped dose-response curve[30] in agreement with the idea that optimal levels of D1 receptor activation are required for efficient working memory formation.[61] Too much DA and hence too much activation of D1 receptors, as during stress, may be deleterious for cortical function.

3.6 REDUCTION OF NMDA RECEPTOR-MEDIATED RESPONSES BY DA OCCURS VIA D2 RECEPTORS

In contrast to the enhancing effects of D1 receptors on NMDAR-mediated responses, D2 receptor activation leads to inhibitory effects.[32] This may be relevant to preventing excessive activation of NMDARs and its consequent Ca^{2+} accumulation that may be deleterious to neurons. For example, DA and the D1 receptor agonist SKF 38393 increased the magnitude of NMDA-induced cell swelling, an index of excitotoxicity.[62,63] This effect was reduced in the presence of SCH 23390 (a D_1 receptor antagonist), demonstrating specificity. In contrast, activation of D2 family receptors with quinpirole (a D2 receptor agonist) resulted in decreased cell swelling.[64] These results provided evidence that DA receptors have the potential to modulate excitotoxicity in the striatum, a process suggested to be responsible for cell dysfunction and ultimately cell death as in Huntington's disease (HD).

Compared to D1–NMDAR interactions, much less is known about the mechanisms by which D2 receptor activation leads to reduction of NMDA currents. Decreased cAMP production and PKA activity are certainly potential mechanisms.

D2 receptors also can modulate neuronal excitability by activating the PLC–IP$_3$–Ca^{2+} cascade.[65] However, at least in cortical pyramidal neurons, D2 attenuation of NMDA responses does not require intracellular Ca^{2+} or PKA inhibition but requires activation of GABA$_A$ receptors, suggesting that this effect is mediated through excitation of GABA interneurons.[46]

D$_4$ receptors are abundant in the prefrontal cortex[66] and may play an important role in schizophrenia and other psychiatric disorders.[67] Mice lacking D$_4$ receptors show signs of hyperexcitability.[68] Application of a D$_4$ receptor agonist produces a decrease of NMDA currents via inhibition of PKA, activation of PP1 and the consequent inhibition of Ca^{2+} calmodulin-dependent kinase II.[69] In CA1 pyramidal neurons, quinpirole depresses excitatory transmission mediated by NMDARs by increasing release of intracellular Ca^{2+}. This depression is dependent on transactivation of platelet-derived growth factor β by D$_4$ receptors.[70] Similar effects were found in prefrontal cortical neurons but they were mediated by D$_{2/3}$ receptors.[71] Physical coupling between D$_2$ receptors and NR2B subunits can also reduce NMDA currents.[72] The mechanism underlying this effect involves disruption of the association between NR2B and CaMKII, thereby reducing subunit phosphorylation. It is believed that the D$_2$–NR2B interaction plays a critical role in the stimulative effect of cocaine.[72]

3.7 GENETIC MANIPULATIONS OF DA–NMDA RECEPTOR INTERACTIONS

The generation of mice lacking specific receptors or receptor subunits via genetic engineering approaches marked a new era in the study of receptor function. These techniques permit the generation of mice deficient in selective DA receptors or NMDAR subunits. Our previous studies demonstrated that in D$_1$ receptor-deficient mice, DA potentiation of striatal NMDA responses was greatly reduced.[73] Similarly, presynaptic modulation of glutamate release along the corticostriatal pathway was enhanced in D$_2$ receptor knock-out animals.[27]

Our laboratory recently examined the enhancement of NMDA currents in mice lacking NR2A subunits.[74] Preliminary observations indicate that D1 modulation of these currents is similar in MSSNs from wild type and NR2A knock-out cohorts. We also examined D2 attenuation of NMDA responses in these mice and again found no statistically significant differences in modulation levels. These results suggest that the presence or absence of the NR2A subunit does not affect D1 or D2 modulation of NMDAR-mediated currents. These studies are relevant to DA–NMDA interactions as modulation of NMDA currents by DA receptors may be mediated by phosphorylation of specific receptor subunits or by physical coupling. Further, recent evidence indicates specific NMDAR subunits may play different roles in synaptic plasticity and excitotoxicity.[75–77]

Mice that express EGFP reporter genes in a variety of cells have been generated.[78] Mice that express specific DA receptor subtypes represent important tools to differentiate neuronal populations within the striatum. DA or its agonists almost always modulate responses induced by NMDAR activation in MSSNs. However, the magnitude of this modulation varies from cell to cell possibly because D1 and D2 receptors are largely segregated in different populations of MSSNs.

Although all MSSNs are GABAergic, they differ in expression of DA receptor subtypes, peptide contents, and projection targets.[79] Two major neuronal subpopulations of MSSNs have been described. One projects primarily to the substantia nigra pars reticulata and the internal segment of the globus pallidus (direct pathway). The other subpopulation projects primarily to the external segment of the globus pallidus (indirect pathway).[80] MSSNs originating the direct pathway mainly express D_1 receptors and colocalize substance P. MSSNs originating the indirect pathway mainly express D_2 receptors and colocalize enkephalin although some overlap exists.[81–83]

We are currently examining DA–NMDAR interactions in acutely dissociated D_1 and D_2 EGFP-positive MSSNs. Application of SKF 812907 (a D1 agonist) dose-dependently and reversibly increased NMDA currents in D_1 but not in D_2 cells. NMDA current enhancement was prevented by SCH 23390 (a D1 antagonist).[84] In contrast, quinpirole, (a D2 agonist), dose-dependently and reversibly decreased NMDA currents in D_2 but not in D_1 cells. The effect was blocked by remoxipride, a D2 antagonist. At the highest concentration, quinpirole induced decreases of NMDA currents in some D_1 cells as well.

3.8 FUNCTIONAL RELEVANCE OF DA–NMDA RECEPTOR INTERACTIONS

The function of DA–NMDAR interactions may vary according to the area in which they occur. In the dorsal striatum, these interactions are important in motor control. In the ventral striatum, they provide mechanisms that may underlie addiction. In the frontal cortex, these interactions are implicated in working memory and cognition. In other areas such as the amygdala, their role is less clear. Overall, the D1–NMDAR interaction, when mediated by second messenger cascades, appears synergistic. The membrane-delimited physical interactions appear antagonistic and have more relevance to neuroprotection, with the caveat that DA agonists and antagonists can also directly modulate the NMDAR channel pore.

In the striatum, electrophysiological studies have shown that high-frequency stimulation of corticostriatal inputs induces LTP in normal physiological conditions or after Mg^{2+} removal.[85–89] Activation of D1 family receptors is required for LTP induction,[40] whereas coactivation of D1 and D2 receptors is required for LTD.[90] The mechanisms by which D1 receptors are permissive to LTP induction are unclear but may involve enhancement of Ca^{2+} influx through L-type channels.[45] In certain conditions, cortical pyramidal neurons and striatal MSSNs oscillate between two preferred (up and down) states.[91] The D1–NMDAR interaction favors the transition to and maintenance of the up state,[92] and thus is more permissive toward synaptic plasticity involving potentiation.

Assuming there is segregation of D_1 (direct pathway) and D_2 (indirect pathway) receptors in striatal MSSNs, plasticity that depends on DA–NMDAR interactions is likely to go in different directions (produce different outcomes). Thus, the D1–NMDAR interaction, by strengthening synapses in the direct pathway (LTP), may function to reinforce a motor program, for example. The D2–NMDAR interaction, by weakening synaptic strength (LTD) along the indirect pathway, may serve to extinguish competitive motor programs.

In the cerebral cortex, D1 receptors and D1–NMDAR interactions play a very important role in working memory and cognitive function. In particular, accumulating evidence indicates that induction and maintenance of persistent activity in prefrontal cortex and related networks is dependent on D1–NMDAR interactions.[23] Alterations of these receptors and their interactions occur in schizophrenia. One important aspect of these interactions is the existence of an optimal level of D1 receptor activation below or above which DA's effects on working memory are deleterious.[93]

D1–NMDAR interactions facilitate the transition and maintenance of the up states in the cerebral cortex and may also initiate these state transitions.[92] One caveat of up and down states in the striatum or cortex is that they are best observed in anesthetized animals or during slow-wave sleep and their relevance or even evidence of their occurrence in the awake state is indirect or unknown. Recent studies of the striatum indicate that these membrane transitions in waking animals do not occur and cell firing is more random than in anesthetized animal preparations.[94]

Synchronous activity (gamma oscillations) may occur in awake animals and this activity may play an important role in cognition.[95] Alterations in gamma oscillations, particularly in the frontal cortex, have been observed in schizophrenia.[96–99] In humans DA D_4 receptor and DA transporter-1 polymorphisms have been shown to modulate gamma activity.[100]

Another form of synchronous activity called neuronal "avalanche" has been demonstrated to occur spontaneously in mature cortical organotypic cultures and in slices after bath application of D1 agonists and NMDA.[101,102] These avalanches may play a role in optimizing information flow across cortical networks. Interestingly, D1 NMDA-induced avalanches displayed a U-shaped pharmacological profile in which moderate DA concentrations maximize spatial correlations in the cortical network; lower or higher concentrations reduce spatial correlations.[103] One speculation is that phasic release of DA as during unexpected reward[104] produces exactly the correct concentration to enhance NMDAR activation and produce an avalanche capable of sustaining working memory. This avalanche may propagate or be replicated in striatal D_1 MSSNs to reinforce specific motor sets conducive to reward.

DA–NMDA interactions also play an important role in neurodegenerative diseases because unregulated enhancement of excitation, particularly excitation mediated by NMDARs, will cause neuronal dysfunction and disturb structural neuronal integrity. For example, the excitotoxicity hypothesis of HD posits that excessive glutamate release at the corticostriatal terminal or altered sensitivity of postsynaptic NMDARs and their signaling systems may induce cell death.[105] Studies in genetic mouse models of HD confirmed increased sensitivity of NMDARs in MSSNs.[106–108] However, the precise location of NMDARs in synaptic or extrasynaptic compartments determines the outcome of receptor activation. In hippocampal neurons, activation of synaptic NMDARs triggers an anti-apoptotic pathway, whereas activation of extrasynaptic NMDARs may cause cell death.[109]

Assuming that activation of NMDARs recruits more functional D_1 receptors in plasma membranes[50,52,53] and that these D_1 receptors in turn recruit more NMDARs,[55] a positive feedback mechanism may be created and the outcome of these interactions can be deleterious for the neuron if the mechanism is not stopped.[110]

Both D_1 and NMDARs independently exert toxic effects on striatal neurons. In addition, D_1 receptor activation also potentiates NMDA toxicity.[64] A number of protective mechanisms must be in place to prevent the deleterious effect of excessive D1–NMDAR stimulation. Activation of D2 receptors may be neuroprotective since it reduces NMDA responses.[64,111] Other mechanisms may be also considered. For example, the interactions of D_1 and NMDARs independent of cAMP production and the D_2–NR2B interaction both reduce NMDA currents and excitotoxicity.[49] The diffusion trap system may represent a fast and efficient way to prevent excessive potentiation of NMDA responses if it makes D_1 receptors less functional—a conclusion that remains to be verified. One drawback is that this mechanism is more or less random; the effectiveness of the trap depends on the availability of the prey. If D_1 receptors are abundant and nearby, the trap will work, but it is nonetheless subject to haphazard encounters.

3.9 CONCLUSIONS

It has been 15 years since the enhancement of NMDA responses by D1 receptor activation was first observed.[31] As generally occurs with any scientific observation or hypothesis, explanations become more complex than initially assumed. The potential mechanisms and even the outcomes of D1–NMDAR interactions continue to multiply. We may speculate that various interactions accomplish different functions. Some may be intended to enhance, whereas others may be designed to inhibit the outcomes of receptor interactions. The traditional pathway involving D1 receptor activation and the cAMP–PKA–DARPP-32 cascade produces various effects that enhance NMDAR function.[47,48]

Physical interactions among these receptors, in the cytoplasm or in membranes, add new levels of complexity. Two pathways in the formation of D_1–NMDA heteroreceptor complexes are envisaged. One is G protein- and Ca^{2+}-dependent, occurs in the cytoplasm, and delivers the complex in a regulated manner to the plasma membrane, in particular the PSD.[57] The other is G protein- and Ca^{2+}-independent, is membrane delimited, and may function as an inhibitory mechanism or brake to prevent and dampen continuous positive feedback.[49,52,53] These interactions in conjunction with the more traditional interactions through signaling pathways fine-tune neuronal function. Alterations of these interactions that occur in some pathological states jeopardize functional and structural neuronal integrity. Understanding these interactions and their possible consequences in normal and diseased states is essential for designing better therapeutic approaches to treat psychiatric and neurological disorders.

ACKNOWLEDGMENTS

This work has benefited from USPHS Grants NS41574 and NS33538 and contracts with the High Q Foundation and CHDI, Inc.

REFERENCES

1. Bloom, F.E., Costa, E., and Salmoiraghi, G.C., Anesthesia and the responsiveness of individual neurons of the caudate nucleus of the cat to acetylcholine, norepinephrine and dopamine administered by microelectrophoresis, *J. Pharmacol. Exp. Ther.*, 150, 244, 1965.

2. McLennan, H. and York, D.H., The action of dopamine on neurones of the caudate nucleus, *J. Physiol.*, 189, 393, 1967.

3. Chiodo, L.A. and Berger, T.W., Interactions between dopamine and amino acid-induced excitation and inhibition in the striatum, *Brain Res.*, 375, 198, 1986.

4. Cepeda, C. and Levine, M.S., Dopamine and N-methyl-D-aspartate receptor interactions in the neostriatum, *Dev. Neurosci.*, 20, 1, 1998.

5. Monaghan, D.T., Bridges, R.J., and Cotman, C.W., The excitatory amino acid receptors: their classes, pharmacology, and distinct properties in the function of the central nervous system, *Annu. Rev. Pharmacol. Toxicol.*, 29, 365, 1989.

6. Hollmann, M. and Heinemann, S., Cloned glutamate receptors, *Annu. Rev. Neurosci.*, 17, 31, 1994.

7. Nakanishi, S., Metabotropic glutamate receptors: synaptic transmission, modulation, and plasticity, *Neuron*, 13, 1031, 1994.

8. Nowak, L., Bregestovski, P., Ascher, P. et al., Magnesium gates glutamate-activated channels in mouse central neurones, *Nature*, 307, 462, 1984.

9. Civelli, O., Bunzow, J.R., and Grandy, D.K., Molecular diversity of the dopamine receptors, *Annu. Rev. Pharmacol. Toxicol.*, 33, 281, 1993.

10. Sibley, D.R. and Monsma, F.J., Jr., Molecular biology of dopamine receptors, *Trends Pharmacol. Sci.*, 13, 61, 1992.

11. McGeer, P.L., McGeer, E.G., Scherer, U. et al., A glutamatergic corticostriatal path?, *Brain Res.*, 128, 369, 1977.

12. Fonnum, F., Storm-Mathisen, J., and Divac, I., Biochemical evidence for glutamate as neurotransmitter in corticostriatal and corticothalamic fibres in rat brain, *Neuroscience*, 6, 863, 1981.

13. Lindvall, O., Bjorklund, A., and Skagerberg, G., Selective histochemical demonstration of dopamine terminal systems in rat di- and telencephalon: new evidence for dopaminergic innervation of hypothalamic neurosecretory nuclei, *Brain Res.*, 306, 19, 1984.

14. Freund, T.F., Powell, J.F., and Smith, A.D., Tyrosine hydroxylase-immunoreactive boutons in synaptic contact with identified striatonigral neurons, with particular reference to dendritic spines, *Neuroscience*, 13, 1189, 1984.

15. Smith, A.D. and Bolam, J.P., The neural network of the basal ganglia as revealed by the study of synaptic connections of identified neurones, *Trends Neurosci.*, 13, 259, 1990.

16. Goldman-Rakic, P.S., Leranth, C., Williams, S.M. et al., Dopamine synaptic complex with pyramidal neurons in primate cerebral cortex, *Proc. Natl. Acad. Sci. USA*, 86, 9015, 1989.

17. Rolls, E.T., Neurophysiology and cognitive functions of the striatum, *Rev. Neurol. (Paris)*, 150, 648, 1994.

18. Graybiel, A.M., Building action repertoires: memory and learning functions of the basal ganglia, *Curr. Opin. Neurobiol.*, 5, 733, 1995.

19. Chesselet, M.F. and Delfs, J.M., Basal ganglia and movement disorders: an update, *Trends Neurosci.*, 19, 417, 1996.

20. Schultz, W., Dopamine neurons and their role in reward mechanisms, *Curr. Opin. Neurobiol.*, 7, 191, 1997.

21. Calabresi, P., Centonze, D., Gubellini, P. et al., Synaptic transmission in the striatum: from plasticity to neurodegeneration, *Prog. Neurobiol.*, 61, 231, 2000.
22. Goldman-Rakic, P.S., Cellular basis of working memory, *Neuron*, 14, 477, 1995.
23. Castner, S.A. and Williams, G.V., Tuning the engine of cognition: a focus on NMDA/ D1 receptor interactions in prefrontal cortex, *Brain Cogn.*, 63, 94, 2007.
24. Rowlands, G.F. and Roberts, P.J., Activation of dopamine receptors inhibits calcium-dependent glutamate release from corticostriatal terminals *in vitro*, *Eur. J. Pharmacol.*, 62, 241, 1980.
25. Kornhuber, J. and Kornhuber, M.E., Presynaptic dopaminergic modulation of cortical input to the striatum, *Life Sci.*, 39, 699, 1986.
26. Maura, G., Giardi, A., and Raiteri, M., Release-regulating D-2 dopamine receptors are located on striatal glutamatergic nerve terminals, *J. Pharmacol. Exp. Ther.*, 247, 680, 1988.
27. Cepeda, C., Hurst, R.S., Altemus, K.L. et al., Facilitated glutamatergic transmission in the striatum of D2 dopamine receptor-deficient mice, *J. Neurophysiol.*, 85, 659, 2001.
28. Cheramy, A., Romo, R., Godeheu, G. et al., In vivo presynaptic control of dopamine release in the cat caudate nucleus II. Facilitatory or inhibitory influence of L-glutamate, *Neuroscience*, 19, 1081, 1986.
29. Konradi, C., Cepeda, C., and Levine, M.S., Dopamine–glutamate interactions, in *Handbook of Experimental Pharmacology* Vol. 154. Di Chiara, G., Ed. Springer Verlag, Berlin, 2002, p. 117.
30. Seamans, J.K. and Yang, C.R., The principal features and mechanisms of dopamine modulation in the prefrontal cortex, *Prog. Neurobiol.*, 74, 1, 2004.
31. Cepeda, C., Radisavljevic, Z., Peacock, W. et al., Differential modulation by dopamine of responses evoked by excitatory amino acids in human cortex, *Synapse*, 11, 330, 1992.
32. Cepeda, C., Buchwald, N.A., and Levine, M.S., Neuromodulatory actions of dopamine in the neostriatum are dependent upon the excitatory amino acid receptor subtypes activated, *Proc. Natl. Acad. Sci. USA*, 90, 9576, 1993.
33. Calabresi, P., De Murtas, M., Pisani, A. et al., Vulnerability of medium spiny striatal neurons to glutamate: role of Na+/K+ ATPase, *Eur. J. Neurosci.*, 7, 1674, 1995.
34. Nicola, S.M. and Malenka, R.C., Modulation of synaptic transmission by dopamine and norepinephrine in ventral but not dorsal striatum, *J. Neurophysiol.*, 79, 1768, 1998.
35. Chergui, K. and Lacey, M.G., Modulation by dopamine D1-like receptors of synaptic transmission and NMDARs in rat nucleus accumbens is attenuated by the protein kinase C inhibitor Ro 32-0432, *Neuropharmacology*, 38, 223, 1999.
36. Wang, J. and O'Donnell, P., D(1) dopamine receptors potentiate nmda-mediated excitability increase in layer V prefrontal cortical pyramidal neurons, *Cereb. Cortex*, 11, 452, 2001.
37. Flores-Hernandez, J., Cepeda, C., Hernandez-Echeagaray, E. et al., Dopamine enhancement of NMDA currents in dissociated medium-sized striatal neurons: role of D_1 receptors and DARPP-32, *J. Neurophysiol.*, 88, 3010, 2002.
38. Chen, G., Greengard, P., and Yan, Z., Potentiation of NMDAR currents by dopamine D_1 receptors in prefrontal cortex, *Proc. Natl. Acad. Sci. USA*, 101, 2596, 2004.
39. Calabresi, P., Gubellini, P., Centonze, D. et al., Dopamine and cAMP-regulated phosphoprotein 32 kDa controls both striatal long-term depression and long-term potentiation, opposing forms of synaptic plasticity, *J. Neurosci.*, 20, 8443, 2000.
40. Kerr, J.N. and Wickens, J.R., Dopamine D-1/D-5 receptor activation is required for long-term potentiation in the rat neostriatum *in vitro*, *J. Neurophysiol.*, 85, 117, 2001.

41. Wickens, J.R., Horvitz, J.C., Costa, R.M. et al., Dopaminergic mechanisms in actions and habits, *J. Neurosci.*, 27, 8181, 2007.
42. Colwell, C.S. and Levine, M.S., Excitatory synaptic transmission in neostriatal neurons: regulation by cyclic AMP-dependent mechanisms, *J. Neurosci.*, 15, 1704, 1995.
43. Blank, T., Nijholt, I., Teichert, U. et al., The phosphoprotein DARPP-32 mediates cAMP-dependent potentiation of striatal N-methyl-D-aspartate responses, *Proc. Natl. Acad. Sci. USA*, 94, 14859, 1997.
44. Snyder, G.L., Fienberg, A.A., Huganir, R.L. et al., A dopamine/D_1 receptor/protein kinase A/dopamine- and cAMP-regulated phosphoprotein (Mr 32 kDa)/protein phosphatase-1 pathway regulates dephosphorylation of the NMDAR, *J. Neurosci.*, 18, 10297, 1998.
45. Cepeda, C., Colwell, C.S., Itri, J.N. et al., Dopaminergic modulation of NMDA-induced whole cell currents in neostriatal neurons in slices: contribution of calcium conductances, *J. Neurophysiol.*, 79, 82, 1998.
46. Tseng, K.Y. and O'Donnell, P., Dopamine-glutamate interactions controlling prefrontal cortical pyramidal cell excitability involve multiple signaling mechanisms, *J. Neurosci.*, 24, 5131, 2004.
47. Greengard, P., Allen, P.B., and Nairn, A.C., Beyond the dopamine receptor: the DARPP-32/protein phosphatase-1 cascade, *Neuron*, 23, 435, 1999.
48. Cepeda, C. and Levine, M.S., Where do you think you are going? The NMDA-D_1 receptor trap, *Sci. STKE*, 2006, 20, 2006.
49. Lee, F.J., Xue, S., Pei, L. et al., Dual regulation of NMDAR functions by direct protein–protein interactions with the dopamine D_1 receptor, *Cell*, 111, 219, 2002.
50. Pei, L., Lee, F.J., Moszczynska, A. et al., Regulation of dopamine D_1 receptor function by physical interaction with the NMDARs, *J. Neurosci.*, 24, 1149, 2004.
51. Cui, C., Xu, M., and Atzori, M., Voltage-dependent block of N-methyl-D-aspartate receptors by dopamine D_1 receptor ligands, *Mol. Pharmacol.*, 70, 1761, 2006.
52. Scott, L., Kruse, M.S., Forssberg, H. et al., Selective up-regulation of dopamine D_1 receptors in dendritic spines by NMDAR activation, *Proc. Natl. Acad. Sci. USA*, 99, 1661, 2002.
53. Scott, L., Zelenin, S., Malmersjo, S. et al., Allosteric changes of the NMDAR trap diffusible dopamine 1 receptors in spines, *Proc. Natl. Acad. Sci. USA*, 103, 762, 2006.
54. Changeux, J.P. and Edelstein, S.J., Allosteric mechanisms of signal transduction, *Science*, 308, 1424, 2005.
55. Dunah, A.W. and Standaert, D.G., Dopamine D1 receptor-dependent trafficking of striatal NMDA glutamate receptors to the postsynaptic membrane, *J. Neurosci.*, 21, 5546, 2001.
56. Dunah, A.W., Sirianni, A.C., Fienberg, A.A. et al., Dopamine D1-dependent trafficking of striatal N-methyl-D-aspartate glutamate receptors requires Fyn protein tyrosine kinase but not DARPP-32, *Mol. Pharmacol.*, 65, 121, 2004.
57. Fiorentini, C., Gardoni, F., Spano, P. et al., Regulation of dopamine D1 receptor trafficking and desensitization by oligomerization with glutamate N-methyl-D-aspartate receptors, *J. Biol. Chem.*, 278, 20196, 2003.
58. Reynolds, J.N., Hyland, B.I., and Wickens, J.R., A cellular mechanism of reward-related learning, *Nature*, 413, 67, 2001.
59. Wickens, J.R., Reynolds, J.N., and Hyland, B.I., Neural mechanisms of reward-related motor learning, *Curr. Opin. Neurobiol.*, 13, 685, 2003.
60. Bamford, N.S., Zhang, H., Schmitz, Y. et al., Heterosynaptic dopamine neurotransmission selects sets of corticostriatal terminals, *Neuron*, 42, 653, 2004.

61. Lidow, M.S., Williams, G.V., and Goldman-Rakic, P.S., The cerebral cortex: a case for a common site of action of antipsychotics, *Trends Pharmacol. Sci.*, 19, 136, 1998.
62. Dodt, H.U., Hager, G., and Zieglgansberger, W., Direct observation of neurotoxicity in brain slices with infrared videomicroscopy, *J. Neurosci. Methods*, 50, 165, 1993.
63. Colwell, C.S. and Levine, M.S., Glutamate receptor-induced toxicity in neostriatal cells, *Brain Res.*, 724, 205, 1996.
64. Cepeda, C., Colwell, C.S., Itri, J.N. et al., Dopaminergic modulation of early signs of excitotoxicity in visualized rat neostriatal neurons, *Eur. J. Neurosci.*, 10, 3491, 1998.
65. Hernandez-Lopez, S., Tkatch, T., Perez-Garci, E. et al., D2 dopamine receptors in striatal medium spiny neurons reduce L-type Ca^{2+} currents and excitability via a novel PLCβ1-IP3-calcineurin-signaling cascade, *J. Neurosci.*, 20, 8987, 2000.
66. Ariano, M.A., Wang, J., Noblett, K.L. et al., Cellular distribution of the rat D4 dopamine receptor protein in the CNS using anti-receptor antisera, *Brain Res.*, 752, 26, 1997.
67. Tarazi, F.I., Zhang, K., and Baldessarini, R.J., Dopamine D4 receptors: beyond schizophrenia, *J. Recept. Signal Transduct. Res.*, 24, 131, 2004.
68. Rubinstein, M., Cepeda, C., Hurst, R.S. et al., Dopamine D4 receptor-deficient mice display cortical hyperexcitability, *J. Neurosci.*, 21, 3756, 2001.
69. Wang, X., Zhong, P., Gu, Z. et al., Regulation of NMDARs by dopamine D4 signaling in prefrontal cortex, *J. Neurosci.*, 23, 9852, 2003.
70. Kotecha, S.A., Oak, J.N., Jackson, M.F. et al., A D2 class dopamine receptor transactivates a receptor tyrosine kinase to inhibit NMDAR transmission, *Neuron*, 35, 1111, 2002.
71. Beazely, M.A., Tong, A., Wei, W.L. et al., D2-class dopamine receptor inhibition of NMDA currents in prefrontal cortical neurons is platelet-derived growth factor receptor-dependent, *J. Neurochem.*, 98, 1657, 2006.
72. Liu, X.Y., Chu, X.P., Mao, L.M. et al., Modulation of D2R-NR2B interactions in response to cocaine, *Neuron*, 52, 897, 2006.
73. Levine, M.S., Altemus, K.L., Cepeda, C. et al., Modulatory actions of dopamine on NMDAR-mediated responses are reduced in D_1A-deficient mutant mice, *J. Neurosci.*, 16, 5870, 1996.
74. Sakimura, K., Kutsuwada, T., Ito, I. et al., Reduced hippocampal LTP and spatial learning in mice lacking NMDAR epsilon 1 subunit, *Nature*, 373, 151, 1995.
75. Li, R., Huang, F.S., Abbas, A.K. et al., Role of NMDAR subtypes in different forms of NMDA-dependent synaptic plasticity, *BMC Neurosci.*, 8, 55, 2007.
76. von Engelhardt, J., Coserea, I., Pawlak, V. et al., Excitotoxicity in vitro by NR2A- and NR2B-containing NMDARs, *Neuropharmacology*, 53, 10, 2007.
77. Liu, Y., Wong, T.P., Aarts, M. et al., NMDAR subunits have differential roles in mediating excitotoxic neuronal death both *in vitro* and *in vivo*, *J. Neurosci.*, 27, 2846, 2007.
78. Gong, S., Zheng, C., Doughty, M.L. et al., A gene expression atlas of the central nervous system based on bacterial artificial chromosomes, *Nature*, 425, 917, 2003.
79. Gerfen, C.R., The neostriatal mosaic: multiple levels of compartmental organization, *Trends Neurosci.*, 15, 133, 1992.
80. Smith, Y., Bevan, M.D., Shink, E. et al., Microcircuitry of the direct and indirect pathways of the basal ganglia, *Neuroscience*, 86, 353, 1998.
81. Kawaguchi, Y., Wilson, C.J., and Emson, P.C., Projection subtypes of rat neostriatal matrix cells revealed by intracellular injection of biocytin, *J. Neurosci.*, 10, 3421, 1990.
82. Surmeier, D.J., Song, W.J., and Yan, Z., Coordinated expression of dopamine receptors in neostriatal medium spiny neurons, *J. Neurosci.*, 16, 6579, 1996.

83. Aizman, O., Brismar, H., Uhlen, P. et al., Anatomical and physiological evidence for D₁ and D2 dopamine receptor colocalization in neostriatal neurons, *Nat. Neurosci.*, 3, 226, 2000.
84. Cepeda, C., Starling, A., Wu, N. et al., Defining electrophysiological properties of subpopulations of striatal neurons using genetic expression of enhanced green fluorescent protein, *Soc. Neurosc. Abstr.*, 30, 2004.
85. Calabresi, P., Pisani, A., Mercuri, N.B. et al., Long-term potentiation in the striatum is unmasked by removing the voltage-dependent magnesium block of NMDAR channels, *Eur. J. Neurosci.*, 4, 929, 1992.
86. Charpier, S. and Deniau, J.M., *In vivo* activity-dependent plasticity at corticostriatal connections: evidence for physiological long-term potentiation, *Proc. Natl. Acad. Sci. USA*, 94, 7036, 1997.
87. Spencer, J.P. and Murphy, K.P., Bi-directional changes in synaptic plasticity induced at corticostriatal synapses *in vitro*, *Exp. Brain Res.*, 135, 497, 2000.
88. Smith, R., Musleh, W., Akopian, G. et al., Regional differences in the expression of corticostriatal synaptic plasticity, *Neuroscience*, 106, 95, 2001.
89. Fino, E., Glowinski, J., and Venance, L., Bidirectional activity-dependent plasticity at corticostriatal synapses, *J. Neurosci.*, 25, 11279, 2005.
90. Calabresi, P., Maj, R., Pisani, A. et al., Long-term synaptic depression in the striatum: physiological and pharmacological characterization, *J. Neurosci.*, 12, 4224, 1992.
91. Wilson, C.J. and Kawaguchi, Y., The origins of two-state spontaneous membrane potential fluctuations of neostriatal spiny neurons, *J. Neurosci.*, 16, 2397, 1996.
92. Tseng, K.Y. and O'Donnell, P., Post-pubertal emergence of prefrontal cortical up states induced by D1–NMDA co-activation, *Cereb. Cortex*, 15, 49, 2005.
93. Williams, G.V. and Castner, S.A., Under the curve: critical issues for elucidating D₁ receptor function in working memory, *Neuroscience*, 139, 263, 2006.
94. Mahon, S., Vautrelle, N., Pezard, L. et al., Distinct patterns of striatal medium spiny neuron activity during the natural sleep-wake cycle, *J. Neurosci.*, 26, 12587, 2006.
95. Buzsaki, G. and Draguhn, A., Neuronal oscillations in cortical networks, *Science*, 304, 1926, 2004.
96. Gallinat, J., Winterer, G., Herrmann, C.S. et al., Reduced oscillatory gamma-band responses in unmedicated schizophrenic patients indicate impaired frontal network processing, *Clin. Neurophysiol.*, 115, 1863, 2004.
97. Spencer, K.M., Nestor, P.G., Perlmutter, R. et al., Neural synchrony indexes disordered perception and cognition in schizophrenia, *Proc. Natl. Acad. Sci. USA*, 101, 17288, 2004.
98. Cho, R.Y., Konecky, R.O., and Carter, C.S., Impairments in frontal cortical gamma synchrony and cognitive control in schizophrenia, *Proc. Natl. Acad. Sci. USA*, 103, 19878, 2006.
99. Basar-Eroglu, C., Brand, A., Hildebrandt, H. et al., Working memory related gamma oscillations in schizophrenia patients, *Int. J. Psychophysiol.*, 64, 39, 2007.
100. Demiralp, T., Herrmann, C.S., Erdal, M.E. et al., DRD4 and DAT1 polymorphisms modulate human gamma band responses, *Cereb. Cortex*, 17, 1007, 2007.
101. Beggs, J.M. and Plenz, D., Neuronal avalanches in neocortical circuits, *J. Neurosci.*, 23, 11167, 2003.
102. Plenz, D. and Thiagarajan, T.C., The organizing principles of neuronal avalanches: cell assemblies in the cortex?, *Trends Neurosci.*, 30, 101, 2007.
103. Stewart, C.V. and Plenz, D., Inverted-U profile of dopamine–NMDA-mediated spontaneous avalanche recurrence in superficial layers of rat prefrontal cortex, *J. Neurosci.*, 26, 8148, 2006.
104. Schultz, W., Getting formal with dopamine and reward, *Neuron*, 36, 241, 2002.

105. DiFiglia, M., Excitotoxic injury of the neostriatum: a model for Huntington's disease, *Trends Neurosci.*, 13, 286, 1990.
106. Levine, M.S., Klapstein, G.J., Koppel, A. et al., Enhanced sensitivity to N-methyl-D-aspartate receptor activation in transgenic and knock-in mouse models of Huntington's disease, *J. Neurosci. Res.*, 58, 515, 1999.
107. Cepeda, C., Ariano, M.A., Calvert, C.R. et al., NMDAR function in mouse models of Huntington disease, *J. Neurosci. Res.*, 66, 525, 2001.
108. Zeron, M.M., Hansson, O., Chen, N. et al., Increased sensitivity to N-methyl-D-aspartate receptor-mediated excitotoxicity in a mouse model of Huntington's disease, *Neuron*, 33, 849, 2002.
109. Hardingham, G.E., Fukunaga, Y., and Bading, H., Extrasynaptic NMDARs oppose synaptic NMDARs by triggering CREB shut-off and cell death pathways, *Nat. Neurosci.*, 5, 405, 2002.
110. Yang, C.R. and Chen, L., Targeting prefrontal cortical dopamine D1 and N-methyl-D-aspartate receptor interactions in schizophrenia treatment, *Neuroscientist*, 11, 452, 2005.
111. Bozzi, Y. and Borrelli, E., Dopamine in neurotoxicity and neuroprotection: what do D2 receptors have to do with it? *Trends Neurosci.*, 29, 167, 2006.
112. Schilstrom, B., Yaka, R., Argilli, E. et al., Cocaine enhances NMDAR-mediated currents in ventral tegmental area cells via dopamine D5 receptor-dependent redistribution of NMDARs, *J. Neurosci.*, 26, 8549, 2006.

4 The NMDA Receptor and Alcohol Addiction

Dorit Ron and Jun Wang

CONTENTS

4.1   INTRODUCTION

Alcohol addiction is a costly and detrimental chronic relapsing disorder, characterized by compulsive alcohol use despite the negative consequences; it is thought to be
associated with aberrant learning and memory processes.[1,2] The NMDA-type glutamate receptor (NMDAR) plays an essential role in synaptic plasticity and learning
and memory.[3,4] Not surprisingly, it is well established that the NMDAR is a major
target of alcohol (ethanol) in the brain and has been implicated in ethanol-associated
phenotypes such as tolerance, dependence, withdrawal, craving, and relapse.[5,6] This
chapter focuses on studies elucidating molecular mechanisms that underlie ethanol's actions on the NMDAR, and discusses the physiological and behavioral consequences of ethanol's actions. Finally, we summarize information regarding the
potential use of modulators of NMDAR function as medication to treat the adverse
effects of alcoholism.

4.2   ACUTE ETHANOL INHIBITION OF
NMDA RECEPTOR FUNCTION

In 1989, Lovinger et al. reported that ethanol (5-100 mM) acutely inhibits NMDA-
activated ion currents in cultured mouse hippocampal neurons.[7] The inhibitory
actions of ethanol on the activity of the channel were further demonstrated by
measuring NMDAR-mediated excitatory postsynaptic potentials/currents (EPSPs/
EPSCs) in slices from many brain regions such as the hippocampus,[8–13] cortex,[8,13–15]
amygdala,[16,17] nucleus accumbens,[18,19] and dorsal striatum,[20–22] as well as in mammalian heterologous expression systems such as HEK cells and *Xenopus* oocytes
expressing recombinant NMDARs. The reduction in NMDAR activity upon acute
exposure to ethanol is concentration-dependent and has a very rapid (less than
100 ms) onset when measured in NMDA-evoked currents using a fast solution
exchange technique.[14,23,24]

Single channel recordings in cultured cortical neurons revealed that ethanol
decreases the open channel probability and mean open time of native NMDARs.[8]
The precise mechanism by which ethanol rapidly inhibits NMDAR function is still
under investigation. However, the very fast reduction of channel activity in response
to ethanol suggests a direct interaction of the NMDAR subunits with ethanol to
regulate channel gating in nonneuronal mammalian cell culture models such as
HEK-293 cells, as well as *Xenopus* oocytes transfected with different combinations
of NMDAR subunits that are commonly used to determine ethanol sensitivity to a
defined subunit composition and/or amino acid substitution of specific amino acids,
as described below.

4.2.1   The NR1 Subunit

The NR1 subunit is encoded from one gene. However, the subunit contains three
sites of alternative splicing, one in the N-terminus and two in the C-terminus.[25] This
results in a total of eight possible splice variants. The C-terminus of the NR1 subunit
is made of four cassettes (C0, C1, C2, and C2'),[26] and the C0 cassette is present in all
splice variants. The C0 cassette is an important mediator of the inhibitory actions of

ethanol on the function of the channel, as deletion of the C0 cassette is reported to reduce the potency of ethanol-mediated inhibition of NMDAR activity.[27] However, this deletion seems to affect only NR1/NR2A, but not NR1/NR2B or 2C combinations.[27,28] In addition, several studies have suggested that amino acids within the third and fourth transmembrane domain of the subunit confer the channel's sensitivity to ethanol.[29,30]

4.2.2 NR2 SUBUNITS

Both NR2A- and NR2B-containing NMDARs are highly sensitive to the inhibitory actions of ethanol and are thought to be more sensitive to ethanol than those containing NR2C or NR2D subunits.[31–33] However, comparison of the degree of ethanol sensitivity of NR2A- and NR2B-containing receptors remains inconclusive. For example, NR2B-containing receptors were found to be more sensitive to ethanol compared to NR2A-containing ones,[34] but opposite results were also reported.[32,33] However, when comparing the sensitivities of different NR2 subunit-containing NMDARs to ethanol, it is important to note that the different NR1 splice variants may also affect the sensitivity of a specific NR2 subunit-containing NMDAR to ethanol. For example, Jin and Woodward tested (in transfected HEK 293 cell) the effect of ethanol exposure on 32 possible NMDARs consisting of 1 of 8 NR1 splice variants and 1 of 4 NR2 subunits. The maximal inhibition of channel activity in the presence of ethanol was observed in NR1-2b/NR2C, while the minimal one was found in NR1-3b/NR2C, NR1-3b/NR2D, and NR1-4b/NR2C.[35] No single NR1 splice variant or NR2 subunit showed a consistently high or low degree of ethanol inhibition when combined with other NR2 subunits or NR1 splice variants.[35] These findings suggest that the overall sensitivity of an individual NMDAR to ethanol depends on specific combinations of NR1 and NR2 subunits. Finally, several amino acids within the third and fourth transmembrane domains of the NR2A subunit have been identified as residues that contribute to the inhibitory actions of ethanol on the activity of the channel.[30,36–38] However, whether or not point mutations in the NR2B subunit affect ethanol sensitivity is yet to be determined.

In summary, the studies described above provide important information on the mechanism of the fast inhibitory action of ethanol on the activity of the channel. However, as these studies were obtained from nonneuronal systems, the results should be further confirmed in more physiologically-relevant systems.

4.2.3 COFACTORS

Cofactors that contribute to the activity of the NMDAR, such as magnesium and zinc ions, as well as the amino acid glycine, may also play a role in the molecular mechanism mediating ethanol's action on the activity of the channel. However, the contribution of cofactors to the modulation of NMDAR activity by ethanol remains unclear.

Mg^{2+} — Extracellular Mg^{2+} voltage-dependently blocks the NMDAR channel by binding to a deep site of the channel pore.[39,40] In hippocampal slices, the IC_{50} of ethanol inhibition of NMDAR activity was reported to be ~50 and ~100 mM in the presence of 1 and 0 mM Mg^{2+}, respectively, suggesting that low Mg^{2+} reduces the sensitivity of NMDARs to ethanol.[10,41] In amygdala slices, 44 mM ethanol inhibits NMDAR-mediated EPSCs in 1 mM Mg^{2+} by 30%, but loses its inhibition in 0.3 mM Mg^{2+}, and increases in NMDAR EPSCs in the presence of ethanol were observed in 0 Mg^{2+}, suggesting again that low Mg^{2+} reduces the sensitivity of NMDARs to ethanol.[16] However, these results remain controversial, as other groups reported that Mg^{2+} does not affect the degree of ethanol inhibition of NMDAR response. For example, in hippocampal slices, ethanol inhibition of NMDAR response was reported to be similar in normal Mg^{2+} (1.5 mM) and low-Mg^{2+} (0.1 mM) solutions.[9] In oocytes expressing NR1/NR2A, NR1/NR2B, or NR1/NR2C, the presence of high (3 mM) or low (0.01 mM) Mg^{2+} does not alter ethanol sensitivity of NMDARs.[31] In cultured hippocampal neurons, Mg^{2+} was also found not to affect ethanol inhibition of NMDA-activated currents.[42]

Glycine — The glycine binding site is located within the NR1 subunit.[43–45] Although some studies reported that glycine modulates ethanol inhibition of NMDAR function,[46–51] other studies have not found evidence to support this finding.[31, 32,42,52–56]

Zn^{2+} — In HEK-293 cells expressing NR1 and NR2A subunits, chelation of Zn^{2+} by EDTA reduced ethanol inhibition of NMDAR activity.[57,58] However, such an effect was not observed in another study.[31]

4.2.4 POSTTRANSLATION MODIFICATIONS

Posttranslation modifications such as phosphorylation–dephosphorylation events, which occur in a time frame of minutes after ethanol exposure, also contribute to the inhibitory actions of ethanol on the NMDAR. For example, we found that exposure of hippocampal slices to ethanol results in the internalization of NR2A-containing receptors via a mechanism that depends on activation of the small G protein H-Ras and inhibition of the tyrosine kinase Src.[59] We further showed that as a result of ethanol-mediated internalization of the channel, the contribution of NR2A to the activity of the channel is decreased.[59]

In addition, Alvestad et al. reported that acute ethanol treatment of hippocampal CA1 slices decreased the basal level of tyrosine phosphorylation of NR2A and NR2B subunits, and that the phosphotyrosine phosphotase inhibitor bpV reduced ethanol inhibition of NMDAR-mediated field EPSPs in hippocampal slices.[60] As tyrosine phosphatases contribute to both the activation and inhibition of the phosphorylation and activity of NMDAR,[61] it is likely that a tyrosine phosphatase contributes to the acute effects of ethanol on NMDAR response.

4.3 FACILITATION OF NMDA RECEPTOR FUNCTION BY ETHANOL

4.3.1 ACUTE TOLERANCE OF NMDA RECEPTORS TO ETHANOL INHIBITION

As stated above, the primary acute effects of ethanol on NMDAR activity is inhibition. However, in some brain regions the inhibitory effect of ethanol is reduced as a function of time. This acute decrease in the sensitivity of NMDARs to ethanol inhibition is termed "acute tolerance," and was first described by Grover et al. in rat hippocampal slices, in which a significant decrease in ethanol inhibition was observed over a 15-minute period of ethanol exposure.[62] Acute tolerance to ethanol's inhibitory actions on NMDAR function was later confirmed in both mouse and rat hippocampal slices.[13,59,63]

In addition, Li et al. showed that in spinal cord slices ethanol depressed NMDAR activity by ~37% at 8–10 minutes, but only ~17% at 20 minutes. Neurons in other brain regions including the nucleus locus creruleus,[64] the basolateral amygdala,[16] and rostral ventrolateral medulla[65] also show similar phenotypes of acute tolerance of NMDAR activity to ethanol inhibition.

An interesting question is whether or not the tolerance will eventually counteract the inhibitory effect of ethanol. The answer appears, at least in some preparations, to be positive. For example, in hippocampal CA1 slices, we previously observed that the NMDAR activity returns to its basal level after 35 minutes of ethanol application.[13] Similarly, in rostral ventrolateral medulla neurons, ethanol inhibition was not detected 40 minutes after ethanol exposure.[65]

4.3.2 REBOUND POTENTIATION AND LONG-TERM FACILITATION OF NMDA RECEPTOR ACTIVITY

We (and others) observed that in various brain regions and in the spinal cord, the activity of the channel is greatly potentiated upon ethanol washout,[13,21,66–69] and even low concentrations of ethanol (10 mM) were shown to induce such potentiation.[69] In addition, we recently observed that in dorsal striatal slices, acute ethanol exposure and withdrawal results in long-term facilitation (LTF) of NMDAR-mediated EPSCs,[21] and a phenomenon similar to LTF was also detected in spinal cord slices,[66–69] in rostral ventrolateral medulla neurons slices,[65] and in locus ceruleus neurons slices.[64]

4.3.3 MOLECULAR MECHANISMS MEDIATING FACILITATION OF NMDA RECEPTOR FUNCTION BY ETHANOL

The NR2 subunits are phosphorylated by the Src family protein tyrosine kinases (PTKs) Fyn and Src,[61] leading to upregulation of channel function.[61] In 1997, Miyakawa et al.[63] observed that NR2B phosphorylation is increased after ethanol administration in Fyn heterozygous (Fyn[+/-]) but not in Fyn deletion (Fyn[-/-]) mice, and that acute tolerance to ethanol inhibition of NMDAR-mediated field EPSPs was observed in hippocampal slices from Fyn[+/-] but not from Fyn[-/-] mice. These results suggest a role for Fyn kinase in acute tolerance of NMDAR activity.

Several years later we identified a molecular mechanism that underlies this phenotype. We found that in the hippocampus, the scaffolding protein RACK1 localizes Fyn kinase to the NR2B subunit;[70] however RACK1 acts as a negative modulator to prevent NR2B phosphorylation by Fyn kinase.[70] Activation of the cAMP/PKA pathway leads to dissociation of the trimolecular complex allowing Fyn kinase to phosphorylate NR2B, which, in turn, leads to an increase in channel function.[71,72] Importantly, we found that exposure of hippocampal slices to ethanol leads to the dissociation of RACK1 from the Fyn/NR2B complex via a mechanism that requires activation of the cAMP/PKA pathway, leading to phosphorylation of NR2B.[13]

Our results further suggest that this mechanism accounts for the development of acute tolerance in the presence of ethanol, and to the rebound potentiation upon ethanol washout. We found that when the Src PTK inhibitor, PP2, was applied to the hippocampal slice preparation prior to ethanol washout, the rebound potentiation was not observed, and when the inhibitor was applied at the peak of the rebound potentiation, a rapid inhibition of NMDAR activity was observed.[13] Furthermore, when recombinant RACK1 was added to the hippocampal slice preparation, ethanol-mediated NR2B phosphorylation was inhibited and acute tolerance was not observed.[13] Taken together, these results suggest that Fyn phosphorylation of the NR2B subunit is, at least in part, the mechanism that accounts for the enhancement of the activity of the channel in the hippocampus in response to ethanol exposure.

As mentioned above, in the dorsal striatum, acute ethanol exposure and withdrawal leads to prolonged enhancement of NMDAR-mediated EPSCs upon ethanol washout.[21] Here, too, treatment with ethanol leads to NR2B phosphorylation both *ex vivo* in slice preparations and *in vivo*, and the corresponding RACK1 dissociation from the trimolecular complex leading to the activation of Fyn kinase.[21] Interestingly, both Fyn activation and NR2B phosphorylation were observed after ethanol washout. Importantly, the LTF of NMDAR-EPSCs was not observed in the presence of the Fyn inhibitor PP2, in dorsal striatal slices from Fyn$^{-/-}$ mice, or upon incubation of dorsal striatal slices with the selective NR2B-containing NMDAR inhibitor Ro 25-6981.[21]

Taken together, these results suggest that this mechanism of Fyn kinase dissociation from RACK1, leading to its activation and to NR2B phosphorylation, accounts for the rebound potentiation and to LTF of NMDAR activity after ethanol exposure in the hippocampus and dorsal striatum, respectively.

Finally, the metabotropic glutamate receptor[67] and PKCγ[68] were found to be required for the development of acute tolerance and withdrawal potentiation, respectively, in the spinal cord. Specifically, Li et al. observed that the extent of ethanol inhibition of NMDAR activity is reduced from ~37% at 8–10 minutes to ~17% at 20 minutes, and that such reduction is enhanced by the metabotropic glutamate receptor agonist ACPD and is attenuated by the antagonist MCPG, suggesting that acute tolerance of NMDAR activity to ethanol in the spinal cord is developed in a metabotropic glutamate receptor-dependent manner.[67]

Also, in spinal cord slices, Li et al. found that the NMDAR activity is potentiated by ~24% at 18 minutes after a 15-minute ethanol application, and that such potentiation is prevented by bath application of a PKCγ inhibitory peptide γV5-3, indicating that withdrawal potentiation in the spinal cord develops via a PKCγ-dependent manner.[68]

4.3.4 BRAIN REGION-SPECIFIC ACTIONS OF ETHANOL

Interestingly, ethanol does not affect NMDAR function at all brain regions identically. For instance, an *in vivo* study showed that in the inferior colliculus and the hippocampus, but not in the lateral septum, ethanol inhibited NMDA-induced neuronal activity.[73] An *ex vivo* imaging study showed that Ca^{2+} influx through NMDARs in neurons from brainstem was not affected by concentrations of ethanol as high as

160 mM.[74] Interestingly, the sensitivity of the NMDAR subunits to posttranslation modifications upon ethanol exposure is also not universal throughout the brain.

We observed that in the hippocampus or the dorsal striatum, acute exposure to ethanol resulted in an increase in the tyrosine phosphorylation of the NR2B subunit of the NMDAR, leading to the upregulation of channel function, but none of these phenotypes were observed in the prefrontal cortex or ventral striatum.[13,21] We found that Fyn is compartmentalized to the NR2B subunit of the NMDAR only in the hippocampus and the dorsal striatum, but not in the ventral striatum or the prefrontal cortex,[13,21] suggesting that brain region specificity to ethanol's actions results, at least in part, from differences in the intracellular compartmentalization of signaling and scaffolding proteins.

4.3.5 Chronic Ethanol and Synaptic Compartmentalization of NMDA Receptor Subunits

Elegant studies conducted by Chandler and colleagues showed that prolonged exposure of hippocampal neurons with moderate doses of ethanol resulted in clustering of NMDARs in dendritic spines, and the clustering was found to be restricted to synaptic but not extrasynaptic pools of the receptor.[75] These changes were blocked by a PKA inhibitor, and by a low dose of a NMDAR antagonist.[75] In addition, the authors observed that the enhanced synaptic localization on NMDARs required the postsynaptic density scaffolding protein, PSD-95.[76] Finally, these changes were correlated with an increase in synaptic NMDAR currents.[75] Interestingly, a recent study by Offenhauser and colleagues showed that acute exposure of cerebellar granule cells to high concentrations of ethanol resulted in the redistribution of F-actin away from postsynaptic sites. The study further suggests that the cytoskeletal remodeling induced by ethanol depends on the NMDAR and actin binding protein Esp8.[77] Although acute ethanol exposure inhibits NMDAR function and thus prevents long-term plasticity such as LTP (see below), prolonged ethanol exposure employs an adaptive process, such as increasing the trafficking of NMDARs to synapses.[75] As calcium influx through NMDARs evokes AMPAR insertion via a calcium/calmodulin-dependent protein kinase II-dependent process,[78] prolonged ethanol exposure would be expected to reduce the clustering of AMPARs in synapses. However, increased trafficking of NMDARs to the synapses may counterbalance the ethanol inhibition of NMDAR activity, leading to no changes in synaptic AMPAR clustering.[75]

4.3.6 Chronic Ethanol Exposure and Withdrawal Alters NMDA Receptor Function

Exposure to ethanol for 24 hours or longer followed by withdrawal has been shown to lead to hyperactivation of the channel in neuronal preparations.[79] For example, withdrawal from chronic exposure of cultured hippocampal slices to ethanol (35 mM or 70 mM) for 5 to 11 days increased the NMDAR activity that occurs within 1 hour after ethanol was removed and lasted for at least 7 hours.[80,81]

In vivo studies showed that withdrawal from exposure of rats to continuous ethanol vapor for at least 2 weeks increased the contribution of the NR2B subunit to

NMDAR function.[82] However, such a change was not observed 1 week after withdrawal from chronic ethanol exposed rats.[83] One explanation for the hyperexcitability of the NMDAR upon withdrawal from chronic ethanol exposure is an increase in number of receptors resulting from an adaptation mechanism that is due to the long lasting inhibition of activity of the channel.

To test this possibility, Ticku et al. and others used primary neurons exposed chronically (several days) to ethanol to determine whether the expression level of the NMDAR subunits was altered. The investigators observed that chronic ethanol treatment upregulated the mRNA level of the NR2B subunit,[84,85] and protein levels of the NR1 and NR2B subunits.[86] More recently, increases in *NR2B* gene AP-1 binding and promotor activity were observed upon chronic exposure of cortical neurons to ethanol, suggesting a mechanism for the increase in *NR2B* expression in response to ethanol.[87] Another possible mechanism mediating the increase in mRNA levels in response to ethanol was reported by Qiang et al., who showed that ethanol exposure leads to a decrease in the mRNA level of the *NR2B* transcriptional repressor NRSF (neuro-restrictive silencer factor).[88] Kumari et al. investigated the molecular mechanism underlying the increase in the expression of the *NR1* subunit and found an enhancement of *NR1* mRNA stability upon chronic exposure to ethanol,[89] possibly via an association with the RNA binding protein GIIβ.[90] They also found that the protein levels of NR1, NR2A, and NR2B were elevated in rat hippocampal and cortical neurons exposed chronically to ethanol *ex vivo* and *in vivo*.[86,91–93] In addition, chronic intermittent exposure of cortical neurons to ethanol resulted in a significant increase in both the message and protein levels of the NR2B subunit.[94] Taken together, these results confirm the hypothesis that hyperexcitability of the NMDAR channel upon chronic ethanol exposure and withdrawal is due to an increase in the mRNA and protein levels of NMDAR subunits. Interestingly, Pawlak et al. showed that the serine protease tissue plasminogen activator (tPA) contributes to the upregulation of NR2B-containing NMDARs upon ethanol exposure and to ethanol withdrawal syndrome. The authors reported that tPA-deficient mice have a decreased severity of seizures upon ethanol withdrawal that corresponds with a reduction in NR2B level. The authors further showed that tPA increased ethanol withdrawal seizures, whereas an NR2B-NMDAR selective antagonist reversed tPA's effect.[95]

4.4 PHYSIOLOGICAL IMPLICATIONS OF MODULATION OF NMDA RECEPTOR FUNCTION BY ETHANOL

4.4.1 ACUTE INHIBITION

It is well known that the NMDAR plays a key role in long-term potentiation (LTP) of the AMPAR-mediated synaptic response, which is a cellular model of learning and memory.[3,4] As expected, inhibition of NMDARs by ethanol was reported to also block hippocampal LTP,[96–98] which was more pronounced in juvenile rats (30 days old) than in adult rats (90 days old).[99] In addition, such inhibition has also been observed *in vivo*. Givens et al. reported that in awake rats, LTP was produced by stimulation of electrodes implanted in the dentate gyrus of the hippocampus, and this LTP was inhibited by intraperitoneal injection of nonintoxicating doses of ethanol (0.5 or 1.0 g/kg) given prior to the LTP induction.[100] Ethanol inhibition of LTP was observed not only in the hippocampus but also in other brain regions. NMDAR-dependent LTP in the dorsomedial striatum[101] was recently reported to be abolished by ethanol at concentrations as low as 10 mM.[20] LTP in the dorsolateral bed nucleus was also shown to be inhibited by ethanol.[102] LTP is a cellular model of learning and memory[4,103], and in humans, ethanol disrupts performance on a variety of short-term memory tasks[104–106] and ethanol inhibition of LTP may be associated with drinking-induced blackouts.[107] Finally, ethanol inhibition of LTP in the hippocampus may underlie episodes of amnesia after alcohol binge drinking.[108]

4.4.2 ACUTE TOLERANCE, REBOUND POTENTIATION, AND LTF

As stated above, prolonged ethanol exposure leads to the development of acute tolerance of NMDARs to ethanol, which is evidenced by a reduction of ethanol inhibition of receptor activity, which may eliminate the depressive effect of ethanol on the induction of NMDAR-dependent LTP. For example, Tokuda et al. observed in hippocampal slices that LTP was abolished by acute application of 60 mM ethanol, but LTP was inducible when ethanol was gradually increased to 60 mM over 75 minutes.[109] They speculated that this slow increase in ethanol concentration induced an acute tolerance of NMDAR activity to ethanol, which preserved NMDAR function and thus LTP induction.

The ability to induce LTP during acute ethanol tolerance further suggests that synaptic plasticity and memory formation may be developed in response to ethanol exposure. In addition, it is intriguing to speculate that the upregulation of NMDAR function in response to ethanol observed *ex vivo* may contribute to the aberrant learning and memory, as well as habit formation associated with alcohol addiction.[110,111]

4.4.3 CHRONIC ETHANOL EXPOSURE AND WITHDRAWAL

Withdrawal from chronic exposure to ethanol leads to excessive activity of the NMDAR. For instance, withdrawal of ethanol following chronic exposure increased the firing rate of hippocampal neurons in both cultures[75] and slices.[81] This effect was abolished by the NMDR agonist APV. In addition, the excessive activity of the NMDAR upon chronic exposure to ethanol, in conjunction with withdrawal and the increase in Ca^{2+} influx, is believed to be the major cause of neurotoxicity and neuronal cell death,[112] which were detected mainly in NMDAR-containing neurons such as cortical pyramidal cells, hippocampal CA1 pyramidal cells, granule cells in the dentate gyrus, and amygdala neurons.[113,114]

Inhibition of NMDARs with an NR2B subunit antagonist was shown to block the neurotoxic actions of ethanol in cultured cortical neurons.[115] The NMDAR-mediated neurotoxicity and cell death resulting from ethanol withdrawal may account for the decrease in the number of neurons in the cortex[116] and in the cerebellum,[117] and a decrease in the number of cortical neuronal dendrites,[118] and dentate gyrus granule cells.[119,120] Hyperactivation of the NMDAR may be the cause of seizures and other symptoms observed in rodents and humans upon ethanol withdrawal, which, if not treated, could be fatal.[121,122]

4.5 NMDA RECEPTORS AND BEHAVIORS ASSOCIATED WITH ETHANOL EXPOSURE

The NMDAR has been linked to many behavioral paradigms associated with ethanol exposure such as intoxication, reward, sensitization, and relapse.[6,123] Below is a summary of some of the studies linking the NMDAR to ethanol-associated behavioral paradigms *in vivo*.

4.5.1 ETHANOL INTOXICATION

Ethanol intoxication is measured in rodents by the length of sleep time upon systemic injection of hypnotic doses (3–4 g/kg) of ethanol. Miyakawa et al. showed that Fyn deletion mice were more sensitive to intoxicating doses of ethanol and therefore their sleep time was longer than the Fyn$^{+/-}$ mice.[63] We found that systemic administration of the NR2B-specific inhibitor, ifenprodil, together with ethanol increased the length of sleep time of the Fyn$^{+/+}$ mice to the same level as the Fyn$^{-/-}$ mice.[124] Taken together, these results suggest that Fyn-mediated phosphorylation of NR2B subunits and the development of acute tolerance reduce the *in vivo* sensitivity to hypnotic doses of ethanol.

Support for a potential role of NR2B-containing NMDARs in the attenuation of the level of intoxication was reported in a recent study in which systemic inhibition of NR2B-containing NMDARs with CGP-37848 or Ro-25-6981 significantly increased sleep time in C57BL/6J mice.[125] These results are also in line with numerous studies by Kalant and colleagues showing that the NMDAR antagonists (+)-MK-801 and ketamine blocked the development of rapid tolerance to ethanol exposure *in vivo*.[126–128]

4.5.2 SENSITIZATION

Sensitization is defined as a progressive increase in the effect of the same dose of a drug when administered repeatedly over time. Sensitization to ethanol's actions is measured in rodents by an increase in the acute stimulating effects of systemic administration of a nonintoxicating dose of ethanol on locomotion. The link between the NMDAR and sensitization stemmed from studies showing that the noncompetitive NMDAR antagonist MK-801 reduced the stimulant effects of ethanol and prevented expression of sensitization.[129,130] However, Meyer and Philips reported that repeated administration of 0.1 mg/kg MK-801 with ethanol potentiated, whereas 0.25 mg/kg attenuated, sensitization to ethanol's locomotor stimulant effect.[131] Interestingly, Broadbent et al. reported that the NR2B-containing NMDAR-specific antagonist, ifenprodil, did not alter expression of sensitization, suggesting the involvement of non–NR2B-containing receptors[130]; this possibility needs to be confirmed in future studies.

4.5.3 REWARD

The level of ethanol reward is measured in mice in a conditioned place preference (CPP) paradigm. Pretreatment with the competitive NMDAR antagonist CGP-37849 reduced the acquisition of ethanol-induced CPP, possibly by impairing the ability of mice to learn the task.[132] Kotlinska et al. found that a noncompetitive NMDAR antagonist neramexane inhibited the acquisition and expression of ethanol-induced CPP.[133] In addition, Biala et al. observed that coapplication of the noncompetitive NMDAR antagonist dizocilpine and the NMDAR antagonist L-701,324 acting on the glycine binding site prevented the acquisition of ethanol-induced CPP.[134] Interestingly, the $NR2A^{-/-}$ and heterozygous mice did not exhibit ethanol-induced CPP, whereas their WT littermates did, suggesting that NR2A-containing NMDARs are important for the rewarding actions of ethanol.[135] Interestingly, we found that Fyn kinase is not required for ethanol-induced CPP,[124] suggesting that the NR2A- and not NR2B-containing NMDAR is required for the rewarding properties of ethanol.

4.5.4 RELAPSE

Alcohol drinking after a period of abstinence can be mimicked in rodents by an alcohol deprivation paradigm in which access of ethanol is renewed after a period of abstinence. This leads to a significant increase in ethanol self-administration. Hotler et al. reported that repeated administration of NMDAR antagonists dose-dependently decreased ethanol consumption in an ethanol deprivation model,[136] suggesting that NMDAR inhibitors could be developed as medications to prevent relapse to alcohol drinking.

4.5.5 ETHANOL WITHDRAWAL SYNDROME

As mentioned above, ethanol withdrawal syndrome is a life-threatening condition and is also a hallmark for physical dependence to ethanol.[122] Other symptoms of ethanol withdrawal syndromes in humans include tachycardia, sweating, tremor, hypertension, anxiety, agitation, auditory and visual hallucinations, and confusion.[121,122] Therefore, ethanol withdrawal symptoms are disabling enough to lead many subjects to resume alcohol consumption at the early stages of withdrawal.[2,112,137,138]

4.6 MODULATORS OF NMDA RECEPTOR FUNCTION AND TREATMENT OF ALCOHOL ABUSE AND DEPENDENCE

During the past 20 years, NMDAR antagonists have been assessed for their potential use as medication for the treatment of various CNS related disorders such as stroke, pain, and Alzheimer's disease.[139,140] Several NMDAR antagonists have been tested in human trials as potential drugs that alleviate adverse phenotypes that are associated with alcoholics. For example, administration of the NMDAR antagonist, ketamine, to recovering alcoholics reduced psychosis, negative symptoms, dysphoric mood, and worsening of cognitive function.[141] A recent study showed that the well-tolerated NMDAR antagonist memantine[142] reduced alcohol-induced cue-induced craving,[143] suggesting that well-tolerated NMDAR antagonists such as memantine could potentially be used as medications for the treatment of alcohol addiction.

Finally, the anticraving and relapse drug, acamprosate, was shown to modulate the activity of the NMDAR, suggesting that the beneficial actions of the drug may be due, at least in part, to its action on the channel. In 2004 the FDA approved acamprosate (Campral) as an anticraving and relapse medication after clinical trials showed efficacy of the drug in maintaining abstinence in recovering alcoholics (FDA 2004-07-29). Interestingly, various studies suggested that acamprosate modulates the activity of the NMDAR. Acamprosate was shown to act as a weak antagonist[144] or a partial "coagonist" at the NMDAR, so that low concentrations enhance activation when receptor activity is low, whereas higher concentrations are inhibitory to high levels of receptor activation.[145] Acamprosate was also shown to decrease NMDAR activity in cortical neurons.[146] However, in neurons of the hippocampal CA1 region and of the nucleus accumbens, the compound enhanced NMDAR function.[147,148] In primary cultured striatal and cerebellar granule cells, acamprosate exposure did not result in alteration of NMDA-induced currents,[149] nor did it alter the inhibitory effects of ethanol (10-100 mM) on receptor function.[149] However, acamprosate was found to cause an up-regulation of the NR1 subunit in the cortex and hippocampus.[144] These data suggest that the actions of acamprosate on the NMDAR are complex and should be further explored.

ACKNOWLEDGMENT

This work was supported by NIAAA (R01/AA/MH13438-O1A1) (D.R.).

REFERENCES

1. Weiss, F. and Porrino, L.J., Behavioral neurobiology of alcohol addiction: recent advances and challenges, *J. Neurosci.*, 22, 3332, 2002.
2. Koob, G.F., Alcoholism: allostasis and beyond, *Alcohol Clin. Exp. Res.*, 27, 232, 2003.
3. Malenka, R.C. and Nicoll, R.A., Long-term potentiation: a decade of progress? *Science*, 285, 1870, 1999.
4. Bliss, T.V. and Collingridge, G.L., A synaptic model of memory: long-term potentiation in the hippocampus, *Nature*, 361, 31, 1993.
5. Trujillo, K.A. and Akil, H., Excitatory amino acids and drugs of abuse: a role for N-methyl-D-aspartate receptors in drug tolerance, sensitization and physical dependence, *Drug Alcohol Depend.*, 38, 139, 1995.
6. Krystal, J.H. et al., N-methyl-D-aspartate glutamate receptors and alcoholism: reward, dependence, treatment, and vulnerability, *Pharmacol. Ther.*, 99, 79, 2003.
7. Lovinger, D.M., White, G., and Weight, F.F., Ethanol inhibits NMDA-activated ion current in hippocampal neurons, *Science*, 243, 1721, 1989.
8. Wright, J.M., Peoples, R.W., and Weight, F.F., Single-channel and whole-cell analysis of ethanol inhibition of NMDA-activated currents in cultured mouse cortical and hippocampal neurons, *Brain Res.*, 738, 249, 1996.
9. Lovinger, D.M., White, G., and Weight, F.F., NMDAR-mediated synaptic excitation selectively inhibited by ethanol in hippocampal slice from adult rat, *J. Neurosci.*, 10, 1372, 1990.
10. Morrisett, R.A. et al., Ethanol and magnesium ions inhibit N-methyl-D-aspartate-mediated synaptic potentials in an interactive manner, *Neuropharmacology*, 30, 1173, 1991.

11. Hendrickson, A.W., Sibbald, J.R., and Morrisett, R.A., Ethanol alters the frequency, amplitude, and decay kinetics of Sr^{2+}-supported, asynchronous NMDAR mEPSCs in rat hippocampal slices, *J. Neurophysiol.*, 91, 2568, 2004.
12. Kolb, J.E., Trettel, J., and Levine, E.S., BDNF enhancement of postsynaptic NMDARs is blocked by ethanol, *Synapse*, 55, 52, 2005.
13. Yaka, R., Phamluong, K., and Ron, D., Scaffolding of Fyn kinase to the NMDAR determines brain region sensitivity to ethanol, *J. Neurosci.*, 23, 3623, 2003.
14. Wirkner, K. et al., Mechanism of inhibition by ethanol of NMDA and AMPA receptor channel functions in cultured rat cortical neurons, *Naun. Schmiede. Arch. Pharmacol.*, 362, 568, 2000.
15. Li, Q., Wilson, W.A., and Swartzwelder, H.S., Differential effect of ethanol on NMDA EPSCs in pyramidal cells in the posterior cingulate cortex of juvenile and adult rats, *J. Neurophysiol.*, 87, 705, 2002.
16. Calton, J.L., Wilson, W.A., and Moore, S.D., Magnesium-dependent inhibition of N-methyl-D-aspartate receptor-mediated synaptic transmission by ethanol, *J. Pharmacol. Exp. Ther.*, 287, 1015, 1998.
17. Calton, J.L., Wilson, W.A., and Moore, S.D., Reduction of voltage-dependent currents by ethanol contributes to inhibition of NMDAR-mediated excitatory synaptic transmission, *Brain Res.*, 816, 142, 1999.
18. Maldve, R.E. et al., DARPP-32 and regulation of the ethanol sensitivity of NMDARs in the nucleus accumbens, *Nat. Neurosci.*, 5, 641, 2002.
19. Nie, Z., Madamba, S.G., and Siggins, G.R., Ethanol inhibits glutamatergic neurotransmission in nucleus accumbens neurons by multiple mechanisms, *J. Pharmacol. Exp. Ther.*, 271, 1566, 1994.
20. Yin, H.H. et al., Ethanol reverses the direction of long-term synaptic plasticity in the dorsomedial striatum, *Eur. J. Neurosci.*, 25, 3226, 2007.
21. Wang, J. et al., Ethanol induces long-term facilitation of NR2B-NMDAR activity in the dorsal striatum: implications for alcohol drinking behavior, *J. Neurosci.*, 27, 3593, 2007.
22. Popp, R.L. et al., Ethanol sensitivity and subunit composition of NMDARs in cultured striatal neurons, *Neuropharmacology*, 37, 45, 1998.
23. Peoples, R.W. and Stewart, R.R., Alcohols inhibit N-methyl-D-aspartate receptors via a site exposed to the extracellular environment, *Neuropharmacology*, 39, 1681, 2000.
24. Criswell, H.E. et al., Macrokinetic analysis of blockade of NMDA-gated currents by substituted alcohols, alkanes and ethers, *Brain Res.*, 1015, 107, 2004.
25. Zukin, R.S. and Bennett, M.V., Alternatively spliced isoforms of the NMDARI receptor subunit, *Trends Neurosci.*, 18, 306, 1995.
26. Wenthold, R.J. et al., Trafficking of NMDARs, *Annu. Rev. Pharmacol. Toxicol.*, 43, 335, 2003.
27. Anders, D.L. et al., Reduced ethanol inhibition of N-methyl-D-aspartate receptors by deletion of NR1 C0 domain or overexpression of alpha-actinin-2 proteins, *J. Biol. Chem.*, 275, 15019, 2000.
28. Mirshahi, T. et al., Intracellular calcium enhances the ethanol sensitivity of NMDARs through an interaction with the C0 domain of the NR1 subunit, *J. Neurochem.*, 71, 10957, 1998.
29. Ronald, K.M., Mirshahi, T., and Woodward, J.J., Ethanol inhibition of N-methyl-D-aspartate receptors is reduced by site-directed mutagenesis of a transmembrane domain phenylalanine residue, *J. Biol. Chem.*, 276, 44729, 2001.
30. Smothers, C.T. and Woodward, J.J., Effects of amino acid substitutions in transmembrane domains of the NR1 subunit on the ethanol inhibition of recombinant N-methyl-D-aspartate receptors, *Alcohol Clin. Exp. Res.*, 30, 523, 2006.

31. Chu, B., Anantharam, V., and Treistman, S.N., Ethanol inhibition of recombinant heteromeric NMDA channels in the presence and absence of modulators, *J. Neurochem.*, 65, 140, 1995.

32. Mirshahi, T. and Woodward, J.J., Ethanol sensitivity of heteromeric NMDARs: effects of subunit assembly, glycine and NMDAR1 Mg^{2+}-insensitive mutants, *Neuropharmacology*, 34, 347, 1995.

33. Masood, K. et al., Differential ethanol sensitivity of recombinant N-methyl-D-aspartate receptor subunits, *Mol. Pharmacol.*, 45, 324, 1994.

34. Blevins, T. et al., Effects of acute and chronic ethanol exposure on heteromeric N-methyl-D-aspartate receptors expressed in HEK 293 cells, *J. Neurochem.*, 69, 2345, 1997.

35. Jin, C. and Woodward, J.J., Effects of eight different NR1 splice variants on the ethanol inhibition of recombinant NMDARs, *Alcohol Clin. Exp. Res.*, 30, 673, 2006.

36. Ren, H., Honse, Y., and Peoples, R.W., A site of alcohol action in the fourth membrane-associated domain of the N-methyl-D-aspartate receptor, *J. Biol. Chem.*, 278, 48815, 2003.

37. Ren, H. et al., Mutations at F637 in the NMDAR NR2A subunit M3 domain influence agonist potency, ion channel gating and alcohol action, *Br. J. Pharmacol.*, 151, 749, 2007.

38. Honse, Y. et al., Sites in the fourth membrane-associated domain regulate alcohol sensitivity of the NMDAR, *Neuropharmacology*, 46, 647, 2004.

39. Mayer, M.L., Westbrook, G.L., and Guthrie, P.B., Voltage-dependent block by Mg^{2+} of NMDA responses in spinal cord neurones, *Nature*, 309, 261, 1984.

40. Nowak, L. et al., Magnesium gates glutamate-activated channels in mouse central neurones, *Nature*, 307, 462, 1984.

41. Martin, D. et al., Ethanol inhibition of NMDA mediated depolarizations is increased in the presence of Mg^{2+}, *Brain Res.*, 546, 227, 1991.

42. Peoples, R.W. et al., Ethanol inhibition of N-methyl-D-aspartate-activated current in mouse hippocampal neurones: whole-cell patch-clamp analysis, *Br. J. Pharmacol.*, 122, 1035, 1997.

43. Schorge, S. and Colquhoun, D., Studies of NMDAR function and stoichiometry with truncated and tandem subunits, *J. Neurosci.*, 23, 1151, 2003.

44. Papadakis, M., Hawkins, L.M., and Stephenson, F.A., Appropriate NR1-NR1 disulfide-linked homodimer formation is requisite for efficient expression of functional, cell surface N-methyl-D-aspartate NR1/NR2 receptors, *J. Biol. Chem.*, 279, 147032, 2004.

45. Cull-Candy, S.G. and Leszkiewicz, D.N., Role of distinct NMDAR subtypes at central synapses, *Sci. STKE*, 2004, 16, 2004.

46. Woodward, J.J. and Gonzales, R.A., Ethanol inhibition of N-methyl-D-aspartate-stimulated endogenous dopamine release from rat striatal slices, *J. Neurochem.*, 54, 712, 1990.

47. Rabe, C.S. and Tabakoff, B., Glycine site-directed agonists reverse the actions of ethanol at the N-methyl-D-aspartate receptor, *Mol. Pharmacol.*, 38, 753, 1990.

48. Popp, R.L., Lickteig, R.L., and Lovinger, D.M., Factors that enhance ethanol inhibition of N-methyl-D-aspartate receptors in cerebellar granule cells, *J. Pharmacol. Exp. Ther.*, 289, 1564, 1999.

49. Hoffman, P.L. et al., N-methyl-D-aspartate receptors and ethanol: inhibition of calcium flux and cyclic GMP production, *J. Neurochem.*, 52, 1937, 1989.

50. Buller, A.L. et al., Glycine modulates ethanol inhibition of heteromeric N-methyl-D-aspartate receptors expressed in Xenopus oocytes, *Mol. Pharmacol.*, 48, 717, 1995.

51. Dildy-Mayfield, J.E. and Leslie, S.W., Mechanism of inhibition of N-methyl-D-aspartate-stimulated increases in free intracellular Ca^{2+} concentration by ethanol, *J. Neurochem.*, 56, 1536, 1991.
52. Bhave, S.V. et al., Mechanism of ethanol inhibition of NMDAR function in primary cultures of cerebral cortical cells, *Alcohol Clin. Exp. Res.*, 20, 934, 1996.
53. Peoples, R.W. and Weight, F.F., Ethanol inhibition of N-methyl-D-aspartate-activated ion current in rat hippocampal neurons is not competitive with glycine, *Brain Res.*, 571, 342, 1992.
54. Cebers, G. et al., Glycine does not reverse inhibitory actions of ethanol on NMDAR functions in cerebellar granule cells, *Naunyn Schmiedebergs Arch. Pharmacol.*, 354, 736, 1996.
55. Gonzales, R.A. and Woodward, J.J., Ethanol inhibits N-methyl-D-aspartate-stimulated [3H]norepinephrine release from rat cortical slices, *J. Pharmacol. Exp. Ther.*, 253, 1138, 1990.
56. Woodward, J.J., A comparison of the effects of ethanol and the competitive glycine antagonist 7-chlorokynurenic acid on N-methyl-D-aspartic acid-induced neurotransmitter release from rat hippocampal slices, *J. Neurochem.*, 62, 987, 1994.
57. Woodward, J.J. and Smothers, C., Ethanol inhibition of recombinant NR1/2A receptors: effects of heavy metal chelators and a zinc-insensitive NR2A mutant, *Alcohol*, 31, 71, 2003.
58. Woodward, J.J., Fyn kinase does not reduce ethanol inhibition of zinc-insensitive NR2A-containing N-methyl-D-aspartate receptors, *Alcohol*, 34, 101, 2004.
59. Suvarna, N. et al., Ethanol alters trafficking and functional N-methyl-D-aspartate receptor NR2 subunit ratio via H-Ras, *J. Biol. Chem.*, 280, 31450, 2005.
60. Alvestad, R.M. et al., Tyrosine dephosphorylation and ethanol inhibition of N-methyl-D-aspartate receptor function, *J. Biol. Chem.*, 278, 11020, 2003.
61. Salter, M.W. and Kalia, L.V., Src kinases: a hub for NMDAR regulation, *Nat. Rev. Neurosci.*, 5, 317, 2004.
62. Grover, C.A., Frye, G.D., and Griffith, W.H., Acute tolerance to ethanol inhibition of NMDA-mediated EPSPs in the CA1 region of the rat hippocampus, *Brain Res.*, 642, 70, 1994.
63. Miyakawa, T. et al., Fyn-kinase as a determinant of ethanol sensitivity: relation to NMDA-receptor function, *Science*, 278, 698, 1997.
64. Poelchen, W., Nieber, K., and Illes, P., Tolerance to inhibition by ethanol of N-methyl-D-aspartate-induced depolarization in rat locus coeruleus neurons *in vitro*, *Eur. J. Pharmacol.*, 332, 267, 1997.
65. Lai, C.C., Chang, M.C., and Lin, H.H., Acute tolerance to ethanol inhibition of NMDA-induced responses in rat rostral ventrolateral medulla neurons, *J. Biomed. Sci.*, 11, 482, 2004.
66. Wong, S.M. et al., Glutamate receptor-mediated hyperexcitability after ethanol exposure in isolated neonatal rat spinal cord, *J. Pharmacol. Exp. Ther.*, 285, 201, 1998.
67. Li, H.F. et al., Ethanol tachyphylaxis in spinal cord motorneurons: role of metabotropic glutamate receptors, *Br. J. Pharmacol.*, 138, 1417, 2003.
68. Li, H.F., Mochly-Rosen, D., and Kendig, J.J., Protein kinase C gamma mediates ethanol withdrawal hyper-responsiveness of NMDAR currents in spinal cord motor neurons, *Br. J. Pharmacol.*, 144, 301, 2005.
69. Wong, S.M. et al., Hyperresponsiveness on washout of volatile anesthetics from isolated spinal cord compared to withdrawal from ethanol, *Anesth. Analg.*, 100, 413, 2005.
70. Yaka, R. et al., NMDAR function is regulated by the inhibitory scaffolding protein, RACK1, *Proc. Natl. Acad. Sci. USA*, 99, 5710, 2002.

71. Yaka, R. et al., Pituitary adenylate cyclase-activating polypeptide (PACAP(1-38)) enhances N-methyl-D-aspartate receptor function and brain-derived neurotrophic factor expression via RACK1, *J. Biol. Chem.*, 278, 9630, 2003.
72. Thornton, C. et al., H-Ras modulates N-methyl-D-aspartate receptor function via inhibition of Src tyrosine kinase activity, *J. Biol. Chem.*, 278, 23823, 2003.
73. Simson, P.E., Criswell, H.E., and Breese, G.R., Inhibition of NMDA-evoked electrophysiological activity by ethanol in selected brain regions: evidence for ethanol-sensitive and ethanol-insensitive NMDA-evoked responses, *Brain Res.*, 607, 9, 1993.
74. Randoll, L.A. et al., N-methyl-D-aspartate-stimulated increases in intracellular calcium exhibit brain regional differences in sensitivity to inhibition by ethanol, *Alcohol Clin. Exp. Res.*, 20, 197, 1996.
75. Carpenter-Hyland, E.P., Woodward, J.J., and Chandler, L.J., Chronic ethanol induces synaptic but not extrasynaptic targeting of NMDARs, *J. Neurosci.*, 24, 7859, 2004.
76. Carpenter-Hyland, E.P. and Chandler, L.J., Homeostatic plasticity during alcohol exposure promotes enlargement of dendritic spines, *Eur. J. Neurosci.*, 24, 3496, 2006.
77. Offenhauser, N. et al., Increased ethanol resistance and consumption in Eps8 knockout mice correlates with altered actin dynamics, *Cell*, 127, 213, 2006.
78. Shi, S.H. et al., Rapid spine delivery and redistribution of AMPA receptors after synaptic NMDAR activation, *Science*, 284, 1811, 1999.
79. Esel, E., Neurobiology of alcohol withdrawal inhibitory and excitatory neurotransmitters, *Turk. Psikiyatri. Derg.*, 17, 129, 2006.
80. Thomas, M.P., Monaghan, D.T., and Morrisett, R.A., Evidence for a causative role of N-methyl-D-aspartate receptors in an *in vitro* model of alcohol withdrawal hyperexcitability, *J. Pharmacol. Exp. Ther.*, 287, 87, 1998.
81. Hendrickson, A.W. et al., Aberrant synaptic activation of N-methyl-D-aspartate receptors underlies ethanol withdrawal hyperexcitability, *J. Pharmacol. Exp. Ther.*, 321, 60, 2007.
82. Roberto, M. et al., Acute and chronic ethanol alter glutamatergic transmission in rat central amygdala: an *in vitro* and *in vivo* analysis, *J. Neurosci.*, 24, 1594, 2004.
83. Roberto, M. et al., Chronic ethanol exposure and protracted abstinence alter NMDARs in central amygdala, *Neuropsychopharmacology*, 31, 988, 2006.
84. Hu, X.J., Follesa, P., and Ticku, M.K., Chronic ethanol treatment produces a selective upregulation of the NMDAR subunit gene expression in mammalian cultured cortical neurons, *Brain Res. Mol. Brain Res.*, 36, 211, 1996.
85. Kumari, M. and Ticku, M.K., Ethanol and regulation of the NMDAR subunits in fetal cortical neurons, *J. Neurochem.*, 70, 1467, 1998.
86. Follesa, P. and Ticku, M.K., Chronic ethanol-mediated up-regulation of the N-methyl-D-aspartate receptor polypeptide subunits in mouse cortical neurons in culture, *J. Biol. Chem.*, 271, 13297, 1996.
87. Qiang, M. and Ticku, M.K., Role of AP-1 in ethanol-induced N-methyl-D-aspartate receptor 2B subunit gene up-regulation in mouse cortical neurons, *J. Neurochem.*, 95, 1332, 2005.
88. Qiang, M., Rani, C.S., and Ticku, M.K., Neuron-restrictive silencer factor regulates the N-methyl-D-aspartate receptor 2B subunit gene in basal and ethanol-induced gene expression in fetal cortical neurons, *Mol. Pharmacol.*, 67, 2115, 2005.
89. Kumari, M. and Anji, A., An old story with a new twist: do NMDAR1 mRNA binding proteins regulate expression of the NMDAR1 receptor in the presence of alcohol?, *Ann. NY Acad. Sci.*, 1053, 311, 2005.
90. Anji, A. and Kumari, M., A novel RNA binding protein that interacts with NMDAR1 mRNA: regulation by ethanol, *Eur. J. Neurosci.*, 23, 2339, 2006.
91. Trevisan, L. et al., Chronic ingestion of ethanol up-regulates NMDAR1 receptor subunit immunoreactivity in rat hippocampus, *J. Neurochem.*, 62, 1635, 1994.

92. Kumari, M., Differential effects of chronic ethanol treatment on N-methyl-D-aspartate R1 splice variants in fetal cortical neurons, *J. Biol. Chem.*, 276, 2976, 2001.
93. Nagy, J. et al., Differential alterations in the expression of NMDAR subunits following chronic ethanol treatment in primary cultures of rat cortical and hippocampal neurones, *Neurochem. Int.*, 42, 35, 2003.
94. Sheela Rani, C.S. and Ticku, M.K., Comparison of chronic ethanol and chronic intermittent ethanol treatments on the expression of GABA(A) and NMDAR subunits, *Alcohol*, 38, 89, 2006.
95. Pawlak, R. et al., Ethanol-withdrawal seizures are controlled by tissue plasminogen activator via modulation of NR2B-containing NMDARs, *Proc. Natl. Acad. Sci. USA*, 102, 443, 2005.
96. Schummers, J. and Browning, M.D., Evidence for a role for GABA(A) and NMDARs in ethanol inhibition of long-term potentiation, *Brain Res. Mol. Brain Res.*, 94, 9, 2001.
97. Sinclair, J.G. and Lo, G.F., Ethanol blocks tetanic and calcium-induced long-term potentiation in the hippocampal slice, *Gen. Pharmacol.*, 17, 231, 1986.
98. Morrisett, R.A. and Swartzwelder, H.S., Attenuation of hippocampal long-term potentiation by ethanol: a patch-clamp analysis of glutamatergic and GABAergic mechanisms, *J. Neurosci.*, 13, 2264, 1993.
99. Pyapali, G.K. et al., Age- and dose-dependent effects of ethanol on the induction of hippocampal long-term potentiation, *Alcohol*, 19, 107, 1999.
100. Givens, B. and McMahon, K., Ethanol suppresses the induction of long-term potentiation *in vivo*, *Brain Res.*, 688, 27, 1995.
101. Partridge, J.G., Tang, K.C., and Lovinger, D.M., Regional and postnatal heterogeneity of activity-dependent long-term changes in synaptic efficacy in the dorsal striatum, *J. Neurophysiol.*, 84, 1422, 2000.
102. Weitlauf, C. et al., High-frequency stimulation induces ethanol-sensitive long-term potentiation at glutamatergic synapses in the dorsolateral bed nucleus of the stria terminalis, *J. Neurosci.*, 24, 5741, 2004.
103. Martin, S.J., Grimwood, P.D., and Morris, R.G., Synaptic plasticity and memory: an evaluation of the hypothesis, *Annu. Rev. Neurosci.*, 23, 649, 2000.
104. Miller, M.E. et al., Effects of alcohol on the storage and retrieval processes of heavy social drinkers, *J. Exp. Psychol. [Hum. Learn.]*, 4, 246, 1978.
105. Lister, R.G. et al., Dissociation of the acute effects of alcohol on implicit and explicit memory processes, *Neuropsychologia*, 29, 1205, 1991.
106. Acheson, S.K., Stein, R.M., and Swartzwelder, H.S., Impairment of semantic and figural memory by acute ethanol: age-dependent effects, *Alcohol Clin. Exp. Res.*, 22, 1437, 1998.
107. Tsai, G. and Coyle, J.T., The role of glutamatergic neurotransmission in the pathophysiology of alcoholism, *Annu. Rev. Med.*, 49, 173, 1998.
108. White, A.M., Matthews, D.B., and Best, P.J., Ethanol, memory, and hippocampal function: a review of recent findings, *Hippocampus*, 10, 88, 2000.
109. Tokuda, K., Zorumski, C.F., and Izumi, Y., Modulation of hippocampal long-term potentiation by slow increases in ethanol concentration, *Neuroscience*, 146, 340, 2007.
110. Dickinson, A., Wood, N., and Smith, J.W., Alcohol seeking by rats: action or habit?, *Q. J. Exp. Psychol. B.*, 55, 331, 2002.
111. Everitt, B.J. and Robbins, T.W., Neural systems of reinforcement for drug addiction: from actions to habits to compulsion, *Nat. Neurosci.*, 8, 1481, 2005.
112. Fadda, F. and Rossetti, Z.L., Chronic ethanol consumption: from neuroadaptation to neurodegeneration, *Prog. Neurobiol.*, 56, 385, 1998.
113. Lovinger, D.M., Excitotoxicity and alcohol-related brain damage, *Alcohol Clin. Exp. Res.*, 17, 19, 1993.

114. Obernier, J.A., Bouldin, T.W., and Crews, F.T., Binge ethanol exposure in adult rats causes necrotic cell death, *Alcohol Clin. Exp. Res.*, 26, 547, 2002.
115. Nagy, J. et al., NR2B subunit selective NMDA antagonists inhibit neurotoxic effect of alcohol-withdrawal in primary cultures of rat cortical neurones, *Neurochem. Int.*, 44, 17, 2004.
116. Harper, C., Kril, J., and Daly, J., Are we drinking our neurones away? *Br. Med. J. (Clin. Res. Ed.)*, 294, 534, 1987.
117. Baker, K.G. et al., Neuronal loss in functional zones of the cerebellum of chronic alcoholics with and without Wernicke's encephalopathy, *Neuroscience*, 91, 429, 1999.
118. Harper, C. and Corbett, D., Changes in the basal dendrites of cortical pyramidal cells from alcoholic patients—a quantitative Golgi study, *J. Neurol. Neurosurg. Psychiatry*, 53, 856, 1990.
119. Cadete-Leite, A. et al., Granule cell loss and dendritic regrowth in the hippocampal dentate gyrus of the rat after chronic alcohol consumption, *Brain Res.*, 473, 1, 1988.
120. Walker, D.W. et al., Neuronal loss in hippocampus induced by prolonged ethanol consumption in rats, *Science*, 209, 711, 1980.
121. Hall, W. and Zador, D., The alcohol withdrawal syndrome, *Lancet*, 349, 1897–900, 1997.
122. De Witte, P. et al., Alcohol and withdrawal: from animal research to clinical issues, *Neurosci. Biobehav. Rev.*, 27, 189, 2003.
123. Kumari, M. and Ticku, M.K., Regulation of NMDARs by ethanol, *Prog. Drug. Res.*, 54, 152, 2000.
124. Yaka, R. et al., Fyn kinase and NR2B-containing NMDARs regulate acute ethanol sensitivity but not ethanol intake or conditioned reward, *Alcohol Clin. Exp. Res.*, 27, 1736, 2003.
125. Boyce-Rustay, J.M. and Holmes, A., Functional roles of NMDAR NR2A and NR2B subunits in the acute intoxicating effects of ethanol in mice, *Synapse*, 56, 222, 2005.
126. Khanna, J.M. et al., Effect of NMDAR antagonists on rapid tolerance to ethanol, *Eur. J. Pharmacol.*, 230, 23, 1993.
127. Khanna, J.M., Shah, G., and Chau, A., Effect of NMDA antagonists on rapid tolerance to ethanol under two different testing paradigms, *Pharmacol. Biochem. Behav.*, 57, 693, 1997.
128. Khanna, J.M., Morato, G.S., and Kalant, H., Effect of NMDA antagonists, an NMDA agonist, and serotonin depletion on acute tolerance to ethanol, *Pharmacol. Biochem. Behav.*, 72, 291, 2002.
129. Camarini, R. et al., MK-801 blocks the development of behavioral sensitization to the ethanol, *Alcohol Clin. Exp. Res.*, 24, 285, 2000.
130. Broadbent, J., Kampmueller, K.M., and Koonse, S.A., Expression of behavioral sensitization to ethanol by DBA/2J mice: the role of NMDA and non-NMDA glutamate receptors, *Psychopharmacology (Berl.)*, 167, 225, 2003.
131. Meyer, P.J. and Phillips, T.J., Bivalent effects of MK-801 on ethanol-induced sensitization do not parallel its effects on ethanol-induced tolerance, *Behav. Neurosci.*, 117, 641, 2003.
132. Boyce-Rustay, J.M. and Cunningham, C.L., The role of NMDAR binding sites in ethanol place conditioning, *Behav. Neurosci.*, 118, 822, 2004.
133. Kotlinska, J. et al., Effect of neramexane on ethanol dependence and reinforcement, *Eur. J. Pharmacol.*, 503, 95, 2004.
134. Biala, G. and Kotlinska, J., Blockade of the acquisition of ethanol-induced conditioned place preference by N-methyl-D-aspartate receptor antagonists, *Alcohol*, 34, 175, 1999.
135. Boyce-Rustay, J.M. and Holmes, A., Ethanol-related behaviors in mice lacking the NMDAR NR2A subunit, *Psychopharmacology (Berl.)*, 187, 455, 2006.

136. Vengeliene, V. et al., The role of the NMDAR in alcohol relapse: a pharmacological mapping study using the alcohol deprivation effect, *Neuropharmacology*, 48, 822, 2005.
137. Duka, T. et al., Consequences of multiple withdrawals from alcohol, *Alcohol Clin. Exp. Res.*, 28, 233, 2004.
138. Bisaga, A. and Popik, P., In search of a new pharmacological treatment for drug and alcohol addiction, *Drug Alcohol Depend.*, 59, 1, 2000.
139. Chizh, B.A., Headley, P.M., and Tzschentke, T.M., NMDAR antagonists as analgesics: focus on the NR2B subtype, *Trends Pharmacol. Sci.*, 22, 636, 2001.
140. Kemp, J.A. and McKernan, R.M., NMDAR pathways as drug targets, *Nat. Neurosci.*, 5 Suppl, 1039, 2002.
141. Krystal, J.H. et al., Altered NMDA glutamate receptor antagonist response in recovering ethanol-dependent patients, *Neuropsychopharmacology*, 28, 2020, 2003.
142. Parsons, C.G., Danysz, W., and Quack, G., Memantine is a clinically well tolerated N-methyl-D-aspartate (NMDA) receptor antagonist—a review of preclinical data, *Neuropharmacology*, 38, 735, 1999.
143. Krupitsky, E.M. et al., Effect of memantine on cue-induced alcohol craving in recovering alcohol-dependent patients, *Am. J. Psychiatry*, 164, 519, 2007.
144. Rammes, G. et al., The anti-craving compound acamprosate acts as a weak NMDA-receptor antagonist, but modulates NMDA-receptor subunit expression similar to memantine and MK-801, *Neuropharmacology*, 40, 749, 2001.
145. Naassila, M. et al., Mechanism of action of acamprosate. Part I. Characterization of spermidine-sensitive acamprosate binding site in rat brain, *Alcohol Clin. Exp. Res.*, 22, 802, 1998.
146. Zeise, M.L. et al., Acamprosate (calciumacetylhomotaurinate) decreases postsynaptic potentials in the rat neocortex: possible involvement of excitatory amino acid receptors, *Eur. J. Pharmacol.*, 231, 47, 1993.
147. Madamba, S.G. et al., Acamprosate (calcium acetylhomotaurinate) enhances the N-methyl-D-aspartate component of excitatory neurotransmission in rat hippocampal CA1 neurons in vitro, *Alcohol Clin. Exp. Res.*, 20, 651, 1996.
148. Berton, F. et al., Acamprosate enhances N-methyl-D-apartate receptor-mediated neurotransmission but inhibits presynaptic GABA(B) receptors in nucleus accumbens neurons, *Alcohol Clin. Exp. Res.*, 22, 183, 1998.
149. Popp, R.L. and Lovinger, D.M., Interaction of acamprosate with ethanol and spermine on NMDARs in primary cultured neurons, *Eur. J. Pharmacol.*, 394, 221, 2000.

5 Transcriptional Regulation of NMDA Receptor Expression

Guang Bai and Peter W. Hoffman

CONTENTS

5.1 INTRODUCTION

The N-methyl-D-aspartate (NMDA) subtypes of glutamate receptors are intimately involved in a number of important neuronal activities in mammalian nervous systems including neuronal migration, synaptogenesis, neuronal plasticity, neuronal survival, and excitotoxicity. Through these activities, NMDA receptors (NRs) play an important role in the development of drug addiction, pain perception, and the pathogenesis of neurological disorders such as schizophrenia and Huntington's disease.[1–10]

It is generally believed that aberrant or pathological NR effects occur mainly via abnormal receptor activity, resulting from altered availability of agonists or modified quality or quantity of membrane-associated receptors. In mammals, functional NRs are heterotetramers of subunits encoded by three gene families, i.e., NMDAR1 (NR1 or *Grin1*), NMDAR2 (NR2 or *Grin2*), and NMDAR3 (NR3 or *Grin3*).[3,4,11] The NR1 family has one gene; the NR2 family has four (designated A through D); and the NR3 family has two (A and B). Structurally, NR1 is an essential component found in all tetramers, while different NR2 members are incorporated based on age and nervous system region. NR3 proteins function as negative components when included in the structures.[3,4,11,12] Eight variants of NR1 protein are produced by alternative splicing and distributed differentially in nervous systems.[13–15] This complex composition of different subunits and splicing variants forms the primary basis of the functional diversity of NRs.

From January 1992 to June 2007, more than 1000 research articles relevant to NR expression were published. In sum, they concluded that the expression of NR genes is cell- or tissue-specific, relatively stable, and regulated differentially by various physiological, pharmacological, and pathological factors. Most of these conclusions were based on assessments of changes of the steady state levels of mRNA and protein that may be driven by numerous sophisticated mechanisms. Transcription is the initial step and generally the most sensitive to cellular needs and environmental cues. Thus, it serves as a major mechanism controlling gene expression.[16]

Precise spatial and temporal expression of a selective set of genes determines phenotypic differences among distinct tissues and cells in higher eukaryotes.[16–18] In the case of the NR gene families, transcription of each subunit gene in a given neuron or cell must be coordinately controlled but differentially responsive to cell type, developmental stage, and environmental signals to maintain healthy cellular function. How this coordinated control takes place is an important and challenging question. This chapter reviews studies that explore the transcriptional control of NR genes. It discusses studies of promoter and regulatory sequences, regulatory units, developmental regulation, cell type specificity, growth factor regulation, neurological disorders, and epigenetic mechanisms.

5.2 IDENTIFICATION OF PROMOTER

The promoter is the gene component that directs transcription.[17,19] It consists of a core or basal promoter that spans the most upstream and nearby transcription start sites (TSSs) to initiate transcription as well as regulatory regions containing enhancers or silencers to regulate transcription rate.[17,20] It had been assumed that

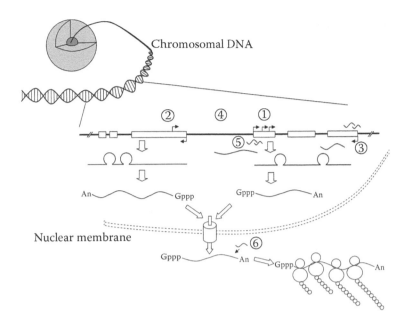

FIGURE 5.1 Current view of transcription and gene expression. The flow of gene expression from the nucleus to the cytoplasm is shown. Emerging concepts of transcription and gene expression are indicated by circled numbers. (1) Multiple TSSs for more than 50% genes, some of which have multiple promoters. (2) TSS of one gene starts from exonic sequence of another. (3) Promoter activity occurs at the end of a gene. (4) Bidirectional promoter for ~11% human genes. (5) Large portion of genomic sequences are transcribed in various sizes for both directions, including promoter-associated short RNA (<200 n.t.).[165] (6) Noncoding transcripts may target translation.

tissue-specific and/or -inducible genes in higher eukaryotes contained a single TSS and an immediate upstream TATA or CAAT box. Multiple TSSs, GC-rich regions, and structures lacking TATA and CAAT were viewed as characteristics of house-keeping genes.[18,21–23] Another early concept was that most eukaryotic genes utilized a single promoter and that the core promoter contained several basic DNA binding sequences.[16,22,24,25] This concept has been modified significantly,[16,20] particularly since large-scale analysis of transcription regulatory regions became possible.[19,26–28]

As shown in Figure 5.1, more than 50% of human genes including many neuronal genes, utilize multiple TSSs from a single exon or from different exons. In the latter case, multiple promoters control a single gene (>20% of human genes).[28] A significant number of promoters start from the exonic sequence of another gene.[19] Many genes exert promoter activities at their 3′ ends, possibly for antisense transcription or for transcription of a downstream gene.[29] Bidirectional promoters exist for many genes (~11% in humans).[30] Supported by additional evidence,[19,26–29,31] GC-rich structures and those lacking TATA and CAAT boxes are associated with many tissue-specific genes including neuronal genes. Based on these observations, our view of promoter structure and function has been simplistic.

Predication of promoters is difficult because no universal consensus may be employed.[16] Despite this difficulty, computer algorithms have been developed and most are available online.[32-36] Identification of *bona fide* promoters still relies largely on experimental mapping TSSs or the 5' ends of mRNA.[16] Several databases have been established to collect promoters identified in experiments.[37-39] Databases of full-length cDNAs are also available although many of the included sequences have artificial or incomplete 5' ends.[40-42]

5.2.1 MAPPING TRANSCRIPTION START SITES

The 5' end of a given mRNA is also the TSS on genomic DNA. In experiments, the TSS is usually defined by mapping the 5' end of the mRNA followed by an alignment to the genomic sequence. Conventional methods include RNase protection assays, primer extension, cell-free *in vitro* transcription, and 5'-rapid amplification of cDNA ends coupled to sequencing of cloned ends. Several novel methods have been developed for a large-scale analysis of the 5' ends, e.g., 5' end serial analysis of gene expression[43] and cap analysis of gene expression.[44]

NR1 was the first glutamate receptor gene to be mapped for its 5' end.[45] Two major clusters of TSSs separated by 40 base pairs (bps) were identified from a GC-rich and TATA and CAAT boxless region. Similar conclusions were reached by later studies of NR1 gene from humans[46] and chicks.[47] Rat NR2A TSSs were identified following systematic RNA mapping.[48] Multiple TSSs were found to spread almost entirely across exon 1. Similar results were obtained for mice[49] and humans.[50] Only one TSS was found for the NR2B gene in mice[51,52] and humans.[53] Two TSSs separated by 18 bps were mapped for the mouse NR2C gene,[54] and the downstream TSS was more highly utilized. All mapped TSSs of NR genes are located within a single exon. No current reports focus on mapping of the NR2D and NR3A/B TSSs.

Recent genome-wide analysis has shown that TSS selection is tissue-dependent for many genes.[55] The proximal TSS cluster of the NR1 gene is mostly recognized in the brain, while the distal TSS cluster is heavily utilized by PC12 cells that express NR1 mRNA but not detectable NR1 protein.[56] Therefore, it has been proposed that the additional 5' untranslated region (5'UTR) sequences transcribed from the distal TSSs interfere with translation initiation.[57] However, whether the transcription from the distal site is part of a *bona fide* control mechanism or simply an aberration of the PC12 cell line is still unclear.

5.2.2 FUNCTIONAL ANALYSIS

All NR genes mapped for TSSs have been functionally tested for minimal sequences that govern transcription initiation. The most sensitive and convenient means to test promoter function is reporter gene technology that allows a putative promoter to drive expression of an easily assayable foreign gene in cultured cells or in transgenic animals.[58]

For the rat NR1 gene, the basal or core promoter has been defined within the −1 to −356 bp fragment by luciferase reporter gene experiments in cultured cell lines and primary neurons.[59,60] The addition of upstream sequence up to −5.4 kb significantly increased promoter activity, suggesting that this region contains enhancer elements.[61]

Activity of the NR2A promoter has been tested in cultured cells and primary neurons in rats,[48] in transgenic mice,[49] and in cultured human neuroblastoma cells.[50] Interestingly, the rat NR2A core promoter elements were restricted within exon 1 (1140 bps). These sequences alone demonstrated cell type selectivity with much stronger activity in neurons than in glial or HEK293 cells.[48] In addition, DNA sequences between the upstream and downstream TSSs retain comparable or even stronger ability to drive luciferase expression in comparison to the upstream genomic sequences.[48] These sequences may represent a novel type of multiple promoter, but the issue remains unexplored. In transgenic mice, an equivalent sequence from the mouse NR2A gene (~1 kb) directed luciferase expression selectively in the brain with activity comparable to a fragment extending −9 kb upstream.[49]

The minimal promoter for the NR2B gene was found within a short fragment of −106 to +158 in NIH3T3 cells but its cell type specificity requires further analysis.[51] However, a larger fragment up to −572 bp of this gene showed neuron-specific activity in the brains of transgenic mice.[52] The NR2C core promoter has been defined within a short sequence (−64 to +203 relative to the most upstream TSS) in cultured cells.[62] No promoter activity has been examined for the NR2D and NR3A/B genes.

5.3 REGULATORY UNITS: *CIS* ELEMENTS AND *TRANS* FACTORS

Nuclear proteins regulate transcription by binding to specific sequences in the regulatory region of the target gene. These sequences were initially identified by their interactions with *trans* factors. Analysis of these sequences indicated that *cis* elements share conserved consensus motifs (typically, 6 to 10 bps).[16,18] To date, five types of *cis* elements have been proposed: enhancers, silencers, insulator/boundary elements, and locus control regions.[63]

Transcription factors interact with sequences in enhancers or silencers to turn transcription on and off or manipulate the efficiency of the cognate promoter. Most of these binding sites are located within the 5′ upstream region in clusters to form enhancers that positively impact transcription or repressors that negatively regulate the promoters. However, some *cis* elements can be found in other regions of the gene or genome.[16,17]

Studies *in vitro* and in living cells demonstrated that one type of *cis* element may interact with different groups of *trans* factors, and conversely one type of *trans* factor may bind different types of *cis* elements.[16] We shall summarize the interactions of *trans* factors and *cis* elements in NR genes, considering each relevant transcription factor or family individually. Figure 5.2 presents the relative positions of the analyzed binding elements on the relevant genes for humans, rats, and mice. Although a number of putative *cis* elements have been proposed to these promoters on the basis of motif searches, their functionality has not been demonstrated and will not be discussed here.

5.3.1 SPECIFIC PROTEIN (SP) FAMILY

Specific protein 1 (Sp1) is the prototype of the Sp family that includes eight additional members (Sp2 to Sp9).[64–67] Sp proteins bind GC (5′-GGGGCGGGG) or GT/CACC (5′-GGTGTGGGG) boxes. Sp1 is expressed during neuronal differentiation[61]

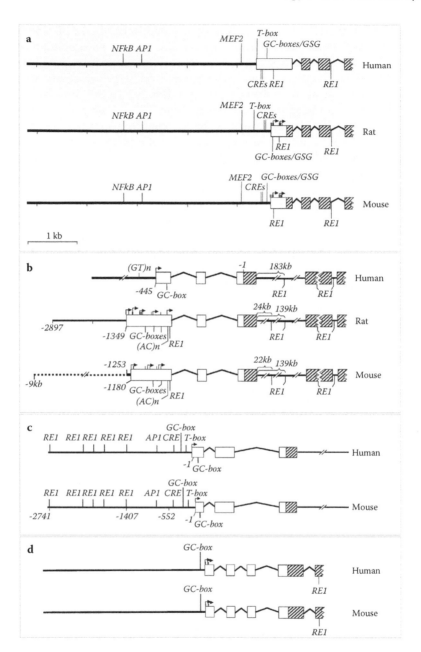

FIGURE 5.2 The 5′ end structures of NR genes. The genomic sequences spanning TSSs, 5′ flanking sequence, and selected exons are shown for promoters of NR genes analyzed experimentally: (a) NR1; (b) NR2A; (c) NR2B; and (d) NR2C. The positions of functional DNA binding sequences and *cis* elements are indicated. Solid lines = nontranscribed or intronic sequences. Open boxes = untranslated exonic sequences. Hatched boxes = coding sequences. Arrow heads = TSSs. Short vertical lines indicate weak TSSs. RE1 sites related to the NR genes are from supplementary data of two recent publications.[108,112]

and deletion of this factor from the genome results in aberrant brain development in the embryo and eventual lethality.[68]

A motif search of functional promoter sequences revealed putative GC boxes near TSSs of the NR1, NR2A, NR2B, and NR2C genes. Two tandem GC boxes immediately upstream of the TSSs of the NR1 gene have been analyzed functionally by a series of sophisticated experiments.[56,59] These elements were shown to be responsible for growth factor upregulation of the promoter.[56] Factors interacting with this sequence include Sp1, Sp3, Egr1, and MAZ.[59,69] Surprisingly, Sp1 and Sp3 also bind an NFκB site 3 kb upstream of the TSSs in the NR1 gene.[70]

Three GC boxes from the core promoter of the NR2A gene were tested by electrophoretic mobility shift assay and reporter gene assay and shown to possess positive regulatory activity.[48] At least three GC boxes were identified from the NR2B promoter. Although all three interacted with nuclear proteins, they may be redundant in upregulating the promoter since a fragment bearing only one such site showed promoter activity comparable to one having all three.[51] One GC-rich sequence harboring two tandem GC boxes was found immediately upstream of the NR2C TSS. This sequence binds Sp proteins and positively regulates the reporter gene.[62]

Sp factor, specifically Sp1, often collaborates with other transcription factors by direct interaction to enhance or reduce its active role on the promoter. On the NR1 promoter, a brain-specific factor known as MEF2C was found to interact directly with Sp1 and synergize promoter activity.[60] Interestingly, this interaction is independent of MEF2C binding to the promoter since forced expression of MEF2C with a promoter lacking the MEF2 site retains a similar impact on the promoter as long as Sp1 factor is coexpressed.

5.3.2 MAZ

Myc-associated zinc finger protein (MAZ) binds the same *cis* element as Sp1.[71] In the NR1 promoter, this protein was shown to compete with Sp1 at the proximal GC boxes.[69] Since GC boxes are common *cis* elements in the regulatory regions of genes in the nervous system, the interactions of transcription factors from different families with this element suggest that an appropriate balance in the interactions of various *trans* factors and these sites is important to maintain neuronal function.[53,72,73]

5.3.3 EARLY GROWTH RESPONSE (EGR) FAMILY

Early growth response (Egr) proteins belong to the immediate early gene family of transcription factors and are encoded by four genes, i.e., Zif268/Egr1/Krox-24, Erg2, Egr3, and Egr4.[74,76] These proteins are inducible and transiently expressed in most tissues including neurons in response to environmental cues such as neurotrophins[74] and glutamate.[75] Egr proteins recognize an Egr response element (GSG, 5′-GCG$_5$CG-3′) proximal to the TSSs of target genes and usually function as activators.

The rat NR1 promoter has a perfect GSG site immediately upstream of the TSSs.[45] Motif searches of genome databases revealed that human and mouse NR1 genes also bear this motif in similar locations. This site responds to growth factor stimulation in PC12 cells, binds recombinant Egr1 and Egr3, and enhances reporter gene expression in response to coexpressed Egr1 and Egr3.[74] Therefore, the GSG site is believed to mediate at least part of the positive effect of nerve growth factor (NGF) on NR1 expression.[56]

5.3.4 JUN AND FOS FAMILIES

Jun and Fos are among the most studied transcription factors.[77] Their expression is associated with many cellular activities such as growth, differentiation, stress, and apoptosis.[1,77] Members of the Fos, Jun, and ATF subfamilies form various homodimers or heterodimers and interact with the activator protein 1 (Ap1) consensus site of 5′-TGA(C/G)TCA to regulate transcription rate. Dimer composition is context-dependent and may significantly influence activity.[78] An active Ap1 site in the distal promoter of the NR1 gene was initially identified by computer alignment and subsequently confirmed by DNA–protein binding assays and reporter gene experiments.[45] An Ap1 site in the NR2B promoter has been suggested to mediate positive effects of ethanol-induced expression of this gene.[79] This Ap1 site is also recognized by factors in the CREB family. Supershift EMSA experiments demonstrated that phosphorylated CREB is increased in DNA–protein complexes bound to the NR2B Ap1 site after ethanol treatment.[79]

5.3.5 MEF2C

MEF2 is a subfamily of the MADS (MCM1 agamous deficiens serum response factor) family of DNA binding proteins consisting of four members (MEF2A, B, C, and D) that bind to the consensus of 5′-YTAW$_4$TAR.[80] MEF2C is highly expressed in the developing brain in parallel with NR expression.[81] A MEF2 site (5′-TTATTTATAG) has been identified approximately 500 bps upstream of the GSG and GC-boxes in the NR1 promoter.[60] This site positively regulates the NR1 promoter in cultured cell lines, primary neurons, and differentiating neurons.[60] MEF2C is the major *trans* factor responsible for this upregulation. Surprisingly, it synergizes this effect with the Sp1 factor via the proximal tandem GC boxes. Considering the expression of MEF2C and Sp1 in developing brains and differentiating neurons,[61] the interaction of these factors may be an important force driving expression of the NR1 gene during development.

5.3.6 CREB

In the nervous system, the cAMP response element (CRE) binding (CREB) protein receives signals primarily from the protein kinase A (PKA) pathway and from other pathways including the Ras-mitogen-activated protein kinase (MAPK) pathway triggered by NR activation.[82] Phosphorylated CREB proteins bind the highly conserved 8-bp palindromic CRE consensus (5′-TKACGTCA), and Ap1 or Ap2 sites.[83] CREs retaining the 3′ half of the palindromic consensus, while still active, are considered atypical and demonstrate less activity than full-size sites.[84,85] CREB has a large number of target genes and a database has been established to collect experimentally proven targets and predict potential CRE sites (http://natural.salk.edu/creb).[86]

The NR1 gene contains several atypical CRE sites around the TSSs and in the distal region.[45,87] In cultured embryonic neural cells, activation of the PKA pathway by forskolin increases both NR1 mRNA and protein. Chromatin immunoprecipitation (CHIP) experiments employing CREB antibody precipitated genomic DNA fragments proximal to the NR1 TSSs.[87] The CRE/CREB signal is often associated with neural activities such as synaptic plasticity[88] and nociception[89] in the mature central nervous system (CNS). The significance of these interactions in developing neurons remains to be explored.

In the NR2B regulatory region, a functional CRE site was found to be responsive to ethanol in cultured cortical neurons.[90] However, the changes observed for mRNA level in the treated neurons and for reporter gene activity in transfected cultures were marginal in comparison to the significant change in DNA binding of the CRE site by nuclear extracts following ethanol treatment.

5.3.7 NFκB Family

The NFκB family is composed of five transcription factors (p50, p52, p65, c-Rel, Rel-B) and associated with mechanisms of neuronal survival, neuronal plasticity, and neuropathology.[91–93] Homodimers and heterodimers of these proteins typically function as activators by binding to the consensus of 5′-GGGRDTYYCC. Analysis of the upstream sequence of the rat NR1 promoter revealed a perfect NFκB site. Unexpectedly, this site did not bind functional NFκB factors found in nuclear extracts of differentiating neurons.[70] More surprisingly, this NFκB site formed complexes with these nuclear extracts independent of any NFκB factor. Sp factors are the protein components of these complexes. The binding of Sp factors to the NR1 NFκB site was further confirmed in living cells by CHIP.[70] The binding strengths of different Sp factors to this site in living cells vary with neuronal differentiation. This kind of Sp factor binding to an NFκB site is the first such example in neuronal gene regulation. Considering the presence of the NFκB sites in a wide spectrum of neuronal genes and the universal expression of Sp factors in the brain, this finding may have broad implications in neuronal gene expression.

5.3.8 Tbr1

Tbr1 is a neuron-specific T box transcription factor expressed during brain development.[94] It binds a palindromic DNA consensus (5′-TSACACCTAGGTGTGAAATT) as well as nonpalindromic sequences homologous to either half side of the consensus such as 5′-YTTCACACCT.[95] NR1 and NR2B promoters both contain nonpalindromic T box elements. A combination of luciferase reporter assays and knock-out mice demonstrated a positive effect of Tbr1 on NR2B and/or NR1 expression.[96] Considering its expression in developing brain, Tbr1 is very likely an activator for the NR gene expression during development.

5.3.9 Estrogen Receptor

Estrogen regulates gene expression by interacting with its nuclear receptors, ERα and ERβ, or by activating signal transduction pathways through unidentified membrane-associated receptor(s).[97] Ligand-bound homodimers of ERα and ERβ recognize an estrogen response element or ERE (5′-GGTCANNNTGACC) and usually upregulate the promoter. ERα or ERβ monomer binding of half ERE sides has been suggested but is still controversial.[98] Increasing evidence suggests that estrogen steroids play roles in several CNS functions such as synaptic plasticity and neuroprotection involving NR activity.[97] Ovarian steroid withdrawal by ovariectomy in rats produced NR hypoactivity, specifically in the hippocampus.[99] Estradiol treatment of the rats recovered hippocampal NR ligand binding preceded by changes in NR1 and NR2B mRNA levels, visualized by *in situ* hybridization, suggesting that estrogen may regulate NR gene transcription.

Although the effects of estrogen on the NR1 and NR2B promoters have not been tested directly, the NR1 promoter is upregulated by the Ras-MAPK pathway that can be activated by estrogen.[97,100,101] In addition, expression of ERα and ERβ in the brains of female mice was well correlated with NR2D mRNA levels by Watanabe et al.[102] These authors identified four half palindromic ERE (5′-TGACC) sites in the 3′UTR of NR2D mRNA. The capability of these half sites to regulate transcription was further tested using a hetero promoter–reporter gene assay in cultured cells.[103] However, the TGACC half site sequence is very common in the genome and therefore the *bona fide* NR2D promoter must be tested to confirm the estrogen effect.

5.3.10 REST/NRSF

An effort to delineate the mechanism underlying neuronal transcription revealed a group of *cis* elements (consensus: 5′DYCAGCACCNNGGACAGNNNC) from the regulatory regions of many neuronal genes—designated repressor element 1 (RE1) or neuron-restrictive silencer element (NRSE).[104–108] The cognate *trans* factor is RE1 silencing transcription factor (REST) or neuron-restrictive silencer factor (NRSF), a zinc finger protein with several short isoforms generated by alternative splicing.[106] REST expression is high in nonneuronal cells and neural progenitors where it acts with cofactors to suppress neuronal genes.[106,109–111]

Putative RE1/NRSE sites have been found bioinformatically and experimentally in the regulatory regions of the NR1, NR2A, and NR2B genes.[52,104,106,107] A global search of REST targets by ChIPSeq uncovered REST bound sequences near the NR2A, 2C, and 3A genes in a human T lymphoblast cell line.[108] Another genome-wide analysis coupling a large-scale CHIP with serial analysis of chromatin occupancy revealed that REST occupied one or more sites linked to each of the NR1, NR2C, and NR2D genes in a cultured mouse kidney cell line.[112]

The functional impacts of REST and identified RE1 sites on *bona fide* promoters in a reporter gene assay have been studied only for the NR1 and NR2B genes.[61,113,114] The NR1 RE1 element was shown to bind REST and negatively regulate the promoter in nonneuronal and neuroprogenitor cells.[61] The RE1 site in the 5′ flanking region of the NR2B gene also demonstrated a negative regulatory effect on the promoters in embryonic neural cells.[114]

5.4 DEVELOPMENTAL EXPRESSION OF NR GENES

Development is the cellular program that produces differentiated tissues and organs from undifferentiated precursors. It has been proposed that a network of interactions between *cis* elements in DNA and grouped *trans* factors expressed following specific temporal and spatial patterns is a key driving force for this program.[115–117]

NR genes exhibit heterogeneous expression in the developing brain. The mRNAs of NR1, NR2A, and 2D become detectable in CNS following embryonic day 14 in rodents.[118,119] During the first 3 postnatal weeks, NR1 is progressively upregulated in the whole brain, followed by NR2A expression. In contrast, NR2B shows high expression in the whole brain within the first 2 postnatal weeks. Subsequently, it decreases in the cerebellum, but remains highly expressed in the forebrain. NR2C is expressed at low levels in the early stages of development and shows progressively higher expression in the cerebellum and olfactory bulb following postnatal day 11. NR2D is highly expressed during the first postnatal week throughout the brain, then declines and becomes restricted mainly to the middle brain in adults.[118–120] NR3A mRNA appears in the rat CNS by E16, becomes robustly expressed in the whole CNS except the forebrain by E19, and peaks by P7 in rats. Its expression then declines significantly, and is limited to a few CNS areas such as the spinal cord and thalamus.[121,122] NR3B expression emerges in motor neurons of the spinal cord by P10 and reaches its maximum by P21.[123]

The impact of *cis* elements on developmental expression of the NR1 promoter has been analyzed in more detail than impacts on other NR genes.[61,70] In P19 embryonic stem cells, changes in NR1 mRNA level and promoter activity are coordinated with neurogenesis and neuronal differentiation. Additionally, REST expression and binding to the NR1 RE1 site are robustly downregulated in the early stages of neurogenesis and differentiation.[61] A 2-day gap between REST downregulation and upregulation of the NR1 gene promoter suggests that other factors are involved. Experiments with *cis* element mutations suggest that the GC box/GSG, MEF2 and NFκB sites and relevant *trans* factors are important for promoter activation following this 2-day gap although the mechanism underlying the gap is unknown.[60,70]

NR2A and 2B promoter fragments have been studied in transgenic mice to determine their total activity in the developing brain. The NR2B promoter up to −9 kb exhibited a 100-fold greater activity in neuronal and glial cocultures than in pure glial cultures. A sequence between −1253 and −1180 is necessary for the developmental expression of NR2A.[49] The activity of these elements during development has not been precisely studied.[49,52]

To follow NR2C expression, Karavanova et al. recently developed a knock-in mouse model by inserting the bacterial LacZ gene into the 5′ end of the coding sequence of the endogenous NR2C gene.[124] They were able to map the regional and developmental expression of NR2C. Unfortunately, this model cannot provide details of how the promoter contributes to this expression. In a separate study, an NR2C promoter–LacZ fusion gene was integrated into the mouse genome and reporter expression was found in layer 4 spiny stellate cells of the adult barrel cortex[125] but no developmental information is available.

The mechanism underlying the decline of the NR2D gene during development remains unexplored. Recent data generated by genome-wide searches for REST targets linked DNA sequences bound by REST to the NR2C and NR2D genes.[108,112] Since the REST–RE1 interaction plays an important role in the developmental expression of neuronal genes, whether these identified RE1 sites participate in developmental expression and restriction of these two genes should be investigated.

5.5   GROWTH FACTOR REGULATION OF NR GENES

Growth factors are important extracellular stimuli that initiate and maintain neuronal differentiation and survival. Brain-derived growth factor treatment of cultured
embryonic cortical neurons increased the NR1 mRNA level about two-fold.[126] Studies with PC12 cells showed that activity of the NR1 promoter is upregulated by several growth factors including NGF, fibroblast growth factor, and epidermal growth
factor.[56,60] A recent report scrutinized several lines of PC12 cells and concluded that
NGF upregulates NR1 mRNA in a cell line-dependent manner.[127]

Based on these studies, NR1 transcription is likely upregulated by growth factors during neuronal differentiation. Further studies demonstrate that NGF utilizes
both MAPK and phosphatidylinositol 3-kinase (PI3K) pathways to regulate the NR1
promoter.[101] Interestingly, MAPK targets included Sp1 and a group of unknown
nuclear proteins specifically interacting with single-stranded DNA of the proximal
region of the NR1 promoter.[101,113]

Using proteomics, DNA binding, and siRNA technologies, we found that the
major components of these complexes are hnRNPs that regulate mRNA transporta-
tion, splicing, and gene transcription. The downstream mediators of the PI3K effect
have not yet been elucidated.[101] NR activation in the nervous system controls the
transcriptional regulation of many genes such as those encoding growth factors
through the Ras-MAPK pathway.[128,129] This reciprocal regulation of growth factor
and NR genes may be part of the mechanism underlying the neurotrophic effect of
NR activation.

5.6   CELL TYPE SPECIFICITY OF NR GENE EXPRESSION

Neurons are the major sites of NR expression. It is surprising that NR expression
and receptor activities are also found in nonneuronal cells and peripheral tissues
including glial cells,[130] pancreatic islands,[131] lungs,[132] bone cells,[131] adrenal medulla
and kidneys,[133] keratinocytes,[134] and heart,[135] although their exact functions in these
tissues are not yet fully understood.

The neuronal specificity of NR1 expression has been shown to reside in the
proximal promoter. Functional analysis disclosed that 356 bps of the NR1 core promoter confer high activity in neuronal PC12 cells and in cultured neurons, low activity in the C6 glioma, and almost no activity in HeLa cells.[45,56,59] As noted, this 356
bp proximal promoter contains a consensus RE1 site that negatively regulates this
promoter in nonneuronal and neuroprogenitor cells.[60,87,113] A number of universal
cis elements from proximal and distal regulatory regions have been found to positively regulate the NR1 promoter. The coordination of these positive and negative
elements determines NR1 expression in different group of neurons.

The neuronal specificity of the NR2A gene is restricted by a fragment consisting of exon 1 and a few upstream sequences in transgenic mice. The addition of the
upstream sequence up to 9 kb did not show additional effects.[49] The same observation has been repeated in cultured cells.[48] The core promoters of the NR2B and
NR2C genes also exhibit neuronal specificity in transgenic mice and cultured neurons.[49,52,62] Since the RE1 sites identified for the NR2 genes in several studies are

distal to these core promoters.[106–108,112,114] and no direct evidence indicates how these sites regulate bona fide promoters, whether RE1/REST is the major player of neuronal specificity of NR2 gene transcription must be reconsidered.

While NR proteins are expressed in several nonneuronal cell types, the mechanisms controlling this expression have not been directly investigated. Only tangential evidence exists. One study employing C6 glioma cells as negative controls found weak NR1 promoter activity in these cells. Nuclear extracts from C6 glioma cells were also found to bind the Spl sites of the NR1 promoter.[59] The control of NR expression in nonneuronal cell types deserves further scrutiny and would certainly provide interesting data on cell-specific transcriptional control and developmental pathways.

5.7   NEUROLOGICAL DISORDERS

NR-mediated glutamate toxicity is considered a common pathological pathway of many acute and chronic neurological disorders such as brain trauma, brain ischemia, schizophrenia, Alzheimer's disease, Parkinson's disease, Huntington's disease, AIDS dementia, and Lou Gehrig's disease.[6–8,10,136] Since NR function and activity are determined by a specific set of NR genes expressed in a given cell and by the level of that expression, abnormal transcription of the NR genes may contribute to NR pathological effects.

5.7.1   REST, HUNTINGTIN, AND HUNTINGTON'S DISEASE

Alteration of NR expression may be one mechanism underlying the pathology of Huntington's disease.[137] NR ligand binding is significantly decreased in the caudate in Huntington's disease.[138] Wild-type huntingtin has been shown to retain REST protein in the cytoplasm, preventing it from silencing neuronal genes.[139] A mutated form of huntingtin genetically associated with Huntington's disease[137] does not retain this function and releases REST that can suppress neuronal genes in mature neurons.[139]

In situ hybridization of a Huntington's disease brain revealed decreases in the NR1 and NR2B mRNAs correlated with severity.[140] Decreases of NR2A and NR2B mRNAs in the hippocampus were also observed in mouse models of Huntington's disease (R6/2).[141] NR2D is upregulated in neuronal nitric oxide synthase-positive interneurons in the caudates of these mice.[142] It is therefore likely that transcription of NR genes is suppressed by free REST in some neuronal nuclei of Huntington's patients, while other cells may express inappropriate subunits.

5.7.2   PROMOTER POLYMORPHISM AND SCHIZOPHRENIA

NR hypofunction has been reported in schizophrenic patients[143] and schizophrenic behaviors have been noted in NR2A-depleted mice.[144] Reduced expression of NR genes including the NR1, NR2B, and NR2C genes in the thalami of schizophrenic patients has also been documented.[145,146]

The possibility that lowered promoter activity due to polymorphisms may contribute to NR hypofunction has been investigated.[147–149] In a case-controlled study (375 schizophrenics, 378 controls, equally divided by sex), the repeat length of a variable GT repeat in the NR2A promoter was correlated with mRNA reduction and the severity of chronic schizophrenia.[50] This initial survey was confirmed in an extended study with twice as many schizophrenics and controls.[147]

This observation was confirmed in an independent study of 122 Han Chinese sibling pair families.[149] The authors believed that the (GT)n polymorphism in the NR2A promoter played a significant role in the etiology of schizophrenia. However, whether a *trans* factor interacts with these repeats is yet to be investigated. In another case-control study, a T-G variant within a proximal GC box of the NR2B promoter was correlated with schizophrenia.[53] In response to NGF treatment in PC12 cells, the T allele showed a 30-fold increase in promoter activity in comparison to the G allele. In a study of an Italian population, a G-C change of the first G in the GGGG sequence of a putative NFκB site in the 5′UTR of the NR1 gene was correlated with human schizophrenia although its impact on promoter activity was not tested.[150] Polymorphisms of other NR subunits in schizophrenic patients have not yet been investigated.

Reduced expression of the NR genes was found in the brains of patients with other neurodegenerative disorders as well. Alzheimer's disease patients showed reduced expression of the NR1 and NR2B gene in the hippocampus.[151] Parkinson's patients exhibited reduced NR1 expression in the striatum and in the superficial layers of the prefrontal cortex.[152] Whether polymorphisms of the regulatory regions also play a role in this altered NR gene expression is an interesting question that remains to be addressed.

5.7.3 ALCOHOLISM

Accumulating data indicate that glutamate neurotransmission may be damaged by ethanol inhibition. Expression of NR2B in the brain was found to be upregulated by ethanol treatment.[153] DNA methylation studies revealed that its promoter is demethylated in cultured neurons and in living animal models following chronic (but not acute) ethanol treatment.[154,155] The demethylation of two CpG islands upstream of the NR2B promoter is well correlated with the increase in NR2B mRNA. However, the functional effect of this demethylation on promoter activity has yet to be defined. Ethanol was also found to interfere with the inhibitory activities of RE1 clusters in the NR2B promoter[114] and enhance interactions of CREB and FosB with the AP1 site in the NR2B promoter,[79] eventually upregulating the NR2B gene.

5.7.4 HYPERACTIVATION BY AGONIST

Overstimulation of NRs by an abnormally high level of glutamate in nervous tissue has been hypothesized to mediate excitotoxicity in neurological disorders. However, hypoactivity of NR and reduced expression of NR genes was observed in a number of chronic neurological disorders and correlated with severity.[140,141,143,151,152] Whether the reductions are caused by glutamate overstimulation at the early stages of the disorders was investigated. Gascon et al. found that treatment of cultured cortical neurons with NMDA or glutamate resulted in a reduction of the NR1 protein and mRNA.[156]

This effect was repeated on the NR1 promoter transfected into these cells. Mao et al. reported similar observations in cultured neurons and proposed that Sp1 protein and the proximal GC boxes were the mediators of this effect.[157] Treatment of cultured striatum neurons with quinolic acid, an endogenous NR agonist, produced the same effect.[158] This negative feedback in NR expression may be a defensive mechanism to avoid overactivation due to the persistent presence of agonist for neurons undergoing chronic pathological changes.

5.8 EPIGENETIC REGULATION

In a broad sense, the term *epigenetics* covers stable modifications of chromatin and DNA not involving DNA sequence change. DNA methylation, chromatin remodeling, and noncoding RNA are considered major epigenetic mechanisms.[159]

5.8.1 DNA METHYLATION

Cytosine in a CpG dinucleotide is the major target of DNA methyltransferase in vertebrates.[159,160] Clusters of CpG sequences often reside in the promoters or their proximal regions to form so-called CpG islands.[159,161] CpG islands may become methylated and negatively regulate transcription. This pathway has been proposed as a critical epigenetic mechanism underlying development and other processes.[159]

The CpG islands have been found within the promoter or proximal regions of the NR1, NR2A, NR2B, and NR2C genes.[45,48,54,162] The role of DNA methylation in promoter regulation has only been studied for the NR2B gene. The NR2B promoter was found to be hypermethylated in primary esophageal squamous cell carcinoma in which the NR2B is not expressed.[162] Demethylation by 5-aza-2'-deoxycytidine unmasked the promoter region and led to expression of NR2B transcripts as measured by reverse transcriptase polymerase chain reaction.

5.8.2 CHROMATIN REMODELING

Whether histones and other nuclear proteins regulate NR gene transcription via chromatin remodeling is largely unexplored. However, many *trans* factors found to directly interact with NR promoters such as REST and CREB are subject to regulation by chromatin remodeling.[110,159] Therefore, it can be hypothesized that this mechanism is also utilized to regulate NR transcription, but detailed direct evidence does not exist.

5.8.3 NONCODING RNA

Initial sequence analysis by TargetScan (http://www.targetscan.org) revealed that mRNAs of NR1, NR2A-D, and NR3A are included in the 30% of human mRNAs considered potential targets of miRNAs.[163] However, no studies have yet addressed directly the involvement of miRNAs in NR gene expression.

5.9 SUMMARY

Studies have revealed the functional activities of 5' flanking sequences of the NR1, NR2A, 2B, and 2C genes using cultured cells and living animals. Interactions of several *cis*-acting regulatory elements with cognate *trans* factors and their impacts on promoter activity have been found to play important roles in NR gene expression under various conditions. Further studies should uncover additional pathways of differential NR expression. These studies will help explain the mechanisms of abnormal NR expression in human diseases and may also reveal potential pharmaceutical targets.

More than 1,000 transcription factors are dynamically expressed in the brain.[164] Thus it is likely that a network of these factors and interactions with a set of *cis* elements under the influence of epigenetic mechanisms ultimately determines NR gene expression under various conditions. A genome-wide approach may help address this question efficiently, particularly for investigating all functional *cis* elements involved in NR gene regulation. Those located in regions other than the 5′ flanking sequences, the 5′UTR, and all functional regions of the NR2D and NR3A/B genes have not been systematically studied. A large-scale search for transcription regulatory regions may allow us to learn whether NR genes are involved in the complex transcription patterns uncovered by recent studies of the functional elements of the human genome (Figure 5.1).[19,44,165,166]

ACKNOWLEDGMENTS

The authors wish to thank Drs. Roland Dubner and Dean Dessem for critical reading of this manuscript and Dong Wei for drawing figures. GB was supported by NIH Grant NS38077 and by start-up funds from the University of Maryland Dental School.

REFERENCES

1. Nestler, E.J., Barrot, M. and Self, D.W., Delta FosB: a sustained molecular switch for addiction, *Proc. Natl. Acad. Sci. U.S.A.*, 98, 11042, 2001.
2. Dubner, R., The neurobiology of persistent pain and its clinical implications, *Suppl. Clin. Neurophysiol.*, 57, 3, 2004.
3. Dingledine, R. et al., The glutamate receptor ion channels, *Pharmacol. Rev.*, 51, 7, 1999.
4. Mori, H. and Mishina, M., Structure and function of the NMDAR channel, *Neuropharmacology*, 34, 1219, 1995.
5. Mikuni, N. et al., NMDA-receptors 1 and 2A/B coassembly increased in human epileptic focal cortical dysplasia, *Epilepsia*, 40, 1683, 1999.
6. Kristiansen, L.V. et al., NMDARs and schizophrenia, *Curr. Opin. Pharmacol.*, 7, 48, 2007.
7. Waxman, E.A. and Lynch, D.R., N-methyl-D-aspartate receptor subtypes: multiple roles in excitotoxicity and neurological disease, *Neuroscientist*, 11, 37, 2005.
8. Olney, J.W., Newcomer, J.W., and Farber, N.B., NMDAR hypofunction model of schizophrenia, *J. Psychiatr. Res.*, 33, 523, 1999.
9. Ikonomovic, M.D. et al., Distribution of glutamate receptor subunit NMDAR1 in the hippocampus of normal elderly and patients with Alzheimer's disease, *Exp. Neurol.*, 160, 194, 1999.
10. Bossy-Wetzel, E., Schwarzenbacher, R., and Lipton, S.A., Molecular pathways to neurodegeneration, *Nat. Med.*, 10, S2, 2004.
11. Das, S. et al., Increased NMDA current and spine density in mice lacking the NMDAR subunit NR3A, *Nature*, 393, 377, 1998.

12. Nishi, M. et al., Motoneuron-specific expression of NR3B, a novel NMDA-type glutamate receptor subunit that works in a dominant-negative manner, *J. Neurosci.*, 21, RC185, 2001.
13. Zukin, R.S. and Bennett, M.V., Alternatively spliced isoforms of the NMDARI receptor subunit, *Trends Neurosci.*, 18, 306, 1995.
14. Prybylowski, K.L. et al., Expression of splice variants of the NR1 subunit of the N-methyl-D-aspartate receptor in the normal and injured rat spinal cord, *J. Neurochem.*, 76, 797, 2001.
15. Tolle, T.R. et al., Cellular and subcellular distribution of NMDAR1 splice variant mRNA in the rat lumbar spinal cord, *Eur. J. Neurosci.*, 7, 1235, 1995.
16. Wray, G.A. et al., The evolution of transcriptional regulation in eukaryotes, *Mol. Biol. Evol.*, 20, 1377, 2003.
17. Levine, M. and Tjian, R., Transcription regulation and animal diversity, *Nature*, 424, 147, 2003.
18. Maniatis, T., Goodbourn, S., and Fischer, J.A., Regulation of inducible and tissue-specific gene expression, *Science*, 236, 1237, 1987.
19. ENCODE, Identification and analysis of functional elements in 1% of the human genome by the ENCODE pilot project, *Nature*, 447, 799, 2007.
20. Smale, S.T., Core promoters: active contributors to combinatorial gene regulation, *Genes Dev.*, 15, 2503–8, 2001.
21. Dynan, W.S., Promoters for housekeeping genes, *Trends Genet.*, 2, 196, 1986.
22. Butler, J.E. and Kadonaga, J.T., The RNA polymerase II core promoter: a key component in the regulation of gene expression, *Genes Dev.*, 16, 2583, 2002.
23. Saltzman, A.G. and Weinmann, R., Promoter specificity and modulation of RNA polymerase II transcription, *FASEB J.*, 3, 1723, 1989.
24. Taylor, M.S. et al., Heterotachy in mammalian promoter evolution, *PLoS Genet.*, 2, e30, 2006.
25. Lee, T.I. and Young, R.A., Regulation of gene expression by TBP-associated proteins, *Genes Dev.*, 12, 1398, 1998.
26. Kimura, K. et al., Diversification of transcriptional modulation: large-scale identification and characterization of putative alternative promoters of human genes, *Genome Res.*, 16, 55, 2006.
27. Sandelin, A. et al., Mammalian RNA polymerase II core promoters: insights from genome-wide studies, *Nat. Rev. Genet.*, 8, 424, 2007.
28. Cooper, S.J. et al., Comprehensive analysis of transcriptional promoter structure and function in 1% of the human genome, *Genome Res.*, 16, 1, 2006.
29. Trinklein, N.D. et al., Integrated analysis of experimental data sets reveals many novel promoters in 1% of the human genome, *Genome Res.*, 17, 720, 2007.
30. Lin, J.M. et al., Transcription factor binding and modified histones in human bidirectional promoters, *Genome Res.*, 17, 818, 2007.
31. Carninci, P. et al., Genome-wide analysis of mammalian promoter architecture and evolution, *Nat. Genet.*, 38, 626, 2006.
32. Murakami, K., Kojima, T., and Sakaki, Y., Assessment of clusters of transcription factor binding sites in relationship to human promoter, CpG islands and gene expression, *BMC Genomics*, 5, 16, 2004.
33. Liu, R. and States, D.J., Consensus promoter identification in the human genome utilizing expressed gene markers and gene modeling, *Genome Res.*, 12, 462, 2002.
34. Xie, X. et al., PromoterExplorer: an effective promoter identification method based on the AdaBoost algorithm, *Bioinformatics*, 22, 2722, 2006.
35. Wasserman, W.W. and Sandelin, A., Applied bioinformatics for the identification of regulatory elements, *Nat. Rev. Genet.*, 5, 276, 2004.

36. Bajic, V.B. et al., Promoter prediction analysis on the whole human genome, *Nat. Biotechnol.*, 22, 1467, 2004.
37. Schmid, C.D. et al., EPD in its twentieth year: toward complete promoter coverage of selected model organisms, *Nucleic Acids Res.*, 34, D82, 2006.
38. Bajic, V.B. et al., Mice and men: their promoter properties, *PLoS Genet.*, 2, e54, 2006.
39. Yamashita, R. et al., DBTSS: data base of human transcription start Sites, progress report 2006, *Nucleic Acids Res.*, 34, D86, 2006.
40. Ota, T. et al., Complete sequencing and characterization of 21,243 full-length human cDNAs, *Nat. Genet.*, 36, 40, 2004.
41. Carninci, P. et al., The transcriptional landscape of the mammalian genome, *Science*, 309, 1559, 2005.
42. Gerhard, D.S. et al., The status, quality, and expansion of the NIH full-length cDNA project: the Mammalian Gene Collection (MGC), *Genome Res.*, 14, 2121, 2004.
43. Hashimoto, S. et al., 5′ end SAGE for the analysis of transcriptional start sites, *Nat. Biotechnol.*, 22, 1146, 2004.
44. Carninci, P., Tagging mammalian transcription complexity, *Trends Genet.*, 22, 501, 2006.
45. Bai, G. and Kusiak, J.W., Cloning and analysis of the 5′ flanking sequence of the rat N-methyl-D-aspartate receptor 1 (NMDAR1) gene, *Biochim. Biophys. Acta*, 1152, 197, 1993.
46. Zimmer, M. et al., Cloning and structure of the gene encoding the human N-methyl-D-aspartate receptor (NMDAR1), *Gene*, 159, 219, 1995.
47. Zarain-Herzberg, A. et al., Cloning and characterization of the chick NMDAR subunit-1 gene, *Brain Res. Mol. Brain Res.*, 137, 235, 2005.
48. Liu, A.G. et al., Functional analysis of the rat N-methyl-D-aspartate receptor 2A promoter: multiple core promoters in exon 1, positive regulation by Sp factors and translational regulation, *J. Biol. Chem.*, 278, 26423, 2003.
49. Desai, A. et al., Analysis of transcriptional regulatory sequences of the N-methyl-D-aspartate receptor 2A subunit gene in cultured cortical neurons and transgenic mice, *J. Biol. Chem.*, 277, 46374, 2002.
50. Itokawa, M. et al., A microsatellite repeat in the promoter of the N-methyl-D-aspartate receptor 2A subunit (GRIN2A) gene suppresses transcriptional activity and correlates with chronic outcome in schizophrenia, *Pharmacogenetics*, 13, 271, 2003.
51. Klein, M. et al., Cloning and characterization of promoter and 5′UTR of the NMDAR subunit epsilon 2: evidence for alternative splicing of 5′-non-coding exon, *Gene*, 208, 259, 1998.
52. Sasner, M. and Buonanno, A., Distinct N-methyl-D-aspartate receptor 2B subunit gene sequences confer neural and developmental specific expression, *J. Biol. Chem.*, 271, 21316, 1996.
53. Miyatake, R., Furukawa, A., and Suwaki, H., Identification of a novel variant of the human NR2B gene promoter region and its possible association with schizophrenia, *Mol. Psychiatr.*, 7, 1101, 2002.
54. Suchanek, B., Seeburg, P.H., and Sprengel, R., Gene structure of the murine N-methyl D-aspartate receptor subunit NR2C, *J. Biol. Chem.*, 270, 41, 1995.
55. Kawaji, H. et al., Dynamic usage of transcription start sites within core promoters, *Genome Biol.*, 7, R118, 2006.
56. Bai, G. and Kusiak, J.W., Nerve growth factor up-regulates the N-methyl-D-aspartate receptor subunit 1 promoter in PC12 cells, *J. Biol. Chem.*, 272, 5936, 1997.
57. Awobuluyi, M., Vazhappilly, R., and Sucher, N.J., Translational activity of N-methyl-D-aspartate receptor subunit NR1 mRNA in PC12 cells, *Neurosignals*, 12, 283, 2003.

58. Alam, J. and Cook, J.L., Reporter genes: application to the study of mammalian gene transcription, *Anal. Biochem.*, 188, 245, 1990.
59. Bai, G. and Kusiak, J.W., Functional analysis of the proximal 5′ flanking region of the N-methyl-D-aspartate receptor subunit gene, NMDAR1, *J. Biol. Chem.*, 270, 7737, 1995.
60. Krainc, D. et al., Synergistic activation of the N-methyl-D-aspartate receptor subunit 1 promoter by myocyte enhancer factor 2C and Sp1, *J. Biol. Chem.*, 273, 26218, 1998.
61. Bai, G. et al., The role of the RE1 element in activation of the NR1 promoter during neuronal differentiation, *J. Neurochem.*, 86, 992, 2003.
62. Pieri, I. et al., Regulation of the murine NMDA-receptor-subunit NR2C promoter by Sp1 and fushi tarazu factor1 (FTZ-F1) homologues, *Eur. J. Neurosci.*, 11, 2083, 1999.
63. Maston, G.A., Evans, S.K., and Green, M.R., Transcriptional Regulatory Elements in the Human Genome, *Annu. Rev. Genomics Hum. Genet.*, 7, 29, 2006.
64. Suske, G., The Sp-family of transcription factors, *Gene*, 238, 291, 1999.
65. Kawakami, Y. et al., Sp8 and Sp9, two closely related buttonhead-like transcription factors, regulate Fgf8 expression and limb outgrowth in vertebrate embryos, *Development.*, 131, 4763, 2004.
66. Safe, S. and Abdelrahim, M., Sp transcription factor family and its role in cancer, *Eur. J. Cancer*, 41, 2438, 2005.
67. Milona, M.A., Gough, J.E., and Edgar, A.J., Expression of alternatively spliced isoforms of human Sp7 in osteoblast-like cells, *BMC Genomics*, 4, 43, 2003.
68. Marin, M. et al., Transcription factor Sp1 is essential for early embryonic development but dispensable for cell growth and differentiation, *Cell*, 89, 619, 1997.
69. Okamoto, S. et al., Effect of the ubiquitous transcription factors, SP1 and MAZ, on NMDAR subunit type 1 (NR1) expression during neuronal differentiation, *Brain Res. Mol. Brain Res.*, 107, 89, 2002.
70. Liu, A.G. et al., NF-kB site interacts with Sp factors and upregulates the NR1 promoter during neuronal differentiation., *J. Biol. Chem.*, 279, 17449, 2004.
71. Song, J. et al., Transcriptional regulation by zinc-finger proteins Sp1 and MAZ involves interactions with the same cis-elements, *Int. J. Mol. Med.*, 11, 547, 2003.
72. Dunah, A.W. et al., Sp1 and TAFII130 transcriptional activity disrupted in early Huntington's disease, *Science*, 296, 2238, 2002.
73. Hansen, T.V., Rehfeld, J.F., and Nielsen, F.C., Function of the C-36 to T polymorphism in the human cholecystokinin gene promoter, *Mol. Psychiatry.*, 5, 443, 2000.
74. Gashler, A. and Sukhatme, V.P., Early growth response protein 1 (Egr-1): prototype of a zinc-finger family of transcription factors, *Prog. Nucleic Acid Res. Mol. Biol.*, 50, 191, 1995.
75. Beckmann, A.M. and Wilce, P.A., Egr transcription factors in the nervous system, *Neurochem. Int.*, 31, 477, 1997.
76. Khachigian, L.M., Early growth response-1: blocking angiogenesis by shooting the messenger, *Cell Cycle*, 3, 10, 2004.
77. Raivich, G. and Behrens, A., Role of the AP-1 transcription factor c-Jun in developing, adult and injured brain, *Prog. Neurobiol.*, 78, 347, 2006.
78. Morgan, J.I. and Curran, T., Immediate-early genes: ten years on, *Trends Neurosci.*, 18, 66, 1995.
79. Qiang, M. and Ticku, M.K., Role of AP-1 in ethanol-induced N-methyl-D-aspartate receptor 2B subunit gene up-regulation in mouse cortical neurons, *J. Neurochem.*, 95, 1332, 2005.
80. McKinsey, T.A., Zhang, C.L., and Olson, E.N., MEF2: a calcium-dependent regulator of cell division, differentiation and death, *Trends Biochem. Sci.*, 27, 40, 2002.

81. Leifer, D. et al., MEF2C, a MADS/MEF2-family transcription factor expressed in a laminar distribution in cerebral cortex, *Proc. Natl. Acad. Sci. U.S.A.*, 90, 1546, 1993.

82. Lonze, B.E. and Ginty, D.D., Function and regulation of CREB family transcription factors in the nervous system, *Neuron*, 35, 605, 2002.

83. Roesler, W.J., Vandenbark, G.R., and Hanson, R.W., Cyclic AMP and the induction of eukaryotic gene transcription, *J. Biol. Chem.*, 263, 9063, 1988.

84. Flammer, J.R., Popova, K.N., and Pflum, M.K., Cyclic AMP response element-binding protein (CREB) and CAAT/enhancer-binding protein beta (C/EBPbeta) bind chimeric DNA sites with high affinity, *Biochemistry*, 45, 9615, 2006.

85. Craig, J.C. et al., Consensus and variant cAMP-regulated enhancers have distinct CREB-binding properties, *J. Biol. Chem.*, 276, 11719, 2001.

86. Zhang, X. et al., Genome-wide analysis of cAMP-response element binding protein occupancy, phosphorylation, and target gene activation in human tissues, *Proc. Natl. Acad. Sci. U.S.A.*, 102, 4459, 2005.

87. Lau, G.C. et al., Up-regulation of NMDAR1 subunit gene expression in cortical neurons via a PKA-dependent pathway, *J. Neurochem.*, 88, 564, 2004.

88. Josselyn, S.A. and Nguyen, P.V., CREB, synapses and memory disorders: past progress and future challenges, *Curr. Drug Targets CNS Neurol. Disord.*, 4, 481, 2005.

89. Wei, F. et al., Genetic elimination of behavioral sensitization in mice lacking calmodulin-stimulated adenylyl cyclases, *Neuron*, 36, 713, 2002.

90. Rani, C.S., Qiang, M., and Ticku, M.K., Potential role of cAMP response element-binding protein in ethanol-induced N-methyl-D-aspartate receptor 2B subunit gene transcription in fetal mouse cortical cells, *Mol. Pharmacol.*, 67, 2126, 2005.

91. Mattson, M.P. et al., Roles of nuclear factor kappaB in neuronal survival and plasticity, *J. Neurochem.*, 74, 443, 2000.

92. Lipton, S.A., Janus faces of NF-kappa B: neurodestruction versus neuroprotection, *Nat. Med.*, 3, 20, 1997.

93. Pizzi, M. and Spano, P., Distinct roles of diverse nuclear factor-kappaB complexes in neuropathological mechanisms, *Eur. J. Pharmacol.*, 545, 22, 2006.

94. Bulfone, A. et al., Expression pattern of the Tbr2 (eomesodermin) gene during mouse and chick brain development, *Mech. Dev.*, 84, 133, 1999.

95. Tada, M. and Smith, J.C., T-targets: clues to understanding the functions of T-box proteins, *Dev. Growth Differ.*, 43, 1, 2001.

96. Wang, T.F. et al., Identification of Tbr-1/CASK complex target genes in neurons, *J. Neurochem.*, 91, 1483, 2004.

97. Brann, D.W. et al., Neurotrophic and neuroprotective actions of estrogen: basic mechanisms and clinical implications, *Steroids*, 72, 381, 2007.

98. Klinge, C.M., Estrogen receptor interaction with estrogen response elements, *Nucleic Acids Res.*, 29, 2905, 2001.

99. Cyr, M. et al., Ovarian steroids and selective estrogen receptor modulators activity on rat brain NMDA and AMPA receptors, *Brain Res. Brain Res. Rev.*, 37, 153, 2001.

100. Wehling, M. and Losel, R., Non-genomic steroid hormone effects: membrane or intracellular receptors?, *J. Steroid Biochem. Mol. Biol.*, 102, 180, 2006.

101. Liu, A. et al., Nerve growth factor uses Ras/ERK and phosphatidylinositol 3-kinase cascades to up-regulate the N-methyl-D-aspartate receptor 1 promoter, *J. Biol. Chem.*, 276, 45372, 2001.

102. Watanabe, T. et al., NMDAR type 2D gene as target for estrogen receptor in the brain, *Brain Res. Mol. Brain Res.*, 63, 375, 1999.

103. Vasudevan, N. et al., Isoform specificity for oestrogen receptor and thyroid hormone receptor genes and their interactions on the NR2D gene promoter, *J. Neuroendocrinol.*, 14, 836, 2002.

104. Schoenherr, C.J., Paquette, A.J., and Anderson, D.J., Identification of potential target genes for the neuron-restrictive silencer factor, *Proc. Natl. Acad. Sci. U.S.A.*, 93, 9881, 1996.
105. Johnson, R. et al., Identification of the REST regulon reveals extensive transposable element-mediated binding site duplication, *Nucleic Acids Res.*, 34, 3862, 2006.
106. Roopra, A., Huang, Y., and Dingledine, R., Neurological disease: listening to gene silencers, *Mol. Interv.*, 1, 219, 2001.
107. Bruce, A.W. et al., Genome-wide analysis of repressor element 1 silencing transcription factor/neuron-restrictive silencing factor (REST/NRSF) target genes, *Proc. Natl. Acad. Sci. U.S.A.*, 101, 10458, 2004.
108. Johnson, D.S. et al., Genome-wide mapping of in vivo protein-DNA interactions, *Science*, 316, 1497, 2007.
109. Lunyak, V.V., Prefontaine, G.G., and Rosenfeld, M.G., REST and peace for the neuronal-specific transcriptional program, *Ann. NY Acad. Sci.*, 1014, 110, 2004.
110. Ballas, N. and Mandel, G., The many faces of REST oversee epigenetic programming of neuronal genes, *Curr. Opin. Neurobiol.*, 15, 500, 2005.
111. Ballas, N. et al., REST and its corepressors mediate plasticity of neuronal gene chromatin throughout neurogenesis, *Cell*, 121, 645, 2005.
112. Otto, S.J. et al., A new binding motif for the transcriptional repressor REST uncovers large gene networks devoted to neuronal functions, *J. Neurosci.*, 27, 6729, 2007.
113. Bai, G. et al., Single-stranded DNA-binding proteins and neuron-restrictive silencer factor participate in cell-specific transcriptional control of the NMDAR1 gene, *J. Biol. Chem.*, 273, 1086, 1998.
114. Qiang, M., Rani, C.S., and Ticku, M.K., Neuron-restrictive silencer factor regulates the N-methyl-D-aspartate receptor 2B subunit gene in basal and ethanol-induced gene expression in fetal cortical neurons, *Mol. Pharmacol.*, 67, 2115, 2005.
115. Istrail, S. and Davidson, E.H., Logic functions of the genomic cis-regulatory code, *Proc. Natl. Acad. Sci. U.S.A.*, 102, 4954, 2005.
116. Levine, M. and Davidson, E.H., Gene regulatory networks for development, *Proc. Natl. Acad. Sci. U.S.A.*, 102, 4936, 2005.
117. Ma, Q., Transcriptional regulation of neuronal phenotype in mammals, *J. Physiol.*, 575, 379, 2006.
118. Monyer, H. et al., Developmental and regional expression in the rat brain and functional properties of four NMDARs, *Neuron*, 12, 529, 1994.
119. Akazawa, C. et al., Differential expression of five N-methyl-D-aspartate receptor subunit mRNAs in the cerebellum of developing and adult rats, *J. Comp. Neurol.*, 347, 150, 1994.
120. Watanabe, M. et al., Developmental changes in distribution of NMDAR channel subunit mRNAs, *Neuroreport*, 3, 1138, 1992.
121. Ciabarra, A.M. et al., Cloning and characterization of chi-1: a developmentally regulated member of a novel class of the ionotropic glutamate receptor family, *J. Neurosci.*, 15, 6498, 1995.
122. Sucher, N.J. et al., Developmental and regional expression pattern of a novel NMDAR-like subunit (NMDAR-L) in the rodent brain, *J. Neurosci.*, 15, 6509, 1995.
123. Fukaya, M., Hayashi, Y., and Watanabe, M., NR2 to NR3B subunit switchover of NMDARs in early postnatal motoneurons, *Eur. J. Neurosci.*, 21, 1432, 2005.
124. Karavanova, I. et al., Novel regional and developmental NMDAR expression patterns uncovered in NR2C subunit-beta-galactosidase knock-in mice, *Mol. Cell. Neurosci.*, 34, 468, 2007.
125. Binshtok, A.M. et al., NMDARs in layer 4 spiny stellate cells of the mouse barrel cortex contain the NR2C subunit, *J. Neurosci.*, 26, 708, 2006.

126. Schratt, G.M. et al., BDNF regulates the translation of a select group of mRNAs by a mammalian target of rapamycin-phosphatidylinositol 3-kinase-dependent pathway during neuronal development, *J. Neurosci.*, 24, 7366, 2004.

127. Edwards, M.A. et al., Lack of functional expression of NMDARs in PC12 cells, *Neurotoxicology*, 5, 5, 2007.

128. West, A.E., Griffith, E.C., and Greenberg, M.E., Regulation of transcription factors by neuronal activity, *Nat. Rev. Neurosci.*, 3, 921, 2002.

129. Wang, J.Q., Fibuch, E.E., and Mao, L., Regulation of mitogen-activated protein kinases by glutamate receptors, *J. Neurochem.*, 100, 1, 2007.

130. Verkhratsky, A. and Kirchhoff, F., NMDARs in glia, *Neuroscientist*, 13, 28, 2007.

131. Hinoi, E. et al., Glutamate signaling in peripheral tissues, *Eur. J. Biochem.*, 271, 1, 2004.

132. Dickman, K.G. et al., Ionotropic glutamate receptors in lungs and airways: molecular basis for glutamate toxicity, *Am. J. Respir. Cell. Mol. Biol.*, 30, 139, 2004.

133. Hinoi, E. et al., Constitutive expression of heterologous N-methyl-D-aspartate receptor subunits in rat adrenal medulla, *J. Neurosci. Res.*, 68, 36, 2002.

134. Fischer, M. et al., N-methyl-D-aspartate receptors influence the intracellular calcium concentration of keratinocytes, *Exp. Dermatol.*, 13, 512, 2004.

135. Leung, J.C. et al., Expression and developmental regulation of the NMDAR subunits in the kidney and cardiovascular system, *Am. J. Physiol. Regul. Integr. Comp. Physiol.*, 283, R964, 2002.

136. Olney, J.W., Wozniak, D.F., and Farber, N.B., Glumate receptor dysfunction and Alzheimer's disease, *Restor. Neurol. Neurosci.*, 13, 75, 1998.

137. Davies, S. and Ramsden, D.B., Huntington's disease, *Mol. Pathol.*, 54, 409, 2001.

138. Dure, L.S.T., Young, A.B., and Penney, J.B., Excitatory amino acid binding sites in the caudate nucleus and frontal cortex of Huntington's disease, *Ann. Neurol.*, 30, 785, 1991.

139. Zuccato, C. et al., Huntingtin interacts with REST/NRSF to modulate the transcription of NRSE-controlled neuronal genes, *Nat. Genet.*, 35, 76, 2003.

140. Arzberger, T. et al., Changes of NMDAR subunit (NR1, NR2B) and glutamate transporter (GLT1) mRNA expression in Huntington's disease: *in situ* hybridization study, *J. Neuropathol. Exp. Neurol.*, 56, 440, 1997.

141. Luthi-Carter, R. et al., Complex alteration of NMDARs in transgenic Huntington's disease mouse brain: analysis of mRNA and protein expression, plasma membrane association, interacting proteins, and phosphorylation, *Neurobiol. Dis.*, 14, 624, 2003.

142. Zucker, B. et al., Transcriptional dysregulation in striatal projection- and interneurons in a mouse model of Huntington's disease: neuronal selectivity and potential neuroprotective role of HAP1, *Hum. Mol. Genet.*, 14, 179, 2005.

143. Coyle, J.T., Tsai, G., and Goff, D., Converging evidence of NMDAR hypofunction in the pathophysiology of schizophrenia, *Ann. NY Acad. Sci.*, 1003, 318, 2003.

144. Miyamoto, Y. et al., Hyperfunction of dopaminergic and serotonergic neuronal systems in mice lacking the NMDAR epsilon1 subunit, *J. Neurosci.*, 21, 750, 2001.

145. Clinton, S.M. et al., Altered transcript expression of NMDAR-associated postsynaptic proteins in the thalamus of subjects with schizophrenia, *Am. J. Psychiatr.*, 160, 1100, 2003.

146. Ibrahim, H.M. et al., Ionotropic glutamate receptor binding and subunit mRNA expression in thalamic nuclei in schizophrenia, *Am. J. Psychiatr.*, 157, 1811, 2000.

147. Iwayama-Shigeno, Y. et al., Extended analyses support the association of a functional (GT)n polymorphism in the GRIN2A promoter with Japanese schizophrenia, *Neurosci. Lett.*, 378, 102, 2005.

148. Iwayama, Y. et al., Analysis of correlation between serum D-serine levels and functional promoter polymorphisms of GRIN2A and GRIN2B genes, *Neurosci. Lett.*, 394, 101, 2006.

149. Tang, J. et al., Significant linkage and association between a functional (GT)n poly-morphism in promoter of the N-methyl-D-aspartate receptor subunit gene (GRIN2A) and schizophrenia, *Neurosci. Lett.*, 409, 80, 2006.

150. Begni, S. et al., Association between the G1001C polymorphism in the GRIN1 gene promoter region and schizophrenia, *Biol. Psychiatr.*, 53, 617, 2003.

151. Mishizen-Eberz, A.J. et al., Biochemical and molecular studies of NMDAR subunits NR1/2A/2B in hippocampal subregions throughout progression of Alzheimer's dis-ease pathology, *Neurobiol. Dis.*, 15, 80, 2004.

152. Meoni, P. et al., NMDA NR1 subunit mRNA and glutamate NMDA-sensitive bind-ing are differentially affected in the striatum and pre-frontal cortex of Parkinson's disease patients, *Neuropharmacology*, 38, 625, 1999.

153. Nagy, J., The NR2B subtype of NMDAR: a potential target for the treatment of alco-hol dependence, *Curr. Drug Targets CNS Neurol. Disord.*, 3, 169, 2004.

154. Marutha Ravindran, C.R. and Ticku, M.K., Changes in methylation pattern of NMDAR NR2B gene in cortical neurons after chronic ethanol treatment in mice, *Brain Res. Mol. Brain Res.*, 121, 19, 2004.

155. Marutha Ravindran, C.R. and Ticku, M.K., Role of CpG islands in the up-regulation of NMDAR NR2B gene expression following chronic ethanol treatment of cultured cortical neurons of mice, *Neurochem. Int.*, 46, 313, 2005.

156. Gascon, S. et al., Transcription of the NR1 subunit of the N-methyl-D-aspartate receptor is down-regulated by excitotoxic stimulation and cerebral ischemia, *J. Biol. Chem.*, 280, 35018, 2005.

157. Mao, X., Moerman, A.M., and Barger, S.W., Neuronal kappa B-binding factors con-sist of Sp1-related proteins. Functional implications for autoregulation of N-methyl-D-aspartate receptor-1 expression, *J. Biol. Chem.*, 277, 44911, 2002.

158. Kumar, U., Characterization of striatal cultures with the effect of QUIN and NMDA, *Neurosci. Res.*, 49, 29, 2004.

159. Levenson, J.M. and Sweatt, J.D., Epigenetic mechanisms in memory formation, *Nat. Rev. Neurosci.*, 6, 108, 2005.

160. Goll, M.G. and Bestor, T.H., Eukaryotic cytosine methyltransferases, *Annu. Rev. Biochem.*, 74, 481, 2005.

161. Gardiner-Garden, M. and Frommer, M., CpG islands in vertebrate genomes, *J. Mol. Biol.*, 196, 261, 1987.

162. Kim, M.S. et al., N-methyl-D-aspartate receptor type 2B is epigenetically inactivated and exhibits tumor-suppressive activity in human esophageal cancer, *Cancer Res.*, 66, 3409, 2006.

163. Lewis, B.P., Burge, C.B., and Bartel, D.P., Conserved seed pairing, often flanked by adenosines, indicates that thousands of human genes are microRNA targets, *Cell*, 120, 15, 2005.

164. Gray, P.A. et al., Mouse brain organization revealed through direct genome-scale TF expression analysis, *Science*, 306, 2255, 2004.

165. Kapranov, P. et al., RNA maps reveal new RNA classes and a possible function for pervasive transcription, *Science*, 316, 1484, 2007.

166. Cheng, J. et al., Transcriptional maps of 10 human chromosomes at 5-nucleotide resolution, *Science*, 308, 1149, 2005.

6 NMDA Receptors and Translational Control

Charles A. Hoeffer and Eric Klann

CONTENTS

6.1 INTRODUCTION

Translation—the synthesis of new polypeptides encoded by messenger RNAs—is an essential process involved in nearly every aspect of cellular survival and growth. In recent years, the importance of translation has been demonstrated to be critical for the manifestation of persistent forms of synaptic plasticity such as long term facilitation (LTF) in *Aplysia* and long term potentiation (LTP) in vertebrate hippocampal neurons.[1–4] Importantly, translation is critical to the formation of long term memory (LTM) in panoplies of phyla.[5–8] N-methyl-D-aspartate (NMDA) receptors (NMDARs) have been critically linked to the regulation of processes both upstream and downstream of neuronal translation. To better understand the role of NMDARs in the regulation of translational machinery, it is important to first overview the many stages and levels of regulation involved in the translation of mRNA into new protein.

6.2 OVERVIEW OF TRANSLATION

An overview of the basic elements that mechanistically define cellular translation is important for understanding the roles of NDMA Receptors in the regulation of neuronal translation. In simplistic terms, eukaryotic translation requires messenger RNA (mRNA), the small (40S) and large (60S) ribosomal subunits, and "charged" (bearing amino-acyl amino acid) transfer RNA (tRNA).[9,10] The regulated assembly of these constituent elements results in peptide synthesis. Translation occurs predominantly in the soma of the cell, but the components of the translational apparatus are also found in more distal compartments of neurons such as axons and dendrites.[11–13]

A considerable body of evidence suggests that extra-somatic translation is critical to long-lasting forms of synaptic plasticity.[11,14–16] Moreover, experimental preparations that physically separate the soma from dendritic regions are fully able to express protein synthesis-dependent forms of synaptic plasticity. Within dendrites, translational machinery is frequently organized into bundles of immature ribosomes, RNA-binding proteins, and mRNAs called polysomes. Following neuronal activity, these polysomes are trafficked into dendritic spines and undergo maturation into active sites of translation. NMDARs are localized primarily to synaptic sites where they are frequently colocalized with translational machinery.[17]

NMDARs are also colocalized with signaling complexes that include upstream signaling elements involved in regulating the translational apparatus.[18] NMDAR activation may play a role in the translocation of translational components to areas of synaptic activity, thereby providing a functional link between protein synthesis and neuronal activity. Whether NMDAR activation is linked to specific classes of polysome maturation or linked to translation in a more general way is not currently understood. The role of extra-synaptic NMDARs in regulation of protein synthesis remains poorly understood.

6.3 THREE STAGES OF TRANSLATION

Eukaryotic translation can be divided into three phases: initiation, elongation, and release (also called termination). Each phase involves the recruitment of specific eukaryotic translation factors to the ribosomal/mRNA complex: initiation factors (eIF)s, elongation factors (eEF)s, and release factors (eRF)s. Initiation involves the assembly of the ribosomal subunits, the "start" methionyl tRNA (Met-tRNA), and the mRNA to be translated. Elongation begins with the release of the initiation complex and maturation of the ribosome/mRNA complex to allow assembly and elongation of polypeptide as directed by tRNA–codon pairing. Release is the process by which the recently translated mRNA and the newly complete peptide are released. Table 6.1 lists the translation factors and their roles in translation.

6.3.1 INITIATION

An early step in initiating translation is generation of the 43S preinitiation complex, a complex composed of a 40S ribosomal subunit and several eIFs including eIF1A, eIF2, and eIF3 (see Table 6.1 for factor descriptions). The primary role of the 43S complex is to bring the Met-tRNA to the early assembly of the functioning translational machinery.

TABLE 6.1

Translation Factors & Associated Proteins	Known Function	NMDA-R Regulation	Note
eIF1	Promotes binding of eIF2-GTP-Met-tRNA to mRNA and 40S ribosomal subunit		
eIF1A	Promotes binding of eIF2-GTP-Met-tRNA to mRNA and 40S ribosomal subunit		
eIF2	Composed of eIF2α and 2B, binds GTP and Met-tRNA to form ternary complex that binds 40S subunit		Major determinant of general protein synthesis rates
eIF2a	Regulatory subunit of eIF2, phosphorylated at Ser 51		Regulated by GCN2, PERK, and other kinases
eIF2B	Encodes GEF activity of eIF2B		Disupted in Vanishing White Matter Disease
eIF3	Promotes eIF2-GTP-Met-tRNA binding to 40S subunit		
eIF4A	DEAD box protein, encodes RNA helicase activity		
eIF4B	Enhances eIF4A activity		
eIF4E	7-methyl-GTP-binding protein or "Cap"- binding protein	via ERK and MNK kinases	Phosphorylation correlated with enhanced translation rates
eIF4F	Cap-binding complex, composed of eIF4A, 4E, 4G, promotes binding of mRNA binding to 43S pre-initiation complex		
eIF4G	Multidomain scaffolding protein, interacts with eIF4E, Maskin, CPEB, PABP, and regulatory kinases		
eIF5	GAP protein that acts on eIF2		
eEF1	Composed of eEF1A, 1B, loads amino-acyl-tRNA onto A-site of ribosomal subunit		
eEF1A	Binds amino-acyl-tRNA, GTPase that allows tRNA release		
eEF1B	Encodes GEF activity for eEF1		

(continued)

TABLE 6.1 (*Continued*)

Translation Factors & Associated Proteins	Known Function	NMDA-R Regulation	Note
eEF2	Translocates ribosome along mRNA undergoing translation	via eEF2 kinase (CaMKIII)	
eRF1	Release factor, hydrolyzes peptide release from 80S ribosome, encodes GAP activity		
eRF3	Release factor, GTPase activity		
4E-BP	Inhibitor of eIF4E, repressed by phosphorylation of Thr-37,46,70 & Ser 65, target of multiple kinases		Large family of translational repressors
PABP	Poly(A)-binding Protein		Binds eIF4G
CPEB	CPE element-binding protein, mediates poly(A) extension of mRNA	via Aurora and CaMKIIα kinases	
Maskin	Binds CPEB, mediates transcript suppression through interaction with eIF4E		Drosophila homolog, *cup*
40S subunit	Small ribosomal unit, composed of rRNA and ribosomal proteins		
43S pre-initiation complex	40S subunit, eIF2-GTP-Met-tRNA, eIF1, eIF1A (possibly other subunits e.g. eIF5)		
48S pre-initiation complex	43S complex plus substrate mRNA		
60S subunit	Large ribosomal subunit, composed of rRNA and ribosomal proteins		
80S ribosome	Completely assembled ribosome plus substrate mRNA, capable of peptide elongation		

The functional states of several eIFs in this preinitiation complex are regulated by upstream signaling that affects their availability to form ternary complexes as well as their functional activities. A major target of regulation in the formation of this complex is eIF2 that exists in GDP (inactive) and GTP (active) states. The ternary structure composed of eIF2-GTP and Met-tRNA associates with the 40S subunit to initiate start codon recognition. Upon recognition, GTP is hydrolyzed, eIF2-GDP is released, and translation can commence.

Inactive eIF2-GDP is restored to its active form via the activity of the guanine nucleotide exchange factor (GEF) eIF2B, a process regulated by the phosphorylation of Ser 51 on the α subunit of eIF2 (pSer51-eIF2 α). pSer-eIF2 α acts as an inhibitor of eIF2B function by preventing the conversion of eIF2-GDP to eIF2-GTP. The phosphorylation of eIF2α is under the regulation of at least four kinases: protein kinase RNA regulated (PKR), heme regulated initiation (HRI) factor 2α, eIF2α kinase 3 (PERK), and general control nonderepressible kinase 2 (GCN). High levels of pSer51-eIF2α inhibit general translation, but lead to increased translation of specific upstream open reading frame (uORF)-bearing mRNAs such as ATF-4 and C/EBP.

Most translated neuronal mRNAs are "capped" (methylated GTP is attached to the 5′ end of the mRNA molecule). The cap is recognized by eIF4E that allows the mRNA to be bound by eIF4G, a large multidomain scaffolding protein that couples the binding of other translation factors (eIF4E, eIF4A) and RNA–binding proteins such as poly(A)-binding protein (PABP). The combined activities of this complex, often referred to as eIF4F, prepare the targeted mRNA for active translation. The formation of the eIF4F complex is controlled by a number of upstream signaling pathways (Figure 6.1A).

The eIF4F cap-binding complex is under the regulation of a family of small modulatory proteins called 4E-binding proteins (4E-BPs) that inhibit protein synthesis by competing with eIF4G for eIF4E. The interaction of 4E-BPs and eIF4E is regulated by phosphorylation of inhibitory residues on 4E-BP (Thr 37, 46, 70 and Ser 65).[19,20] Phosphorylation of these residues disrupts the interaction between 4E-BP and eIF4E, allowing protein synthesis to proceed via the formation of eIF4F. Phosphorylation of 4E-BP is governed by several kinase signaling pathways, including the mammalian target of rapamycin (mTOR), extracellular signal regulated kinase (ERK), and phosphoinositide-3 kinase (PI3K)[19] (Figure 6.2).

mRNA bound to the eIF4F ternary structure forms an aggregate with the 43S preinitiation complex to form the 48S pre-initiation complex. Upon binding the mRNA, the ribosomal complex "scans" the mRNA for the AUG start codon. The 5′ untranslated regions (UTRs) often contain complex secondary structures that can inhibit this process. In these cases, scanning is facilitated via the RNA helicase activity of eIF4A. Upon recognition of the start AUG, eIF2B within the 48S structure hydrolyzes its bound GTP, releasing it and several other factors from the 48S preinitiation complex. To complete the initiation process, GTP-bound eIF5B brings the 60S ribosomal subunit to the complex. After the GTP hydrolysis event, eIF5B is released and the final assembly forms the translationally active 80S ribosome. At this stage, the initiation factors associated with the mRNA cap and the 40S ribosomal subunits are released and the process of peptide elongation comes under the regulation of elongation factors.

6.3.2 ELONGATION

Translational elongation is a widely conserved process that, similar to initiation, is regulated in neurons. Elongation begins with a peptidyl tRNA in the ribosomal P site and an adjacent vacant A site. An amino-acyl tRNA is brought to the empty A site as part of a ternary complex comprised of the amino-acyl tRNA and the GTP-bound elongation factor, eEF1A. GTP is exchanged for GDP onto eEF1A via the GEF activity of a trimeric protein complex termed eEF1B. The four peptides

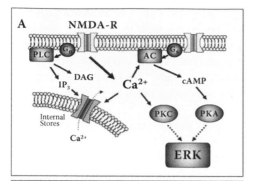

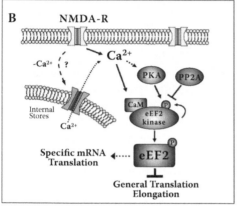

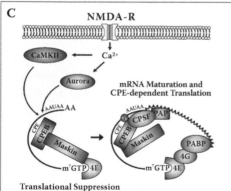

FIGURE 6.1 (See color insert following page 212) (A) NMDAR activation is linked to numerous intracellular signaling pathways. N-methyl-D-aspartate (NMDA) receptor (NMDAR) activation is the major source of activity-dependent calcium (Ca^{2+}) entry into the neuron. In addition, NMDAR activation may promote the generation of other second messengers [cAMP; diacylglyercol (DAG); and inositol-3,4,5-trisphosphate (IP_3)] through NMDAR association with membrane bound G-protein (Gs, Go) signaling to adenylate cyclase (AC) and calcium activated phospholipase (PLC). These second messengers may promote the activation of signal kinases such as cAMP-dependent kinase (PKA) and protein kinase C (PKC). Increased intracellular Ca^{2+} and IP_3 can trigger additional calcium influx via stimulation of calcium-release channels regulating calcium release from internal stores. (B) Signaling pathways activated by NMDARs involved in translational elongation. Activation of NMDARs results in extracellular calcium (Ca^{2+}) entry that eventually activates cAMP-dependent protein kinase (PKA). PKA-dependent phosphorylation and calcium-bound calmodulin promote the activity of eEF2 kinase (also known as CaMKIII). eEF2 kinase exhibits autophosphorylation activity that allows it to remain active after upstream signaling ceases. eEF2 kinase phosphorylates eEF2 (p-eEF2), which suppresses translational elongation. This has the effect of repressing general protein synthesis but can also produced enhanced translation of some mRNAs (i.e., 5′ TOP mRNA) that under normal conditions are translated with low efficiency. Protein phosphatase 2A (PP2A) can dephosphorylate eEF2 kinase to reduce its activity and de-repress translation elongation. NMDAR activation may also lead to eEF2 kinase activation via a mechanism independent of extracellular Ca^{2+} entry. (C) NMDAR regulation of mRNA maturation through CPEB. Activation of NMDARs leads to extracellular Ca^{2+} influx that activates Ca^{2+}/calmodulin-dependent protein kinase II (CaMKII) and Aurora kinase. Specific mRNAs contain 3′ untranslated region sequences (UUUUUAU) called CPEs. Immature mRNAs with CPEs are bound by cytoplasmic polyadenylation element-binding protein (CPEB), which binds maskin that also associates with eIF4E (4E) bound to the m⁷GTP cap of the mRNA transcript. Translation of the transcript is inhibited by short poly(A) tails and the sequestration of 4E. Transcript de-repression is achieved via the phosphorylation of CPEB (p-CPEB) by CaMKII and Aurora. This promotes increased interaction between CPEB and polyadenylation specificity factor (CPSF). CPEB–CPSF association results in the recruitment of poly(A) polymerase (PAP) to lengthen the poly(A) tail of the immature mRNA transcript. Poly(A)-binding protein (PABP) then binds to the extended poly(A) tail and in turn interacts with eIF4G (4G). 4G displaces maskin binding with 4E, which permits the bound transcript to be translated. m⁷GTP = 7-methyl GTP. AAUAA = polyadenylation signal.

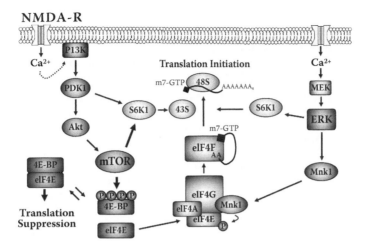

FIGURE 6.2 (See color insert following page 212) Signaling pathways activated by NMDARs and involved in translational initiation. NMDAR activation in turn activates the phosphatidylinositol 3-kinase (PI3K) and the mitogen-activated protein kinase (MEK)/extracellular signal-regulated kinase (ERK) signaling pathways. NMDAR activation produces sequential activation of PI3K, phosphoinositide-dependent kinase 1 or 2 (PDK1), protein kinase-B (Akt), and mammalian target of rapamycin (mTOR). mTOR activation leads to activation of S6 kinase 1 (S6K1) and phosphorylation (P) of 4E-binding proteins (4E-BPs). This phosphorylation causes disassociation of 4E-BPs from initiation factor 4E (eIF4E). Released eIF4E interacts with initiation factor 4G (eIF4G) and forms the active eIF4F (eIF4E-eIF4A-eIF4G) complex. eIF4F promotes mRNA binding to the 43S prcinitiation complex to form the 48S preinitiation complex. ERK-dependent phosphorylation of both MAPK-interacting serine/threonine kinase 1 (Mnk1) that can phosphorylate eIF4E and S6K1 that can phosphorylate ribosomal protein S6 is correlated with enhanced translation initiation. m7G = 7-methyl-GTP. $AAAAAAA_n$ = poly(A) tail.

that compose the eEF1A/B complex are targets of phosphoregulation. The primary kinases involved in regulating the eEF1A/B complex are protein kinase C (PKC) and casein kinase (CK). Cognate binding of the codon and anticodon activates eEF1A GTPase activity, thereby releasing it from the amino-acyl tRNA that then permits peptide elongation.

The 80S ribosome must translocate along the mRNA to continue elongation that is accomplished via the activity of eEF2 that hydrolyzes GTP to facilitate the three-base pair movement of the ribosome along the mRNA. Similar to eEF1A/B, the activity of eEF2 is regulated by phosphorylation of a C terminal threonine-56 (Thr 56) that reduces eEF2 activity, thus slowing elongation and protein synthesis. Phosphorylation of this site is regulated by protein phosphatase 2A (PP2A), believed to be the major link between the mTOR signaling pathway and elongation. Interestingly, only a single kinase known as eEF2 kinase has been identified as regulator of eEF2 activity (Figure 6.1B).

Calcium is a major regulator of eEF2 kinase activity. In the presence of calmodulin and high Ca^{2+} concentrations, eEF2 kinase exhibits autophosphorylation that not only increases its nascent activity, but liberates it from transient Ca^{2+} regulation. eEF2 kinase also is regulated by cAMP-dependent protein kinase (PKA), providing additional regulatory links from eEF2 kinase to channels at the membrane and neurotransmitter receptors coupled to G-proteins. The eEF2 GTPase-driven translocation of mRNA during the elongation is repeated until a stop codon is reached, at which point elongation is terminated and the ribosomal subunits are released from the mRNA.

6.3.3 TERMINATION

Translational termination occurs when the elongating ribosome encounters a consensus stop codon (i.e., UAG) in the ribosomal A site. The resulting interaction between the presence of the stop codon and eRFs results in the release of the newly translated peptide following a hydrolysis reaction from the translating ribosomal complex. The hydrolysis reaction that permits peptide release is catalyzed by eRF1. A second release factor, eRF3, then removes eRF1 from the ribosomal A site, permitting the ribosome to return to the pool of available subunits. eRF3 possesses GTPase activity, but the specific role and timing of GTP hydrolysis involved in the process of termination is poorly understood.[9] In contrast to initiation or elongation, very little is known about regulation of translation termination.

6.4 RNA BINDING PROTEINS AND mRNA LOCALIZATION

Translation also is governed by the availability and stability of mRNA templates. In many ways, these factors are as important as the process of translation. mRNA availability for translation is regulated by number of RNA binding proteins (RBPs)[21–23] that accomplish this through a variety of mechanisms. Some proteins such as fragile X mental retardation (FMRP) protein and Staufen (Stau 1 and 2) proteins bind multiple types of mRNAs because they contain several different RNA binding domains that permit binding to a multitude of specific RNA sequences such as G quartets and polyuridine tracts or via interactions with microRNAs or double-stranded RNAs. Other RBPs are more sequence-specific. RBPs such as cytoplasmic polyadenylation element (CPE) element binding protein and Hus bind only specific elements contained in the 3′ UTRs of specific mRNA transcripts. Because these sequences are found only in subsets of mRNA populations, their activities are restricted to small segments of the RNA population.

How may RBPs affect translational expression? It is generally believed that they "silence" mRNAs by occluding their interactions with translational machinery such as eIF4E. RBPs can also positively regulate translation by trafficking mRNAs to sites of active translation. They may play a critical role in the manifestation of an important protein synthesis-dependent phenomenon called synaptic tagging—a process by which specific synapses in a myriad population of synapses are tagged or marked by synaptic activity for the specific recruitment of cellular products that promote long-lasting synaptic change (such as newly synthesized peptides). Reasonably strong evidence indicates that the regulation of RBP activity may be a component of the mechanism by which tagging takes place.[24–27]

6.5 ADMINISTERING TRANSLATION: NMDA RECEPTOR-DEPENDENT SIGNALING TO TRANSLATIONAL MACHINERY

Translating an mRNA into a peptide is a very complex process involving numerous factors that are regulated at nearly every step. The evidence now available places most translation control at the level of initiation via the integration of upstream signaling cascades. NMDARs are critical neuronal signaling molecules involved in regulating neuronal translation. They exert control over translation via two general processes: calcium entry from the extracellular space and activation of signaling cascades linked to the NMDAR protein complexes.[18,28]

Calcium entry via the NMDAR activation represents perhaps the most potent element of NMDAR-dependent control of signaling cascades. A number of signaling cascades (Figure 6.1A) are activated in response to calcium entry at the NMDAR.[18,28] This entry via the NMDAR may result in the production of other second messenger molecules including cyclic adenosine monophosphate (cAMP), inositol triphosphate (IP3), and diacylglycerol (DAG).[29–31] The ability of NMDARs to promote second messenger signaling through such a wide variety of cellular mechanisms makes the receptor a potent regulator of neuronal metabolic activity.

6.5.1 NMDA RECEPTORS AND TRANSLATIONAL INITIATION

Prominently coupled to NMDAR activation is the ERK (extracellular signal-regulated kinase) signal transduction pathway. NMDAR activation promotes the activation of several signaling kinases such as Ras and Rap upstream of ERK activation.[32,33] ERK regulates several translational regulatory proteins upstream of translational initiation including the MAP kinase signal interacting kinases (Mnk1, Mnk2).[34] Mnk1 and 2 phosphorylate eIF4E in an ERK-dependent manner. eIF4E *in vivo*[35,36] is localized to postsynaptic densities and dendritic lipid rafts.[37,38] It is generally thought that eIF4E phosphorylation enhances general translation,[39] although the physiological function of this phosphorylation event has been debated.[40–42]

Nevertheless, a direct connection of NMDAR activation, Mnk activation, and enhanced eIFE4E phosphorylation has been demonstrated in the mouse hippocampus.[43] Also, it has been demonstrated that NMDAR-dependent L-LTP is associated with an ERK-dependent increase eIEF4E phosphorylation.[44] These findings demonstrate a direct role for NMDA activation in promoting activation of translation factors via the ERK signaling cascade.

NMDAR activation can influence translation through other signaling cascades and has been shown to trigger the translation of ribosomal S6 kinase 2 (Rsk2) via an ERK-independent signaling pathway. Critical to the stimulation of Rsk2 translation is the activation of the mTOR signaling cascade.[45] Activation of mTOR is vital for translation initiation (Figure 6.2) and NMDAR activation has been implicated in the activation of both upstream and downstream components of the mTOR signaling cascade. Numerous studies also have demonstrated that NMDAR activation results in activation of PI3-kinase, PDK1, and Akt/PKB, all of which have been shown to regulate the activity of mTOR.[46–53]

More direct evidence linking NMDAR activation to mTOR-dependent regulation of translation arose from studies in hippocampal neurons demonstrating that stimulation of NMDARs results in dendritic protein synthesis that is sensitive to rapamycin, a potent inhibitor of mTOR signaling.[16,54] NMDAR-dependent L-LTP is rapamycin-sensitive[16] and results in increased phosphorylation of p70 S6 kinase (S6K) at threonine 389, the mTOR target residue.[55] It is generally accepted that mTOR activation promotes translation,[56] but mTOR activation may function to suppress translation of certain mRNAs.[57] In a surprising study, NMDAR-dependent signaling through mTOR led to suppressed dendritic translation of the potassium channel, Kv1.1.[58] These studies provide strong evidence linking NMDAR activity to regulation of mTOR signaling and protein synthesis, although this regulation may not always follow the accepted paradigm of mTOR activation resulting in increased rates of translation.

6.5.2 NMDA Receptors and Translational Elongation

NMDARs have been shown also to be involved in regulating another aspect of translational regulation: peptide elongation. As noted, NMDARs are major sources of activity-dependent calcium entry into the postsynaptic compartment. Calcium entry via the NMDAR channel regulates eEF2 kinase activity via calcium-bound calmodulin that binds near the catalytic domain of the kinase. Interestingly, this is not the only route through which NMDAR activation promotes calcium regulation of eEF2 kinase. Exposure of cortical neurons to NMDA in absence of extracellular calcium results in increased eEF2 phosphorylation but not eIF2α phosphorylation.

The eEF2 activation arising from calcium released from intracellular stores also resulted in inhibition of protein synthesis.[59] Several studies have shown that NMDAR activation leads to the phosphorylation of eEF2 and the subsequent inhibition of protein synthesis.[60–65] In light of the evidence coupling NMDARs to activation of translational initiation pathways, concomitant repression of translation elongation presents a paradoxical scenario in which NMDA simultaneously stimulates and represses protein synthesis (Figure 6.1B).

A closer examination of the studies linking NMDAR activation to the repression of protein synthesis reveals that NMDAR regulation of translational elongation may in fact be representative of a more finely regulated mechanism that contributes to the proper timing and spatial localization of neuronal translational events. For example, NMDAR-dependent regulation of eEF2 is spatially restricted to sites of synaptic activity; in cultured hippocampal neurons, blocking the NMDAR components of miniature synaptic events reduced eEF2 phosphorylation in a synapse-specific fashion and site-directed blockade of eEF2 kinase resulted in increased protein synthesis.[66] In addition, stimulation of NMDARs did not appear to lead to the persistent repression of elongation because prolonged application of hippocampal neurons to NMDA (30 min) resulted in only transient eEF2 phosphorylation that peaked 10 min before returning to basal levels within 60 min. Briefer periods of stimulation resulted in even more transient eEF2 phosphorylation (peaking <5 min).[63]

In a study of rat synaptoneurosomes (biochemical preparations containing enriched pre-and postsynaptic compartments), stimulation with NMDA led to repressed protein synthesis. The repression of translation initially was pronounced (within 5 min), then followed by an increase in protein synthesis that persisted 60 min.[67] This initial repression of translation in response to NMDAR stimulation may represent a mechanism to delay protein synthesis until specific mRNA transcripts arrive at sites of synaptic activity.

NMDAR-dependent regulation of eEF2 activity may also regulate translation via a mechanism of specificity, that is, NMDAR-dependent signaling may increase translation of a specific mRNA even though general protein synthesis is inhibited. For example, NMDAR stimulation that results in the suppression of general protein synthesis at the same time *increases* the synthesis of CaMKIIα.[67] In addition, slowing the rate of translation elongation may permit less efficiently translated messages to be utilized to a greater extent.

Many components of the translational machinery belong to a class of mRNAs containing 5′ terminal oligopyrimidines (TOPs). These mRNAs contain minimal 5′ structures and are not normally translated with great efficiency.[68,69] Upon cellular stimulation, these mRNAs are translated at increased levels.[70,71] Following the induction of NMDAR-dependent L-LTP in rat hippocampal slices, persistent increases in dendritic eEF1A levels have been observed.[72] This finding suggests the possibility that NMDAR activity may alter translation along a temporal gradient.

The increased eEF1A associated with L-LTP was rapamycin-sensitive during an early temporal window shortly after stimulation, but was resistant to rapamycin applied at later points.[72] Thus, NMDAR activation may initially drive translation initiation via mTOR activation, but also affect translation in a way that permits critical pools of mRNA to be specifically translated at later times.

Finally, although not demonstrated in neurons, eEF2 kinase can be regulated via a rapamycin-sensitive pathway at Ser 78 (adjacent to the Ca^{2+}–calmodulin binding site). This type of signaling requires the PI3 kinase signaling cascade, raising the possibility that activation of NMDARs may trigger cross-talk between eEF2 kinase and mTOR.[73] All this evidence points to a role for NMDAR-dependent regulation of translation elongation in order to alter the translation of specific mRNAs. Moreover, the translation of 5′ TOP-containing mRNAs may be regulated by this mechanism that may allow NMDARs to be involved in increasing overall neuronal translational capacity. Further evidence linking NMDAR signaling to increasing the general translational capacity of a neuron comes from recent studies involving the AIDA-1d postsynaptic protein. Following NMDAR stimulation, AIDA-1d was trafficked in a retrograde fashion into the nucleus where it mediated increased nucleolar numbers, the cellular sites for ribosomal maturation.[74] Whether the ribosomes are then trafficked back to synapses remains to be determined.

6.5.3 NMDA RECEPTORS AND TRANSLATION TERMINATION

Translation termination is the final critical step in protein synthesis. The nascent peptide is released and the ribosomal synthetic machinery is recycled. Surprisingly, little is known about the general cellular mechanisms governing this process in eukaryotes. No direct evidence points to NMDAR-dependent regulation of translation termination. However, based on the broad influence of NMDAR signaling on translation initiation and elongation, it may simply a matter of time before such regulation is revealed.

6.6 NMDA RECEPTORS AND REGULATION OF TRANSLATIONAL SUBSTRATES

The precise regulation of the availability, localization, and stability of mRNA, the translational substrate, is as at least as important as the core machinery in serving as a component of the cellular protein synthesis apparatus. Neurons are highly organized and polarized cells with specialized compartments. The proper trafficking of mRNAs to distal dendritic sites is critical to normal neuronal function. Messenger RNA is exported to neuronal processes in large RNA "granules" composed of mRNAs, RBPs, ribosomes (polyribosomes), and translation factors. Neuronal activity modulates the trafficking and subsequent synthesis of specific mRNAs in dendritic regions where synaptic activity occurs.[75–80] Based on the extensive roles of NMDARs in triggering the activation of signaling pathways that regulate protein synthesis, the question is whether NMDARs also play a role in the regulation of the mRNAs availability for translation.

6.6.1 NMDA Receptor Regulation of CPEB

The association between mRNAs and RBPs is largely dictated by specific sequences within the mRNA that convey binding specificity to the RBPs. One type of RBP is the CPE–binding protein (CPEB). CPEB binds a consensus 3′ UTR sequence (UUUUUAU) where it regulates both translational repression and activation through direct binding and the polyadenylation of the mRNAs.[81] CPEB is phosphorylated by at least two kinases, Aurora kinase and CaMKIIα, at Thr 171.[82,83] Phosphorylated CPEB interacts with cleavage and polyadenylation specificity factor (CPSF) whose activity extends the poly(A) tail of the bound mRNA[84] (Figure 6.1C).

In oocytes, translation of dormant mRNAs is activated through CPEB binding and subsequent polyadenylation of the dormant mRNAs. Initially, CPEB silences the immature mRNA through recruitment of another protein known as maskin that binds eIF4E which, when bound to the mRNA cap, prevents translation initiation. Phosphorylation of CPEB induces polyadenylation of the mRNA and binding of the poly(A)-binding protein (PABP) that brings eIF4G to the mRNA–protein complex. This association displaces the maskin interaction with eIF4E, allowing translation to begin.[85] A similar mechanism in *Drosophila* involving a maskin homolog known as *Cup* has been described.[86] Neurons may utilize similar mechanisms to regulate mRNA substrate availability for translation. One caveat in this model of NMDAR CPEB-dependent regulation of synaptic translation is that a functional mammalian homolog for maskin has yet to be conclusively demonstrated. Several homologs for mammalian maskin have been proposed, including a protein called neuroguidin.[87]

In addition to its presence in the soma of neurons, CPEB is present in dendrites.[88] It is particularly enriched in biochemical preparations that isolate the postsynaptic density (PSD), identifying the synapse as the major site for CPEB localization. Importantly, following stimulation with visual activity, synaptic CaMKIIα mRNA was polyadenylated and its protein expression was upregulated.[88]

A direct link between CPEB and NMDARs was first demonstrated in cultured cortical neurons in which NMDAR activation stimulated Aurora kinase to phosphorylate CPEB and subsequently enhance CaMKIIα mRNA polyadenylation.[82,89] In addition to stimulating CaMKIIα translation, NMDAR activation may also promote CPEB activity via stimulation of CaMKII phosphotransferase activity. CaMKIIα has been shown to phosphorylate CPEB at Thr 171 to promote polyadenylation.[83,90] This finding highlights convergent signaling to CPEB from the NMDAR. Thus, it is possible that NMDAR activation can result in differential CPEB activation based on the distinct kinetic profiles of the kinase activity of Aurora and CaMKII. Of particular interest in this regard is the ability of CaMKII to auto-phosphorylate and remain active independent of calcium long after calcium signals that initially promote its activity are gone.[91,92]

Although it is clear that NMDAR activation regulates CPEB to promote translation, whether it mimics the signaling pathway that exists in amphibians and insects remains to be determined.[86,88] Nonetheless, NMDAR-dependent regulation of CPEB activity provides a mechanism for the rapid transition of CPE-containing transcripts from a state translational repression to a state of translational activation. Finally, because CPEB also interacts with microtubule motor proteins to regulate the transport of mRNA in vertebrate neurons, NMDAR activity may regulate both the competency of translational substrates and also their transport to sites of neuronal activity.[93] These findings clearly demonstrate a role for NMDARs in regulating the availability of specific mRNA transcripts for neuronal translation.

6.6.2 FRAGILE X MENTAL RETARDATION PROTEIN (FMRP) AND TRANSLATIONAL SUPPRESSION

FMRP is another RBP critically involved in synaptic plasticity, cognitive function, and mental retardation.[23,94] Like CPEB, FMRP is involved in translational repression. It can repress translation via a number of mechanisms. First, it possesses several RNA–binding domains and can sequester mRNAs, thereby removing their availability to act as translational substrates. Second, high levels of FMRP have been shown to inhibit translation by disrupting formation of the 80S ribosomal complex.[95] Finally, FMRP can inhibit synaptic translation via its involvement in the micro RNA (miRNA) pathway.[96]

FMRP is associated with many components of the miRNA system including miRNA synthetic biosynthetic machinery (Dicer, Argonaut) and several species of miRNA.[97,98] Translational repression by miRNA is achieved through two mechanisms. It binds imperfectly to short 3′ UTR segments on mRNA and inhibits translation through an unknown mechanism. miRNA binding to mRNA also can destabilize mRNA, targeting it for degradation.[99] These features distinguish FMRP as a powerful translational suppressor.

The best evidence linking regulation of neuronal translation to FMRP function comes from studies of mGluR-dependent LTD, a protein synthesis-dependent form of LTD. Studies by Huber et al. and Hou et al. demonstrated that mGluR-dependent LTD is enhanced in mice deficient for FMRP and that FMRP is rapidly translated during mGluR-dependent LTD.[15,100] The best data linking NMDARs to FMRP function comes from studies of the cortex. FMRP protein levels increase in rat barrel cortex following whisker stimulation due to elevated rates of FMRP translation of existing FMRP transcript rather than *de novo* gene expression. Importantly, this increase was prevented by NMDAR activity blockade.[101] Similarly, FMRP protein levels increase in response to visual stimulation. This increase is also dependent on NMDAR activity.[102] These data are consistent with the idea that NMDAR activity acts to suppress translation via increased FMRP biosynthesis. However, this idea is at odds with other information we have about NMDAR, FMRP, and protein synthesis-dependent forms of synaptic plasticity.

NMDAR activity is linked to activation of transcriptional initiation and enhanced mRNA availability for translation. The finding that FMRP is translationally induced by neuronal activity suggests that it is involved in at least some forms of translational activation. This concept is supported by studies demonstrating that FMRP is required for translation of PSD-95 following mGluR activation.[103] One possibility is that NMDARs act through FMRP to induce a period of translational slowing to allow the preferential translation of certain classes of mRNA over others, such as those that contain internal ribosomal entry sites.[104]

6.7 CONCLUDING THOUGHTS

NMDARs are involved in coordinated signaling immediately downstream of synaptic activity during neuronal development, synaptic plasticity, and long-term memory. Based on the broad requirement for protein synthesis in these processes, it is not surprising to find evidence that NMDARs are intimately involved in many facets of translational control in neurons.

Despite the tremendous progress in the elucidation of the signaling pathways that couple NMDARs to protein synthesis machinery, many fascinating questions remain. Two of the most intriguing and difficult questions in this regard are determining the identities of the proteins synthesized in response to NMDAR activation and understanding how these new proteins are incorporated into the proteome to promote long-term functional changes in neurons. Another critical question is whether alterations in NMDAR-dependent protein synthesis are present in disorders triggered by mutations or deletions in translational control molecules such as fragile X mental retardation. We anticipate that answers to these questions will provide a more clear understanding of the relationships of NMDARs and translation.

ACKNOWLEDGMENTS

We thank Marie Monfils, Dana Leventhal, Kiriana Cowansage, Monique Beadouin, Johanna Withers, and J. Lebowski for useful comments and discussions.

REFERENCES

1. Montarolo, P.G. et al., A critical period for macromolecular synthesis in long-term heterosynaptic facilitation in *Aplysia*, *Science*, 234, 1249, 1986.
2. Frey, U. and Morris, R.G., Synaptic tagging and long-term potentiation, *Nature*, 385, 533, 1997.
3. Nguyen, P.V. and Kandel, E.R., Brief theta-burst stimulation induces a transcription-dependent late phase of LTP requiring cAMP in area CA1 of the mouse hippocampus, *Learn. Mem.*, 4, 230, 1997.
4. Kandel, E.R., The molecular biology of memory storage: a dialog between genes and synapses, *Biosci. Rep.*, 21, 565, 2001.
5. Bailey, C.H., Bartsch, D., and Kandel, E.R., Toward a molecular definition of long-term memory storage, *Proc. Natl. Acad. Sci. USA*, 93, 13445, 1996.
6. Davis, H.P. and Squire, L.R., Protein synthesis and memory: a review, *Psychol. Bull.*, 96, 518, 1984.
7. Schafe, G.E. and LeDoux, J.E., Memory consolidation of auditory Pavlovian fear conditioning requires protein synthesis and protein kinase A in the amygdala, *J. Neurosci.*, 20, RC96, 2000.
8. Nader, K., Schafe, G.E., and Le Doux, J.E., Fear memories require protein synthesis in the amygdala for reconsolidation after retrieval, *Nature*, 406, 722, 2000.
9. Kapp, L.D. and Lorsch, J.R., The molecular mechanics of eukaryotic translation, *Annu. Rev. Biochem.*, 73, 657, 2004.
10. Klann, E. and Dever, T.E., Biochemical mechanisms for translational regulation in synaptic plasticity, *Nat. Rev. Neurosci.*, 5, 931, 2004.
11. Kang, H. and Schuman, E.M., A requirement for local protein synthesis in neurotrophin-induced hippocampal synaptic plasticity, *Science*, 273, 1402, 1996.
12. Steward, O. and Schuman, E.M., Protein synthesis at synaptic sites on dendrites, *Annu. Rev. Neurosci.*, 24, 299, 2001.
13. Raymond, C.R., Redman, S.J., and Crouch, M.F., The phosphoinositide 3-kinase and p70 S6 kinase regulate long-term potentiation in hippocampal neurons, *Neuroscience*, 109, 531, 2002.
14. Martin, K.C. et al., Synapse-specific, long-term facilitation of aplysia sensory to motor synapses: a function for local protein synthesis in memory storage, *Cell*, 91, 927, 1997.
15. Huber, K.M., Kayser, M.S., and Bear, M.F., Role for rapid dendritic protein synthesis in hippocampal mGluR-dependent long-term depression, *Science*, 288, 1254, 2000.
16. Tang, S.J. and Schuman, E.M., Protein synthesis in the dendrite, *Philos. Trans. R. Soc. Lond. B. Biol. Sci.*, 357, 521, 2002.
17. Liu, L. et al., Role of NMDAR subtypes in governing the direction of hippocampal synaptic plasticity, *Science*, 304, 1021, 2004.
18. Husi, H. et al., Proteomic analysis of NMDAR-adhesion protein signaling complexes, *Nat. Neurosci.*, 3, 661, 2000.
19. Raught, B., Gingras, A.C., and Sonenberg, N., *Translational Control of Gene Expression*, ed. N. Sonenberg, J.W. Hershey, and M.B. Matthews. 2000: Cold Spring Harbor Press. 245-294.
20. Gingras, A.C. et al., Hierarchical phosphorylation of the translation inhibitor 4E-BP1, *Genes Dev.*, 15, 2852, 2001.
21. Bolognani, F. and Perrone-Bizzozero, N.I., RNA-protein interactions and control of mRNA stability in neurons, *J. Neurosci. Res.*, 86, 481, 2008.
22. Deschenes-Furry, J. et al., Role of ELAV-like RNA-binding proteins HuD and HuR in the post-transcriptional regulation of acetylcholinesterase in neurons and skeletal muscle cells, *Chem. Biol. Interact.*, 157–158, 43, 2005.

23. Zalfa, F., Achsel, T., and Bagni, C., mRNPs, polysomes or granules: FMRP in neuronal protein synthesis, *Curr. Opin. Neurobiol.*, 16, 265, 2006.
24. Steward, O. et al., Synaptic activation causes the mRNA for the IEG Arc to localize selectively near activated postsynaptic sites on dendrites, *Neuron*, 21, 741, 1998.
25. Rook, M.S., Lu, M., and Kosik, K.S., CaMKIIalpha 3′ untranslated region-directed mRNA translocation in living neurons: visualization by GFP linkage, *J. Neurosci.*, 20, 6385, 2000.
26. Tiruchinapalli, D.M. et al., Activity-dependent trafficking and dynamic localization of zipcode binding protein 1 and beta-actin mRNA in dendrites and spines of hippocampal neurons, *J. Neurosci.*, 23, 3251, 2003.
27. Smith, W.B. et al., Dopaminergic stimulation of local protein synthesis enhances surface expression of GluR1 and synaptic transmission in hippocampal neurons, *Neuron*, 45, 765, 2005.
28. Hardingham, G.E., Arnold, F.J., and Bading, H., A calcium microdomain near NMDARs: on switch for ERK-dependent synapse-to-nucleus communication, *Nat. Neurosci.*, 4, 565, 2001.
29. Sabatini, B.L., Oertner, T.G., and Svoboda, K., The life cycle of Ca(2+) ions in dendritic spines, *Neuron*, 33, 439, 2002.
30. Vanhoutte, P. and Bading, H., Opposing roles of synaptic and extrasynaptic NMDARs in neuronal calcium signalling and BDNF gene regulation, *Curr. Opin. Neurobiol.*, 13, 366, 2003.
31. Dell'Acqua, M.L. et al., Regulation of neuronal PKA signaling through AKAP targeting dynamics, *Eur. J. Cell. Biol.*, 85, 627, 2006.
32. Coogan, A.N., O'Leary, D.M., and O'Connor, J.J., P42/44 MAP kinase inhibitor PD98059 attenuates multiple forms of synaptic plasticity in rat dentate gyrus *in vitro*, *J. Neurophysiol.*, 81, 103, 1999.
33. Cullen, P.J. and Lockyer, P.J., Integration of calcium and Ras signalling, *Nat. Rev. Mol. Cell. Biol.*, 3, 339, 2002.
34. Fukunaga, R. and Hunter, T., MNK1, a new MAP kinase-activated protein kinase, isolated by a novel expression screening method for identifying protein kinase substrates, *EMBO J.*, 16, 1921, 1997.
35. Waskiewicz, A.J. et al., Mitogen-activated protein kinases activate the serine/threonine kinases Mnk1 and Mnk2, *EMBO J.*, 16, 1909, 1997.
36. Waskiewicz, A.J. et al., Phosphorylation of the cap-binding protein eukaryotic translation initiation factor 4E by protein kinase Mnk1 *in vivo*, *Mol. Cell. Biol.*, 19, 1871, 1999.
37. Asaki, C. et al., Localization of translational components at the ultramicroscopic level at postsynaptic sites of the rat brain, *Brain. Res.*, 972, 168, 2003.
38. Tanoue, T. and Nishida, E., Molecular recognitions in the MAP kinase cascades, *Cell. Signal.*, 15, 455, 2003.
39. Sonenberg, N. and Dever, T.E., Eukaryotic translation initiation factors and regulators, *Curr. Opin. Struct. Biol.*, 13, 56, 2003.
40. Scheper, G.C. and Proud, C.G., Does phosphorylation of the cap-binding protein eIF4E play a role in translation initiation?, *Eur. J. Biochem.*, 269, 5350, 2002.
41. Kleijn, M. et al., Regulation of translation initiation factors by signal transduction, *Eur. J. Biochem.*, 253, 531, 1998.
42. Raught, B. and Gingras, A.C., eIF4E activity is regulated at multiple levels, *Int. J. Biochem. Cell. Biol.*, 31, 43, 1999.
43. Banko, J.L., Hou, L., and Klann, E., NMDAR activation results in PKA- and ERK-dependent Mnk1 activation and increased eIF4E phosphorylation in hippocampal area CA1, *J. Neurochem.*, 91, 462, 2004.

44. Kelleher, R.J., et al., Translational control by MAPK signaling in long-term synaptic plasticity and memory, *Cell*, 116, 467, 2004.
45. Kaphzan, H., Doron, G., and Rosenblum, K., Co-application of NMDA and dopamine-induced rapid translation of RSK2 in the mature hippocampus, *J. Neurochem.*, 103, 388, 2007.
46. Yano, S., Tokumitsu, H., and Soderling, T.R., Calcium promotes cell survival through CaM-K kinase activation of the protein-kinase-B pathway, *Nature*, 396, 584, 1998.
47. Sutton, M.A. et al., Miniature neurotransmission stabilizes synaptic function via tonic suppression of local dendritic protein synthesis, *Cell*, 125, 785, 2006.
48. Perkinton, M.S. et al., Phosphatidylinositol 3-kinase is a central mediator of NMDAR signalling to MAP kinase (Erk1/2), Akt/PKB and CREB in striatal neurones, *J. Neurochem.*, 80, 239, 2002.
49. Daw, M.I. et al., Phosphatidylinositol 3 kinase regulates synapse specificity of hippocampal long-term depression, *Nat. Neurosci.*, 5, 835, 2002.
50. Zhu, D., Lipsky, R.H., and Marini, A.M., Co-activation of the phosphatidylinositol-3-kinase/Akt signaling pathway by N-methyl-D-aspartate and TrkB receptors in cerebellar granule cell neurons, *Amino Acids*, 23, 11, 2002.
51. Opazo, P. et al., Phosphatidylinositol 3-kinase regulates the induction of long-term potentiation through extracellular signal-related kinase-independent mechanisms, *J. Neurosci.*, 23, 3679, 2003.
52. Schmitt, J.M. et al., Calmodulin-dependent kinase kinase/calmodulin kinase I activity gates extracellular-regulated kinase-dependent long-term potentiation, *J. Neurosci.*, 25, 1281, 2005.
53. Yoshii, A. and Constantine-Paton, M., BDNF induces transport of PSD-95 to dendrites through PI3K-AKT signaling after NMDAR activation, *Nat. Neurosci.*, 10, 702, 2007.
54. Gong, R. et al., Roles of glutamate receptors and the mammalian target of rapamycin (mTOR) signaling pathway in activity-dependent dendritic protein synthesis in hippocampal neurons, *J. Biol. Chem.*, 281, 18802, 2006.
55. Cammalleri, M. et al., Time-restricted role for dendritic activation of the mTOR-p70S6K pathway in the induction of late-phase long-term potentiation in the CA1, *Proc. Natl. Acad. Sci. USA*, 100, 14368, 2003.
56. Klann, E. et al., Synaptic plasticity and translation initiation, *Learn. Mem.*, 11, 365, 2004.
57. Pirola, L. et al., Phosphoinositide 3-kinase-mediated reduction of insulin receptor substrate-1/2 protein expression via different mechanisms contributes to the insulin-induced desensitization of its signaling pathways in L6 muscle cells, *J. Biol. Chem.*, 278, 15641, 2003.
58. Raab-Graham, K.F. et al., Activity- and mTOR-dependent suppression of Kv1.1 channel mRNA translation in dendrites, *Science*, 314, 144, 2006.
59. Gauchy, C. et al., N-methyl-D-aspartate receptor activation inhibits protein synthesis in cortical neurons independently of its ionic permeability properties, *Neuroscience*, 114, 859, 2002.
60. Nairn, A.C. et al., Nerve growth factor treatment or cAMP elevation reduces Ca^{2+}/calmodulin-dependent protein kinase III activity in PC12 cells, *J. Biol. Chem.*, 262, 14265, 1987.
61. Ryazanov, A.G., Shestakova, E.A., and Natapov, P.G., Phosphorylation of elongation factor 2 by EF-2 kinase affects rate of translation, *Nature*, 334, 170, 1988.
62. Ryazanov, A.G. and Davydova, E.K., Mechanism of elongation factor 2 (EF-2) inactivation upon phosphorylation: phosphorylated EF-2 is unable to catalyze translocation, *FEBS Lett.*, 251, 187, 1989.

63. Marin, P. et al., Glutamate-dependent phosphorylation of elongation factor-2 and inhibition of protein synthesis in neurons, *J. Neurosci.*, 17, 3445, 1997.

64. Scheetz, A.J., Nairn, A.C., and Constantine-Paton, M., N-methyl-D-aspartate receptor activation and visual activity induce elongation factor-2 phosphorylation in amphibian tecta: a role for N-methyl-D-aspartate receptors in controlling protein synthesis, *Proc. Natl. Acad. Sci. USA*, 94, 14770, 1997.

65. Nairn, A.C. et al., Elongation factor-2 phosphorylation and the regulation of protein synthesis by calcium, *Prog. Mol. Subcell. Biol.*, 27, 91, 2001.

66. Sutton, M.A. et al., Postsynaptic decoding of neural activity: eEF2 as a biochemical sensor coupling miniature synaptic transmission to local protein synthesis, *Neuron*, 55, 648, 2007.

67. Scheetz, A.J., Nairn, A.C., and Constantine-Paton, M., NMDAR-mediated control of protein synthesis at developing synapses, *Nat. Neurosci.*, 3, 211, 2000.

68. Levy, S. et al., Oligopyrimidine tract at the 5' end of mammalian ribosomal protein mRNAs is required for their translational control, *Proc. Natl. Acad. Sci. USA*, 88, 3319, 1991.

69. Davuluri, R.V. et al., CART classification of human 5' UTR sequences, *Genome Res.*, 10, 1807, 2000.

70. Aloni, R., Peleg, D., and Meyuhas, O., Selective translational control and nonspecific post-transcriptional regulation of ribosomal protein gene expression during development and regeneration of rat liver, *Mol. Cell. Biol.*, 12, 2203, 1992.

71. Jefferies, H.B. et al., Rapamycin suppresses 5' TOP mRNA translation through inhibition of p70s6k, *EMBO J.*, 16, 3693, 1997.

72. Tsokas, P. et al., Local protein synthesis mediates a rapid increase in dendritic elongation factor 1A after induction of late long-term potentiation, *J. Neurosci.*, 25, 5833, 2005.

73. Browne, G.J. and Proud, C.G., A novel mTOR-regulated phosphorylation site in elongation factor 2 kinase modulates the activity of the kinase and its binding to calmodulin, *Mol. Cell. Biol.*, 24, 2986, 2004.

74. Jordan, B.A. et al., Activity-dependent AIDA-1 nuclear signaling regulates nucleolar numbers and protein synthesis in neurons, *Nat. Neurosci.*, 10, 427, 2007.

75. Lyford, G.L. et al., Arc, a growth factor and activity-regulated gene, encodes a novel cytoskeleton-associated protein that is enriched in neuronal dendrites, *Neuron*, 14, 433, 1995.

76. Knowles, R.B. et al., Translocation of RNA granules in living neurons, *J. Neurosci.*, 16, 7812, 1996.

77. Weiler, I.J. et al., Fragile X mental retardation protein is translated near synapses in response to neurotransmitter activation, *Proc. Natl. Acad. Sci. USA*, 94, 5395, 1997.

78. Ostroff, L.E. et al., Polyribosomes redistribute from dendritic shafts into spines with enlarged synapses during LTP in developing rat hippocampal slices, *Neuron*, 35, 535, 2002.

79. Shan, J. et al., A molecular mechanism for mRNA trafficking in neuronal dendrites, *J. Neurosci.*, 23, 8859, 2003.

80. Kanai, Y., Dohmae, N., and Hirokawa, N., Kinesin transports RNA: isolation and characterization of an RNA-transporting granule, *Neuron*, 43, 513, 2004.

81. Richter, J.D., CPEB: a life in translation, *Trends. Biochem. Sci.*, 32, 279, 2007.

82. Huang, Y.S. et al., N-methyl-D-aspartate receptor signaling results in Aurora kinase-catalyzed CPEB phosphorylation and alpha CaMKII mRNA polyadenylation at synapses, *EMBO J.*, 21, 2139, 2002.

83. Atkins, C.M. et al., Cytoplasmic polyadenylation element binding protein-dependent protein synthesis is regulated by calcium/calmodulin-dependent protein kinase II, *J. Neurosci.*, 24, 5193, 2004.

84. Mendez, R. and Richter, J.D., Translational control by CPEB: a means to the end, *Nat. Rev. Mol. Cell. Biol.*, 2, 521, 2001.

85. Cao, Q. and Richter, J.D., Dissolution of the maskin-eIF4E complex by cytoplasmic polyadenylation and poly(A)-binding protein controls cyclin B1 mRNA translation and oocyte maturation, *EMBO J.*, 21, 3852, 2002.

86. Nakamura, A., Sato, K., and Hanyu-Nakamura, K., Drosophila cup is an eIF4E binding protein that associates with Bruno and regulates oskar mRNA translation in oogenesis, *Dev. Cell.*, 6, 69, 2004.

87. Jung, M.Y., Lorenz, L., and Richter, J.D., Translational control by neuroguidin, a eukaryotic initiation factor 4E and CPEB binding protein, *Mol. Cell. Biol.*, 26, 4277, 2006.

88. Wu, L. et al., CPEB-mediated cytoplasmic polyadenylation and the regulation of experience-dependent translation of alpha-CaMKII mRNA at synapses, *Neuron*, 21, 1129, 1998.

89. Wells, D.G. et al., A role for the cytoplasmic polyadenylation element in NMDAR-regulated mRNA translation in neurons, *J. Neurosci.*, 21, 9541, 2001.

90. Atkins, C.M. et al., Bidirectional regulation of cytoplasmic polyadenylation element-binding protein phosphorylation by Ca^{2+}/calmodulin-dependent protein kinase II and protein phosphatase 1 during hippocampal long-term potentiation, *J. Neurosci.*, 25, 5604, 2005.

91. Ouyang, Y. et al., Visualization of the distribution of autophosphorylated calcium/calmodulin-dependent protein kinase II after tetanic stimulation in the CA1 area of the hippocampus, *J. Neurosci.*, 17, 5416, 1997.

92. Giese, K.P. et al., Autophosphorylation at Thr286 of the alpha calcium-calmodulin kinase II in LTP and learning, *Science*, 279, 870, 1998.

93. Huang, Y.S. et al., Facilitation of dendritic mRNA transport by CPEB, *Genes Dev.*, 17, 638, 2003.

94. Martin, K.C. and Zukin, R.S., RNA trafficking and local protein synthesis in dendrites: an overview, *J. Neurosci.*, 26, 7131, 2006.

95. Laggerbauer, B. et al., Evidence that fragile X mental retardation protein is a negative regulator of translation, *Hum. Mol. Genet.*, 10, 329, 2001.

96. Filipowicz, W. et al., Post-transcriptional gene silencing by siRNAs and miRNAs, *Curr. Opin. Struct. Biol.*, 15, 331, 2005.

97. Zalfa, F. et al., The fragile X syndrome protein FMRP associates with BC1 RNA and regulates the translation of specific mRNAs at synapses, *Cell*, 112, 317, 2003.

98. Jin, P. et al., Biochemical and genetic interaction between the fragile X mental retardation protein and the microRNA pathway, *Nat. Neurosci.*, 7, 113, 2004.

99. Bagga, S. et al., Regulation by let-7 and lin-4 miRNAs results in target mRNA degradation, *Cell*, 122, 553, 2005.

100. Hou, L. et al., Dynamic translational and proteasomal regulation of fragile X mental retardation protein controls mGluR-dependent long-term depression, *Neuron*, 51, 441, 2006.

101. Todd, P.K., Malter, J.S., and Mack, K.J., Whisker stimulation-dependent translation of FMRP in the barrel cortex requires activation of type I metabotropic glutamate receptors, *Brain Res. Mol. Brain. Res.*, 110, 267, 2003.

102. Gabel, L.A. et al., Visual experience regulates transient expression and dendritic localization of fragile X mental retardation protein, *J. Neurosci.*, 24, 10579, 2004.

103. Todd, P.K., Mack, K.J., and Malter, J.S., The fragile X mental retardation protein is required for type-I metabotropic glutamate receptor-dependent translation of PSD-95, *Proc. Natl. Acad. Sci. USA*, 100, 14374, 2003.

104. Pinkstaff, J.K. et al., Internal initiation of translation of five dendritically localized neuronal mRNAs, *Proc. Natl. Acad. Sci. USA*, 98, 2770, 2001.

7 Regulation of NMDA Receptors by Kinases and Phosphatases

*Michael W. Salter, Yina Dong, Lorraine V. Kalia,
Xue Jun Liu, and Graham Pitcher*

CONTENTS

7.1 INTRODUCTION

Phosphorylation is a fundamental and pervasive mechanism widely known to regulate the functions of proteins,[94,133] and lipids.[8] Phosphorylation of specific amino acid residues is a reversible process controlled enzymatically by the competing activities of protein kinases that catalyze phosphorylation and phosphoprotein phosphatases that catalyze dephosphorylation. Several years before the cloning of glutamate receptors, phosphorylation was found to increase NMDA currents, and dephosphorylation to decrease these currents in neurons from the hippocampus.[76] Since then, two principal protein kinase/phosphatase families have been studied extensively related to regulation of NMDA receptors (NMDARs) in the central nervous system: those that act at serine/threonine residues and those that act at tyrosine residues.

Phosphorylation and dephosphorylation may regulate the gating or cell surface expression of NMDARs. Recently, an additional mechanism, alteration of the relative permeability of the NMDAR channel to Ca^{2+}, has been suggested to be subject to regulation by phosphorylation.[112] The simplest biochemical event that may underlie the regulation of NMDARs is phosphorylation of a single amino acid in one of the core NR subunit proteins. This phosphorylation may then be reversed by the action of phosphoprotein phosphatase, and thus the relative levels of phosphorylation and dephosphorylation are determined by the competing actions of those enzymes, i.e., those that are most proximate in the regulatory pathways.

While such direct phosphorylation on serine/threonine and tyrosine residues has been demonstrated, whether such phosphorylation alone is necessary or sufficient for the subsequent increase in NMDAR currents is not known definitively and remains an open question. Alternative mechanisms that are nearly as simple, such as phosphorylation of regulatory or trafficking proteins in NMDAR complexes or of cytoskeletal or other elements also may contribute to changes in NMDAR currents.

The kinase and phosphatase enzymes most proximate in the regulatory control of NMDARs are typically held within NMDAR complexes through anchoring proteins (Figure 7.1) that allow the strategic localization of each enzyme in proximity to its substrate in the complex. This may enhance the efficiency and specificity of the signaling pathways. Based on the key role of NMDARs in many forms of synaptic plasticity, that signaling complexes containing both kinases and their counterpart phosphatases are specifically targeted to the receptor complex facilitates bidirectional regulation of NMDARs during synaptic plasticity. Moreover, these enzymes are also subject to complex regulation by intracellular biochemical signaling networks, leading to multiple levels of control that are dynamic in time and space in certain neurons. Adding to the complexity, the enzymes regulating NMDARs may be differentially expressed in different neuronal populations in the CNS, and the expression may change during development or under physiological or pathological conditions. This chapter provides an overview of the current state of knowledge about NMDAR regulation by serine/threonine and tyrosine phosphorylation, and the cross-talk between these kinase/phosphatase signaling pathways.

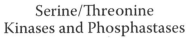

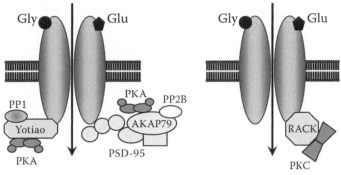

Serine/Threonine
Kinases and Phosphastases

Tyrosine
Kinases and Phosphastases

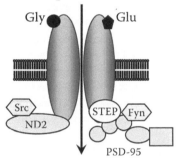

FIGURE 7.1 Comparison of anchoring of serine/threonine versus tyrosine kinases and phosphatases that regulate NMDARs.

7.2 NMDA RECEPTOR REGULATION BY SERINE AND THREONINE KINASES AND PHOSPHATASES

Phosphorylation by serine/threonine kinases is a mechanism for functionally regulating a range of ligand-gated ion channels including GABA$_A$,[87] glycine,[116] nicotinic cholinergic,[42] AMPA,[33,80,99,141] and NMDA[19,52,128] receptors. The intracellular domains of NMDAR subunits contain consensus phosphorylation sites for serine/threonine kinases. Protein kinase A (PKA) and protein kinase C (PKC) are the two that have been most extensively studied in terms of regulating NMDARs. Several other kinases including casein kinase II (CK2) and cyclin-dependent kinase 5 have been found to regulate NMDAR function. In addition, calcium-calmodulin-dependent kinase II (CAMKII) is known to translocate to NMDARs in an activity-dependent manner.[6,7,81]

7.2.1 PROTEIN KINASE A REGULATION OF NMDA RECEPTORS

Protein kinase A (PKA) has been shown to increase NMDAR currents as indicated through elevation of PKA activity by forskolin or cAMP analogs or direct intracellular administration of PKA.[16] Moreover, NMDAR currents are increased by activation of PKA through stimulating G-protein coupled receptors (GPCRs) including β-adrenergic receptors with agonists norepinephrine and isoproterenol.[129] The increase in NMDAR current appears to occur through increased gating as indicated by increased channel open probability (P_o) and currents evoked by exogenously administrated NMDAR agonists. NMDAR-mediated excitatory postsynaptic currents (EPSCs) are increased by PKA. The enhancements may be due in part to PKA-mediated suppression of the desensitization of synaptic NMDARs.[98]

PKA is held in association with NMDAR complexes through binding to two main scaffolding proteins or A kinase anchoring proteins (AKAPs). AKAP 79/150[54] interacts indirectly with NMDARs via PSD-95[23] and yotiao binds directly to the NR1 subunit.[68,109,145] A critical feature of these scaffolds is that they anchor both PKA and the phosphoprotein phosphatases—calcineurin and protein phosphatase 1 (PP1), respectively—that oppose the action of PKA on NMDAR function (see below). This allows highly localized, tightly balanced, and complex regulation of NMDAR function by the interplay of PKA and these phosphatases. Under basal conditions in one prominent model, constitutively active PP1 may keep NMDARs in a state of dephosphorylation and low activity. Upon activation by cAMP, PKA may phosphorylate PP1, decreasing its activity and thereby, with direct phosphorylation of the channel, lead to a shift in the balance of NMDARs to a higher phosphorylation state and thus a higher activity state.[82,145]

While most studies focused on NMDAR currents, recent evidence suggests that, in addition to changes in channel gating, permeability of the channel to Ca^{2+} may be regulated by PKA phosphorylation.[112] Most compellingly, Ca^{2+} entry through the channels, as assayed by an indicator dye overload technique, was suppressed by inhibitors of PKA to a much greater extent than would be predicted by the decrease in NMDAR current. Consistent with this, PKA blockers reduced the $P_{Ca}/P_{monovalent}$ ratio, as determined from reversal potential shifts of NMDAR currents. It was also found that NMDAR-mediated Ca^{2+} increases in dendritic spines were suppressed by these blockers, with little or no decrease in NMDAR EPSCs. Notably this effect was more prominent in neurons from young than in those from mature animals. While this work relied exclusively on blockers of PKA, it raises the possibility that phosphorylation may regulate channel permeability in addition to changes in channel gating and receptor trafficking, and that this regulation may be developmentally controlled.

MacDonald and colleagues[75] described an additional PKA-mediated regulatory mechanism by which PKA may conversely decrease NMDAR currents. They found that activation of GPCRs such as the PDGF receptor leads to activation of PKA and subsequent phosphorylation of the Csk tyrosine kinase. Csk is a major negative regulator of Src family kinases. PKA-mediated phosphorylation of Csk activates this kinase, inhibiting Src kinases, and thereby suppressing NMDAR currents.

7.2.2 PROTEIN KINASE C UPREGULATION OF NMDA RECEPTORS

The serine/threonine PKC family is ubiquitously expressed and involved in multiple neuronal functions including neurotransmitter release, receptor regulation, and synaptic remodeling. Based on its structure and selective sensitivity to second messenger activators Ca^{2+} and diacylglycerol (DAG), PKC can be classified into three major groups. The conventional or calcium-dependent cPKCs (α, βI, βII, and γ) are activated by Ca^{2+} and DAG. The novel or calcium-independent nPKCs (δ, θ, η, and ε) lack a Ca^{2+} binding domain but are still activated by DAG. Finally, the atypical, aPKCs (ζ and λ/ι), are both Ca^{2+}- and DAG-independent but sensitive to other phospholipids.[4] PKC activation is associated with its translocation from the cytosol to the different intracellular compartments including plasma and nuclear membranes, where it is held close to the pertinent substrates by interacting with PKC anchoring proteins or RACKs (receptors for activated C kinase).

Each PKC isozyme may have a specific RACK or anchoring protein that directs the relocation of PKC after its activation and in part mediates isozyme-specific function.[108] PKC isozymes (β, γ, and ε) were found in the NMDAR complex in the PSD.[44] Thus, upon activation, translocated membrane-bound PKC may phosphorylate NMDARs and other proteins in the PSD.

Evidence supporting a role for PKC in regulating NMDAR function is abundant. Electrophysiological studies showed that activation of PKC by application of 4β-phorbol 12–myristate 13-acetate (4β-PMA) enhanced peak NMDA-evoked currents recorded from isolated CA1 hippocampal neurons. This potentiation is prevented by PKC inhibitors chelerythrine or calphostin C, confirming the role of endogenous PKC in response to 4β-PMA.[72] The constitutively active fragment of PKC (PKM) also potentiates peak NMDA currents in hippocampal neurons.[148] PKC modulation of peak NMDA currents depends primarily on the NR2 subunits expressed. Enhancement by PKC is pronounced for receptors containing the NR2A or NR2B subunits, but absent for receptors containing the NR2C or NR2D subunits.[135,47] Instead, activation of PKC inhibits NR2C and NR2D responses.[31] PKC also enhances the functions of synaptically located NMDARs as indicated by the increased NMDAR-mediated components of spontaneous miniature currents by PKM.[72]

In *Xenopus* oocytes expressing native NMDARs, stimulation of PKC using 12-*O*-tetradecanoyl phorbol-13-acetate (TPA) potentiates NMDA channel activity, with no change in single-channel conductance, reversal potential, or mean open time.[69] Activation of GPCRs including phosphoinositol-coupled metabotropic glutamate receptors,[113] muscarinic receptors, and lysophosphatidic acid receptors[72] also potentiates NMDAR currents via activation of PKC.

PKC potentiation of NMDAR function may be mediated through direct phosphorylation of NMDARs. Biochemical studies show that NR1, NR2A, and NR2B subunits may be phosphorylated by PKC both *in vitro* and *in vivo*.[62,127] Phosphorylation sites have been identified for NR1 (Ser-890 and Ser-896),[127] NR2A (Ser-1416),[28] and NR2B (Ser-1303 and Ser-1323).[65] Using specific inhibitors of PKC isoforms and antibodies recognizing specifically phosphorylated serine, PKC sites of NR1 have been found to be phosphorylated by different PKC isoforms, with Ser-896 phosphorylated by PKC α and Ser-890 phosphorylated by PKCγ.[104]

Earlier studies have shown PKC-induced potentiation of NR1/NR2A receptor currents in mutant receptors lacking the entire intracellular domains of both NR1 and NR2A[152], which contain consensus PKC phosphorylation sites. It is therefore unlikely that the potentiation is caused by direct phosphorylation of NMDAR subunits. Instead, PKC could modify associated proteins, involved in signaling and/or trafficking of NMDARs.

Evidence supports the role of direct phosphorylation in PKC-mediated potentiation of NMDAR currents. In *Xenopus* oocytes expressing NR1/NR2B receptors, direct phosphorylation of NR2B (Ser-1303 and Ser-1323) is involved in the PKC-mediated potentiation of NMDAR currents, as mutation of either of these residues severely reduces PKC potentiation.[65] A comparable effect was seen for insulin that potentiates NMDAR currents through PKC. Mutation of Ser-1303 and Ser-1323 of NR2B significantly reduces the potentiation effect of insulin on NR1/NR2B receptors. Similarly, mutating of homologous sites in NR2A (Ser-1291 and Ser-1312) abolishes the insulin potentiation of NR1/NR2A receptors.[48]

Evidence indicates that PKC potentiation of NMDARs may involve cross-talk with tyrosine kinase pathways. Studies of GPCRs have shown that the Src tyrosine kinase is involved in the PKC-induced potentiation and functions as the downstream of PKC. Activation of muscarinic receptors and lysophosphatidic acid receptors potentiates NMDARs current in isolated CA1 hippocampal neurons via activation of PKC. This potentiation is blocked by tyrosine kinase inhibitors genistein or lavendustin A, Src unique domain peptide fragments (40-58), and the anti-cst1 antibody that selectively inhibits the Src family of kinases.[72] The intermediary between PKC and Src is focal adhesion kinase cell adhesion kinase-β/proline-rich tyrosine kinase 2 (CAKβ/Pyk2).[40] PKC activates CAKβ/Pyk2 in hippocampal neurons[63,40] that in turn binds to and activates Src kinase.[40] Thus, PKC potentiation may also be mediated via the sequential activation of CAKβ/Pyk2 and the nonreceptor tyrosine kinase Src (see section below).

In addition to increasing NMDAR currents per se in neurons in the trigeminal nucleus caudalis, activating PKC causes relief of the voltage-dependent blockade of NMDARs by Mg^{2+}.[18] While this relief is prominent in these neurons, the degree of relief of Mg^{2+} blockade is markedly less in neurons in the hippocampus.[139] Thus, this mechanism for enhancing NMDAR currents in physiological conditions may have limited relevance to the functions of NMDARs in the orofacial processing region of the brain stem.

PKC thus regulates multiple properties of NMDARs. In addition to potentiating peak NMDAR current, PKC also enhances NMDAR desensitization. In CA1 pyramidal neurons of the hippocampus, activation of PKC potentiates peak NMDAR currents and also enhances inactivation of steady-state NMDAR currents. This Ca^{2+}-dependent inactivation is mediated via the competitive binding of Ca^{2+} CaM to a site located on the C terminus of the NR1 subunit that also binds α-actinin 2.[71] Unlike potentiation of peak currents by PKC that occurs through the sequential activation of the tyrosine kinases CAKβ/Pyk2 and Src[72], enhancement of Ca^{2+}-dependent inactivation of NMDAR currents is independent of CAKβ/Pyk2 and Src activity.[71] Activation of PKC also enhances glycine-insensitive desensitization of NR1/NR2A receptors expressed in HEK-293 cells, which is independent of previously identified

PKC sites in NR1 and NR2A but may depend on the unidentified PKC sites.[47] Thus, PKC exerts a combined effect on NMDARs, namely enhancement of peak currents and suppression of steady-state currents. By regulating NMDAR desensitization kinetics, PKC may allow more precise control over the time course of Ca^{2+} entry following NMDAR activation, thereby preventing excessive and potentially damaging ionic influxes.

7.2.3 OTHER SERINE AND THREONINE KINASES

7.2.3.1 Casein Kinase II (CK2)

CK2 was shown to regulate NMDAR currents through studies using cell-attached and excised patch recordings of single NMDA channels from acutely dissociated adult hippocampal dentate granule cells.[67] Applying purified CK2 enzyme increased NMDAR channel function that was conversely decreased by a selective inhibitor of CK2, 5,6-dichloro-1-β-D-ribofuranosyl benzimidazole (DRB). DRB also inhibited NMDAR-mediated synaptic transmission. Subsequently, CK2 was found to phosphorylate the Ser-1480 serine residue within the C terminal PDZ ligand (IESDV) of the NR2B subunit of NMDAR *in vitro* and *in vivo*. This phosphorylation of Ser-1480 disrupted the interaction of NR2B with the PDZ domains of PSD-95 and SAP102 and led to decreased surface NR2B expression in neurons. This decrease in surface expression appears to be in opposition to the increase in channel gating seen in electrophysiological studies. This apparent paradox likely points to complex regulatory effects of CK2 on NMDARs.

7.2.3.2 Cyclin-Dependent Kinase 5 (Cdk5)

Cdk5 is a serine/threonine kinase activated by neuron-specific p35 and p39 proteins.[17,130] It exists as a large, multimeric complex associated with cytoskeletal proteins in neurons and has been shown to phosphorylate a wide variety of proteins, including a number of synaptic proteins.[78,111] Cdk5 phosphorylates the NR2A subunit on Ser-1232 both *in vitro* and in intact cells.[64] This phosphorylation may be inhibited by roscovitine, a Cdk5 inhibitor that also suppresses NMDA-evoked currents in hippocampal neurons. In a recent conditional knock-out of Cdk5, NMDAR EPSCs were increased compared with controls.[39] This increase was attributed to increased synaptic NR2B-containing receptors arising from inhibition of calpain-mediated degradation. Thus, Cdk5 may phosphorylate both NR2A- and NR2B-containing NMDARs with opposing functional effects. Phosphorylation increases NR2A but conversely decreases NR2B.

7.2.4 PHOSPHATASES OPPOSING UPREGULATION BY SERINE AND THREONINE KINASES

The major phospho-serine/threonine protein phosphatases 1 (PP1), 2A (PP2A), and 2B (PP2B or calcineurin) suppress the activities of NMDARs, presumably opposing the actions of the kinases described above. In excised patches from hippocampal neurons, applying exogenous PP1 or PP2A depressed open probability of NMDAR single channels.[140] Conversely, selective inhibitors of PP1 and PP2A, calyculin A and okadaic acid (at low concentrations), exerted the opposite effect, increasing NMDAR currents. This implies that these phosphatases endogenously regulate NMDAR channel activity. Subsequently, the regulation by PP1 was determined to be due to localization of this enzyme at NMDARs through anchoring to yotiao.[145]

Likewise for calcineurin, by using cell-attached recordings in acutely dissociated adult rat dentate gyrus granule cells, inhibitors of this phosphatase (high concentration okadaic acid or FK-506) prolonged the duration of single NMDA channel openings, bursts, clusters, and superclusters.[66]

These inhibitors were ineffective when Ca^{2+} entry through NMDA channels was prevented, indicating that calcineurin, activated by calcium entry through native NMDA channels, shortens the duration of channel openings. Subsequently, calcineurin-mediated feedback was shown to regulate synaptic NMDAR currents[129] and depend on anchoring of calcineurin to the scaffold AKAP 79/150.[54] By measuring NMDAR currents from NR1/ NR2A expressing HEK-293 cells, Ser-900 and Ser-929 were identified as residues dephosphorylated by calcineurin.[58]

The phosphatases are subject to regulation that may vary in different neuronal types leading to cell-type specific regulation. For example, in striatal neurons, stimulating D1 dopamine receptors activated adenylate cyclase, increased cAMP, and activated PKA that then phosphorylated and activated the protein phosphatase inhibitor, DARPP-32. Phosphorylated DARPP-32 is a potent inhibitor of PP1.[11] In hippocampal neurons, PP1 is inhibited by a different protein known as inhibitor 1, which is also a substrate of PKA.[110] Thus, PKA activation leads to inhibition of PP1 and decreased dephosphorylation (i.e., enhanced phosphorylation) of downstream substrates including NMDARs.[122] In CA1 neurons, PKA-phosphorylated inhibitor 1 likely interacts with and inhibits PP1 causing enhanced phosphorylation of NMDARs.[110]

7.3   REGULATION OF NMDA RECEPTORS BY PROTEIN TYROSINE KINASES AND PHOSPHATASES

Over the past decade, tyrosine phosphorylation has emerged as a key form of regulation of NMDARs.[3,75,103] Central for the regulation of NMDARs by tyrosine phosphorylation are members of the Src family of protein tyrosine kinases (PTKs) that upregulate NMDAR function.

7.3.1   ENHANCEMENT OF NMDA RECEPTOR FUNCTION BY SRC

In the mammalian CNS, five members of the Src family of nonreceptor PTKs are expressed: Src, Fyn, Yes, Lck, and Lyn. These kinases were initially thought to be involved in regulating cell proliferation and differentiation because Src, the prototype member, was initially identified as a proto-oncogene.[118] However, Src family kinases (SFKs) are expressed in neurons of the adult CNS,[24,119] suggesting additional functions for SFKs since neurons are differentiated postmitotic cells.

SFKs are now known to be expressed widely throughout the CNS and involved in a range of cellular functions. One major function of SFKs in the developed CNS is regulating the activities of ion channels. In the CNS, the first type of channel found to be subject to regulation by SFKs was the NMDAR subtype of ionotropic glutamate receptor.[142] Subsequently, SFKs have been shown to regulate other types of channels in CNS neurons including voltage-gated ion channels, such as potassium[25] and calcium channels,[14] as well as ionotropic neurotransmitter receptors, including GABAA (γ-aminobutyric acid type A) receptors[86,136] and nicotinic acetylcholine receptors.[138]

Electrophysiological recordings from neurons showed that NMDAR currents are governed by a balance between tyrosine phosphorylation and dephosphorylation. Inhibiting endogenous PTK activity[142,143] or increasing phosphotyrosine phosphatase (PTP) activity by introducing exogenous PTP[143] leads to suppression of NMDAR currents. Conversely, inhibiting endogenous PTP activity or increasing PTK activity by introducing exogenous Src causes enhancement of NMDAR currents.[142] Exogenous Src and Fyn were found to potentiate currents mediated by recombinant NMDARs expressed in HEK-293 cells[56] and in *Xenopus* oocytes.[20] As shown on recordings of NMDAR single channel currents, the predominant effect of PTK activity or inhibiting PTPs was to increase NMDAR channel gating with no effect on NMDAR single channel conductance.[143] Moreover, because the effects of manipulating PTKs and PTPs were present with NMDARs in excised membrane patches, it was inferred that PTK and PTP must be intimately associated with the NMDAR complex.

While these studies also showed that exogenous SFKs are sufficient to enhance NMDAR channel gating, further work is needed to identify the endogenous PTK and PTP compounds that mediate NMDAR upregulation and downregulation, respectively. Kinases in the Src family were implicated as endogenous enzymes upregulating NMDAR activity via a phosphopeptide SFK activator (pYEEI peptide) that increased the activities of synaptic NMDAR-mediated currents in cultured neurons[150] and in CA1 pyramidal neurons in hippocampal slices.[73] When applied to the cytoplasmic aspects of inside-out membrane patches, the activating peptide produced an increase in gating of NMDARs without affecting single channel conductance. Conversely an SFK inhibitory antibody known as anti-cst1 exerted an opposite effect, depressing NMDAR channel gating.

Are all five Src family members expressed in the CNS responsible for the upregulation of NMDAR function or do specific SFKs regulate NMDARs? Src,[40] Fyn,[121] Lck, Lyn, and Yes[50] are found in the postsynaptic density (PSD), the main postsynaptic structural component of glutamatergic synapses. Furthermore, Src,[150] Fyn,[149], Lyn, and Yes[50] were shown to be components of the NMDAR complex. Thus, Src, Fyn, Lyn, and Yes are all at appropriate locations to potentially regulate NMDAR function.

Src was implicated through the use of reagents—an inhibitory antibody (anti-src1)[100] and an inhibitory peptide (Src40-58)[150]—that selectively inhibit this kinase but not other members of the Src kinase family. Each of these Src-specific inhibitors decreases synaptic NMDAR-mediated currents and produces a decrease in NMDAR channel gating, the same changes caused by the general SFK inhibitor, the anti-cst1 antibody. Src40-58 is the immunogen for anti-src1 and corresponds to amino acids 40 to 58 within the unique domain of Src. These reagents were hypothesized to block Src-mediated upregulation of NMDAR activity by disrupting a protein–protein interaction of the Src unique domain that allows Src to interact with NMDARs to modify receptor function.

As mentioned above, exogenous Fyn was shown to increase currents mediated by recombinant NMDARs[56] but whether endogenous Fyn or other SFKs present in the CNS regulates native NMDARs remains to be tested directly because inhibitors that selectively block the activities of these SFKs have yet to be developed. However, the Src-specific inhibitors prevent the increase in channel activity produced by the SFK activating pYEEI peptide,[150] implying that endogenous Src plays a critical role in the upregulation of NMDAR activity by SFKs. Src may cause upregulation of NMDAR channel gating via direct tyrosine phosphorylation of the NR2A and NR2B subunits. As discussed elsewhere,[103] it is possible that phosphorylation by Src of proteins in the NMDAR complex, other than NMDAR subunits, may be responsible for Src-mediated changes in NMDAR function.

7.3.2 ANCHORING SRC IN NMDA RECEPTOR COMPLEX VIA UNIQUE DOMAIN

That Src coimmunoprecipitates as part of the NMDAR complex[150] implies that it is held there by binding to an anchoring protein or proteins. The main attributes of such a protein are as follows[30]: (1) it must bind directly to the unique domain of Src through amino acids 40 to 58; (2) this binding must be prevented by a peptide with the sequence of amino acids 40 to 58 of Src (Src40-58); (3) the protein must be present at excitatory synapses and must be a component of the NMDAR complex; and (4) lack of the protein must prevent upregulation of NMDAR activity by endogenous Src. A candidate protein identified by yeast two-hybrid screening using bait constructs containing the Src unique domain was NADH dehydrogenase subunit 2 (ND2), a 347-amino acid protein known to be a subunit of the inner mitochondrial membrane enzyme NADH dehydrogenase (Complex I).

Direct interaction of the Src unique domain and ND2 was confirmed through *in vitro* binding assays. Results from these experiments also identified the ND2.1 region as necessary and sufficient for interacting with the Src unique domain. ND2.1 bound directly to the Src40-58 peptide and the *in vitro* binding of the Src unique domain to ND2.1 was prevented by Src40-58. Src and ND2 coimmunoprecipitated from tissue extracts and, importantly, from PSD preparations from brain. The coimmunoprecipitation was prevented by Src40-58, implying that the Src–ND2 interaction identified *in vitro* may occur *in vivo*.

In addition to finding ND2 in PSD protein preparations, ND2 immunoreactivity was found by immunogold electron microscopy in PSDs in the CA1 hippocampus. Coimmunoprecipitation experiments indicated that ND2 is a component of the NMDAR complex and that the Src–ND2 interaction is required for the association of Src, but not ND2, with NMDARs. Finally, we found that depleting ND2 suppresses Src association with the NMDAR complex and prevents the upregulation of NMDAR function by activating endogenous Src at excitatory synapses. These multiple and converging lines of evidence lead to the conclusion that ND2 mediates interactions between NMDARs and the unique domain of Src. Surprisingly, ND2 acts as an adaptor protein that anchors Src within the NMDAR complex, where it allows Src to upregulate NMDAR activity.

The finding that ND2 binds to Src through the unique domain establishes that, like the SH2 and SH3 domains, this part of Src is a protein–protein interaction region. ND2 binds to Src through a sequence that is not conserved among members of the Src kinase family. Because the unique domains of several Src family kinases have potential binding partners,[29,134] a unifying principle for the role of this region may be in mediating protein–protein interactions. However, unlike the highly-conserved SH2 and SH3 domains that mediate interactions shared by Src family members, the weakly conserved unique domains readily allow distinct interactions for each kinase. Differences in unique domain binding partners may contribute to the functions of the various members of the Src family of kinases, including Fyn that is also known to be held within the NMDAR complex but does not interact with ND2.[30]

7.3.3 REGULATION OF SRC WITHIN NMDA RECEPTOR COMPLEX

As in other systems, the activity of Src within the CNS is tightly regulated. At excitatory synapses in the adult CNS, the basal activity of Src is normally maintained in a low state but can be enhanced by upstream signaling events. Src family kinases serve as molecular hubs through which numerous signaling cascades converge to regulate NMDARs.[103] Some of the same molecules identified in other systems that regulate Src activity also play important roles in the regulation of Src within the NMDAR complex and include the activating enzymes tyrosine kinase CAKβ/Pyk2[40] and the protein tyrosine phosphatase PTPα[61] along with the inhibitory kinase Csk.[144] In addition to these well-characterized regulators of Src, three PSD proteins were recently identified to modulate Src within the NMDAR complex: RACK1,[149] H-Ras,[126] and PSD-95.[49]

7.3.4 STEP OPPOSITION TO SRC UPREGULATION OF NMDA RECEPTOR FUNCTION

NMDAR function is not regulated by Src alone but by the balance of the activities of Src and a PTP that depresses NMDAR gating, reversing the effects of Src. Inhibiting PTPs pharmacologically increases NMDAR channel gating in excised membrane patches[143] and PTP activity coimmunoprecipitates with NMDARs,[3] indicating that endogenous PTP is intrinsic to the NMDAR complex. One family of PTPs observed at the PSD of glutamatergic synapses are the STEPs (striatal-enriched tyrosine phosphatases),[93] a family of brain-specific, nonreceptor type PTPs.[9]

The STEP61 isoform has been found to be a component of the NMDAR complex in spinal cord and hippocampus[95] and therefore is located appropriately to downregulate NMDAR function. Applying recombinant STEP to the cytoplasmic aspects of inside-out membrane patches suppresses NMDAR channel gating, mimicking the effect of inhibiting Src. Similarly, recombinant STEP applied intracellularly reduces NMDAR EPSCs. Conversely, intracellular application of a function-blocking STEP antibody or a dominant-negative STEP produced an increase in NMDAR-mediated EPSCs, implying that NMDAR activity is regulated by endogenous STEP. Both the reduction of NMDAR currents produced by exogenous STEP and the increase of NMDAR currents arising from inhibiting endogenous STEP required Src since both were prevented by blocking Src activity.[95] Thus, it was concluded that STEP is the endogenous PTP that regulates the function of NMDARs in opposition to Src.

Two additional roles for STEP in the regulation of NMDARs have been elucidated. First, STEP-mediated dephosphorylation suppresses the constitutive trafficking of NMDARs, leading to a decrease in NMDAR cell surface expression.[10,114] Second, STEP has been found to dephosphorylate Fyn, reducing its activity[90] and possibly indirectly and directly suppressing tyrosine phosphorylation of NMDARs.

7.4 NMDA RECEPTOR PHOSPHORYLATION IN SYNAPTIC PLASTICITY

NMDARs are pivotal for several types of lasting forms of synaptic plasticity in the CNS required for physiological events including learning and memory and pathological processes such as pain. Long-term potentiation (LTP) is a prominent form of lasting enhancement of synaptic transmission and the predominant cellular model of learning and memory.[77] Clearly, the induction of one main form of LTP exemplified by the tetanus-induced potentiation at Schaffer collateral-CA1 synapses in the hippocampus requires substantially enhanced entry of Ca^{2+} through NMDARs. Depolarization-induced reduction of Mg^{2+}-inhibition of NMDAR currents is a commonly accepted mechanism, but NMDAR currents may be enhanced in other ways, e.g., stimulating signaling cascades. When such cascades are activated through synaptic activity, they provide a form of coincidence detection, a hallmark of synaptic theories of learning and memory analogous to that proposed to arise from postsynaptic depolarization.

Kinase-mediated upregulation of NMDARs may participate directly in mediating synaptic plasticity. Also, the bidirectional control of NMDARs through a balance of kinase and phosphatase activity may be critical in synaptic metaplasticity, i.e., in the "plasticity of plasticity."[2]

7.4.1 Src Upregulation of NMDA Receptors in LTP at CA1 Synapses

SFKs have been implicated from physiological and pharmacological approaches as critical for the induction of LTP in CA1.[32,73] PTKs were first implicated on the basis that tetanus-induced LTP in CA1 neurons was prevented by broad spectrum inhibitors that did not alter preexisting potentiation, implying involvement of PTKs in induction rather than in maintenance of LTP. Upregulation of NMDAR activity by Src is required for the induction of LTP at Schaffer collateral CA1 synapses in the hippocampus.[40,73,95] Intracellular administration of Src40-58 or anti-Src1 directly into postsynaptic neurons by means of a patch pipette electrode prevents LTP induction in CA1 neurons. Because Src40-58 and anti-Src1 prevent the upregulation of NMDAR activity by endogenous Src but do not affect excitatory synaptic transmission,[40,73,150] the most parsimonious explanation for the suppression of LTP induction is that these reagents disrupt the Src–ND2 interactions at synaptic NMDARs. Thus, it was inferred that the interaction between the Src unique domain and ND2 is essential for induction of LTP at CA1 synapses.

The SFK activator pYEEI peptide increased synaptic AMPAR responses, an increase prevented by Src40-58. This increase occludes that produced by tetanus, implying common signaling steps. The pYEEI-induced increase in AMPAR responses is prevented by chelating intracellular Ca^{2+}, but this has no effect on pYEEI-induced increases in NMDAR currents. Because blocking NMDARs prevents the potentiation of AMPAR responses by pYEEI, the simplest model is that Src-mediated upregulation of NMDARs is necessary for tetanus-induced LTP in CA1 neurons.[3] Consistent with this model is the finding that the level of phosphorylation of Y1472 of NR2B increases following tetanic stimulation in CA1 hippocampus.[89] Tyrosine phosphorylation of NR2B increased after LTP induction in the dentate gyrus of the hippocampus.[101,102]

STEP has also been implicated in the induction of LTP.[95] In hippocampal slices, inhibiting endogenous STEP activity with an inhibitory antibody delivered into CA1 neurons enhanced transmission and occluded LTP induction through a mechanism dependent on NMDARs, Ca^{2+}, and Src.[95] Conversely, administering recombinant STEP into CA1 neurons prevented induction of LTP. Neither administering STEP nor inhibiting Src affected basal synaptic transmission or NMDAR currents in CA1 neurons and hence no suppression of NMDARs resulting from these experimental maneuvers might otherwise account for the blockade of LTP induction. STEP does reverse the enhancement of NMDAR currents produced by activating Src. Thus, STEP acts tonically as a brake on Src-mediated synaptic potentiation.

Consistent with the role of Src-mediated upregulation of NMDARs in LTP induction, recent studies implicated PTPα, a well-characterized activator of Src, in LTP.[61,96] Induction of LTP in hippocampal CA1 neurons was prevented by inhibiting endogenous PTPα activity through intracellular application of an inhibitory antibody.[61] LTP induction in CA1 hippocampus was impaired in mice with targeted deletions of PTPα. The impairment was associated with a reduction in phosphorylation levels of Y1472 in the NR2B C tails in the PTPα$^{-/-}$ mice.[96]

Induction of LTP in hippocampal CA1 neurons is prevented by blocking CAKß using the dominant negative mutant described above.[40] Conversely, administering CAKß into CA1 neurons produces a lasting enhancement of AMPAR synaptic responses, mimicking and occluding LTP. This CAKß-stimulated enhancement of synaptic AMPAR responses is prevented by blocking NMDARs, chelating intracellular Ca^{2+}, or blocking Src. Synaptic NMDAR currents in CA1 neurons are not tonically upregulated by CAKß-Src signaling, but CAKß becomes activated and recruited to Src by stimulation that produces LTP.[40,59] Thus, activation of CAKß leading to stimulation of Src is essential for the induction of tetanus-evoked LTP.

Figure 7.2a illustrates a simple model for induction of LTP based on the work described above. It is hypothesized that tetanic stimulation rapidly activates CAKß that associates with and thereby activates Src, allowing tonic suppression of NMDAR function by STEP to be overcome. This kinase-dependent upregulation may be further amplified by a rise in intracellular Na^+ that occurs during high levels of activity since Src kinases increase NMDAR function and also sensitize the channels to potentiation by intracellular Na^+.[151] Coupled with depolarization-induced reduction of Mg^{2+} inhibition, a dramatic boost in the influx of Ca^{2+} through NMDARs sets in motion the downstream cascade[77] that ultimately results in potentiation of synaptic AMPAR responses by recruiting new AMPARs to the synapse and/or by phosphorylating existing AMPARs.

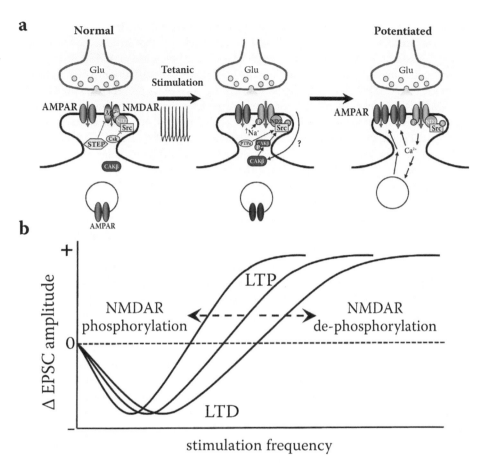

FIGURE 7.2 (a) Model for the role of Src, STEP, CAKβ, PTPα, Csk, and ND2 in the induction of LTP at synapses, e.g., in CA1 hippocampus. Left: under basal conditions, NMDAR activity is suppressed by partial blockade of the channel by Mg^{2+} and activity of STEP and Csk. ND2 acts as an adaptor protein for Src at the NMDAR complex. Middle: tetanic stimulation causes increased current through NMDARs by relief of Mg^{2+} inhibition by activation of Src (Src∗) via the actions of PTPα and activated CAKβ (CAKβ-P) that overcomes suppression by STEP and sensitizes the NMDARs to raised Na^+_i. The protein–protein interaction of ND2 and the Src unique domain allow activated Src to interact with NMDARs to upregulate receptor function. Right: upregulated NMDARs allow greatly increased entry of Ca^{2+} into CA1 neurons, which binds to calmodulin (CaM), causing activation of CaMKII. Expression of LTP arises from increased numbers of AMPARs in postsynaptic membrane or enhanced AMPAR channel activity. (b) Hypothesis for metaplasticity by altering the balance of phosphorylation and dephosphorylation of NMDARs.

7.4.2 KINASE AND PHOSPHATASE REGULATION OF NMDA RECEPTORS IN SYNAPTIC METAPLASTICITY

Metaplasticity is defined as regulation of the processes that underlie synaptic plasticity.[2] This can be seen in thresholds for the induction of LTP or long term depression (LTD) that may be influenced by prior activity or conditioning stimuli that alone do not alter the amplitudes of basal EPSCs or the efficacy of synaptic transmission. For NMDAR-dependent LTP or LTD, one mechanism for metaplasticity may alter the gating or expression of synaptic NMDARs. This is precisely what protein kinases and phosphatases do, as described above.

LTP and LTD can be considered in terms of the relationship of stimulation frequency and resultant change in synaptic efficacy (Figure 7.2b). From this perspective, metaplasticity is reflected by a right or left shift in the frequency–plasticity relationship. Thus, it is hypothesized that kinase-induced enhancement of NMDAR function or number results in production of LTP at lower stimulation frequencies. Conversely, when the kinase/phosphatase balance is shifted toward dephosphorylation and suppression of NMDARs, higher stimulation frequencies may be required to elicit LTP. Changes in NMDARs caused by several serine/threonine and tyrosine kinases or phosphatases have already been implicated in producing metaplasticity.[75] Because these regulatory enzymes are convergence points for many signaling pathways involving diverse upstream receptors in various CNS regions, it is anticipated that complex metaplasticity relationships existing in various neuronal types depend on the unique biochemical networks that are active at a given time.

An interesting example of the complexity by which kinase regulation of NMDARs may affect synaptic plasticity comes from recent work on Cdk5 regulation of NMDARs.[39] As noted, the conditional knockout of Cdk5 in the adult mouse brain increases NR2B expression at synapses. In hippocampal slices from these animals, thresholds for induction of LTP were lowered. Consistent with incorporation of functions of NR2B receptors, LTP induction became sensitive to the ifenprodil NR2B-selective blocker. Furthermore, the Cdk5 conditional knockouts showed improved performance in spatial learning tasks, also sensitive to ifenprodil. Thus, the control of synaptic plasticity by kinase/phosphatase regulation of NMDARs appears to depend on the ultimate effect of the particular kinase/phosphatase on the level of function or expression of synaptic NMDARs and may be highly dependent upon the enzymes, their control by cell signaling pathways, and the NMDAR subunits affected.

7.5 NMDA RECEPTOR PHOSPHORYLATION IN CENTRAL NERVOUS SYSTEM PATHOLOGY

Adaptive plasticity that underlies physiological processes such as learning and memory requires a correct level of NMDAR activity to appropriately modify synaptic transmission. Conversely, maladaptive plasticity underlying disorders characterized by pathological hyperexcitability such as epilepsy or pain may result from excessive activity of NMDARs. Excessive NMDAR activity may also contribute to neuronal cell loss in CNS ischemia and neurodegenerative disorders. However, pathologically suppressed activity of NMDARs is a prominent hypothesis for schizophrenia. Emerging evidence indicates roles for imbalance in kinase/phosphatase regulation of NMDARs in disorders characterized by hyper- or hypofunctioning of these receptors.

7.5.1 PAIN

Upregulation of NMDARs appears crucial for the initiation and maintenance of the enhanced responsiveness of nociceptive neurons in the dorsal horn of the spinal cord in experimental pain models.[146] In spinal cord slices, peripheral inflammation[34] and nerve injury[45] alter NMDAR-mediated currents in superficial dorsal horn neurons. Peripheral nerve injury increases the amplitude, slows the decay phase of NMDA EPSCs,[46] and produces prolonged facilitation of membrane currents and calcium transit induced by bath application of NMDA,[45] thus potentiating glutamatergic transmission.

In the dorsal horn, glutamatergic transmission may be potentiated homosynaptically, as in CA1, although the predominant form of enhancement of synaptic transmission is heterosynaptic.[147] As in CA1, NMDARs in dorsal horn neurons are regulated by CAKβ Src signaling balanced by STEP activity *in vitro*. *In vivo*, tyrosine phosphorylation of NR2B in the spinal cord increases with models of inflammatory[35,36] and neuropathic pain.[1] Inhibition of SFKs *in vivo* delays the onset of inflammatory hyperalgesia[106] and inhibition of SFKs, PKC or group I mGluRs prevents the increase in NR2B tyrosine phosphorylation.[35,36] This indicates that a GPCR signaling cascade upstream of SFK-mediated NMDAR upregulation may be required for pain-related maladaptive changes in synaptic transmission. Peripheral nerve injury activates SFKs in lumbar spinal cord.[51] Intrathecal administration of PP2, a nonselective SFK inhibitor, suppresses mechanical hypersensitivity in nerve-injured mice,[51] suggesting a role of SFK in neuropathic pain.

Studies of mice with deletions of specific SFK genes indicate that Src,[70] Fyn,[1] and Lyn[132] are essential for the development of neuropathic pain. Mice lacking each of these genes exhibited deficits in peripheral nerve injury-induced mechanical hypersensitivity. However, the role of these SFKs in neuropathic pain may be different. Spinal cord dorsal horn NR2B phosphorylation induced by peripheral nerve injury is reduced in both Src and Fyn mutant mice, indicating that NMDARs are downstream of Src and Fyn. However, Lyn is predominantly activated in microglia following PNI, and the upregulation of the ionotropic purinoceptor P2X4 in microglia[131] is deficient in Lyn null mutant mice.

As multiple signaling pathways converge on SFKs in synaptic transmission,[103] SFK-dependent NMDAR upregulation may also serve as a convergence point in the development and maintenance of chronic pain. For example, activation of EphB in the spinal cord with ephrinB2 resulted in prolonged hyperalgesia,[5] while inhibition of EphB reduced chronic inflammatory[5] and neuropathic pain.[55] EphB activation induced phosphorylation of SFKs,[5] leading to phosphorylation of NR2B and amplifying NMDAR responses.[125] The convergence of multiple signaling pathways on SFKs allows both homosynaptic and heterosynaptic plasticity in the dorsal horn that is likely mediated through upregulation of NMDARs by these kinases.

Accumulating evidence indicates that PKA- and PKC-mediated NMDAR phosphorylation participates in generation of pain hypersensitivity. Increased phosphorylation of NR1 proteins in spinal dorsal horn neurons was observed in spinal cords of rats following noxious heat,[12] capsaicin injection,[153,155] formalin injection,[53] and peripheral nerve injury.[26,27] This increase in phosphorylation was detected in both a PKC-dependent site (Ser-896)[12] and a PKA-dependent site (Ser-897).[27,53,154] Pharmacological studies of selective protein kinase inhibitors also suggest that both PKA and PKC are involved in this increased phosphorylation of NR1 following noxious stimulation.[154,155] Thus, serine phosphorylation of NR1 subunits through PKA- and PKC-mediated pathways may contribute to both acute and persistent pain.

7.5.2 EPILEPSY

Some forms of epilepsy such as that produced by kindling are also thought to depend on upregulation of NMDAR function.[83] Kindling shares physiological and pharmacological properties with LTP, so they may share common signaling pathways. Targeted disruption of Fyn in mice delayed the induction of kindling.[13] Introduction of native Fyn into Fyn$^{-/-}$ mice resulted in accelerated kindling.[57] Moreover, transgenic mice expressing a constitutively active form of Fyn showed higher seizure activity.[57] Consistent with this, tyrosine phosphorylation of NR2A and NR2B increased after seizure activity in kainate-induced status epilepticus, another model of epilepsy.[43,88] In an *in vitro* model of epileptiform activity in the CA3 region of hippocampus, Src activity increased with epileptiform activity and the frequency of epileptiform discharge was reduced by pharmacological blockade of SFKs.[105]

7.5.3 ISCHEMIA-INDUCED NEURONAL CELL DEATH

Excitotoxicity mediated by NMDARs is implicated in neuronal death in many pathological conditions including CNS ischemia, trauma, and neurodegenerative diseases. Studies of SFK signaling in models of cerebral ischemia revealed that transient ischemia induces increases in tyrosine phosphorylation of NR2A and NR2B.[21,22,123,124] This is associated with the recruitment of Src, Fyn, and CAKβ to the PSD[21,123] and with the activation of Src and CAKβ[22]. Y1472 in the NR2B C terminal tail is hyperphosphorylated in postischemic rats. Phosphorylation of Y1472 is reduced by inhibition of SFKs.[22] Furthermore, SFK inhibitors suppress NMDA-evoked excitotoxicity *in vitro*.[38] These results implicate a Src-mediated pathway and tyrosine phosphorylation of NMDARs in the pathophysiological mechanism of neuronal death caused by ischemia.

Upregulation of NMDARs by the serine/threonine kinase Cdk5 may also contribute to ischemia-induced neuronal cell loss. In rat hippocampal CA1 neurons, forebrain ischemia induced Cdk5-mediated phosphorylation of the NR2A subunit at Ser-1232.[137] Inhibiting endogenous Cdk5 or perturbing interactions of Cdk5 and NR2A subunits abolished NR2A phosphorylation at Ser-1232 and protected CA1 pyramidal neurons from ischemic insult. Thus, both serine/threonine and tyrosine phosphorylation of NMDARs may be critical in deaths of neurons produced by acute ischemic injury.

7.5.4 HUNTINGTON'S DISEASE

NMDARs are considered to play a role in neuronal loss in several neurodegenerative conditions. The potential involvement of SFK-mediated NMDAR phosphorylation in

Huntington's disease is well investigated. This disease is a progressive neurodegenerative disorder with autosomal-dominant inheritance. The gene for Huntington's disease encodes the huntingtin protein that has an expanded polyglutamine stretch near the 5' end of the gene.[15]

Expression of polyglutamine-expanded huntingtin in a hippocampal cell line activates Src and increases tyrosine phosphorylation of the NR2B subunit of recombinant NMDARs.[115] Expression of mutant huntingtin sensitizes NMDARs and promotes neuronal death induced by glutamate.[120] Inhibition of SFKs decreases glutamate-induced neuronal death mediated by mutant huntingtin, as does co-expression of a mutant NR2B subunit in which Y1252, Y1336, and Y1472 are substituted to phenylalanine. These results indicate involvement of an SFK-mediated signaling pathway upstream in the NMDAR-dependent degeneration of neurons in Huntington's disease.

7.5.5 ALZHEIMER'S DISEASE

Dysregulation of NMDAR phosphorylation is also implicated in Alzheimer's disease. A central causative factor of this disease is the accumulation of a small secreted peptide known as amyloid-β.[79] One effect of a toxic fragment of amyloid-β known as amyloid-β1-42 is activation of the α7 nicotinic acetylcholine receptor resulting in α7-mediated Ca^{2+} influx and activation of calcineurin.[114] PP2B dephosphorylates and activates STEP, which dephosphorylates the NR2B subunit at Tyr-1472 and promotes internalization of NR2B-containing NMDARs. STEP may also depress Fyn and hence the Fyn-mediated phosphorylation of Tyr-1472.[10] It is hypothesized that high levels of amyloid-β reduce NMDA EPSCs and inhibit synaptic plasticity.[114]

7.5.6 SCHIZOPHRENIA

NMDAR hypofunction is implicated in a number of the behavioral manifestations of schizophrenia—social withdrawal, increased motor stereotypy, cognitive deficits, and locomotor activity—in humans and animal models.[84,85,107] Dysregulation of NMDAR phosphorylation may thus contribute to the etiology of schizophrenia. Schizophrenia has one of the highest heritabilities among neuropsychiatric disorders. In several human association and linkage studies, ErbB4 has been identified as a key risk gene that confers susceptibility to schizophrenia.[37,60,91,92] ErbB4 encodes a receptor tyrosine kinase, the ErbB4 receptor, that is expressed in the adult CNS. The cognate ligand for ErbB4 is the peptide neuregulin 1 (NRG1), also strongly linked to schizophrenia in humans.[117] In post-mortem prefrontal cortex tissues of patients with schizophrenia, marked increases of NRG1-induced activation of ErbB4 attributable to increased association of ErbB4 with PSD-95 were observed.[37]

This overactivation of ErbB4 by NRG1 suppressed tyrosine phosphorylation of NR2A in human samples. In rodents, NRG1 ErbB4 signaling blocks induction of LTP at CA1 synapses,[41,74] likely by suppressing Src-mediated enhancement of NMDARs. These studies lead to the hypothesis that cognitive deficits in schizophrenia may be consequences of hyperfunction of NRG1 ErbB4 signaling, leading to suppressed NMDAR-dependent synaptic plasticity.[97]

7.6 CONCLUSIONS

Our understanding of the regulation of NMDARs by protein kinases and phosphatases has increased at an accelerating pace in recent years. We now have abundant evidence about the molecular mechanisms by which these enzymes regulate the functions and cell surface expression of NMDARs. These molecular insights provided a number of tools that are beginning to reveal roles for phosphorylation and dephosphorylation of NMDARs in a diversity of CNS processes. We anticipate that the convergent regulation of NMDARs by serine/threonine and tyrosine kinases will be widely relevant to various states of health and disease.

ACKNOWLEDGMENTS

The work of the authors is supported by the Canadian Institutes of Health Research (CIHR). MWS holds a Canada Research Chair (Tier I) in Neuroplasticity and Pain, and is an International Research Scholar of the Howard Hughes Medical Institute. Thanks to Janice Hicks for assistance preparing this manuscript.

REFERENCES

1. Abe, T. et al., Fyn kinase-mediated phosphorylation of NMDAR NR2B subunit at Tyr1472 is essential for maintenance of neuropathic pain, *Eur. J. Neurosci.*, 22, 1445, 2005.
2. Abraham, W.C. and Bear, M.F., Metaplasticity: the plasticity of synaptic plasticity, *Trends Neurosci.*, 19, 126, 1996.
3. Ali, D.W. and Salter, M.W., NMDAR regulation by Src kinase signalling in excitatory synaptic transmission and plasticity, *Curr. Opin. Neurobiol.*, 11, 336, 2001.
4. Amadio, M., Battaini, F., and Pascale, A., The different facets of protein kinases C: old and new players in neuronal signal transduction pathways, *Pharmacol. Res.*, 54, 317, 2006.
5. Battaglia, A.A. et al., EphB receptors and ephrin-B ligands regulate spinal sensory connectivity and modulate pain processing, *Nat. Neurosci.*, 6, 339, 2003.
6. Bayer, K.U. et al., Interaction with the NMDAR locks CaMKII in an active conformation, *Nature*, 411, 801, 2001.
7. Bayer, K.U. et al., Transition from reversible to persistent binding of CaMKII to postsynaptic sites and NR2B, *J. Neurosci.*, 26, 1164, 2006.
8. Blume-Jensen, P. and Hunter, T., Oncogenic kinase signalling, *Nature*, 411, 355, 2001.
9. Boulanger, L.M. et al., Cellular and molecular characterization of a brain-enriched protein tyrosine phosphatase., *J. Neurosci.*, 15, 1532, 1995.
10. Braithwaite, S.P. et al., Synaptic plasticity: one STEP at a time, *Trends Neurosci.*, 29, 452, 2006.
11. Brautigan, D.L., Phosphatases as partners in signaling networks, *Adv. Second Messenger Phosphoprotein Res.*, 31, 113, 1997.
12. Brenner, G.J. et al., Peripheral noxious stimulation induces phosphorylation of the NMDAR NR1 subunit at the PKC-dependent site, serine-896, in spinal cord dorsal horn neurons, *Eur. J. Neurosci.*, 20, 375, 2004.
13. Cain, D.P. et al., Fyn tyrosine kinase is required for normal amygdala kindling, *Epilepsy Res.*, 22, 107, 1995.

14. Cataldi, M. et al., Protein-tyrosine kinases activate while protein-tyrosine phosphatases inhibit L-type calcium channel activity in pituitary GH_3 cells, *J. Biol. Chem.*, 271, 9441, 1996.

15. Cattaneo, E. et al., Loss of normal huntingtin function: new developments in Huntington's disease research, *Trends Neurosci.*, 24, 182, 2001.

16. Cerne, R., Rusin, K.I., and Randic, M., Enhancement of the N-methyl-D-aspartate response in spinal dorsal horn neurons by cAMP-dependent protein kinase, *Neurosci. Lett.*, 161, 124, 1993.

17. Chae, T. et al., Mice lacking p35, a neuronal specific activator of Cdk5, display cortical lamination defects, seizures, and adult lethality, *Neuron*, 18, 29, 1997.

18. Chen, L. and Huang, L.Y.M., Protein kinase C reduces Mg block of NMDA-receptor channels as a mechanism of modulation, *Nature*, 356, 521, 1992.

19. Chen, L. and Huang, L.-Y.M., Sustained potentiation of NMDAR-mediated glutamate responses through activation of protein kinase C by a μ opioid, *Neuron*, 7, 319, 1991.

20. Chen, S.J. and Leonard, J.P., Protein tyrosine kinase-mediated potentiation of currents from cloned NMDARs, *J. Neurochem.*, 67, 194, 1996.

21. Cheung, H.H. et al., Altered association of protein tyrosine kinases with postsynaptic densities after transient cerebral ischemia in the rat brain, *J. Cereb. Blood Flow Metab.*, 20, 505, 2000.

22. Cheung, H.H. et al., Inhibition of protein kinase C reduces ischemia-induced tyrosine phosphorylation of the N-methyl-D-aspartate receptor, 410, *J. Neurochem.*, 86, 1441, 2003.

23. Colledge, M. et al., Targeting of PKA to glutamate receptors through a MAGUK-AKAP complex, *Neuron*, 27, 107, 2000.

24. Cotton, P.C. and Brugge, J.S., Neural tissues express high levels of the cellular src gene product pp60c-src, *Mol. Cell Biol.*, 3, 1157, 1983.

25. Fadool, D.A. et al., Tyrosine phosphorylation modulates current amplitude and kinetics of a neuronal voltage-gated potassium channel, *J. Neurophysiol.*, 78, 1563, 1997.

26. Gao, X. et al., Enhancement of NMDAR phosphorylation of the spinal dorsal horn and nucleus gracilis neurons in neuropathic rats, *Pain*, 116, 62, 2005.

27. Gao, X. et al., Reactive oxygen species (ROS) are involved in enhancement of NMDA-receptor phosphorylation in animal models of pain, 385, *Pain*, 131, 262, 2007.

28. Gardoni, F. et al., Protein kinase C activation modulates alpha-calmodulin kinase II binding to NR2A subunit of N-methyl-D-aspartate receptor complex, *J. Biol. Chem.*, 276, 7609, 2001.

29. Gervais, F.G. and Veillette, A., The unique amino-terminal domain of p56lck regulates interactions with tyrosine protein phosphatases in T lymphocytes, *Mol. Cell Biol.*, 15, 2393, 1995.

30. Gingrich, J.R. et al., Unique domain anchoring of Src to synaptic NMDARs via the mitochondrial protein NADH dehydrogenase subunit 2, *Proc. Natl. Acad. Sci. USA*, 101, 6237, 2004.

31. Grant, E.R. et al., Opposing contributions of NR1 and NR2 to protein kinase C modulation of NMDARs, *J. Neurochem.*, 71, 1471, 1998.

32. Grant, S.G. et al., Impaired long-term potentiation, spatial learning, and hippocampal development in fyn mutant mice, *Science*, 258, 1903, 1992.

33. Greengard, P. et al., Enhancement of the glutamate response by cAMP-dependent protein kinase in hippocampal neurons, *Science*, 253, 1135, 1991.

34. Guo, H. and Huang, L.Y., Alteration in the voltage dependence of NMDAR channels in rat dorsal horn neurones following peripheral inflammation, *J. Physiol*, 537, 115, 2001.

35. Guo, W. et al., Group I metabotropic glutamate receptor NMDAR coupling and signaling cascade mediate spinal dorsal horn NMDAR 2B tyrosine phosphorylation associated with inflammatory hyperalgesia, *J. Neurosci.*, 24, 9161, 2004.
36. Guo, W. et al., Tyrosine Phosphorylation of the NR2B Subunit of the NMDAR in the Spinal Cord during the Development and Maintenance of Inflammatory Hyperalgesia, *J. Neurosci.*, 22, 6208, 2002.
37. Hahn, C.G. et al., Altered neuregulin 1-erbB4 signaling contributes to NMDAR hypofunction in schizophrenia, *Nat. Med.*, 12, 824, 2006.
38. Hashimoto, R. et al., Lithium-induced inhibition of Src tyrosine kinase in rat cerebral cortical neurons: a role in neuroprotection against N-methyl-D-aspartate receptor-mediated excitotoxicity, *FEBS Lett.*, 538, 145, 2003.
39. Hawasli, A.H. et al., Cyclin-dependent kinase 5 governs learning and synaptic plasticity via control of NMDAR degradation, *Nat. Neurosci.*, 10, 880, 2007.
40. Huang, Y. et al., CAKß/Pyk2 kinase is a signaling link for induction of long-term potentiation in CA1 hippocampus, *Neuron*, 29, 485, 2001.
41. Huang, Y.Z. et al., Regulation of neuregulin signaling by PSD-95 interacting with ErbB4 at CNS synapses, *Neuron*, 26, 443, 2000.
42. Huganir, R.L. et al., Phosphorylation of the nicotinic acetylcholine receptor regulates its rate of desensitization, *Nature*, 321, 774, 1986.
43. Huo, J.Z., Dykstra, C.M., and Gurd, J.W., Increase in tyrosine phosphorylation of the NMDAR following the induction of status epilepticus, *Neurosci. Lett.*, 401, 266, 2006.
44. Husi, H. et al., Proteomic analysis of NMDAR-adhesion protein signaling complexes, *Nat. Neurosci.*, 3, 661, 2000.
45. Isaev, D. et al., Facilitation of NMDA-induced currents and Ca^{2+} transients in the rat substantia gelatinosa neurons after ligation of L5-L6 spinal nerves, *NeuroReport*, 11, 4055, 2000.
46. Iwata, H. et al., NMDAR 2B subunit-mediated synaptic transmission in the superficial dorsal horn of peripheral nerve-injured neuropathic mice, *Brain Res.*, 1135, 92, 2007.
47. Jackson, M.F. et al., Protein kinase C enhances glycine-insensitive desensitization of NMDARs independently of previously identified protein kinase C sites, *J. Neurochem.*, 96, 1509, 2006.
48. Jones, M.L. and Leonard, J.P., PKC site mutations reveal differential modulation by insulin of NMDARs containing NR2A or NR2B subunits, *J. Neurochem.*, 92, 1431, 2005.
49. Kalia, L.V. et al., PSD-95 is a negative regulator of the tyrosine kinase Src in the NMDAR complex, *EMBO J.*, 25, 4971, 2006.
50. Kalia, L.V. and Salter, M.W., Interactions between Src family protein tyrosine kinases and PSD-95, *Neuropharmacology*, 45, 720, 2003.
51. Katsura, H. et al., Activation of Src-family kinases in spinal microglia contributes to mechanical hypersensitivity after nerve injury, *J. Neurosci.*, 26, 8680, 2006.
52. Kelso, S.R., Nelson, T.E., and Leonard, J.P., Protein kinase C-mediated enhancement of NMDA currents by metabotropic glutamate receptors in *Xenopus* oocytes, *J. Physiol.*, 449, 705, 1992.
53. Kim, H.W. et al., Intrathecal treatment with sigma1 receptor antagonists reduces formalin-induced phosphorylation of NMDAR subunit 1 and the second phase of formalin test in mice, *Br. J. Pharmacol.*, 148, 490, 2006.
54. Klauck, T.M. et al., Coordination of three signaling enzymes by AKAP79, a mammalian scaffold protein, *Science*, 271, 1589, 1996.
55. Kobayashi, H. et al., Involvement of EphB1 receptor/EphrinB2 ligand in neuropathic pain, *Spine*, 32, 1592, 2007.

56. Kohr, G. and Seeburg, P.H., Subtype-specific regulation of recombinant NMDAR-channels by protein tyrosine kinases of the src family, *J. Physiol.*, 492 (Pt 2), 445, 1996.
57. Kojima, N. et al., Higher seizure susceptibility and enhanced tyrosine phosphorylation of N-methyl-D-aspartate receptor subunit 2B in fyn transgenic mice, *Learn. Mem.*, 5, 429, 1998.
58. Krupp, J.J. et al., Calcineurin acts via the C-terminus of NR2A to modulate desensitization of NMDARs, *Neuropharmacology*, 42, 593, 2002.
59. Lauri, S.E., Taira, T., and Rauvala, H., High-frequency synaptic stimulation induces association of fyn and c- src to distinct phosphorylated components, *NeuroReport*, 11, 997, 2000.
60. Law, A.J. et al., Disease-associated intronic variants in the ErbB4 gene are related to altered ErbB4 splice-variant expression in the brain in schizophrenia, *Hum. Mol. Genet.*, 16, 129, 2007.
61. Lei, G. et al., Gain control of N-methyl-D-aspartate receptor activity by receptor-like protein tyrosine phosphatase alpha, *EMBO J.*, 21, 2977, 2002.
62. Leonard, A.S. and Hell, J.W., Cyclic AMP-dependent protein kinase and protein kinase C phosphorylate N-methyl-D-aspartate receptors at different sites, *J. Biol. Chem.*, 272, 12107, 1997.
63. Lev, S. et al., Protein tyrosine kinase PYK2 involved in Ca(2+)-induced regulation of ion channel and MAP kinase functions, *Nature*, 376, 737, 1995.
64. Li, B.S. et al., Regulation of NMDARs by cyclin-dependent kinase-519, *Proc. Natl. Acad. Sci. USA*, 98, 12742, 2001.
65. Liao, G.Y. et al., Evidence for direct protein kinase-C mediated modulation of N-methyl-D-aspartate receptor current, *Mol. Pharmacol.*, 59, 960, 2001.
66. Lieberman, D.N. and Mody, I., Regulation of NMDA channel function by endogenous Ca^{2+}- dependent phosphatase, *Nature*, 369, 235, 1994.
67. Lieberman, D.N. and Mody, I., Casein kinase-II regulates NMDA channel function in hippocampal neurons, *Nat. Neurosci.*, 2, 125, 1999.
68. Lin, J.W. et al., Yotiao, a novel protein of neuromuscular junction and brain that interacts with specific splice variants of NMDAR subunit NR1, *J. Neurosci.*, 18, 2017, 1998.
69. Lin, Y. et al., PSD-95 and PKC converge in regulating NMDAR trafficking and gating, *Proc. Natl. Acad. Sci. USA*, 103, 19902, 2006.
70. Liu, X. et al., Inhibition of neuropathic pain behaviour by a tyrosine kinase Src unique domain peptide, *Soc. Neurosci. Abstr.*, 32, 803.20, 2006.
71. Lu, W.Y. et al., In CA1 pyramidal neurons of the hippocampus protein kinase C regulates calcium-dependent inactivation of NMDARs, *J. Neurosci.*, 20, 4452, 2000.
72. Lu, W.Y. et al., G-protein-coupled receptors act via protein kinase C and Src to regulate NMDARs, *Nat. Neurosci.*, 2, 331, 1999.
73. Lu, Y.M. et al., Src activation in the induction of long-term potentiation in CA1 hippocampal neurons, *Science*, 279, 1363, 1998.
74. Ma, L. et al., Ligand-dependent recruitment of the ErbB4 signaling complex into neuronal lipid rafts, *J. Neurosci.*, 23, 3164, 2003.
75. MacDonald, J.F., Jackson, M.F., and Beazely, M.A., G protein-coupled receptors control NMDARs and metaplasticity in the hippocampus, *Biochim. Biophys. Acta*, 1768, 941, 2007.
76. MacDonald, J.F., Mody, I., and Salter, M.W., Regulation of N-methyl-D-aspartate receptors revealed by intracellular dialysis of murine neurones in culture, *J. Physiol.*, 414, 17, 1989.
77. Malenka, R.C. and Nicoll, R.A., Long-term potentiation: a decade of progress? *Science*, 285, 1870, 1999.

78. Matsubara, M. et al., Site-specific phosphorylation of synapsin I by mitogen-activated protein kinase and Cdk5 and its effects on physiological functions, *J. Biol. Chem.*, 271, 21108, 1996.
79. Mattson, M.P., Pathways towards and away from Alzheimer's disease, *Nature*, 430, 631, 2004.
80. McGlade-McCulloh, E. et al., Phosphorylation and regulation of glutamate receptors by calcium/calmodulin-dependent protein kinase II, *Nature*, 362, 640, 1993.
81. Merrill, M.A. et al., Activity-driven postsynaptic translocation of CaMKII, *Trends Pharmacol. Sci.*, 26, 645, 2005.
82. Michel, J.J. and Scott, J.D., AKAP-mediated signal transduction, *Annu. Rev. Pharmacol. Toxicol.*, 42, 235, 2002.
83. Mody, I., Synaptic plasticity in kindling, *Adv. Neurol.*, 79, 631, 1999.
84. Mohn, A.R. et al., Mice with reduced NMDAR expression display behaviors related to schizophrenia, *Cell*, 98, 427, 1999.
85. Morrison, P.D. and Pilowsky, L.S., Schizophrenia: more evidence for less glutamate, *Expert. Rev. Neurother.*, 7, 29, 2007.
86. Moss, S.J. et al., Modulation of GABAA receptors by tyrosine phosphorylation, *Nature*, 377, 344, 1995.
87. Moss, S.J. et al., Functional modulation of GABA$_A$ receptors by cAMP-dependent protein phosphorylation, *Science*, 257, 661, 1992.
88. Moussa, R.C. et al., Seizure activity results in increased tyrosine phosphorylation of the N-methyl-D-aspartate receptor in the hippocampus, *Brain Res. Mol. Brain Res.*, 95, 36, 2001.
89. Nakazawa, T. et al., Characterization of Fyn-mediated tyrosine phosphorylation sites on GluR epsilon 2 (NR2B) subunit of the N-methyl-D-aspartate receptor, *J. Biol. Chem.*, 276, 693, 2001.
90. Nguyen, T.H., Liu, J., and Lombroso, P.J., Striatal enriched phosphatase 61 dephosphorylates Fyn at phosphotyrosine 420, *J. Biol. Chem.*, 277, 24274, 2002.
91. Nicodemus, K.K. et al., Further evidence for association between ErbB4 and schizophrenia and influence on cognitive intermediate phenotypes in healthy controls, *Mol. Psychiatr.*, 11, 1062, 2006.
92. Norton, N. et al., Evidence that interaction between neuregulin 1 and its receptor erbB4 increases susceptibility to schizophrenia, *Am. J. Med. Genet. B Neuropsy chiatr. Genet.*, 141, 96, 2006.
93. Oyama, T. et al., Immunocytochemical localization of the striatal enriched protein tyrosine phosphatase in the rat striatum: a light and electron microscopic study with a complementary DNA-generated polyclonal antibody, *Neuroscience*, 69, 869, 1995.
94. Pawson, T. and Scott, J.D., Protein phosphorylation in signalling: 50 years and counting, *Trends Biochem. Sci.*, 30, 286, 2005.
95. Pelkey, K.A. et al., Tyrosine phosphatase STEP is a tonic brake on induction of long-term potentiation, *Neuron*, 34, 127, 2002.
96. Petrone, A. et al., Receptor protein tyrosine phosphatase alpha is essential for hippocampal neuronal migration and long-term potentiation, *EMBO J.*, 22, 4121, 2003.
97. Pitcher, G.M. et al., Schizophrenia risk gene *ErbB4* is a suppressor of synaptic long-term potentiation in the adult hippocampus, *NeuroReport,* in press, 2007.
98. Raman, I.M., Tong, G., and Jahr, C.E., β-Adrenergic regulation of synaptic NMDARs by cAMP-dependent protein kinase, *Neuron*, 16, 415, 1996.
99. Raymond, L.A., Blackstone, C.D. and Huganir, R.L., Phosphorylation and modulation of recombinant GluR6 glutamate receptors by cAMP-dependent protein kinase, *Nature*, 361, 637, 1993.
100. Roche, S. et al., DNA synthesis induced by some but not all growth factors requires src family protein tyrosine kinases, *Mol. Cell Biol.*, 15, 1102, 1995.

101. Rosenblum, K., Dudai, Y., and Richter-Levin, G., Long-term potentiation increases tyrosine phosphorylation of the N-methyl-D-aspartate receptor subunit 2B in rat dentate gyrus in vivo, *Proc. Natl. Acad. Sci. USA*, 93, 10457, 1996.
102. Rostas, J.A. et al., Enhanced tyrosine phosphorylation of the 2B subunit of the N-methyl-D-aspartate receptor in long-term potentiation, *Proc. Natl. Acad. Sci. USA*, 93, 10452, 1996.
103. Salter, M.W. and Kalia, L.V., Src kinases: a hub for NMDAR regulation, *Nat. Rev. Neurosci.*, 5, 317, 2004.
104. Sanchez-Perez, A.M. and Felipo, V., Serines 890 and 896 of the NMDAR subunit NR1 are differentially phosphorylated by protein kinase C isoforms, *Neurochem. Int.*, 47, 84, 2005.
105. Sanna, P.P. et al., A role for Src kinase in spontaneous epileptiform activity in the CA3 region of the hippocampus, *Proc. Natl. Acad. Sci. U. S. A*, 97, 8653, 2000.
106. Sato, E. et al., Involvement of spinal tyrosine kinase in inflammatory and N-methyl-D-aspartate-induced hyperalgesia in rats, *Eur. J. Pharmacol.*, 468, 191, 2003.
107. Sawa, A. and Snyder, S.H., Schizophrenia: diverse approaches to a complex disease, *Science*, 296, 692, 2002.
108. Schechtman, D. and Mochly-Rosen, D., Isozyme-specific inhibitors and activators of protein kinase C, *Methods Enzymol.*, 345, 470, 2002.
109. Sheng, M. and Kim, M.J., Post-synaptic signaling and plasticity mechanisms, *Science*, 298, 776, 2002.
110. Shenolikar, S., Protein serine/threonine phosphatises: new avenues for cell regulation, *Annu. Rev. Cell Biol.*, 10, 55, 1994.
111. Shuang, R. et al., Regulation of Munc-18/syntaxin 1A interaction by cyclin-dependent kinase 5 in nerve endings, *J. Biol. Chem.*, 273, 4957, 1998.
112. Skeberdis, V.A. et al., Protein kinase A regulates calcium permeability of NMDARs, *Nat. Neurosci.*, 9, 501, 2006.
113. Skeberdis, V.A. et al., mGluR1-mediated potentiation of NMDARs involves a rise in intracellular calcium and activation of protein kinase C, *Neuropharmacology*, 40, 856, 2001.
114. Snyder, E.M. et al., Regulation of NMDAR trafficking by amyloid-beta, *Nat. Neurosci.*, 8, 1051, 2005.
115. Song, C. et al., Expression of polyglutamine-expanded huntingtin induces tyrosine phosphorylation of N-methyl-D-aspartate receptors, *J. Biol. Chem.*, 278, 33364, 2003.
116. Song, Y.M. and Huang, L.-Y.M., Modulation of glycine receptor chloride channels by cAMP- dependent protein kinase in spinal trigeminal neurons, *Nature*, 348, 242, 1990.
117. Stefansson, H. et al., Neuregulin 1 and schizophrenia, *Ann. Med.*, 36, 62, 2004.
118. Stehelin, D., Varmus, H.E., and Bishop, J.M., DNA related to the transforming gene(s) of avian sarcoma viruses is present in normal avian DNA, *Nature*, 260, 170, 1976.
119. Sudol, M., Expression of proto-oncogenes in neural tissues, *Brain Res.*, 472, 391, 1988.
120. Sun, Y. et al., Polyglutamine-expanded huntingtin promotes sensitization of N-methyl-D-aspartate receptors via post-synaptic density 95, *J. Biol. Chem.*, 276, 24713, 2001.
121. Suzuki, T. and Okamura-Noji, K., NMDAR subunits ε1 (NR2A) and ε2 (NR2B) are substrates for Fyn in the postsynaptic density fraction isolated from the rat brain, *Biochem. Biophys. Res. Commun.*, 216, 582, 1995.
122. Svenningsson, P. et al., DARPP-32: an integrator of neurotransmission, *Annu. Rev. Pharmacol. Toxicol.*, 44, 269, 2004.
123. Takagi, N. et al., The effect of transient global ischemia on the interaction of Src and Fyn with the N-methyl-D-aspartate receptor and postsynaptic densities: possible involvement of Src homology 2 domains, *J. Cereb. Blood Flow Metab.*, 19, 880, 1999.

124. Takagi, N. et al., Transient ischemia differentially increases tyrosine phosphorylation of NMDAR subunits 2A and 2B, *J. Neurochem.*, 69, 1060, 1997.
125. Takasu, M.A. et al., Modulation of NMDAR-dependent calcium influx and gene expression through EphB receptors, *Science*, 295, 491, 2002.
126. Thornton, C. et al., H-Ras modulates N-methyl-D-aspartate receptor function via inhibition of Src tyrosine kinase activity, *J. Biol. Chem.*, 278, 23823, 2003.
127. Tingley, W.G. et al., Characterization of protein kinase A and protein kinase C phosphorylation of the N-methyl-D-aspartate receptor NR1 subunit using phosphorylation site-specific antibodies, *J. Biol. Chem.*, 272, 5157, 1997.
128. Tingley, W.G. et al., Regulation of NMDAR phosphorylation by alternative splicing of the C-terminal domain, *Nature*, 364, 70, 1993.
129. Tong, G., Shepherd, D., and Jahr, C.E., Synaptic desensitization of NMDARs by calcineurin, *Science*, 267, 1510, 1995.
130. Tsai, L.H. et al., p35 is a neural-specific regulatory subunit of cyclin-dependent kinase 5, *Nature*, 371, 419, 1994.
131. Tsuda, M. et al., P2X4 receptors induced in spinal microglia gate tactile allodynia after nerve injury, *Nature*, 424, 778, 2003.
132. Tsuda, M. et al., Lyn tyrosine kinase is required for P2X(4) receptor upregulation and neuropathic pain after peripheral nerve injury, *Glia*, 406, 2007.
133. Ubersax, J.A. and Ferrell, J.E., Jr., Mechanisms of specificity in protein phosphorylation, *Nat. Rev. Mol. Cell Biol.*, 8, 530, 2007.
134. Veillette, A. et al., The CD4 and CD8 T cell surface antigens are associated with the internal membrane tyrosine-protein kinase p56lck, *Cell*, 55, 301, 1988.
135. Wagner, D.A. and Leonard, J.P., Effect of protein kinase-C activation on the Mg(2+)-sensitivity of cloned NMDARs, *Neuropharmacology*, 35, 29, 1996.
136. Wan, Q. et al., Modulation of GABAA receptor function by tyrosine phosphorylation of beta subunits, *J. Neurosci.*, 17, 5062, 1997.
137. Wang, J. et al., Cdk5 activation induces hippocampal CA1 cell death by directly phosphorylating NMDARs, *Nat. Neurosci.*, 6, 1039, 2003.
138. Wang, K. et al., Regulation of the neuronal nicotinic acetylcholine receptor by SRC family tyrosine kinases, *J. Biol. Chem.*, 279, 8779, 2004.
139. Wang, L.Y. and MacDonald, J.F., Modulation by magnesium of the affinity of NMDARs for glycine in murine hippocampal neurones, *J. Physiol.*, 486 (Pt 1), 83, 1995.
140. Wang, L.Y. et al., Regulation of NMDARs in cultured hippocampal neurons by protein phosphatases 1 and 2A, *Nature*, 369, 230, 1994.
141. Wang, L.Y., Salter, M.W., and MacDonald, J.F., Regulation of kainate receptors by cAMP-dependent protein kinase and phosphatases, *Science*, 253, 1132, 1991.
142. Wang, Y.T. and Salter, M.W., Regulation of NMDARs by tyrosine kinases and phosphatases, *Nature*, 369, 233, 1994.
143. Wang, Y.T., Yu, X.M., and Salter, M.W., Ca(2+)-independent reduction of N-methyl-D-aspartate channel activity by protein tyrosine phosphatase, *Proc. Natl. Acad. Sci. USA*, 93, 1721, 1996.
144. Weerapura, M. et al., C-terminal Src kinase (Csk) associates with N-methyl-D-aspartate (NMDA) receptors and controls excitatory synaptic transmission, *Soc. Neurosci. Abstr.*, 32, 624.11, 2006.
145. Westphal, R.S. et al., Regulation of NMDARs by an associated phosphatase-kinase signaling complex, *Science*, 285, 93, 1999.
146. Woolf, C.J. and Salter, M.W., Neuronal plasticity: increasing the gain in pain, *Science*, 288, 1765, 2000.
147. Woolf, C.J. and Salter, M.W., Plasticity and pain: role of the dorsal horn, in *Wall and Melzack's Textbook of Pain*, 5th ed., McMahon, S.B. and Kotzenburg, M., Eds., Elsevier Churchill Livingston, London, 2006, p. 91.

148. Xiong, Z.G. et al., Regulation of N-methyl-D-aspartate receptor function by constitutively active protein kinase C, *Mol. Pharmacol.*, 54, 1055, 1998.
149. Yaka, R. et al., NMDAR function is regulated by the inhibitory scaffolding protein, RACK1, *Proc. Natl. Acad. Sci. USA*, 99, 5710, 2002.
150. Yu, X.M. et al., NMDA channel regulation by channel-associated protein tyrosine kinase Src, *Science*, 275, 674, 1997.
151. Yu, X.M. and Salter, M.W., Gain control of NMDA-receptor currents by intracellular sodium, *Nature*, 396, 469, 1998.
152. Zheng, X. et al., Protein kinase C potentiation of N-methyl-D-aspartate receptor activity is not mediated by phosphorylation of N-methyl-D-aspartate receptor subunits, *Proc. Natl. Acad. Sci. USA*, 96, 15262, 1999.
153. Zou, X., Lin, Q., and Willis, W.D., Enhanced phosphorylation of NMDAR 1 subunits in spinal cord dorsal horn and spinothalamic tract neurons after intradermal injection of capsaicin in rats, *J. Neurosci.*, 20, 6989, 2000.
154. Zou, X., Lin, Q., and Willis, W.D., Role of protein kinase A in phosphorylation of NMDAR 1 subunits in dorsal horn and spinothalamic tract neurons after intradermal injection of capsaicin in rats, *Neuroscience*, 115, 775, 2002.
155. Zou, X., Lin, Q., and Willis, W.D., Effect of protein kinase C blockade on phosphorylation of NR1 in dorsal horn and spinothalamic tract cells caused by intradermal capsaicin injection in rats, *Brain Res.*, 1020, 95, 2004.

8 Trafficking and Targeting of NMDA Receptors

Ronald S. Petralia, Rana A. Al-Hallaq,
and Robert J. Wenthold

CONTENTS

8.1 INTRODUCTION

At glutamatergic synapses, NMDA receptors (NMDARs) are localized with other ionotropic glutamate receptors [AMPA receptors (AMPARs) and kainate receptors] and
with metabotropic glutamate receptors. Targeting the necessary number of NMDARs
to the proper sites at synapses is critical for normal glutamatergic neurotransmission
and synaptic plasticity. Additional diversity of NMDAR responses arises from the complexity of subunit composition and variations in localization. Thus, the mechanisms
of NMDAR trafficking and targeting must address the complex needs of neurons. For

example, NMDARs must be transported to different subcellular sites because some receptors are localized to synaptic sites (pre- and postsynaptic) while others are localized extrasynaptically.[1–4]

NMDAR subunit expression differs as a function of brain region and developmental age.[5–7] Subunit composition may determine subcellular localization, i.e., in adults, NR2A-containing receptors are enriched at synapses while extrasynaptic receptors are predominantly NR2B-containing complexes.[1,2,4] Furthermore, evidence indicates the incorporation of more than one type of NR2 subunit in each complex (tri-heteromeric NR1/NR2X/NR2Y) such as NR1/NR2A/NR2B receptors in hippocampal neuron synapses, NR1/NR2A/NR2C in cerebellar granule cell synapses, and NR1/NR2B/NR2D in substantia nigra dopaminergic neurons.[8] Finally, different NMDARs may be expressed at different synapses within the same neuron.[4] Thus, the mechanisms of NMDAR trafficking must be varied and well regulated to meet the needs of neurons.

8.2   NMDA RECEPTOR TRANSPORT THROUGH THE BIOSYNTHETIC PATHWAY, AT THE SYNAPSE, AND FOLLOWING INTERNALIZATION

8.2.1   Processing in the Endoplasmic Reticulum

Generally, membrane proteins are made in the endoplasmic reticulum (ER) and then subjected to various kinds of quality control that inhibit export of misfolded or otherwise imperfect protein molecules. For multimeric proteins such as ion channels, assembly of the monomers occurs in the ER in which mechanisms prevent export of monomers and incompletely assembled complexes. Thus, subunits of NR1, NR2, and NR3 come together in various combinations to form tetramers that generate the NMDAR ion channel complex. These subunits are retained in the ER until they assemble.

An ER retention and retrieval factor with a 3-amino acid residue motif (RXR) was identified in the intracellular C terminal region of NR1. Similar retention factors in other subunits are likely present but have not been identified (NR2B[9] and NR3B[10]; see below). When subunits join to form a complex, the ER retention must be negated by some mechanism such as by steric masking of the ER retention site or the presence of an export signal that somehow overrides the ER retention function. Four of the eight variants of NR1 contain C1 cassettes (alternatively spliced exons of C terminal regions of NR1). This cassette includes the RXR retention motif (RRR[10–14]). When the NR1-1 variant containing this C1 cassette is expressed in heterologous cells, it is retained in the ER.[11] In contrast, expression of any of the other three variants (NR1-2, NR1-3, or NR1-4) causes trafficking of these transfected proteins to the cell surface.

Trafficking to the cell surface is expected for both NR1-2 and NR1-4 because they lack C1 cassettes (with ER retention motifs). NR1-3 contains C1 but still travels to the surface, because it also carries a C2′ cassette that contains a signal that can mask or override the ER retention mechanism. The last six amino acids of C2′ including the PDZ (PSD-95/Dlg/ZO-1) binding domain STVV are sufficient to suppress ER retention. In addition, soluble fusion proteins that contain the PDZ binding domain of C2' block the surface expression of NR1-4 (presumably because the PDZ proteins are saturated by the mutant molecule). This suggests that some kind of PDZ-containing protein normally binds to the C2'-containing NR1 subunits (NR1-3 and NR1-4) very early in the secretory pathway, perhaps at ER exit sites.[15]

Other factors also may affect release of NR1 subunits from ER retention. PKC phosphorylation of serines near the RXR ER retention motif of the C1 cassette relieves ER retention and elicits robust surface expression of NR1.[12] This also may involve phosphorylation of an adjacent serine by PKA and coordination of the actions of PKA and PKC.[16] However, relief from ER retention elicited by the C2' cassette may not be due to the binding of a PDZ protein to the STVV C terminus but may involve an alternative mechanism. The valine residues in the C terminus may bind COPII proteins that are found at ER exit sites; this also may be a common mechanism for ER exit of integral membrane proteins with type I PDZ-binding motifs (T(S)XV).[17]

Exit mechanisms of NMDARs from the ER probably involve the association of the NR1 with NR2 and/or NR3 subunits; in neurons, these combinations may exit the ER and traffic to the plasma membrane. For example, in CA1 pyramidal neurons in mice with NR1 deletions restricted to the hippocampal CA1 region, an aggregation of NR2 subunits in intracisternal granules of the ER[18] was noted. This supports other studies that show that NR2 subunits are retained in the ER in the absence of NR1.[19]

In expression studies in heterologous cells, homomeric NR3A complexes and NR2A/NR3A complexes are retained in the ER; only heteromeric complexes that contain NR1 can reach cell surfaces (they used the NR1-1a variant, which is retained in the ER when transfected singly in heterologous cells).[20] NR3B interactions with other NMDAR subunits are probably similar.[10] Thus, although lone subunits can manifest effective ER retention mechanisms, these mechanisms somehow are suppressed or overridden when subunits combine in proper heteromeric complexes. Presumably, retention signals may be masked and/or exit signals may be enhanced due to sustained conformational changes that occur with quaternary folding of the complex.

One potential exit signal for NR1/NR2 complexes is the HLFY motif found in the proximal C terminal region immediately following the last transmembrane domain of the NR2 subunit. This motif mediates exit of assembled NMDARs from the ER.[9] Even if this motif is mutated, NR1 and NR2 subunits still assemble into functional complexes in the ER, although they cannot exit. Yang et al.[21] demonstrated that the HLFY motif is not necessary for ER exit; surface expression occurs even when the motif is replaced by alanines as long as the remainder of the C terminus is absent. Perhaps HLFY provides only a structural role to ensure proper orientation of the C terminus, rather than serving as an export motif involved in overriding ER retention.

8.2.2 Transport in the Golgi, Trans-Golgi Network and Dendrites

When NMDAR-containing complexes exit the ER, they are modified in the Golgi apparatus and then sorted in the trans-Golgi network (TGN) where they are packaged into different kinds of vesicular or tubulovesicular carriers (Figure 8.1). From there, NMDARs may be transported directly to plasma membranes or may enter

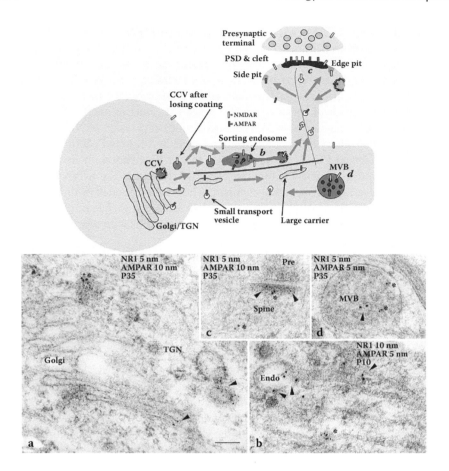

FIGURE 8.1 Trafficking of NMDARs from Golgi/TGN to synapse (top) with corresponding EM figures (bottom; a–d matches *a–d* in diagram). Newly synthesized NMDARs and AMPARs are sequestered in carriers that traffic from Golgi/TGN and originate as clathrin-coated vesicles (CCVs) or non-coated carriers (vesicles or large tubulovesicular carriers). Some selective separation of NMDARs and AMPARs may occur at this point (a). These carriers may fuse directly with plasma membranes or sorting endosomes (where they may mix with recycling receptors) for further selective processing (b). Carriers originating from TGN or endosomes travel on microtubule pathways (thick line) in dendrites and then along actin pathways (thin line) into the spine (c). These receptors are also present in extrasynaptic areas of plasma membranes and may traffic along the membranes. Ultimately, receptors targeted for degradation are sequestered in vesicles inside late endosome and multivesicular bodies (MVBs; d). The EM examples are double immunogold labeled sections of the CA1 stratum pyramidale/radiatum area of the hippocampus, labeled for NR1 (5 nm gold in a and c and 10 nm gold in b and d) and AMPARs (10 nm gold in a and c and 5 nm gold in b and d). NR1 labeling (arrowheads) is associated with CCVs at the TGN in micrograph a and in sorting endosomes in b. AMPARs (asterisks) are associated with a large, tubulovesicular organelle near the TGN in a and probably a smaller vesiculate structure in b. Pre = presynaptic terminal. Scale bar = 100 nm.

endosomes. Transport of NMDAR-containing carrier structures occurs via kinesin motor proteins on polarized microtubules. Nascent NMDARs that reach endosomes may mix with NMDARs recycled from the surface, and additional sorting occurs here as in the TGN. Eventually, NMDARs reach the surfaces of neurons; the final stage probably involves myosin motors that move along actin filaments. Nascent NMDARs may first enter the extrasynaptic membranes and then proceed to synapses, although they also may enter the postsynaptic membrane directly (see Section 8.2.3.1).

For NMDARs and neuronal proteins in general, many of these trafficking steps are not clear. Studies in young cortical neuron cultures elucidated some of the intermediate steps in trafficking, but many steps remain poorly understood.[22–24] Both NMDARs and AMPARs travel via mobile transport packets (various vesicular or tubulovesicular structures) in neurons before and during synaptogenesis. Typically mobile NMDAR clusters lack AMPARs and move rapidly as compared to those that contain AMPARs but lack NMDARs. At least two kinds of carrier vesicles or tubulovesicular organelles probably exist and one may be specialized for rapid trafficking of NMDARs. In these young cortical neurons, prior to synapse formation, NMDARs travel along dendrites and go through cycles of exocytosis to (dependent on SNARE protein SNAP-23) and endocytosis from dendrite surfaces.[23] Transport in the dendrites is via large tubulovesicular organelles that move along microtubules and contain the early endosomal antigen 1 (EEA1) marker. While this distinctive sequential cycling of nascent NMDARs to and from the surfaces occurs specifically for a presynapse stage, it has not been demonstrated for trafficking of nascent NMDARs following synapse formation or in adults. Many of these findings are supported by preliminary immunogold studies in the developing and adult hippocampus.[25,26]

These studies indicate that selective cargo sorting of NMDARs and AMPARs occurs at the TGN and perhaps also at endosomes. In fact, some recycling endosomes contain AMPARs along with their associated PDZ proteins, GRIP and PICK, but no NMDARs.[27] Little evidence indicates that NR2A-containing NMDARs sort differently from NR2B-containing NMDARs at the TGN. Comparisons of binding of NR2A and NR2B C termini with adaptor proteins that may be involved in this transport show only minor differences between the two NR2 subunits.[28] The adaptor subunits that bind to NR2 subunits include $\mu1$, $\mu3\alpha$, and $\mu4$, and are found in adaptor proteins AP-1, AP-3, and AP-4, respectively.[29] Both $\mu1$ and $\mu3\alpha$ are involved in clathrin-mediated cargo selection and transport from the TGN and endosomes; $\mu4$ is involved in similar functions at the TGN, but clathrin is not involved. NR2A and NR2B C termini bind strongly and equally to $\mu1$ and $\mu4$, while they bind relatively weakly to $\mu3\alpha$. Binding of $\mu3\alpha$ is more prevalent for NR2B than for NR2A, suggesting sorting differences.

Trafficking of NMDARs to cell surfaces and then to synapses employs a large complex of proteins.[30,31] One protein of the complex is synapse-associated protein 102 (SAP102), a common membrane-associated guanylate kinase (MAGUK; see Section 8.2.3.3.1) well known for its association with major NMDARs of the early postnatal forebrain NR1/NR2B complexes. NMDARs travel in dendrites in large tubulovesicular organelles, probably early endosomes, containing SAP102.[23] Sans et al.[30,31] showed that NMDARs are carried on membranes of transport vesicles or tubulovesicular carriers in which the NMDAR is bound to SAP102 (or other MAGUKs) that in turn binds the exocyst component Sec8. SAP102 also binds mPins (mammalian homologue of *Drosophila melanogaster* partner of inscuteable) that binds to the G protein subunit, $G\alpha_i$. Thus mPins and $G\alpha_i$ may exist as a complex with SAP102 and NR2B and influence NMDAR trafficking.[30,31] Binding of mPins to $G\alpha_i$ may mediate G protein signaling by inhibiting binding of $G\alpha_i$ to $G\beta\gamma$ (other G-protein subunits) and thus enhancing $G\beta\gamma$ signaling. This complex of proteins consisting of NR1, NR2B, SAP102, Sec8, mPins, and perhaps also $G\alpha_i$ may form in early stages of the secretory pathway such as in the ER, Golgi, or TGN.

The exocyst or Sec6/8 complex consists of eight proteins and has been studied in both yeast and mammalian cells. It may direct intracellular membrane vesicles to their sites of fusion with plasma membranes. Both Sec8 and the NR2 subunits of NMDARs bind to the same region of SAP102, primarily the first and second PDZ domains.[30] A dominant-negative form of Sec8 that lacks a PDZ-binding domain may be used to block the interaction of Sec8 with SAP102; this prevents delivery of NMDARs to the cell surface. Thus, surface delivery of NMDARs requires both SAP102 and Sec8. However, when the PDZ-binding domain of the NR2B subunit is deleted, this NMDAR can be delivered to the cell surface by a mechanism independent of the exocyst and MAGUKs, although these mutated NR2B-containing NMDARs cannot enter synapses.[30,32] At least for NR2B-containing NMDARs, such a phenomenon may allow extrasynaptic NMDARs to be sorted independently from synaptic NMDARs, possibly requiring a MAGUK association for entry into synapses.

mPins is the mammalian form of *Drosophila* Pins that plays a role in cell polarity and cell division. It interacts with the Src-homology-3 (SH3)–guanylate kinase (GK) domains of SAP102, and the interaction influences NMDAR trafficking in neurons. Expression of dominant-negative constructs of mPins in hippocampal neurons in culture decreases native SAP102 in dendrites. For transfected NR2B constructs, the dominant-negative constructs and short interfering RNA (siRNA)-mediated knockdown of mPins reduce the density of surface NMDAR puncta and the intensity of staining per labeled surface punctum. Thus, the mPins–SAP102 complex may promote efficient targeting of NMDARs to cell surfaces. SAP102 and mPins probably bind together in a closed or inactive state; this complex then may be opened via binding of NMDARs. Interaction of $G\alpha_i$ with mPins in this complex also may help stabilize the complex, allowing it to be properly folded and thus to reach the cell surface.

Trafficking of the complex containing NR1, NR2B, SAP102, Sec8, and mPins likely associates with kinesin motors to travel along microtubules in dendrites.[33] NR2B can bind to a complex of other proteins (mLin-7, mLin-2, mLin-10) linking the NMDAR to the kinesin KIF17.[34,35] This complex then mediates the transport of the NMDAR bound in the membrane of the carrier vesicle or tubulovesicular organelle along the dendrite. Knockdown or blockage of KIF17 impairs the expression (as determined by Western blot) and synaptic localization of NR2B, and is followed by a corresponding increase in NR2A at synapses. Thus, replacement of

NR2B- with NR2A-containing receptors may be subject to reciprocal control with NR2A-containing receptors replacing NR2B-containing ones at the synapse.

Presumably, this involves a different mechanism of transport for NR2A than for NR2B. However, it seems unlikely that such a large complex may be joined to the complex of NR1, NR2B, SAP102, Sec8, and mPins to mediate transport of all NMDARs. This suggests more direct links between motors and the latter complex. Another kinesin known as KIF1Bα is associated with several MAGUKs including PSD-95 and SAP97 and the related S-SCAM protein. Its role in NMDAR transport is unknown.[36] KIF1Bα may be a good candidate for motor movement of the complex containing NR1, NR2B, SAP102, Sec8, and mPins because of its direct association with NMDAR-binding MAGUKs. As noted above Section 8.2.2, NMDARs traveling via kinesins on microtubule tracks may switch to myosin motors on actin filaments for final transport to cell surfaces[37] and transports within postsynaptic spines.[38] NMDARs also are regulated by myosin light chain kinase, probably indirectly via effects on actomyosin,[39] and directly interact with myosin regulatory light chains.[40]

Interestingly, while PSD-95 and PSD-93 appear to be mainly postsynaptic, both SAP97 and SAP102 are found in both pre- and postsynaptic structures. This is based on several studies[7,41–43] and the authors' unpublished data. Furthermore, chimeras of PSD-95 containing SAP102 N termini can traffic to both dendrites and axons, thus imitating the distribution of SAP102.[42] This suggests that NMDAR trafficking may be controlled to an extent by the selection of different associated MAGUKs. This may be related to the developmental switch in MAGUKs (mainly SAP102 to PSD-95/PSD-93) associated with the switch in NR2 subunits (NR2B to NR2A), as noted in Section 8.2.3.3.4. Also, NR2B may be one of the major presynaptic NMDAR subunits (Section 8.2.3.5). This may also relate to the association of SAP102 and NR2B-containing NMDARs.

Typically, nascent NMDARs traffic to synapses by traveling from ER export sites, through the Golgi/TGN, and into carriers, beginning typically in cell bodies, then continuing along the dendrites and then to the synapses. However, some NMDARs may be released from ER export sites within dendrites[44,45] or even synthesized locally near synapses from mRNAs transported from cell bodies.[46] Perforant path transection induces trafficking of NR1 mRNA into the dendrites of dentate gyrus granule cells, probably in response to increased terminal proliferation and sprouting.[47] Substantial evidence indicates that some cytoplasmic proteins associated with NMDAR function such as CaMKII or Arc may be synthesized locally to achieve more precise control at individual synapses.[46]

8.2.3 TRAFFICKING TO AND FROM SYNAPSES

8.2.3.1 Exocytosis and Lateral Movement

Exocytosis of glutamate receptors, including NMDARs, occurs at or near synapses. The exact locations are not clear.[48] Pit-like structures labeled for AMPARs and lacking evident clathrin coats are seen often on the sides of spines in adults.[25,49,50] They resemble noncoated pits described in structural studies[51] of what may be exocytotic or noncoated endocytotic sites. Preliminary data indicate that SNARE, synapse-associated protein 23 (SNAP-23), is concentrated in these areas, suggesting that these are sites of exocytosis or contain lipid rafts or caveoli involved in receptor regulation.[52–54]

While these sites of possible exocytosis may be specific for AMPARs,[55] our preliminary studies indicate that vesicles and pits labeled with immunogold for AMPARs and NMDARs are also in the perisynaptic regions at the sides of PSDs and possibly within PSDs.[25,26,56] Newly exocytosed NMDARs initially may form extrasynaptic clusters,[57] or may be incorporated more directly into synapses, presumably via actin/myosin-mediated transport (Section 8.2.2 discusses actin and myosin).[35] NMDARs do not need to enter synapses directly since they are very mobile in surface membranes.[58–60] Changes in neuronal activity modify AMPAR mobility but not NMDAR mobility, although activation of PKC modifies mobility for both.[59,60]

Studies of lateral movements and mobility between synapses and extrasynaptic sites are problematic due to technical constraints. All these studies utilized light microscope techniques to track single molecule movements on cell surfaces and are thus subject to limitations in accuracy of delineation of synaptic versus extrasynaptic localizations. Such problems can be resolved with electron microscope (EM) analysis. The authors used EM to examine extrasynaptic localization of AMPARs[49] and NMDARs (unpublished data) in neurons. However, EM studies may present problems in localization of labels due to "bleeding" of DAB or the relatively large sizes of the gold particles that may have difficulty entering synaptic clefts, as noted for quantum dots in the light microscope studies cited above. Of course, EM studies do not allow for live imaging as light microscope studies do.

8.2.3.2 Extrasynaptic NMDA Receptors

Typically, extrasynaptic NMDARs in the vicinities of mature synapses tend to be mainly NR2B-containing receptors. The synapses mainly have NR2A-containing NMDARs.[2,61,62] The compositions of synaptic NMDARs vary throughout the brain. Thus, cerebellar granule cells lose all their NR2B by later stages of postnatal maturation; adults have NR2A and NR2C. In these cells, even the extrasynaptic NR2B-containing NMDARs must eventually be replaced by other kinds of NMDARs.[63,64] Generally, NR2A expression increases during development and/or activity, and NR2B-containing NMDARs tend to be more readily removed from synapses than NR2A-containing NMDARs.[62,65]

NR2B-containing NMDARs appear to be more mobile than NR2A-containing NMDARs.[66] NR2A-containing NMDARs also show preferential binding to PSD-95 that may further limit their internalization from synapses.[62] Nevertheless, some NR2B still is found in synapses in adults,[67] as NR1/NR2B or NR1/NR2A/NR2B. These findings are supported by functional studies of developing neurons *in vitro*.[68] Conversely, NR2A-containing NMDARs may be localized extrasynaptically.[63,68]

Nevertheless, this predominant separation of NMDARs into synaptic NR2A-containing and extrasynaptic NR2B-containing receptors in adults must have some function. Synaptic receptors may be activated by precise release of glutamate at synapses, while extrasynaptic receptors should be activated only after extensive release of glutamate followed by spillover into extrasynaptic spaces. Thus, extrasynaptic

NMDARs may be adapted to elicit plastic changes to compensate for synapse overactivity.

NR2A was proposed to be associated mainly with long-term potentiation (LTP) and NR2B mainly with long-term depression (LTD)[69,70] although some studies show that both may induce LTP.[71,72] Results for NR2A are not clear due to problems in specificity of the antagonists used in these studies. Interestingly, in rat olfactory bulb granule cells, activation of extrasynaptic NMDARs generates inhibitory currents via BK-type calcium-activated potassium channels.[73] Also, synaptic and extrasynaptic NMDARs can mediate opposite long-term changes in neuronal gene expression, probably due to differences in their local associated signaling complexes.[74–76] The signaling pathways of synaptic NMDARs selectively activate a regulatory cascade involving Ras, ERK, and CREB (cAMP-responsive element-binding protein), while the pathways of extrasynaptic NMDARs inactivate them. Thus, synaptic NMDARs promote CREB activation and induction of BDNF gene expression, while extrasynaptic ones shut off CREB and inhibit BDNF gene expression.[74]

Activation of synaptic NMDARs may upregulate pro-survival genes and down-regulate pro-death genes, while activation of extrasynaptic NMDARs does the exact opposite.[75,76] Activation of extrasynaptic NMDARs by glutamate spillover at cerebellar parallel fiber stellate cell synapses induces the switch from GluR2-lacking calcium-permeable AMPARs to GluR2-containing calcium-impermeable ones. This requires activation of protein kinase C (PKC).[77] An extreme case involves the synapses of retinal ganglion cells in which most NMDARs are extrasynaptic; indeed, mEPSCs are mediated solely by AMPARs.[78] Finally, a more general function of extrasynaptic NMDARs is formation of a mobile reserve pool of receptors for interchange with those in synapses (see Section 8.2.3.1).[60] Organization of extrasynaptic NMDARs is not well understood but may involve associations with other proteins that affect NMDAR trafficking, as indicated in preliminary studies of GIPC (see Section 8.2.3.6).[79–81]

8.2.3.3 Synaptic Localization

In the PSD—an electron-dense structure that lies along the postsynaptic membrane—NMDARs are associated directly or indirectly with a dense network of proteins.[14,82,83] One prevalent group in the PSD is the MAGUK family that includes PSD-95, PSD-93, SAP97, and SAP102.[84] NMDARs also interact directly and indirectly with adhesion molecules that play a role in synapse formation, maturation, function, and plasticity.[85] Interactions with actin-associated proteins are important for proper trafficking, anchoring, and stabilization of NMDARs and PSD components in the spine. This protein complex in the PSD has a role in normal synaptic neurotransmission and synaptic plasticity, and its disruption may contribute to synaptic dysfunction and even cell death.

8.2.3.3.1 PDZ Domain Proteins That Interact with NMDA Receptors
The MAGUKs are among the most abundant components of PSDs.[84] The PSD-95 family composed of PSD-95, PSD-93, SAP97, and SAP102, has three PDZ domains, an SH3 domain, and a GK domain. Other large, PDZ domain-containing scaffolding molecules including synaptic scaffolding molecules (S-SCAM)[86] and

channel-interacting PDZ domain protein (CIPP) were reported to associate similarly with NMDARs.[87]

The PSD-95-related MAGUKs are abundant at glutamatergic synapses although the types vary with age (Section 8.2.3.3.4), suggesting that they play important roles at synapses. Two proposed functions include (1) anchoring and clustering NMDARs at synapses and (2) serving as large scaffolding molecules that link proteins and signaling molecules in and around the PSD (see Section 8.2.3.3.2).[14,82,88] MAGUKs can bind directly to NMDARs via interaction of the MAGUK PDZ domains with the PDZ-binding domains located at the extreme C termini of the NR2 subunits and the NR1 isoforms containing C2′ alternatively spliced cassettes.[11,14,82,88] Interestingly, NMDARs are less affected by changes in MAGUK expression than AMPARs. Knock-down of PSD-95 by small hairpin RNA (shRNA) produces no change in NMDARs with decreases in AMPARs[89] or leads to a moderate reduction in NMDA EPSCs.[90]

While NMDARs can bind to MAGUKs, the significance of this interaction for synapses, especially mature synapses, is unclear. Animal model studies have shown that the NR2 C terminus is required for synaptic localization of NMDARs.[91,92] The entire C terminus was deleted, and the reported effects may be attributed to loss of components of the C terminus other than the PDZ-binding domain. The role of the MAGUKs, particularly SAP102, in the trafficking of NR2B-containing receptors, is better established (Section 8.2.2).[30]

Binding of NR2B to SAP102 or PSD-95 is required for synaptic localization of NR2B-containing NMDARs (Sections 8.2.2 and 8.2.3.2).[30,93] Transfection studies of wild-type and mutant NMDAR subunits demonstrated that NR2B-containing receptors depend on the PDZ-binding domain for entry and/or retention at synapses.[93] This interaction may be regulated by phosphorylation of residues within the PDZ-binding domain of NR2B by casein kinase II (CK2) in an activity-dependent manner (see Section 8.2.3.4.4).[94]

The role of the direct association of MAGUKs with NR2A in trafficking, particularly in more mature synapses, has not been clearly established. In contrast to NR2B, NR2A-containing receptors do not require the PDZ-binding domain for entry and/or retention at synapses.[68,93,95] Overexpression of PSD-95 in hippocampal neurons does not affect synaptic clustering or functioning of NMDARs[96,97] or change NMDAR EPSCs in hippocampal slices.[97] In contrast, overexpression of PSD-95 in cerebellar granule cells promotes the synaptic insertion of NR2A-containing receptors, produces faster NMDAR EPSCs, and appears to favor the changeover from NR2B to NR2A generally seen during maturation.[98] The discrepancies among studies may arise from higher expression of PSD-95 in the hippocampus compared to the cerebellum at the ages studied.[98]

The significance of the MAGUKs as related to NMDAR function has been investigated in transgenic animals.[14] PSD-95 and PSD-93 and SAP102[101] knockout mice exhibited no significant losses of synaptic NMDARs. However, mice lacking both PSD-95 and SAP102 are not viable.[101] Double knockdowns of PSD-95 and PSD-93 resulted in reductions in NMDAR-mediated currents.[89] These results indicate that compensatory mechanisms may stabilize NMDARs at synapses and make analysis of MAGUK function more challenging.

SAP102 expression is elevated in PSD-95 mutant mice, and more PSD-95 co-immunoprecipitates with NR1 in SAP102 knockout mice.[101] This further supports the view that MAGUK proteins act in a compensatory manner. NMDAR localization at synapses in PSD-95 mutant mice is not affected, but these mice have impaired spatial learning and show enhanced LTP at different frequencies of synaptic activation than those found in normal mice. These results support the proposed role of PSD-95 as an anchor bringing certain molecules in close proximity to facilitate particular mechanisms of plasticity.

A role for PSD-95 in LTP is supported by PSD-95-overexpression studies in which the ability to generate LTP is occluded and induction of LTD is enhanced.[102] As noted, the effects of overexpression of PSD-95 on NMDARs are less conclusive and therefore may arise from changes in AMPARs. Single knockout of SAP102 results in impaired spatial learning memory, altered LTP, and altered MAP kinase-mediated signaling.[101] Mice lacking PSD-93 show impairments in some systems but not in others, possibly due to compensation by other MAGUKs. In cerebellar Purkinje cells, in which the only definitive PSD-95-related MAGUK is PSD-93, PSD-93 knockout did not affect the development or function of parallel fiber synapses, Purkinje cells, or cerebellum-dependent behaviors, consistent with the presence of compensatory mechanisms.[100,103] In contrast, in mice lacking PSD-93, surface expression of NR2A and NR2B was reduced and NMDAR-mediated postsynaptic function was impaired in spinal dorsal horn and forebrain.[104] Interestingly, these mice exhibited blunted NMDAR-dependent persistent pain.[104]

S-SCAM, a member of the membrane-associated guanylate kinase with inverted orientation (MAGI) family, links NMDARs to the MAP kinase pathway by binding to neural GDP/GTP exchange protein for Rap1 small G protein (nRAP GEP).[105] S-SCAM (or PSD-95) also binds to membrane-associated guanylate kinase interacting protein (MAGUIN)[106] which affects cell polarity.[107] Similarly, CIPP may serve as a scaffolding protein, as it has been shown to bind the Kir 4.0 potassium channels, neuregulin, neurexins, NMDARs,[87] serotonin 5-HT_{2A} receptor,[108] and acid-sensing ion channel 3 (ASIC3).[109] CIPP has a limited distribution that resembles that of NR2C and NR2D; it is abundant only in the thalamus, colliculus, cerebellum, and brain stem.[87]

8.2.3.3.2 NMDA Receptors and MAGUK-Associated Proteins of the PSD

While the role of a direct association between PSD-95 with NR2A in mature synapses has yet to be determined, it is likely that PSD-95 is a major component of the scaffold of interlinked proteins that comprise the PSD. This protein complex includes kinases, GTPase activators and inhibitors, and cytoskeleton-associated proteins (CAPs) that link the surface receptors and cytoplasmic proteins.

MAGUKs bring NMDARs into contact with signaling molecules such as neuronal nitric oxide synthase (nNOS), which can mediate NMDAR-induced excitotoxicity[110] and synaptic Ras-GTPase-activating protein (SynGAP) which couples NMDARs to the MAP kinase pathway.[111] PSD-95 can bind nNOS via its PDZ2 domain and the NMDAR via its PDZ1 domain; calcium influx through the NMDAR can thus lead to specific activation of nNOS and the production of NO that may modulate NMDAR signaling and underlie neuronal excitotoxicity.[110]

Several proteins associated with the MAP kinase pathway bind to PDZ domains of scaffolding proteins in NMDAR-containing complexes. SynGAP binds to PSD-95,[111,112] nRap GEP binds to S-SCAM,[105] and MAGUIN binds to both PSD-95 and S-SCAM.[106] SynGAP maintains a low steady-state level of active Ras near synapses by catalyzing the hydrolysis of Ras-GTP to Ras-GDP. Calcium entry via NMDARs can activate CaMKII, which phosphorylates and thus inactivates Syn-GAP.[112] As a result of SynGAP inactivation, Ras-GTP accumulates and increases activation of the MAP kinase cascade associated with LTP.[112]

A proposed alternative model indicates that SynGAP may affect synaptic function through the following chain of interacting proteins: SynGAP→PDZ domain protein→MUPP1→CaMKII→NMDAR.[113] NMDAR-mediated calcium influx causes dissociation of CaMKII from this complex, leading to dephosphorylation of SynGAP, inactivation of Rap, decreased p38 MAP kinase activity, and subsequent AMPAR removal from synapses.[113]

SynGAP is selectively associated with NR2B-containing NMDARs,[114] and its expression is high at excitatory synapses in the hippocampus at postnatal day (P) 2, when most NMDARs contain NR2B (see Section 8.2.3.3.4).[67] The SynGAP–NR2B association couples NR2B to inhibition of the Ras-ERK pathway that may underlie the removal of AMPARs from synapses, causing a weakening of synaptic transmission.[114] Interestingly, NR2A-containing NMDARs have the opposite effect: activating the Ras-ERK pathway and promoting surface delivery of GluR1.[114] Consistent with these findings, synaptic NMDARs activate ERK while extrasynaptic NMDARs inactivate it (see Section 8.2.3.2).

In addition to tethering NMDARs with signaling proteins, MAGUKs bring NMDARs in contact with regulatory proteins. For example, Fyn association with MAGUKs enhances Fyn phosphorylation at NR2B Tyr-1472, disrupting the interaction of NR2B with AP-2, which ultimately inhibits internalization of NR2B-containing NMDARs (see Section 8.2.3.6).[93,115] Also, a PSD-95 molecule in close proximity to an NR2A-containing NMDAR can regulate phosphorylation of NR2A via Fyn, Src, and other kinases, regulating receptor function and/or trafficking. Selective localization of NR2A-containing NMDARs at synapses, in the absence of direct NR2A–MAGUK binding (assuming diheteromeric NR1/NR2A complexes[116]) may rely on other proteins that can bind to the NMDAR such as adhesion factors that bind to NR1 (see Section 8.2.3.3.3) and CaMKII (see Section 8.2.3.4.1).[117,118] CaMKII, like the MAGUKs, is one of the most abundant proteins of and interacts with multiple components of the PSD.

The scaffold created by PDZ-containing proteins can link NMDARs to other glutamate receptors and ion channels. Thus, guanylate kinase-associated protein (GKAP) binds to the GK domains of MAGUKs and S-SCAM.[119] GKAP binds to Shank that binds to Homer dimers that may be associated with perisynaptic metabotropic glutamate receptors and TRPC cation channels.[38,67,120] Shank is a multifaceted scaffolding protein; it can bind through Homer dimers to inositol 1,4,5-triphosphate (IP3) receptors in smooth ER cisternae that extend into the spine, directly to metabotropic and delta glutamate receptors, and indirectly to AMPARs via the PDZ protein GRIP.[121,122]

Proteins also link NMDARs and AMPARs to MAGUKs, including PSD-95, which links to AMPARs via TARP members such as stargazin.[123,124] However, coimmunoprecipitation does not reveal evidence of an NMDAR–AMPAR association.[7] Additional possible links of AMPARs and NMDARs occur via CaMKII and an assembly of AMPAR-associated proteins, including SAP97 (binds to GluR1), 4.1N protein, actinin, and actin.[117]

NMDAR complexes are linked to the actin cytoskeleton of the synaptic spine.[38] Actin filaments form pathways for transport in the spine to and from the postsynaptic membrane and are responsible for the structural integrity of the postsynaptic spine. At least four such types of connections exist: NMDARs–actinin–actin,[125] GKAP–Shank–cortactin–actin, PSD-95–SPAR–actin, and PSD-95–citron.[126–128] SPAR is one of the Rap-specific GTPase-activating proteins (GAPs) implicated in regulation of MAP kinase cascades, cell adhesion, and activation of integrins.[129] It binds to the GK domain of PSD-95 and regulates spine morphology via direct interaction with F-actin and possibly also via Rap signaling.[129]

Citron is a target of Rho, which can regulate actin cytoskeleton organization. Citron is limited in distribution to certain specialized neurons and may mediate forms of NMDAR-dependent synaptic plasticity in these neurons. NMDARs can also associate with α1-chimerin, an inhibitor of Rac1,[130] that may indirectly regulate actin in spines. Thus, after the binding of α1-chimerin to the NR2A subunit, α1-chimerin inactivates local Rac1 (Rho GTPase family member) via its GAP domain. Rac1 can promote actin polymerization, increase dendrite arbor complexity, and stimulate spine formation.[130] These interactions suggest a mechanism for NMDAR activity to control overall spine structure (see Section 8.2.3.3.4). Although actin–protein associations play important roles in synaptic structure and function, anchoring of NMDAR–PSD-95 complexes at synapses is only partially dependent on actin, indicating that an actin-independent component also is involved in NMDAR synaptic localization.[131]

8.2.3.3.3 Adhesion Proteins Associated with NMDA Receptors

A number of adhesion proteins are localized to the pre- and postsynaptic membranes and regulate NMDAR-containing complexes. NMDARs interact directly and indirectly with these proteins, including some that act transsynaptically to link NMDARs to presynaptic terminals. Adhesion molecules play a role in the formation, maturation, function, and plasticity of synapses.[85] Similar to PDZ proteins, one of their functions may be clustering NMDARs at synapses. Recent evidence suggests that PDZ proteins and adhesion molecules work together,[132,133] perhaps in a compensatory manner.

Relatively few of these adhesion proteins bind directly to NMDARs. The Eph family of receptor tyrosine kinases may function like adhesion factors at synapses,[134] with presynaptic ephrin-B binding to postsynaptic EphB receptors. This may promote a direct interaction of EphB receptors with the extracellular domains of NR1 subunits,[135] enhancing NMDAR-mediated synaptic function.[136,137] Activated EphB receptors function as tyrosine kinases and may also indirectly potentiate NMDAR-mediated calcium influx.[138] Activated EphB2 may also modulate NMDAR calcium fluxes by linking NMDARs to tyrosine kinases such as Src and Fyn that bind to Eph receptors. These calcium fluxes may control the Rho-family GEFs (kalirin 7 and Tiam) that activate Rac—a regulator of actin polymerization and subsequent dendrite arborization.[139–141]

Activated EphB2 leads to synapse formation.[135] EphBs 1 through 3 are critical for the formation of the PSD and dendritic spines.[142]

A family of synaptic adhesion-like molecules (SALMs) was recently discovered on the basis of its interactions with MAGUKs.[143,144] Of the five members, SALMs 1 through 3 have PDZ-binding domains. SALM1 has been shown to bind to PSD-95, SAP102, and SAP97, serving as a mechanism of indirectly linking SALMs to NMDARs at synapses.[143–146] SALM1 can also bind directly to the extracellular domains of NR1 subunits when expressed in heterologous cells.[144] Although no presynaptic ligand has yet been found, SALM1 and SALM2 can induce synapse formation and neurite outgrowth. SALM1 has been shown to enhance surface expression of transfected NR2A.[143,144]

Other families of transsynaptic adhesion molecules may indirectly connect to NMDAR-containing complexes and act to affect synaptic NMDARs (see Section 8.2.3.3.4).[85] Presynaptic neurexin that is bound to PDZ domain-containing proteins in presynaptic terminals binds to postsynaptic neuroligin that interacts with a PDZ domain of PSD-95.[67,147,148] Neuroligin localization at synapses may help recruit NMDARs to synapses and determine whether developing synapses become excitatory or inhibitory.[67,147–149] Neuroligin binds to the PDZ domains of S-SCAM.[86] The PDZ domains of CIPP can bind to both neuroligins and neurexins,[87] providing other mechanisms by which neuroligin and neurexin may be localized to synapses.

Both cadherins and catenins are associated with the NMDAR-containing complex.[150,151] Homophilic interactions between pre- and postsynaptic cadherins induce interactions with postsynaptic β-catenin that can bind directly to the S-SCAM PDZ scaffold and may control its synaptic targeting.[86] Stability of the synaptic contact may be regulated directly by association, as dimerization of cadherins is associated with NMDAR activation.[151] NMDARs and associated PSD-95 also are found in cadherin-based attachment plaques in cerebellar glomeruli.[152] Thus, glutamate spillover from adjacent synapses in the glomerulus may control the overall stability of the glomerulus.

Adhesion proteins of the L1/NrCAM and NCAM families also are linked to NMDAR-containing synapses.[67] NCAM mediates synaptic plasticity[153] and modulates neuronal positioning and dendritic orientation.[154] In the hippocampus, NCAM180 found in the central region of the PSD associates with NR2A-containing NMDARs and undergoes distribution changes following LTP.[155] NrCAM is found on both the pre- and postsynaptic sides of glutamatergic synapses[67] and can link directly to SAP102 via its PDZ-binding domain.[156] This suggests that NrCAM may be associated with the NMDAR-containing complex at the PSD, and like neuroligin, may help link NMDARs to presynaptic components of synapses.

Integrins mediate the developmental switch from NR2B-containing receptors to NR2A-containing receptors at synapses[157] and potentiate NMDAR-mediated currents by activating Src tyrosine kinases.[158,159] Tyrosine receptor kinase B (TrkB) is found at glutamatergic synapses and may act as an adhesion factor across the synaptic cleft via homophilic binding of two TrkB molecules linked by a dimer of their ligand, brain-derived neurotrophic factor (BDNF).[67] TrkB may control expression and function of NMDARs.[160,161] In visual cortical neurons of young mice, NMDAR activation may activate BDNF–TrkB signaling to recruit more PSD-95 to synapses and thus promote the AMPAR-mediated increase in synaptic current.[162]

8.2.3.3.4 Developmental Changes in NMDA Receptors and Associated Proteins in PSD

During embryonic development and early into postnatal development, NMDARs containing mainly NR1, NR2B, NR2D, and NR3A are present throughout the brain. NR2B, NR2D, and NR3A decrease during maturation, while NR2A and/or NR2C become more abundant in maturing brains.[5,6,163–166] The relative levels of NR1 isoforms also change with age.[167] These changes have been best studied in the forebrain and cerebellum, and exceptions occur in other areas.

Immunoblot analyses and ultrastructural immunogold studies examined postnatal development in the CA1 region of the hippocampus.[7,67] Synaptic labeling for NR2B is highest at P2 and decreases gradually to approximately half by P35. Studies examining NR2A mRNA levels in mice reported that NR2A signals are first detected in the CA1 region of the hippocampus at P1, followed by a substantial increase throughout the brain over the next 2 postnatal weeks.[5] Similarly, immunogold labeling for NR2A at synapses is present but very low at P2; about 12 times as much is present by P35.[7,67]

Another major switch in NR2 subunit expression occurs in cerebellar granule cells in which NR2C replaces NR2B. After the granule cells complete migration and are innervated by mossy fibers, they downregulate NR2B and begin to express NR2C.[168] NR2A is present during this period.[63,64] Neuregulin secreted by mossy fiber terminals can interact with ErbB2 and ErbB4 receptors on granule cells, inducing NR2C expression, perhaps via an indirect structural link of neuregulin and NMDARs.[169,170]

Arguably, the most studied developmental change in NMDARs is the switch from NR2B-containing NMDARs to NR2A-containing NMDARs noted during maturation of excitatory synapses (Figure 8.2). This switch was visualized in ultrastructural studies in the thalamus and cerebral cortex[69,171] and hippocampus.[67] The turnover from NR2B-containing to NR2A-containing NMDARs has been tied to learning experiences[62] such as visual exposure during postnatal development[172] and rule learning for odor discrimination in adults.[173] In cerebellar granule cells, a

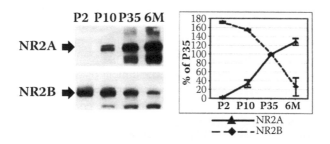

FIGURE 8.2 Patterns of developmental expression of NR2A and NR2B in same homogenates from postnatal (P) days 2, 10, 35, and 6-mo rat hippocampus analyzed by SDS-PAGE and immunoblotting with affinity-purified antibodies. Histogram shows relative amount of protein displayed as percent of that of P35, measured by densitometric scanning. Note how NR2A and NR2B show opposite patterns of expression through development. (Reprinted from Sans, N. et al., *Neuroscience*, 20, 1260, 2000. With permission.)

similar switch occurs when NR2C replaces NR2B after cell migration and innervation by mossy fibers,[168] developing in a caudal-to-rostral progression through the cerebellum.[174]

A change of the major MAGUKs at the synapse from SAP102 to PSD-95 parallels this change from NR2B to NR2A.[7,67,175] In the adult synapse, NR2A is the most prevalent form although NR2B is still found; PSD-95 is the prevalent MAGUK even though SAP102 remains.[7,62,176] In developing neurons of the visual cortices of NR2A knockout mice, PSD-95 does not seem to form effective scaffolds for NR1 and NR2B receptors that persist in adults based on a selective loss of spontaneous NMDAR currents seen in NR2A knockout mice.[61] The resulting model suggests that early synaptic NR2B-containing NMDARs bound to SAP102 are replaced normally by NR2A-containing NMDARs bound to PSD-95 in the centers of synapses. Biochemical studies isolating NR1–NR2A and NR1–NR2B diheteromeric receptors indicate no preference for NR2B with SAP102 and NR2A with PSD-95 in adult hippocampus.[116] PSD-93 and SAP97 also increase with development, as seen with PSD-95.[7,43]

During glutamatergic synapse development, NMDARs are thought to be absent from the earliest nascent excitatory synapses and are recruited after other components initiate synaptogenesis.[67,177,178] Adhesion molecules (Section 8.2.3.3.3) are likely the earliest components of nascent synapses. For example, pre- and postsynaptic NCAMs,[179,180] presynaptic ephrins, postsynaptic EphB receptors,[134] and neuroligin[148] are thought to be recruited early in synaptogenesis and may play a major role in synaptogenesis and synapse maturation. Interestingly, the targeting of neuroligin to the developing synapse is thought to be independent of the postsynaptic anchoring of neuroligin to PSD-95, suggesting that PSD-95 may be subsequently recruited.

As mentioned in Section 8.2.3.3.3, in visual cortical neurons of young mice, NMDAR-mediated activation of TrkB results in recruitment of PSD-95 to synapses.[162] In the CA1 stratum radiatum of the hippocampus, immunogold labeling for catenin shows a preference for the more immature-appearing synaptic contacts, while in comparison, labeling for NR2B, SAP102, and Homer 1b and c is more prevalent on the mature-appearing synapses, again suggesting that adhesion molecules precede entry of other proteins at early synaptic contacts (Figure 8.3).[67]

Subsequently, presynaptic components in packets and then postsynaptic MAGUKs and NMDARs are delivered to developing synapses.[177,181] Discrepancies have been reported regarding whether synapse formation is initiated by the development of pre- or postsynaptic structures. Recruitment of NMDAR clusters has been reported to precede synaptic vesicle proteins.[22] Others indicate that presynaptic differentiation precedes PSD-95 recruitment.[178,181,182] In accordance with both proposed models, mobile postsynaptic transport packets were reportedly recruited to both nascent and existing presynaptic contacts.[183] PSD-95 was identified as an essential component of the postsynaptic scaffolding complex because small interfering RNA (siRNA) knockdown of PSD-95 interfered with postsynaptic scaffold development and excitatory synapse formation.

After the earliest nascent synapse is formed, other proteins are incorporated into the developing PSD. Trans-synaptic signaling is important for coordinated development of pre- and postsynaptic structures, as seen with SAP97 overexpression

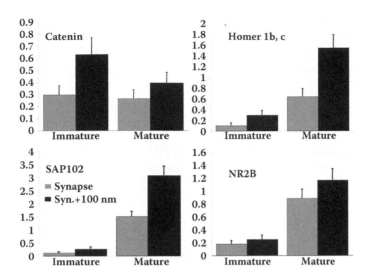

FIGURE 8.3 Comparison of immunogold labeling in immature versus mature synaptic contacts at P2 in hippocampus CA1 stratum radiatum. Immature synapses lacked a substantial PSD and had few synaptic vesicles near the active zone.[49,67] No significant change in catenin labeling was noted. A highly significant increase ($P \leq 0.001$) in labeling for NR2B, SAP102, and Homer 1b, c occurred. The Y axis indicates gold per synapse or per synapse + 100 nm (total within 100 nm perpendicular to postsynaptic membrane). The synapse category is a subset. (Reprinted from Petralia, R.S. et al., *Mol. Cell Neurosci.*, 29, 436, 2005. With permission.)

in postsynaptic cultured hippocampal neurons.[133] This overexpression results in increased presynaptic function, active zone size, and presynaptic protein content that are dependent on adhesion protein activity. Additionally, SAP97 overexpression recruits a complex of postsynaptic proteins.

Synapse formation, as measured by the presence of a distinctive PSD with multiple proteins linked directly or indirectly to NMDARs, occurs less than 2 hr after initial axodendritic contact.[178,181] Thus, in the CA1 region of the hippocampus at P2, relatively mature PSDs have been reported and label for many proteins found in adults.[67] Additionally, the spatial arrangements of these proteins are similar to those found in adults, e.g., Shank and Homer antibodies produce distinctive clusters of labeling below the PSD (in addition to labeling within the PSD) that look similar in synapses at P2, P10, and P35. This position for Shank and Homer below the PSD may correspond to their potential roles as central links in a chain, i.e., NMDARs, MAGUKs, and GKAP are linked to the deeper Shank and Homer and linked to other proteins in internal stores and on the perisynaptic membranes surrounding synapses.

Distinctive differences in the patterns of proteins over development seem to relate mainly to the developmental switch in NMDARs from NR2B-containing to NR2A-containing forms.[67] The change from SAP102 to PSD-95 at synapses parallels this change from NR2B to NR2A,[7,67,175] as mentioned above. At P2, the major GKAP is the higher molecular weight form (130 kD), while both forms (95 and 130 kD) are common in adults.[67,175,176] Note, however, that in adults, NR2B and SAP102 are still fairly common at synapses in the hippocampus, but the overall pattern of NMDAR–MAGUK–GKAP changes dramatically during maturation. The importance of these changes is reflected in studies of NR2A knockout mice.[61] Their visual neurons exhibited selective losses of spontaneous NMDAR currents during development, presumably because PSD-95 cannot act as an effective scaffold for NR2B-containing NMDARs, as noted above. However, it is not clear how selective this NMDAR–MAGUK association may be; at least in adults, NR2A and NR2B appear to associate equally with SAP102, PSD-95, and PSD-93.[116]

Another change during synapse maturation is the appearance of CaMKIIα, which is usually absent from early postnatal synapses,[67] although other forms such as CaMKIIβ may appear early on.[184] NR2B-containing NMDARs bind CaMKII with higher affinity than do NR2A-containing NMDARs; therefore, the NR2B-to-NR2A switch may result in a substantial reduction in LTP and may be associated with the reduced plasticity of glutamatergic synapses of the mature forebrain.[95] The increase in CaMKIIα with maturation may also be related to the switch from NR2B-containing to NR2A-containing NMDARs. In addition, the adult synapse has higher levels of AMPARs.[185] Thus, many proteins appear in the NMDAR-associated complex in early postnatal and adult brains. The final stage in the maturation of excitatory synapses, especially those on principal output neurons, is the development of the postsynaptic spine. In spite of studies indicating that NMDARs play roles in spine formation, mice with CA1-targeted NR1-knockouts (NR1 is lost gradually, mainly during late postnatal development) nevertheless can maintain spines and even may produce new spines in the adult after environmental enrichment.[18,186]

8.2.3.4   Activity-Dependent Alterations in NMDA Receptors

NMDARs are modulated by posttranslational modifications such as ubiquitination, S-nitrosylation, and phosphorylation (see Section 8.2.3.6). Phosphorylation of NMDARs produces a variety of effects[187] and this indicates the importance of this mechanism of regulation in NMDAR function and trafficking. Binding of the kinase directly to the substrate or to an NMDAR-associated protein allows more selectivity in phosphorylation of the appropriate substrate and adds to the complex nature of this regulatory mechanism. Additionally, dephosphorylation of NMDARs by phosphatases has been implicated in downregulation of NMDAR.[188,189] Less is known about this mechanism of regulation. The discussion of these mechanisms of regulation will be limited to their role in NMDAR localization and protein interactions.

8.2.3.4.1   Phosphorylation of NMDA Receptors by CaMKII
NMDAR activation results in calcium entry that, with calmodulin, activates CaMKII. CaMKII activation subsequently results in autophosphorylation at Thr-286, increased CaMKII localization at the PSD, and increased association of CaMKII with NMDARs.[190–192] Autophosphorylation of CaMKII locks it in a constitutively active, calmodulin-trapping state that cannot be reversed by phosphatases, and suppresses inhibitory autophosphorylation of Thr-305 and -306. This activated form may subsequently induce autophosphorylation of neighboring CaMKII molecules.[193]

Thus, through the interaction of CaMKII with NMDARs, activated CaMKII is brought into close proximity with AMPARs,[117,194] enabling it to phosphorylate AMPARs and mediate LTP.[95,195,196]

The subsequent phosphorylation of AMPARs causes synapse potentiation by inducing synaptic insertion and increasing single-channel conductance of AMPARs.[192,193] In summary, CaMKII is recruited to synapses early after potentiation, and through association with NMDARs in a subunit-specific manner, synapses are potentiated via increased AMPAR synaptic insertion and conductance.

CaMKII binds the NR1, NR2A, and NR2B subunits of NMDARs[192,197] with two binding sites identified in NR2B.[118] CaMKII binding to NR2A and NR2B does not occur via direct binding of CaMKII to the major phosphorylation sites on NR2A (Ser-1289)[198] or NR2B (Ser-1303).[199] The interaction of CaMKII with NR2B is stronger than with NR2A, and stable complexes of CaMKII–NR1 and CaMKII–NR2B (but not CaMKII–NR2A) may form,[199] indicating subunit-specific determinations of CaMKII function. Binding of CaMKII to NR2B does not require autophosphorylation of Thr-286 on the CaMKII,[117] but stimulation of NMDARs can increase the association of CaMKII with NR2B.[192]

The high affinity CaMKII–NR2B interaction may be crucial for activity-dependent plasticity.[95,200] CaMKII can regulate NMDARs as a result of various downstream effects. For example, in a PSD preparation from hippocampus, CaMKII can compete with PSD-95 for binding to NR2A.[198] In hippocampal neurons, CaMKII regulates the interaction of NR2A with SAP97 by phosphorylating SAP97 at Ser-232 in an NMDAR-dependent manner,[201] resulting in increased NR2A surface expression.[202] CaMKII also regulates the casein kinase 2 phosphorylation (CK2) of Ser-1480 in the C terminal PDZ-binding domain of NR2B, controlling the binding of NR2B to SAP102 or PSD-95 at synapses (see Section 8.2.3.4.4).[94]

In hippocampal slices potentiated by an experimental LTP protocol, the association of CaMKII with NR2A and B increases, as does CaMKII-dependent activity, and a decrease in association of PSD-95 and NR2A/B is seen.[198] Phosphorylation of NR2A by PKC at Ser-1416 (with phorbol ester or the t-ACPD metabotropic glutamate receptor-specific agonist) inhibits CaMKII binding and results in the dissociation of the CaMKII-NR2A complex.[198] Activation of PKC can also induce translocation of CaMKII to synapses in cultured hippocampal neurons and rapid dispersal of NMDARs from synapses to extrasynaptic membranes, perhaps downregulating synaptic NMDARs.[203] Thus, the role of CaMKII in NMDAR regulation is complex and varied, allowing for subunit-specific effects along with broader PSD-related effects.

8.2.3.4.2 Phosphorylation of NMDA Receptors by PKAs and PKCs

NMDAR-dependent calcium influx results in the generation of cyclic AMP (cAMP) and subsequent activation of cAMP-dependent protein kinases (PKAs). PKAs can phosphorylate NR1, NR2A, and NR2B, increasing NMDAR activity and synaptic targeting.[204] PKAs also regulate NMDAR calcium permeability and directly modulate the induction of NMDAR-dependent LTP in Schaffer collateral CA1 synapses in the hippocampus[205] and spatial long-term memory.[206] PKAs indirectly associate with NMDARs and modulate NMDAR function. For example, PKAs are found in a complex with A kinase anchoring protein (AKAP) 79/150 associated in a complex with PSD-95 and NMDARs.[207]

PKAs also associate with Yotiao, which binds to the NR1 C1 exon cassette. Yotiao binds type 1 protein phosphatase (PP1), which is constitutively active. However, PKA can overcome constitutive PP1 activity and subsequent rapid enhancement of NMDAR currents occurs.[188] PKA can also phosphorylate the CREB immediate early gene that then activates other signaling cascades such as the MAP kinase pathway or can induce new protein synthesis.[208]

Like PKA, PKC phosphorylates NR1, NR2A, and NR2B subunits; however, studies investigating the role of PKC in NMDAR regulation yielded conflicting results. Direct phosphorylation of NMDARs by PKC induces rapid dispersal of NMDARs from synaptic to extrasynaptic sites.[203,209] This may contribute to downregulation of synaptic NMDARs upon PKC activation. In other studies, PKC activation increased the amplitude of NMDAR-mediated currents and channel open probability,[210–212] increasing NMDAR insertion into synaptic membranes[213] and increasing NMDAR sensitivity to inactivation by intracellular calcium.[214] These enhancements are likely indirect through second messenger cascades.[215]

PKC and PSD-95 potentiate NMDAR currents by modulating channel gating and increasing receptor numbers at cell surfaces in *Xenopus* oocytes.[216] These effects are dependent on Ser-1462 in NR2A; mutation at this serine abolishes PSD-95 potentiation of NMDAR currents. PKC activation also regulates NMDAR plasticity by regulating other NMDAR modulators, including CaMKII (Section 8.2.3.4.1) and Src (Section 8.2.3.4.2). Additionally, PKAs and PKCs can work together, suppressing NMDAR ER retention (Section 8.2.1).[16]

8.2.3.4.3 Phosphorylation of NMDA Receptors by PTKs

The Src protein tyrosine kinase (PTK) family includes five members in the CNS: Src, Fyn, Lyn, Lck, and Yes.[217] Phosphorylation of tyrosine residues on NR2A and NR2B[218] potentiates NMDAR-mediated currents.[219] In Fyn knockout mouse striatum, NR2A and NR2B basal tyrosine phosphorylation is reduced; dopamine-dependent NMDAR redistribution to synapses is blocked.[115] Similarly, LTP at Schaffer collateral CA1 synapses in the hippocampus requires upregulation of NMDAR activity by Src.[220]

PSD-95 acts as a linker and promotes Fyn phosphorylation of NR2A.[217,221] Lyn, Lck, and Yes also associate with PSD-95, and Src-mediated potentiation of NR1 and NR2A receptor currents in *Xenopus* oocytes is dependent on the presence of PSD-95.[222] Proline-rich tyrosine kinase (Pyk2), a calcium-dependent tyrosine kinase that activates Src, is associated with PSD-95 and SAP102.[223] Similarly, upregulation of NMDARs by Src is dependent on Src anchoring to the NMDAR-containing complex by the adaptor protein NADH dehydrogenase subunit 2 (ND2).[220] Fyn may be linked to NR2B via the RACK1 scaffolding protein that inhibits NR2B phosphorylation by Fyn, resulting in decreased NMDAR-mediated currents in the CA1 hippocampus region *in vitro*.[224]

In striatal neurons, NR2A and NR2B in the synaptosomal membrane compartment are more highly tyrosine phosphorylated, indicating that tyrosine phosphorylation is an important mechanism that determines subcellular localization of NMDARs.[225] The YEKL internalization motif on the NR2B subunit contains the tyrosine residue Tyr-1472 that is a major substrate for Fyn kinase[226] (see Section 8.2.3.6). In cerebellar granule cells, phosphorylation of this residue by Fyn increases the synaptic localization of NMDARs by preventing their internalization.[93] Along with Tyr-1472, Tyr-1336 is one of the most readily phosphorylated residues on NR2B in cell culture systems.[226] Calpain-mediated proteolysis of NR2B (see Section 8.2.3.6)[37] can be decreased by inhibition of NR2B Tyr-1336 phosphorylation by Fyn via Fyn-directed siRNA or by blocking Tyr-1336 phosphorylation.[227] Fyn and other PTKs may be important for mediating the actions of ephrins and their receptors, the Eph tyrosine kinases, that directly bind NMDARs (see Sections 8.2.3.3.3 and 8.2.3.3.4) and are involved in establishing axon–dendrite connections during development.[138]

8.2.3.4.4 Phosphorylation of NMDA Receptors by Other Kinases

Activation of CK2 potentiates NMDAR activity[228] and may contribute to the induction of LTP.[229] CK2 phosphorylates serine residue Ser-1480 in the PDZ-binding domains of NR2B subunits, thereby blocking the interaction of NR2B-containing receptors with PDZ proteins and decreasing their surface expression.[94] This occurs in an activity-dependent manner and thus may facilitate the replacement of NR2B-containing receptors with NR2A-containing receptors. Cyclin-dependent kinase 5 (Cdk5) phosphorylates NR2A subunits at Ser-1232.[230] This modification is important for mediating NMDA-evoked synaptic currents during the induction of LTP[230] and may be involved in ischemic degeneration of hippocampal CA1 pyramidal neurons.[231]

Cdk5 also facilitates the degradation of NR2B by interacting directly with both NR2B and its protease calpain; thus, Cdk5 knockout mice have enhanced synaptic plasticity due to an increase in NR2B.[232] Glycogen synthase kinase 3 (GSK-3), implicated in neuronal development, mood disorders, and neurodegeneration,[233] modulates NMDAR trafficking.[234] GSK-3 inhibitors and siRNA directed at GSK-3 reduce NMDAR synaptic currents in cultured cortical neurons through increasing Rab5-mediated and PSD-95-regulated NMDAR internalization.

8.2.3.4.5 Modulation of NMDA Receptors by Phosphatases

As discussed above, NMDARs are extensively regulated by complex mechanisms of phosphorylation, and it follows that kinase activity must be reversed by phosphatases. NMDAR regulation by type I protein phosphatase (PP1) is mediated by the Yotiao linker protein that binds to the NR1 subunit and to PKA (see Section 8.2.3.4.2).[188] PKA activation is necessary for enhancing NMDAR currents by inhibiting constitutively active PP1. Furthermore, the catalytic subunit of protein phosphatase 2A (PP2A) is associated with the carboxyl domain of NR3A.[189] This association increases the phosphatase activity of PP2A, resulting in dephosphorylation of Ser-897 of NR1; however, this complex may be disrupted by activation of the NMDAR, reducing PP2A activity.

Additionally, striatal-enriched tyrosine phosphatase (STEP) modulates NMDAR-mediated glutamatergic transmission and LTP.[235] STEP may indirectly modulate NMDARs by dephosphorylating Fyn and reducing its activity.[236] In the hippocampus, STEP binds directly to the NR2A and NR2B subunits, and knockdown of STEP by siRNA results in increased NMDAR-mediated response and increased levels of NMDARs at cell surfaces.[237]

8.2.3.4.6 Modulation of NMDA Receptors by Other Receptors

Cross-talk between the glutamatergic system and other receptor systems has been identified. For example, dopamine increases NMDAR-mediated calcium influx in a PSD-95-dependent manner,[238] further supporting the role of PSD-95 as a scaffold linking various signaling pathways. In the nucleus accumbens, the dopamine-induced increase in NMDAR-mediated calcium influx is mediated by PKA and PKC and can be reversed by activity of the pathway of dopamine- and cAMP-regulated phosphoprotein (DARPP-32; M_r 32 kD) and PP-1,[239] indicating that the effect of dopamine is mediated through phosphorylation. In contrast to the effects of dopamine, activation of the 5-HT_{1A} receptor by serotonin inhibits NMDAR-mediated synaptic currents in cortical pyramidal neurons.[33] This inhibition is associated with a decrease in surface NR2B-containing NMDARs through a mechanism dependent on KIF17 dendritic transport of NR2B-containing NMDARs (see Section 8.2.2).

Insulin can potentiate the activity of NMDARs in *Xenopus* oocytes and enhance NMDAR-mediated synaptic transmission in the hippocampus.[240] The mechanism by which this potentiation occurs differs with respect to the specific NR2 subunit incorporated into the receptor, although NR2A and NR2B are both tyrosine phosphorylated following insulin stimulation.[241] Insulin potentiation of NR1 and NR2B receptors is mediated by both PKCs and tyrosine kinases, while NR1–NR2A potentiation is mediated mainly by PKC.[242] Interestingly, PSD-95 coexpression in *Xenopus* oocytes eliminates insulin-mediated potentiation of NR1/NR2A receptors with no effect on NR1–NR2B potentiation.[222]

8.2.3.4.7 Modulation of NMDA Receptors by Activity

Alterations in activity levels modulate NMDAR trafficking, likely through the phosphorylation and internalization noted above. Chronic blockade of activity in hippocampal neurons *in vitro* results in accumulation of NMDARs at synapses.[243–245] Furthermore, synaptic scaling, a homeostatic form of plasticity that restores neuronal activity to a baseline level, results in increased NMDAR currents.[246] These effects arise from increased NMDARs on cell surfaces, with no changes in single-channel conductance or decay kinetics, implying no change in subunit composition. Furthermore, the effects of activity are subunit-specific, as NR2B-containing receptors are constitutively inserted at the synapse, whereas NR2A-containing receptors require synaptic activity to replace NR2B-containing receptors.[65]

8.2.3.4.8 Modulation of NMDA Receptor-Associated MAGUKs

MAGUKs are responsible for clustering and linking NMDARs (see Section 8.2.3.3.1), providing an additional level of potential regulation. Recent studies identified posttranslational modifications of MAGUKs that effectively regulate NMDAR binding to MAGUKs and, in some cases, receptor clustering. Such modifications include palmitoylation, phosphorylation, and ubiquitination.

Trafficking of some MAGUKs may involve palmitoylation of their N terminal cysteines. Whether palmitoylation is involved in trafficking NMDAR–MAGUK complexes is unknown.[42,247] Of the four PSD-95 family MAGUKs, only PSD-95 and the major isoforms of PSD-93 can be palmitoylated. Dual palmitoylation of PSD-95 may control cellular trafficking of PSD-95, including its initial association with perinuclear organelles and subsequent transfer to the surface via tubulovesicular organelles.[247,248]

The major role of palmitoylation in glutamate receptor trafficking may be transport of AMPARs rather than NMDARs; it may involve palmitoylation of MAGUKs, AMPARs, and their associated proteins (GRIP, ABP), and control trafficking of AMPARs both at the Golgi and cell surface.[248,249] In contrast to PSD-95 and PSD-93, SAP97 and SAP102 are not palmitoylated; SAP97 lacks N terminal cysteines and SAP102 has a specialized cysteine- and histidine-rich N terminal motif that binds zinc and is not palmitoylated.

Cdk5 phosphorylates PSD-95 at three sites in the N terminal region and disrupts the ability of PSD-95 to multimerize.[250] PSD-95 phosphorylated at these sites maintains its ability to bind NMDARs and Kv1.4 potassium channels and cluster the potassium channels in heterologous cells. The N terminal domain is important for its clustering.[251] Disrupting the multimerization of PSD-95 likely affects its role as a scaffolding protein. Phosphorylated PSD-95 is present in the PSD,[250] suggesting that Cdk5 phosphorylation regulates PSD-95 localized at the synapse.

CaMKII phosphorylation of PSD-95 at Ser-73 in the first PDZ domain has also been identified.[252] In contrast to Cdk5 phosphorylation of PSD-95, phosphorylation of Ser-73 disrupts the interaction of PSD-95 with NR2A, although NR2B binding is not affected. Additionally, mass spectrometric analysis of isolated PSD fractions identified phosphorylated serine residues of PSD-95 (Ser-295) and PSD-93 (Ser-365).[253] These residues are located between the second and third PDZ domains, and their functional roles have yet to be determined. Furthermore, PSD-93 is a highly tyrosine-phosphorylated protein in the NMDAR-associated complex.[254] It is phosphorylated on Tyr-384, located between the second and third PDZ domains and may be phosphorylated by Fyn kinase *in vitro*. Fyn phosphorylation promotes the interaction of PSD-93 with the SFK-negative regulator Csk, suggesting that it plays a role in linking NMDARs to the SFK regulatory machinery.

Ubiquitination of MAGUKs has also been identified as a mechanism by which they may be regulated (see Section 8.2.3.6).[255] PSD-95 is associated with the ubiquitin E3 ligase Mdm2 in synaptic membranes. Following NMDA treatment, PSD-95 is ubiquitinated and subsequently degraded by the 26S proteasome. This process requires activity of the calcium-sensitive phosphatase PP2B and phosphorylation of PKA substrates, and is essential for NMDA-induced AMPAR internalization and LTD.

8.2.3.5 Presynaptic NMDA Receptors

While NMDARs generally are assumed to be postsynaptic, several studies in the early 1990s showed evidence of presynaptic NMDARs. Petralia et al.[256,257] found immunocytochemical evidence for presynaptic and axonal localizations of NMDARs in scattered populations throughout the brain. Other studies provided detailed evidence for presynaptic NMDARs associated with specific synapse populations. These were specialized glutamatergic terminals that released or co-released neuropeptides or hormones, e.g., noradrenergic terminals,[258] mossy terminals of the CA3 region of the hippocampus,[259] specific types of primary afferent terminals of the spinal cord dorsal horn,[260,261] and hypothalamic terminals in the pituitary.[262]

It is now apparent that presynaptic NMDARs have broader functions and may underlie common presynaptic mechanisms for regulating releases of the major glutamate and GABA neurotransmitters. By analyzing sectioned or sliced spinal cord preparations with attached dorsal roots (P3–15), Bardoni et al.[263] found that activation of NMDARs in preterminal axons and presynaptic terminals caused an inhibition of glutamate release from the terminals. In the cerebellum, a single glutamatergic granule cell parallel fiber can develop LTD via stimulation of its presynaptic NMDARs, presumably by direct action of glutamate released by its terminal.[264] In contrast, retrograde activation of presynaptic NMDARs on terminals of GABAergic inhibitory interneurons can enhance GABA release at these cerebellar interneuron-Purkinje cell synapses. These presynaptic NMDARs may be activated by local release of glutamate from Purkinje cells.[265]

In GABAergic inputs to tectal neurons of the developing retinotectal system of *Xenopus*, the opposite effect of LTD is caused by a reduction in GABA release mediated by presynaptic NMDARs responding to high frequency visual stimulation. Presumably the glutamate acting on these NMDARs comes from spillover from adjacent glutamatergic terminals.[266] Timing-dependent LTD of neocortical pyramidal cell glutamatergic synapses requires simultaneous activation of presynaptic NMDARs and CB1 cannabinoid receptors.[267] In contrast, simultaneous activation of converging cortical and thalamic afferents in the lateral amygdala (mimicking the effect of fear conditioning) induces associative LTP in the cortical inputs that may depend on activation of presynaptic NMDARs on the cortical afferent terminals caused by release of glutamate from the thalamic afferents.[268]

Pre- and postsynaptic NMDARs may be composed differently. Wang and Thukral[269] presented pharmacological evidence suggesting that presynaptic NMDARs preferentially contain NR1 subunits with exon 5 inserts in the N terminal domains. A number of studies suggest that presynaptic NMDARs contain NR2D (cerebellar interneurons,[270] spinal cord,[263] hippocampus[271]), e.g., because of low sensitivity to magnesium block. These studies concentrate on postnatal stages when NR2D is still common in the brain. However, Thompson et al.[272] presented immunohistochemical evidence for NR2D in presynaptic NMDARs in regions of the hippocampus of adult mice. According to Mameli et al.,[271] CA1 pyramidal neurons in hippocampal slices from rats up to P5 show a short-term increase in probability of glutamate release involving presynaptic NR2D-containing NMDARs and the pregnenolone sulfate excitatory neurosteroid. This triggered a long-term enhancement of AMPARs mediated by postsynaptic NR2B-containing NMDARs.

In contrast, Jourdain et al.[273] showed functional and immunocytochemical evidence for NR2B-containing presynaptic NMDARs in glutamatergic perforant path terminals that form synapses on granule cells of the dentate gyrus of the hippocampus. These are concentrated on the extrasynaptic portions of the terminals and respond to glutamate released from adjacent astrocytes activated by initial glutamate release from the terminal and by activation of P2Y1 purinoreceptors by ATP. Activation of these presynaptic NMDARs enhances synaptic strength at synapses. Yang et al.[274] revealed that the primary presynaptic NMDAR in the entorhinal cortex is NR1/NR2B. This autoreceptor is abundant in young adult rats but appears to decline in older animals. Interestingly, the decline may be reversed in chronic epileptic seizure.

8.2.3.6 NMDA Receptor Internalization

NMDARs are cycled to and from synapses via both constitutive and regulated pathways. Regulation involves a number of factors and depends largely on synapse activity.[62,275–277] Regulated internalization of NMDARs involves direct effects of agonist binding during activity.[278,279] In addition to internalization due to direct activation of the NMDAR, activation-induced internalization may occur indirectly via activation of metabotropic glutamate receptors.[280] Activity-based internalization may also be controlled selectively, e.g., by differential regulation of synaptic and extrasynaptic NMDARs.[3]

Internalization generally requires specific internalization motifs on NMDAR C termini.[14,275,276] NR2B contains a tyrosine-based endocytotic C terminus motif (YEKL) that binds to the $\mu2$ subunit of the AP-2 adaptor protein[28,281,282] involved in clathrin-mediated endocytosis from cell surfaces. NR2A has a similar tyrosine-based motif (YKKM) that is not involved in this endocytosis function. Instead, endocytosis of NR2A seems to involve a dileucine motif in the C terminus. Both NR2A and NR2B can bind to PSD-95 via the ends of their C termini (ESDV), and this interaction with PSD-95 may regulate internalization, i.e., the binding of PSD-95 to the C terminus of NR2A or NR2B may inhibit clathrin-mediated endocytosis of NMDARs.

The importance of this binding to NR2A is not clear because its binding to a PDZ protein is not required for synaptic localization, as discussed below.[65,68,93] Endocytosed NR2A-containing and NR2B-containing NMDARs enter early endosomes, but then they diverge. Generally, NR2A enters late endosomes for degradation, while NR2B enters recycling endosomes for recycling to cell surfaces.[28,282] Interestingly, activation of the NMDAR–CaMKII pathway regulates CK2 phosphorylation of the serine of the PDZ-binding motif at the end of the NMDAR subunit C termini.[94] This phosphorylation decreases surface expression of NR2B in neurons via disruption of the interaction of NR2B with PSD-95 and SAP102, subsequently permitting internalization of the NMDAR from the surface.[31]

This may lead to a natural disruption in forward trafficking of NMDARs to the surface.[94] While phosphorylation of the serine in the PDZ-binding domain (ESDV) of NR2B (but not that of NR2A) may induce internalization of the NMDAR, Fyn-mediated phosphorylation of the tyrosine in the AP-2 binding site (YEKL) of NR2B may prevent AP-2 binding and thus promote retention of NR2B-containing NMDARs at synapses.[93] This suggests that MAGUKs are not simply mechanical scaffolds for anchoring of receptors at synapses. Instead, the major function of MAGUK binding is to keep Fyn kinase in close proximity to the AP-2 binding site of NR2B. The role of Fyn is complex. Activation of dopamine D1 receptors induces a Fyn-dependent redistribution of NMDARs between intracellular and postsynaptic subcellular compartments in the striatum.[115] Also, RACK1 binds to both NR2B and Fyn and prevents Fyn from phosphorylating NR2B.[224] In contrast to NR2B, synaptic localization of NR2A does not require interactions with PDZ proteins or AP-2 binding to its YKKM motif,[93] although PSD-95 promotes Fyn-mediated tyrosine phosphorylation of NR2A (but the specific tyrosine residues involved were not identified[221]). As noted above, coexpression with PSD-95 inhibits NR2A-mediated endocytosis.[28] The regulation of NR2A-containing receptors must be very different from that of NR2B-containing receptors.[68,93]

In addition to the endocytotic motif near the distal C terminus of NR2B, internalization of NMDARs also involves another tyrosine residue-containing endocytotic motif in the proximal C terminus near the last trans-membrane domains in NR1 and NR2 subunits.[283] The three residues following the tyrosine vary in this motif. NR1 has an additional endocytotic motif (VWRK) near the first (YKRH). These proximal motifs appear to be involved in targeting NMDARs to degradation via late endosomes, unlike the distal motif of NR2B that may target endocytosed NR2B-containing NMDARs to recycling endosomes and back to the surface.[283]

The mechanism for this is not clear, but may involve dephosphorylation of this proximal tyrosine residue (possibly phosphorylated by Src kinase), followed by binding of the AP-2 adaptor and clathrin-mediated endocytosis, as shown for NR1 and NR2A NMDARs.[278] This NMDAR internalization is primed by agonist binding and is independent of ion flow.[278,279] Glycine binding enhances the association of AP-2 with the NMDAR and may be sufficient to prime the receptor for internalization, although both glycine and glutamate binding are required for endocytosis.[279]

A number of other proteins involved in clathrin-dependent endocytosis associate with NMDARs. RNAi knockdown of CPG2 (candidate plasticity gene 2), which is localized specifically to the postsynaptic endocytotic zones of excitatory synapses, may increase the number of postsynaptic clathrin-coated vesicles including some that contain NMDARs, and increase the number of surface NR1 and AMPAR GluR2 molecules.[284]

GIPC is associated with endocytosis and contains a PDZ domain that binds to the C terminal ESDV domain of NR2B. Preliminary studies indicate that GIPC helps regulate surface stabilization, endocytosis, and recycling of NMDARs.[79,80] Selective endocytosis of NR3A-containing NMDARs during postnatal development is mediated by the PACSIN1–syndapin 1 adaptor that binds to the C terminal of the NR3A, as well as to dynamin and the N-WASP actin-organizing protein. This mechanism is activity-dependent and may help regulate synaptic maturation.[285]

The site of clathrin-mediated endocytosis of NMDARs and other glutamate receptors is commonly found on the side of the spine.[49,128,286] This location for endocytosis is typical for mature synaptic spines but is more variable in immature synapses and at early postnatal ages. This site is often located in the perisynaptic membrane at the border of the young PSD, typically formed directly on a dendrite shaft.[49] It is difficult to find immunolabeling for NMDARs within identified clathrin-coated pits, but a few examples have been found,[49,284] including some near synapses seen during CPG2 knockdown *in vitro*[284] and in normal neurons (authors' unpublished data).

NMDARs and other glutamate receptors may be internalized by methods other than clathrin-mediated endocytosis. Clathrin-independent endocytosis is seen for a number of proteins including some receptors. Some proteins can be endocytosed by clathrin-dependent or clathrin-independent mechanisms.[287,288] For example, exposure to low levels of epidermal growth factor (EGF) induces endocytosis of EGF receptors

(EFGRs) via clathrin-coated pits, while higher levels of ligand cause EGFRs to be ubiquitinated and endocytosed via a lipid raft-associated, clathrin-independent pathway (both low and high levels of EGF are physiologically relevant).[289]

Ubiquitination is involved in regulation of a number of PSD proteins including PSD-95, Shank, GKAP, and AKAP79 and 109, and may be involved in activity-dependent changes in NMDARs at synapses, particularly the switch from NR2B- to NR2A-containing NMDARs,[62,244] although the mechanism is not fully understood. Ubiquitination is important for direct regulation of NMDARs, although again the mechanism is not clear. F-box protein 2 (Fbx2) can bind to high-mannose glycans of the N terminal extracellular domain of NR1, presumably following retrotranslocation of the N terminal to the cytoplasm by an unknown mechanism.[290] Fbx2 induces ubiquitination of NR1 via linkage of ubiquitin-transferring enzymes, resulting in degradation by the proteasome. Presumably such a mechanism requires additional, unidentified proteins that can direct the N terminal extracellular domain of NR1 (and the other attached subunits of the NMDAR) through the membrane and into the cytoplasm[291]; this mechanism remains speculative.

A ubiquitin-based degradation mechanism may regulate activity-dependent recycling of NMDARs between the cell surface and internal compartments, as overexpression of an Fbx2 mutant accompanied by augmented activity (using a $GABA_A$ receptor antagonist) increases the density of extrasynaptic NMDARs. In addition, prolonged activation of NMDARs leads to calpain-dependent downregulation in NMDAR currents, which involves the degradation of NR2A and NR2B. This is independent of dynamin,[37] a component of clathrin-dependent and some forms of clathrin-independent endocytosis. Overactivity probably leads to calpain-mediated cleavage of NR2 subunits of NMDARs that are on cell surfaces, leading to their destruction and loss via a form of clathrin-independent mechanism; this may protect neurons from excitotoxicity. Susceptibility to calpain-mediated cleavage varies with neuronal maturity, and this process may be blocked by the association of NMDARs with PSD-95 at synapses.[292,293]

It is likely that lipid rafts and associated proteins such as flotillin and caveolin interact under some conditions with NMDARs and other glutamate receptors[52,53,294–297] potentially related to clathrin-independent endocytosis.[294] The presence of caveolin in neurons is controversial, but evidence continues to support significant function for caveolin in neurons.[298,299]

S-nitrosylation of cysteines of NR2A via nitric oxide (NO) can downregulate ion channel activity.[300] NO may be produced in neurons after NMDAR activation, and this may be facilitated by coupling of both the NMDAR and nNOS to synapse-localized PSD-95 (see Section 8.4.1).

8.3 NMDA RECEPTOR TRAFFICKING AND DISEASE

The excessive calcium influx through NMDARs triggers excitotoxic cell death,[8,301] implying that NMDAR-mediated calcium influx mediates specific downstream signaling cascades that may ultimately lead to cell death.[302] Mounting evidence suggests that disruptions in NMDAR trafficking and targeting may play a role in disease states distinct from NMDAR-mediated excitotoxic cell death.[277] NMDAR

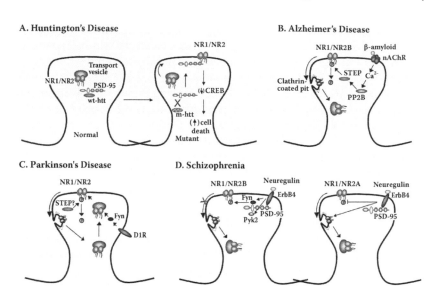

FIGURE 8.4 NMDAR trafficking is altered in neurological disorders. A: Huntington's disease. Wild-type huntingtin (htt) binds the SH3 domain of PSD-95 (left).[332] In a proposed model of HD (right), mutant htt disrupts the association of PSD-95 and htt, increasing the availability of PSD-95 to bind NMDARs[332], and stabilize them on the surface.[281] Mice overexpressing mutant htt exhibit decreased phosphorylation of NR1 Ser-897,[335] resulting in decreased dopamine D1-mediated CREB activation[338] and increased NMDAR-mediated cell death.[310] B: Alzheimer's disease. Calcium influx induced by β-amyloid binding to α7-containing AChRs results in activation of PP2B, which dephosphorylates and activates STEP. STEP dephosphorylates NR2B at Tyr-1472,[356] resulting in increased clathrin-mediated endocytosis of NMDARs.[93] C: Parkinson's disease. D1 dopamine receptor agonist increases NMDAR phosphorylation by Fyn kinase,[115] causing increased NMDAR stabilization on cell surfaces.[93] Constitutive dephosphorylation of NMDARs normally occurs, possibly mediated by STEP, resulting in clathrin-mediated endocytosis of NMDARs. The ratio of phosphorylated to dephosphorylated NMDARs may be disrupted in Parkinson's disease, producing increased NMDAR localization on cell surfaces. D: Schizophrenia. Neuregulin-mediated alterations in NMDAR trafficking are presented in two alternative models. In the first (left), neuregulin-1 signaling activates Fyn and Pyk kinases and causes phosphorylation of NR2B-containing NMDARs.[388] This prevents subsequent clathrin-mediated endocytosis of the NMDAR.[93] In the second model, consistent with hypofunctional NMDARs, neuregulin-1 activates ErbB4 (right) and promotes actin-dependent NMDAR internalization.[386] Activation of ErbB4 by neuregulin suppresses tyrosine phosphorylation of NR2A,[170] the effects of which have not been determined.

dysfunction may lead to loss of synaptic function that may ultimately mediate a collapse of the synaptic functional unit in the absence of calcium-mediated cell death. Furthermore, loss of synaptic function may occur before cell death and in some diseases appears presymptomatically.[303–306] Thus, the complex nature of NMDAR function in disease states has expanded beyond the traditional view of excitotoxic cell death to include NMDAR deregulation. The next section briefly reviews recent developments in elucidating NMDAR mistargeting in the context of neuropathological conditions (Figure 8.4).

8.3.1 NMDA Receptor Subunit-Specific Determinants of Excitotoxic Cell Death

Certain common features of NMDAR deregulation may apply to various diseases. For example, studies suggest that NMDAR-mediated excitoxicity is dependent on NMDAR subunit composition. Cells transfected with NR1/NR2B or NR1/NR2A subunits are more susceptible to cell death than those transfected with NR1/NR2C or NR1/NR2D,[301] possibly due to higher calcium permeability of NR1/NR2A and NR1/NR2B NMDARs. Furthermore, increases in toxicity occur with development as NR2B expression decreases and NR2A expression increases,[307] consistent with subunit-specific vulnerability to excitotoxic cell death. A recent study investigating the effects of NR2B-specific and NR2A-preferring antagonists revealed that while NR2B appears to mediate cell death mechanisms in cortical neurons cultured 14 days *in vitro*, NR2A- and NR2B-containing receptors contribute at 21 days *in vitro*.[308] Interestingly, at 21 days, NR2A exhibited neuroprotective properties at submaximal NMDA concentrations, the effects of which were dependent on the presence of its C terminal tail.

Related to subunit composition, subcellular localization may determine subunit-specific effects. NR2A- and NR2B-containing receptors are coupled to distinct downstream signaling cascades as a result of their predominantly synaptic or extrasynaptic localizations, respectively (see Section 8.2.3.2).[309] Extrasynaptic NMDARs have been associated with the induction of cell death, and synaptic receptors mediate induction of synaptic plasticity.[310,311]

A recent study investigated expression profiles in response to synaptic (induced by blocking GABA$_A$-mediated inhibition with a bicuculline antagonist) and whole-cell (induced by bath application of glutamate) activity.[76] The researchers identified subsets of transcripts induced by NMDAR-dependent and -independent mechanisms, leading to the expression of pro- or antiapoptotic transcripts. Thus, along with subunit composition, synaptic or extrasynaptic NMDAR localization determines whether receptor activation produces synaptic plasticity or neurotoxicity.[312]

8.3.2 Huntington's Disease

Huntington's disease (HD) is an inherited neurodegenerative disorder resulting in cognitive deficits, motor decline, and mood dysfunction.[313] Expansion of the polyglutamine repeat in the gene encoding the huntingtin (htt) protein has been identified as the cause.[314] The disease involves extensive cell death in GABAergic projection medium-sized spiny neurons of the neostriatum and to a lesser extent, cell death in the cortex.[315]

NMDARs have been implicated in HD neuropathology based on ample evidence.[306] For example, HD patients exhibit disproportionate losses of NMDAR binding sites in the striata compared to GABAergic or cholinergic binding sites, suggesting that NMDAR expression increases vulnerability of the neurons to cell death.[316] In addition, in cultured striatal neurons, NMDAR agonists induce neurotoxicity more effectively than agonists for other glutamate receptor subclasses.[317] Furthermore, HD-like behavioral and neuropathological symptoms may be mimicked by administration of NMDAR agonists such as quinolinic acid into the striata of rodents and primates.[318,319]

A role for the NMDARs in HD independent of excitotoxicity has been proposed. Mouse models of HD displayed behavioral and electrophysiological alterations in the absence of cell loss.[320] In HD, cognitive deficits often precede motor symptoms.[321–323] Similarly, presymptomatic alterations in glutamatergic transmission occur in dissociated cortical neurons in mouse model of HD.[324] In a transgenic HD mouse model, NMDAR physiological properties were altered in populations of medium spiny neurons, exhibiting increased response to glutamate and decreased sensitivity to magnesium.[325] These findings support a role for NMDARs in HD neuropathology distinct from the induction of excitotoxic cell death.

Changes in NMDAR subunit expression in HD have been reported[306], and NR2B has been identified as a contributing factor to HD neuropathology.[326] NR2B is enriched in the striatum,[327,328] which exhibits a significantly higher ratio of NR2B to NR2A than the cortex or hippocampus.[329] Most NMDARs in medium spiny neurons are sensitive to the NR2B-specific antagonist ifenprodil, which is as effective at blocking NMDAR-mediated cell death as the NMDAR antagonist MK-801.[330] Thus, the current model suggests that striatal neurons expressing predominantly NR2B are more susceptible to excitotoxic cell death (see Section 8.3.1).[330] This model is supported by findings in HEK-293 cells, in which coexpression of NR2B with mutant htt led to larger NMDAR currents than with wild-type htt. No differences were seen when NR2A was coexpressed with mutant versus wild-type htt.[326] Apoptotic cell death is also significantly greater in cells cotransfected with mutant htt and NR2B, but not NR2A.[331] As noted, extrasynaptic NR2B-containing receptors are suggested to more effectively mediate cell death than synaptic NR2A-containing receptors.[310,311]

In addition to subunit composition and receptor localization, alterations in NMDAR trafficking may contribute to HD neuropathology. NMDARs may be linked to htt by cytoskeletal proteins. NR1 surface localization is increased in cultured medium spiny neurons from transgenic mice expressing htt with moderately expanded polyglutamine lengths.[306] Additionally, in transfected HEK-293 cells and human cortical tissue, NMDARs are linked to htt via its association with the SH3 domain of PSD-95, which plays a role in NMDAR synaptic localization (see Section 8.2.3.3.2).[332] The PSD-95–htt association decreases in the presence of mutant htt, presumably increasing the availability of unbound PSD-95 to bind NMDARs[332] and stabilizing NMDARs on the surface.[281] Consistent with this model is the finding that wild-type htt serves a neuroprotective role, possibly by blocking excessive NMDAR activity.[333] PSD-95 and membrane-associated nNOS are decreased in transgenic models of HD,[334,335] suggesting that the NMDAR–PSD-95–nNOS complex may be disrupted (see Section 8.2.3.3.2).[336] Therefore, the presence of mutant htt may disrupt NMDAR function by increasing NMDAR-mediated calcium influx and by disrupting normal synaptic function.

NMDAR are extensively regulated by phosphorylation (see Section 8.2.3.4). Phosphorylation of NMDARs and their downstream signaling partners is altered in HD. Phosphorylated Src and targeting of activated phospho-Src and PSD-95 to membrane fractions is increased by mutant htt relative to wild-type htt.[337] Changes in PKC and PKA phosphorylation may be affected by mutant htt.[306] Decreased Ser-897 phosphorylation on NR1 was reported in transgenic mouse models of

HD,[335] resulting in decreased dopamine D1-mediated CREB activation[338] linked to increased NMDAR-mediated cell death.[310] Mice overexpressing mutant htt exhibited increased CRE-mediated transcription compared with mice overexpressing wild-type htt,[339] although inhibition of CRE-mediated gene transcription has been linked to HD.[314]

8.3.3 ALZHEIMER'S DISEASE

Alzheimer's disease (AD) is a progressive neurodegenerative disorder character-ized by impairments in memory and cognition, likely caused by the accumulation of β-amyloid peptide. While earlier work focused on the role of the cholinergic system, mounting evidence suggests a glutamatergic synaptic dysfunction in early stages.[304,305] Memantine, a noncompetitive NMDAR antagonist,[340] has been approved for treatment of AD.[341]

Early observations in biopsied and postmortem AD brains reported losses of synaptic spines accompanied by decreases in the numbers of synapses per cortical neuron; these findings have since been confirmed.[342–344] Subsequent studies implicated the NMDAR in the synapse losses.[345] Furthermore, β-amyloid oligomers were found to interrupt synaptic plasticity *in vivo* and in slices[346] and transiently impair learned behaviors in rats.[347–349] Interestingly, nontoxic amounts of β-amyloid reduce synaptic plasticity and glutamatergic transmission without inducing cell death.[350]

Analyses of postmortem AD brains reported altered NMDAR subunit expres-sion. N1-containing NR1 isoforms are decreased in AD-affected patients in brain regions most susceptible to pathological damage.[351,352] While levels of NR2C and NR2D mRNA are not altered, NR2A and NR2B mRNA and protein show decreases in susceptible brain regions.[353] Within the hippocampus, NR1 and NR2B mRNA and protein levels are reduced, while NR2A mRNA and protein levels are unchanged,[354] perhaps producing altered neuronal vulnerability. Discrepant results regarding alter-ations in NMDAR subunit expression suggest that additional mechanisms may be involved in the neuropathological changes.[355]

One such mechanism is NMDAR protein trafficking that is reportedly altered in AD. β-amyloid decreases NMDAR localization at synaptic sites and increases endo-cytosis of NMDARs in cultured cortical neurons and in transgenic mice expressing mutant amyloid precursor protein (APP), with no affect on $GABA_A$ β2 and 3 sub-units.[356] Furthermore, an increase in the tyrosine phosphatase function of STEP results in decreased phosphorylation of NR2B at Tyr-1472,[356] leading to increased endocytosis of the receptor.[93] Prolonged depression of NMDAR-mediated transmission may initi-ate pathological changes or, alternatively, may be neuroprotective based on the find-ing that preconditioning with β-amyloid reduces glutamate-induced neurotoxicity by promoting endocytosis of NMDARs.[357]

8.3.4 PARKINSON'S DISEASE

Parkinson's disease (PD) is a progressive disorder characterized by degeneration of the dopaminergic neurons projecting from the substantia nigra to the striatum. This dopamine deficit is mimicked in animal models that reveal nigrostriatal denerva-tion induced by 6-hydroxydopamine (6-OHDA) injections.[358–360] The characteristic symptoms of PD—resting tremors, rigidity, and other motor symptoms—are alleviated by treatment with L-DOPA (levodopa). However, chronic L-DOPA treatment results in abnormal involuntary spasmodic movements (dyskinesia).[359] Evidence for NMDAR involvement in PD neuropathology comes from effective reduction of neuropathology by NMDAR antagonists, including the NR2B antagonists ifenprodil and CP-101,606, in experimental models[361,362] and in humans.[363]

NMDAR subcellular localization and function are reportedly altered in animal models of PD and following chronic L-DOPA treatment.[358–360] In a unilateral 6-OHDA model of PD, alterations in synaptic function included reorganization of spines,[364] decreases in synapse numbers,[365] and reductions of LTP and LTD.[366,367] While NR1 expression is not altered in postmortem brains of PD patients,[368,369] NR2A is upregulated in a nonhuman primate model of dyskinesia.[358] Altered NMDAR binding was reported in experimental models of PD and human postmortem PD brains.[359]

Changes in subunit composition, subcellular localization, and phosphorylation states of NMDARs have also been reported in 6-OHDA mouse models of PD.[370–372] Total NMDAR subunit expression is reportedly not altered.[370,371,373] However, synaptic localization of NR1 and NR2B decreased in the lesioned striatum relative to unlesioned striatum, with no change in NR2A.[371] Similarly, in L-DOPA-induced dyskinesia, NR2B was reported to redistribute from synaptic to extrasynaptic sites.[360] Significantly, chronic L-DOPA treatment of 6-OHDA-lesioned rats restored the abundance of NR1, NR2A, and NR2B subunits in homogenate and synaptosomal membrane fractions.[371,373]

The exact mechanisms by which NR2B-containing receptors redistribute following 6-OHDA lesions have not been determined, but phosphorylation plays a role. NR2A and NR2B subunits localized in synaptosomal membranes are tyrosine phosphorylated, but those in the light membranes or synaptic vesicle-enriched fractions (containing trafficking vesicles) are not.[225] In the PD 6-OHDA lesion model, no changes were measured in the phosphorylation of NR2A or NR2B in total homogenates, but NR2B phosphorylation decreased in the subpopulations of NMDARs in membrane fractions,[370,371,373] suggesting that changes in phosphorylation alter subcellular localization in PD. Consistent with this model, the dopamine D1 agonist increases NR1, NR2A, and NR2B localization to synaptosomal membrane fractions and tyrosine phosphorylation of the subunits in those fractions.[225]

Dopamine depletion alters serine and threonine phosphorylation of NMDAR subunits. In a unilateral 6-OHDA rat model of PD, phosphorylation of NR1 at Ser-890 and Ser-896 decreased.[371] Ser-897 was not affected. However, increased phosphorylation at this site was reported in another study utilizing the 6-OHDA model and also in MPTP primate models of PD.[374] These modifications have downstream functional effects, as phosphorylation of Ser-890 results in the dispersion of NMDAR clusters.[209] Increased phosphorylation at Ser-897 led to increased dopamine D1-mediated CREB activation.[338] Increased serine phosphorylation of NR2A, but not NR2B, was reported in striatal tissue following unilateral 6-OHDA lesions and subsequent chronic L-DOPA treatment.[375] Interestingly, chronic L-DOPA treatment enhances serine phosphorylation of NR1 at Ser-890, Ser-896, Ser-897, and NR2A along with tyrosine phosphorylation of both NR2A and NR2B.[371,373,375] These studies strongly support a model for regulation of surface delivery and/or stabilization of NMDARs by subunit-specific phosphorylation that may contribute to the neuropathology of PD.

8.3.5 SCHIZOPHRENIA AND MOOD DISORDERS

The glutamate dysfunction hypothesis received much attention in recent years as a mechanism involved in schizophrenia and mood disorders.[376] Reduction of NMDAR function by noncompetitive antagonists such as ketamine and phencyclidine led to dopaminergic hyperactivity and behavioral changes characteristic of schizophrenia.[377] In postmortem tissues of schizophrenic patients, altered NR1, NR2A, NR2B, and NR3A expression was reported.[378–381] Mice expressing only 5% of normal levels of NR1 displayed behavioral abnormalities that correlated with schizophrenia and could have been reversed by treatment with antipsychotics such as haloperidol or clozapine.[382] Possible changes in trafficking or localization of NMDARs were suggested and are supported by a report of decreased striatal expression of SAP102.[383] Interestingly, in many neuropathologies, glutamatergic hyperactivity was implicated; however, schizophrenia is associated with hypofunction of the glutamatergic system.[376,384] This lends further support to the model implicating NMDARs in neuropathologies such as schizophrenia by mechanisms other than excitotoxic cell death.

Neuregulins are widely expressed growth and differentiation factors that have been genetically linked to schizophrenia.[385] Neuregulin function has been associated with changes in glutamatergic synapse function,[170,386] linking neuregulins, glutamatergic hypofunction, and schizophrenia. Neuregulin-1 activity suppresses NMDAR activation, an effect more pronounced in schizophrenic subjects than in normal controls.[170] The neuregulin-β isoform induces NR2C mRNA expression in cultured cerebellar granule cells in an NMDAR activity-dependent manner.[168] In the prefrontal cortex, neuregulin-1 promotes actin-dependent NMDAR internalization and decreases NMDAR-mediated EPSCs.[386]

In hippocampal neurons, neuregulin-1 induces decreases in AMPAR-mediated EPSCs, but not in NMDAR EPSCs, along with a corresponding increase in AMPAR endocytosis.[387] A more tightly coupled complex of the neuregulin receptor ErbB4, PSD-95, and NMDARs was reported in postmortem brain tissue of schizophrenic patients compared to normal controls.[170] The functional effects of this complex were elucidated more recently.[388] ErbB4 binds Fyn, and neuregulin signaling stimulates NR2B Tyr-1472 phosphorylation through the actions of Pyk2 and Fyn kinases. Transgenic mice with neuregulin-1 or ErbB4 knockdowns are hypophosphorylated at Tyr-1472 NR2B and may be rescued by treatment with clozapine. These data suggest that the glutamatergic hypofunction model of schizophrenia may be mediated in part by the effects of neuregulins on NMDAR function.

Major depression and bipolar disorder may be effectively treated with antagonists of NMDARs, implicating the glutamatergic system in the diseases.[389] A single dose of ketamine, an NMDAR-specific antagonist, produced significant improvements in subjects with major depressive disorders.[390] Mice lacking NR2A subunits exhibited anxiolytic and antidepressant-like effects.[391] Exercise did not enhance neurogenesis as occurred in wild-type mice.[392] These findings agree with suggestions

182   Biology of the NMDA Receptor

that depression results from impaired neurogenesis.[393] Treatment of rats with lithium—an effective remedy for patients with bipolar disorder—resulted in block of NMDAR-mediated signaling via phospholipase A_2 activation and arachadonic acid release.[394] GSK-3, one of the main targets of lithium,[395] was shown to induce NMDAR internalization (see Section 8.2.3.4.4).[234]

Interestingly, sleep deprivation is an effective and rapid short-term treatment for depression.[396] Studies indicate that sleep deprivation induces changes in NMDAR function in the hippocampus.[397–399] Longer sleep deprivation (24 hr) produced impaired LTP and spatial learning, reduced NMDAR-mediated currents, and decreased NR1 surface expression with no changes in other NMDAR subunits.[397] Studies investigating shorter periods of sleep deprivation reported different results based on the sleep deprivation protocols used. After 4 hr of sleep deprivation, NR2A protein levels in synaptosomal fractions increased with no changes in NR2B.[399] The effects of sleep deprivation were reversible following 3 hr of recovery. After 72 hr of sleep deprivation, NMDAR-mediated currents decreased in amplitude with no changes in decay kinetics as a result of decreased NR1 and NR2A (but not NR2B) surface expression.[398]

As in schizophrenia studies, bipolar disorder and depression studies reported alterations in NR1, NR2A, NR2B, and, interestingly, NR3A subunit expression.[381,383,400,401] Decreased striatal expression of PSD-95 and SAP102 or SAP102 only was reported in postmortem tissues of patients with bipolar disorder and depression, respectively.[383]

8.3.6 Other Disorders

NMDARs have also been implicated in other neuropathological conditions including lupus erythematosus,[402,403] neuropathic pain,[404] ischemia,[405] epilepsy,[360] amyotrophic lateral sclerosis,[406] and HIV dementia.[407]

ACKNOWLEDGMENTS

We thank Dr. Mark E. Schneider for reading and commenting on the manuscript.

REFERENCES

1. Stocca, G. and Vicini, S., Increased contribution of NR2A subunit to synaptic NMDARs in developing rat cortical neurons, *J. Physiol.*, 507, 13, 1998.
2. Tovar, K.R. and Westbrook, G.L., The incorporation of NMDARs with a distinct subunit composition at nascent hippocampal synapses in vitro, *J. Neurosci.*, 19, 4180, 1999.
3. Li, B. et al., Differential regulation of synaptic and extrasynaptic NMDARs, *Nat. Neurosci.*, 5, 833, 2002.
4. Kohr, G., NMDAR function: Subunit composition versus spatial distribution, *Cell Tissue Res.*, 326, 439, 2006.
5. Watanabe, M. et al., Developmental changes in distribution of NMDAR channel subunit mrnas, *Neuroreport*, 3, 1138, 1992.
6. Wenzel, A. et al., *N*-methyl-D-aspartate receptors containing the NR2D subunit in the retina are selectively expressed in rod bipolar cells, *Neuroscience*, 78, 1105, 1997.

7. Sans, N. et al., A developmental change in NMDAR-associated proteins at hippocampal synapses, *J. Neurosci.*, 20, 1260, 2000.
8. Cull-Candy, S., Brickley, S., and Farrant, M., NMDAR subunits: diversity, development and disease, *Curr. Opin. Neurobiol.*, 11, 327, 2001.
9. Hawkins, L. et al., Export from the endoplasmic reticulum of assembled *N*-methyl-D-aspartic acid receptors is controlled by a motif in the C terminus of the NR2 subunit, *J. Biol. Chem.*, 279, 28903, 2004.
10. Matsuda, K. et al., Specific assembly with the NMDAR 3B subunit controls surface expression and calcium permeability of NMDARs, *J. Neurosci.*, 23, 10064, 2003.
11. Standley, S. et al., PDZ domain suppression of an ER retention signal in NMDAR NR1 splice variants, *Neuron*, 28, 887, 2000.
12. Scott, D. et al., An NMDAR ER retention signal regulated by phosphorylation and alternative splicing, *J. Neurosci.*, 21, 3063, 2001.
13. Xia, H., Hornby, Z.D., and Malenka, R., An ER retention signal explains differences in surface expression of NMDA and AMPA receptor subunits, *Neuropharmacology*, 41, 714, 2001.
14. Wenthold, R. et al., Trafficking of NMDARs, *Annu. Rev. Pharmacol. Toxicol.*, 43, 335, 2003.
15. Holmes, K. et al., The *N*-methyl-D-aspartate receptor splice variant NR1-4 C-terminal domain. Deletion analysis and role in subcellular distribution, *J. Biol. Chem.*, 277, 1457, 2002.
16. Scott, D.B., Blanpied, T.A., and Ehlers, M.D., Coordinated PKA and PKC phosphorylation suppresses RXR-mediated ER retention and regulates the surface delivery of NMDARs, *Neuropharmacology*, 45, 755, 2003.
17. Mu, Y. et al., Activity-dependent mRNA splicing controls ER export and synaptic delivery of NMDARs, *Neuron*, 40, 581, 2003.
18. Fukaya, M. et al., Retention of NMDAR NR2 subunits in the lumen of endoplasmic reticulum in targeted NR1 knockout mice, *Proc. Natl. Acad. Sci. USA*, 100, 4855, 2003.
19. Mcilhinney, R. et al., Assembly intracellular targeting and cell surface expression of the human *N*-methyl-D-aspartate receptor subunits NR1a and NR2A in transfected cells, *Neuropharmacology*, 37, 1355, 1998.
20. Perez-Otano, I. et al., Assembly with the NR1 subunit is required for surface expression of NR3A-containing NMDARs, *J. Neurosci.*, 21, 1228, 2001.
21. Yang, W. et al., A three amino acid tail following the tm4 region of the *N*-methyl-D-aspartate receptor (NR) 2 subunits is sufficient to overcome endoplasmic reticulum retention of NR1-1a subunit, *J. Biol. Chem.*, 282, 9269, 2007.
22. Washbourne, P., Bennett, J., and Mcallister, A., Rapid recruitment of NMDAR transport packets to nascent synapses, *Nat. Neurosci.*, 5, 751, 2002.
23. Washbourne, P. et al., Cycling of NMDARs during trafficking in neurons before synapse formation, *J. Neurosci.*, 24, 8253, 2004.
24. Bressloff, P., Stochastic model of protein receptor trafficking prior to synaptogenesis, *Phys. Rev. E Stat. Nonlin. Soft Matter Phys.*, 74, 031910, 2006.
25. Petralia, R. et al., Morphological correlates of glutamate receptor trafficking to synapses, *Soc. Neurosci. Abs.*, 843.2, 2004.
26. Petralia, R. et al., Specific endosomal associations during the trafficking of synaptic glutamate receptors, *Soc. Neurosci. Abs.*, 949.1, 2005.
27. Lee, S. et al., Biochemical and morphological characterization of an intracellular membrane compartment containing AMPA receptors, *Neuropharmacology*, 41, 680, 2001.
28. Lavezzari, G. et al., Subunit-specific regulation of NMDAR endocytosis, *J. Neurosci.*, 24, 6383, 2004.

29. Rodriguez-Boulan, E., Kreitzer, G., and Musch, A., Organization of vesicular trafficking in epithelia, *Nat. Rev. Mol. Cell Biol.,* 6, 233, 2005.
30. Sans, N. et al., NMDAR trafficking through an interaction between PDZ proteins and the exocyst complex, *Nat. Cell Biol.,* 5, 520, 2003.
31. Sans, N. et al., Mpins modulates PSD-95 and SAP102 trafficking and influences NMDAR surface expression, *Nat. Cell Biol.,* 7, 1179, 2005.
32. Prybylowski, K. et al., Relationship between availability of NMDAR subunits and their expression at the synapse, *J. Neurosci.,* 22, 8902, 2002.
33. Yuen, E. et al., Serotonin 5-HT1A receptors regulate NMDAR channels through a microtubule-dependent mechanism, *J. Neurosci.,* 25, 5488, 2005.
34. Setou, M. et al., Kinesin superfamily motor protein KIF17 and mLin-10 in NMDAR-containing vesicle transport, *Science,* 288, 1796, 2000.
35. Guillaud, L., Setou, M., and Hirokawa, N., KIF17 dynamics and regulation of NR2B trafficking in hippocampal neurons, *J. Neurosci.,* 23, 131, 2003.
36. Mok, H. et al., Association of the kinesin superfamily motor protein KIF1balpha with postsynaptic density-95 (PSD-95), synapse-associated protein-97, and synaptic scaffolding molecule PSD-95/discs large/zona occludens-1 proteins, *J. Neurosci.,* 22, 5253, 2002.
37. Wu, H.Y. et al., Regulation of *N*-methyl-D-aspartate receptors by calpain in cortical neurons, *J. Biol. Chem.,* 280, 21588, 2005.
38. Petralia, R. et al., Glutamate receptor targeting in the postsynaptic spine involves mechanisms that are independent of myosin Va, *Eur. J. Neurosci.,* 13, 1722, 2001.
39. Lei, S. et al., Regulation of NMDAR activity by f-actin and myosin light chain kinase, *J. Neurosci.,* 21, 8464, 2001.
40. Amparan, D. et al., Direct interaction of myosin regulatory light chain with the NMDAR, *J. Neurochem.,* 92, 349, 2005.
41. Muller, B. et al., Molecular characterization and spatial distribution of SAP97, a novel presynaptic protein homologous to SAP90 and the drosophila discs-large tumor suppressor protein, *J. Neurosci.,* 15, 2354, 1995.
42. El-Husseini, A.E. et al., Ion channel clustering by membrane-associated guanylate kinases. Differential regulation by n-terminal lipid and metal binding motifs, *J. Biol. Chem.,* 275, 23904, 2000.
43. Sans, N. et al., Synapse-associated protein 97 selectively associates with a subset of AMPA receptors early in their biosynthetic pathway, *J. Neurosci.,* 21, 7506, 2001.
44. Aridor, M. et al., Endoplasmic reticulum export site formation and function in dendrites, *J. Neurosci.,* 24, 3770, 2004.
45. Horton, A.C. and Ehlers, M., Secretory trafficking in neuronal dendrites, *Nat. Cell Biol.,* 6, 585, 2004.
46. Steward, O. and Schuman, E., Protein synthesis at synaptic sites on dendrites, *Annu. Rev. Neurosci.,* 24, 299, 2001.
47. Gazzaley, A. et al., Differential subcellular regulation of NMDAR1 protein and mRNA in dendrites of dentate gyrus granule cells after perforant path transection, *J. Neurosci.,* 17, 2006, 1997.
48. Kneussel, M., Postsynaptic scaffold proteins at non-synaptic sites. The role of postsynaptic scaffold proteins in motor-protein-receptor complexes, *EMBO Rep.,* 6, 22, 2005.
49. Petralia, R.S., Wang, Y.X., and Wenthold, R.J., Internalization at glutamatergic synapses during development, *Eur. J. Neurosci.,* 18, 3207, 2003.
50. Tomita, S. et al., Functional studies and distribution define a family of transmembrane AMPA receptor regulatory proteins, *J. Cell Biol.,* 161, 805, 2003.
51. Spacek, J. and Harris, K., Three-dimensional organization of smooth endoplasmic reticulum in hippocampal CA1 dendrites and dendritic spines of the immature and mature rat, *J. Neurosci.,* 17, 190, 1997.

52. Desouza, S. and Ziff, E., AMPA receptors do the electric slide, *Sci. STKE,* 2002, PE45, 2002.

53. Hering, H., Lin, C., and Sheng, M., Lipid rafts in the maintenance of synapses, dendritic spines, and surface AMPA receptor stability, *J. Neurosci.,* 23, 3262, 2003.

54. Kanzaki, M. and Pessin, J., Insulin signaling: glut4 vesicles exit via the exocyst, *Curr. Biol.,* 13, R574, 2003.

55. Passafaro, M., Piech, V., and Sheng, M., Subunit-specific temporal and spatial patterns of AMPA receptor exocytosis in hippocampal neurons, *Nat. Neurosci.,* 4, 917, 2001.

56. Gerges, N. et al., Dual role of the exocyst in AMPA receptor targeting and insertion into the postsynaptic membrane, *EMBO J.,* 25, 1623, 2006.

57. Rao, A. et al., Heterogeneity in the molecular composition of excitatory postsynaptic sites during development of hippocampal neurons in culture, *J. Neurosci.,* 18, 1217, 1998.

58. Tovar, K.R. and Westbrook, G.L., Mobile NMDARs at hippocampal synapses, *Neuron,* 34, 255, 2002.

59. Groc, L. et al., Differential activity-dependent regulation of the lateral mobilities of AMPA and NMDARs, *Nat. Neurosci.,* 7, 695, 2004.

60. Triller, A. and Choquet, D., Surface trafficking of receptors between synaptic and extrasynaptic membranes: and yet they do move! *Trends Neurosci.,* 28, 133, 2005.

61. Townsend, M. et al., Developmental loss of miniature N-methyl-D-aspartate receptor currents in NR2A knockout mice, *Proc. Natl. Acad. Sci. USA,* 100, 1340, 2003.

62. Perez-Otano, I. and Ehlers, M.D., Homeostatic plasticity and NMDAR trafficking, *Trends Neurosci.,* 28, 229, 2005.

63. Rumbaugh, G. and Vicini, S., Distinct synaptic and extrasynaptic NMDARs in developing cerebellar granule neurons, *J. Neurosci.,* 19, 10603, 1999.

64. Fu, Z., Logan, S., and Vicini, S., Deletion of the NR2A subunit prevents developmental changes of NMDA-mEPSCs in cultured mouse cerebellar granule neurones, *J. Physiol.,* 563, 867, 2005.

65. Barria, A. and Malinow, R., Subunit-specific NMDAR trafficking to synapses, *Neuron,* 35, 345, 2002.

66. Groc, L. et al., NMDAR surface mobility depends on NR2A-2B subunits, *Proc. Natl. Acad. Sci. USA,* 103, 18769, 2006.

67. Petralia, R.S. et al., Ontogeny of postsynaptic density proteins at glutamatergic synapses, *Mol. Cell Neurosci.,* 29, 436, 2005.

68. Thomas, C.G., Miller, A.J., and Westbrook, G.L., Synaptic and extrasynaptic NMDAR NR2 subunits in cultured hippocampal neurons, *J. Neurophysiol.,* 95, 1727, 2006.

69. Liu, L. et al., Role of NMDAR subtypes in governing the direction of hippocampal synaptic plasticity, *Science,* 304, 1021, 2004.

70. Woo, N. et al., Activation of p75NTR by proBDNF facilitates hippocampal long-term depression, *Nat. Neurosci.,* 8, 1069, 2005.

71. Berberich, S. et al., Lack of NMDAR subtype selectivity for hippocampal long-term potentiation, *J. Neurosci.,* 25, 6907, 2005.

72. Weitlauf, C. et al., Activation of NR2A-containing NMDARs is not obligatory for NMDAR-dependent long-term potentiation, *J. Neurosci.,* 25, 8386, 2005.

73. Isaacson, J. and Murphy, G.J., Glutamate-mediated extrasynaptic inhibition: Direct coupling of NMDARs to Ca(2+)-activated K+ channels, *Neuron,* 31, 1027, 2001.

74. Vanhoutte, P. and Bading, H., Opposing roles of synaptic and extrasynaptic NMDARs in neuronal calcium signalling and BDNF gene regulation, *Curr. Opin. Neurobiol.,* 13, 366, 2003.

75. Medina, I., Extrasynaptic NMDARs reshape gene ranks, *Sci. STKE,* 2007, pe23, 2007.
76. Zhang, S. et al., Decoding NMDAR signaling: Identification of genomic programs specifying neuronal survival and death, *Neuron,* 53, 549, 2007.
77. Sun, L. and Liu, S.J., Activation of extrasynaptic NMDARs induces a PKC-dependent switch in AMPA receptor subtypes in mouse cerebellar stellate cells, *J. Physiol.,* 583, 537, 2007.
78. Zhang, J. and Diamond, J.S., Distinct perisynaptic and synaptic localization of NMDA and AMPA receptors on ganglion cells in rat retina, *J. Comp. Neurol.,* 498, 810, 2006.
79. Yi, Z. et al., NMDARs interact with GIPC, *Soc. Neurosci. Abs.,* 487.10, 2005.
80. Yi, Z. et al., GIPC, a single PDZ domain-containing protein, interacts with the NMDAR and regulates its surface expression, *Soc. Neurosci. Abs.,* 31.14, 2006.
81. Petralia, R. et al., Organization of synaptic and extrasynaptic NMDARs, *Soc. Neurosci. Abs.,* 2007.
82. Lim, I. et al., Disruption of the NMDAR-PSD-95 interaction in hippocampal neurons with no obvious physiological short-term effect, *Neuropharmacology,* 45, 738, 2003.
83. Sheng, M. and Hoogenraad, C., The postsynaptic architecture of excitatory synapses: a more quantitative view, *Annu. Rev. Biochem.,* 76, 823, 2007.
84. Funke, L., Dakoji, S., and Bredt, D.S., Membrane-associated guanylate kinases regulate adhesion and plasticity at cell junctions, *Annu. Rev. Biochem.,* 74, 219, 2005.
85. Dalva, M.B., Mcclelland, A.C., and Kayser, M.S., Cell adhesion molecules: Signalling functions at the synapse, *Nat. Rev. Neurosci.,* 8, 206, 2007.
86. Nishimura, W. et al., Interaction of synaptic scaffolding molecule and beta-catenin, *J. Neurosci.,* 22, 757, 2002.
87. Kurschner, C. et al., CIPP, a novel multivalent PDZ domain protein, selectively interacts with Kir 4.0 family members, NMDAR subunits, neurexins, and neuroligins, *Mol. Cell Neurosci.,* 11, 161, 1998.
88. Kim, E. and Sheng, M., PDZ domain proteins of synapses, *Nat. Rev. Neurosci.,* 5, 771, 2004.
89. Elias, G.M. et al., Synapse-specific and developmentally regulated targeting of AMPA receptors by a family of MAGUK scaffolding proteins, *Neuron,* 52, 307, 2006.
90. Ehrlich, I. et al., PSD-95 is required for activity-driven synapse stabilization, *Proc. Natl. Acad. Sci. USA,* 104, 4176, 2007.
91. Sprengel, R. et al., Importance of the intracellular domain of NR2 subunits for NMDAR function *in vivo, Cell,* 92, 279, 1998.
92. Steigerwald, F. et al., C-terminal truncation of NR2A subunits impairs synaptic but not extrasynaptic localization of NMDARs, *J. Neurosci.,* 20, 4573, 2000.
93. Prybylowski, K. et al., The synaptic localization of NR2B-containing NMDARs is controlled by interactions with PDZ proteins and AP-2, *Neuron,* 47, 845, 2005.
94. Chung, H.J. et al., Regulation of the NMDAR complex and trafficking by activity-dependent phosphorylation of the NR2B subunit PDZ ligand, *J. Neurosci.,* 24, 10248, 2004.
95. Barria, A. and Malinow, R., NMDAR subunit composition controls synaptic plasticity by regulating binding to CaMKII, *Neuron,* 48, 289, 2005.
96. El-Husseini, A.E. et al., PSD-95 involvement in maturation of excitatory synapses, *Science,* 290, 1364, 2000.
97. Schnell, E. et al., Direct interactions between PSD-95 and stargazin control synaptic AMPA receptor number, *Proc. Natl. Acad. Sci. USA,* 99, 13902, 2002.

98. Losi, G. et al., PSD-95 regulates NMDARs in developing cerebellar granule neurons of the rat, *J. Physiol.*, 548, 21, 2003.
99. Migaud, M. et al., Enhanced long-term potentiation and impaired learning in mice with mutant postsynaptic density-95 protein, *Nature*, 396, 433, 1998.
100. Mcgee, A.W. et al., Psd-93 knock-out mice reveal that neuronal MAGUKs are not required for development or function of parallel fiber synapses in cerebellum, *J. Neurosci.*, 21, 3085, 2001.
101. Cuthbert, P. et al., Synapse-associated protein 102/dlgh3 couples the NMDAR to specific plasticity pathways and learning strategies, *J. Neurosci.*, 27, 2673, 2007.
102. Stein, V. et al., Postsynaptic density-95 mimics and occludes hippocampal long-term potentiation and enhances long-term depression, *J. Neurosci.*, 23, 5503, 2003.
103. Sonoda, T. et al., Binding of glutamate receptor δ2 to its scaffold protein, delphilin, is regulated by PKA, *Biochem. Biophys. Res. Comm.*, 350, 748, 2006.
104. Tao, Y. et al., Impaired NMDAR-mediated postsynaptic function and blunted NMDAR-dependent persistent pain in mice lacking postsynaptic density-93 protein, *J. Neurosci.*, 23, 6703, 2003.
105. Ohtsuka, T. et al., nRap GEP: A novel neural GDP/GTP exchange protein for Rap1 small G protein that interacts with synaptic scaffolding molecule (S-SCAM), *Biochem. Biophys. Res. Commun.*, 265, 38, 1999.
106. Yao, I. et al., MAGUIN, a novel neuronal membrane-associated guanylate kinase-interacting protein, *J. Biol. Chem.*, 274, 11889, 1999.
107. Ohtakara, K. et al., Densin-180, a synaptic protein, links to PSD-95 through its direct interaction with MAGUIN-1, *Genes Cells*, 7, 1149, 2002.
108. Becamel, C. et al., The serotonin 5-HT2A and 5-HT2C receptors interact with specific sets of PDZ proteins, *J. Biol. Chem.*, 279, 20257, 2004.
109. Anzai, N. et al., The multivalent PDZ domain-containing protein CIPP is a partner of acid-sensing ion channel 3 in sensory neurons, *J. Biol. Chem.*, 277, 16655, 2002.
110. Cao, J. et al., The PSD95–nNOS interface: a target for inhibition of excitotoxic p38 stress-activated protein kinase activation and cell death, *J. Cell Biol.*, 168, 117, 2005.
111. Kim, J.H. et al., Syngap: a synaptic RasGAP that associates with the PSD-95/SAP90 protein family, *Neuron*, 20, 683, 1998.
112. Chen, H. et al., A synaptic ras-GTPase activating protein (p135 SynGAP) inhibited by cam kinase ii, *Neuron*, 20, 895, 1998.
113. Krapivinsky, G. et al., Syngap-mupp1-CaMKII synaptic complexes regulate p38 map kinase activity and NMDAR-dependent synaptic AMPA receptor potentiation, *Neuron*, 43, 563, 2004.
114. Kim, M. et al., Differential roles of NR2A- and NR2B-containing NMDARs in ras-erk signaling and AMPA receptor trafficking, *Neuron*, 46, 745, 2005.
115. Dunah, A.W. et al., Dopamine d1-dependent trafficking of striatal N-methyl-D-aspartate glutamate receptors requires fyn protein tyrosine kinase but not darpp-32, *Mol. Pharmacol.*, 65, 121, 2004.
116. Al-Hallaq, R.A. et al., NMDA di-heteromeric receptor populations and associated proteins in rat hippocampus, *J. Neuro. sci.*, in press, 2007.
117. Lisman, J., Schulman, H., and Cline, H., The molecular basis of CaMKII function in synaptic and behavioural memory, *Nat. Rev. Neurosci.*, 3, 175, 2002.
118. Merrill, M.A. et al., Activity-driven postsynaptic translocation of CaMKII, *Trends Pharmacol. Sci.*, 26, 645, 2005.
119. Hirao, K. et al., A novel multiple PDZ domain-containing molecule interacting with *N*-methyl-D-aspartate receptors and neuronal cell adhesion proteins, *J. Biol. Chem.*, 273, 21105, 1998.
120. Kim, S. et al., Activation of the TRPC1 cation channel by metabotropic glutamate receptor mglur1, *Nature*, 426, 285, 2003.

121. Tu, J.C. et al., Coupling of mGluR/homer and PSD-95 complexes by the shank family of postsynaptic density proteins, *Neuron,* 23, 583, 1999.
122. Uemura, T., Mori, H., and Mishina, M., Direct interaction of GluRdelta2 with shank scaffold proteins in cerebellar Purkinje cells, *Mol. Cell Neurosci.,* 26, 330, 2004.
123. Chen, L. et al., Stargazin regulates synaptic targeting of AMPA receptors by two distinct mechanisms, *Nature,* 408, 936, 2000.
124. Rouach, N. et al., TARP gamma-8 controls hippocampal AMPA receptor number, distribution and synaptic plasticity, *Nat. Neurosci.,* 8, 1525, 2005.
125. Wyszynski, M. et al., Competitive binding of alpha-actinin and calmodulin to the NMDAR, *Nature,* 385, 439, 1997.
126. Furuyashiki, T. et al., Citron, a rho-target, interacts with PSD-95–SAP-90 at glutamatergic synapses in the thalamus, *J. Neurosci.,* 19, 109, 1999.
127. Zhang, W. et al., Citron binds to PSD-95 at glutamatergic synapses on inhibitory neurons in the hippocampus, *J. Neurosci.,* 19, 96, 1999.
128. Blanpied, T.A., Scott, D.B., and Ehlers, M.D., Dynamics and regulation of clathrin coats at specialized endocytic zones of dendrites and spines, *Neuron,* 36, 435, 2002.
129. Pak, D. et al., Regulation of dendritic spine morphology by SPAR, a PSD-95-associated rapgap, *Neuron,* 31, 289, 2001.
130. Van De Ven, T.J., Vandongen, H., and Vandongen, A., The nonkinase phorbol ester receptor alpha 1-chimerin binds the NMDAR NR2A subunit and regulates dendritic spine density, *J. Neurosci.,* 25, 9488, 2005.
131. Allison, D. et al., Role of actin in anchoring postsynaptic receptors in cultured hippocampal neurons: differential attachment of NMDA versus AMPA receptors, *J. Neurosci.,* 18, 2423, 1998.
132. Futai, K. et al., Retrograde modulation of presynaptic release probability through signaling mediated by PSD-95-neuroligin, *Nat. Neurosci.,* 10, 186, 2007.
133. Regalado, M. et al., Trans-synaptic signaling by postsynaptic synapse-associated protein 97, *J. Neurosci.,* 26, 2343, 2006.
134. Murai, K. and Pasquale, E., Eph receptors, ephrins, and synaptic function, *Neuroscientist,* 10, 304, 2004.
135. Dalva, M.B. et al., EphB receptors interact with NMDARs and regulate excitatory synapse formation, *Cell,* 103, 945, 2000.
136. Grunwald, I.C. et al., Kinase-independent requirement of EphB2 receptors in hippocampal synaptic plasticity, *Neuron,* 32, 1027, 2001.
137. Henderson, J.T. et al., The receptor tyrosine kinase EphB2 regulates NMDA-dependent synaptic function, *Neuron,* 32, 1041, 2001.
138. Takasu, M.A. et al., Modulation of NMDAR-dependent calcium influx and gene expression through EphB receptors, *Science,* 295, 491, 2002.
139. Misra, C. and Ziff, E., EphB2 gets a GRIP on the dendritic arbor, *Nat. Neurosci.,* 8, 848, 2005.
140. Tolias, K. et al., The Rac1-GEF tiam1 couples the NMDAR to the activity-dependent development of dendritic arbors and spines, *Neuron,* 45, 525, 2005.
141. Tolias, K. et al., The Rac1 guanine nucleotide exchange factor tiam1 mediates EphB receptor-dependent dendritic spine development, *Proc. Natl. Acad. Sci. USA,* 104, 7265, 2007.
142. Henkemeyer, M. et al., Multiple EphB receptor tyrosine kinases shape dendritic spines in the hippocampus, *J. Cell Biol.,* 163, 1313, 2003.
143. Ko, J. et al., SALM synaptic cell adhesion-like molecules regulate the differentiation of excitatory synapses, *Neuron,* 50, 233, 2006.
144. Wang, C.Y. et al., A novel family of adhesion-like molecules that interacts with the NMDAR, *J. Neurosci.,* 26, 2174, 2006.

145. Seabold, G. et al., A family of adhesion-like molecules associated with the NMDAR complex, *Soc. Neurosci. Abs.*, 487.11, 2005.
146. Seabold, G.K. et al., The SALM family of adhesion-like molecules forms heteromeric and homomeric complexes, *Soc. Neurosci. Abs.*, 31.5, 2006.
147. Levinson, J.N. et al., Neuroligins mediate excitatory and inhibitory synapse formation: involvement of PSD-95 and neurexin-1beta in neuroligin-induced synaptic specificity, *J. Biol. Chem.*, 280, 17312, 2005.
148. Dean, C. and Dresbach, T., Neuroligins and neurexins: Linking cell adhesion, synapse formation and cognitive function, *Trends Neurosci.*, 29, 21, 2006.
149. Chubykin, A. et al., Activity-dependent validation of excitatory versus inhibitory synapses by neuroligin-1 versus neuroligin-2, *Neuron*, 54, 919, 2007.
150. Husi, H. et al., Proteomic analysis of NMDAR-adhesion protein signaling complexes, *Nat. Neurosci.*, 3, 661, 2000.
151. Tanaka, H. et al., Molecular modification of n-cadherin in response to synaptic activity, *Neuron*, 25, 93, 2000.
152. Petralia, R., Wang, Y., and Wenthold, R., NMDARs and PSD-95 are found in attachment plaques in cerebellar granular layer glomeruli, *Eur. J. Neurosci.*, 15, 583, 2002.
153. Bukalo, O. et al., Conditional ablation of the neural cell adhesion molecule reduces precision of spatial learning, long-term potentiation, and depression in the CA1 subfield of mouse hippocampus, *J. Neurosci.*, 24, 1565, 2004.
154. Demyanenko, G.P. et al., Close homolog of L1 modulates area-specific neuronal positioning and dendrite orientation in the cerebral cortex, *Neuron*, 44, 423, 2004.
155. Fux, C.M. et al., NCAM180 and glutamate receptor subtypes in potentiated spine synapses: an immunogold electron microscopic study, *Mol. Cell Neurosci.*, 24, 939, 2003.
156. Davey, F. et al., Synapse associated protein 102 is a novel binding partner to the cytoplasmic terminus of neurone-glial related cell adhesion molecule, *J. Neurochem.*, 94, 1243, 2005.
157. Chavis, P. and Westbrook, G., Integrins mediate functional pre- and postsynaptic maturation at a hippocampal synapse, *Nature*, 411, 317, 2001.
158. Lin, B. et al., Integrins regulate NMDAR-mediated synaptic currents, *J. Neurophysiol.*, 89, 2874, 2003.
159. Bernard-Trifilo, J.A. et al., Integrin signaling cascades are operational in adult hippocampal synapses and modulate NMDAR physiology, *J. Neurochem.*, 93, 834, 2005.
160. Yamada, K. and Nabeshima, T., Brain-derived neurotrophic factor/TrkB signaling in memory processes, *J. Pharmacol. Sci.*, 91, 267, 2003.
161. Elmariah, S. et al., Postsynaptic TrkB-mediated signaling modulates excitatory and inhibitory neurotransmitter receptor clustering at hippocampal synapses, *J. Neurosci.*, 24, 2380, 2004.
162. Yoshii, A. and Constantine-Paton, M., BDNF induces transport of PSD-95 to dendrites through PI3K-AKT signaling after NMDAR activation, *Nat. Neurosci.*, 10, 702, 2007.
163. Laurie, D.J. et al., Regional, developmental and interspecies expression of the four NMDAR2 subunits, examined using monoclonal antibodies, *Brain Res. Mol. Brain Res.*, 51, 23, 1997.
164. Ciabarra, A. et al., Cloning and characterization of chi-1: a developmentally regulated member of a novel class of the ionotropic glutamate receptor family, *J. Neurosci.*, 15, 6498, 1995.
165. Kirson, E. et al., Early postnatal switch in magnesium sensitivity of NMDARs in rat CA1 pyramidal cells, *J. Physiol.*, 521 Pt 1, 99, 1999.

166. Al-Hallaq, R.A. et al., Association of NR3A with the *N*-methyl-D-aspartate receptor NR1 and NR2 subunits, *Mol. Pharmacol.*, 62, 1119, 2002.

167. Laurie, D. et al., The distribution of splice variants of the NMDAR1 subunit mRNA in adult rat brain, *Brain Res. Mol. Brain Res.*, 32, 94, 1995.

168. Ozaki, M. et al., Neuregulin-beta induces expression of an NMDA-receptor subunit, *Nature*, 390, 691, 1997.

169. Garcia, R., Vasudevan, K., and Buonanno, A., The neuregulin receptor erbB-4 interacts with PDZ-containing proteins at neuronal synapses, *Proc. Natl. Acad. Sci. USA*, 97, 3596, 2000.

170. Hahn, C. et al., Altered neuregulin 1-erbB4 signaling contributes to NMDAR hypofunction in schizophrenia, *Nat. Med.*, 12, 824, 2006.

171. Erisir, A. and Harris, J., Decline of the critical period of visual plasticity is concurrent with the reduction of NR2B subunit of the synaptic NMDAR in layer 4, *J. Neurosci.*, 23, 5208, 2003.

172. Quinlan, E.M., Olstein, D.H., and Bear, M.F., Bidirectional, experience-dependent regulation of *N*-methyl-D-aspartate receptor subunit composition in the rat visual cortex during postnatal development, *Proc. Natl. Acad. Sci. USA*, 96, 12876, 1999.

173. Quinlan, E.M. et al., A molecular mechanism for stabilization of learning-induced synaptic modifications, *Neuron*, 41, 185, 2004.

174. Karavanova, I. et al., Novel regional and developmental NMDAR expression patterns uncovered in NR2C subunit-beta-galactosidase knock-in mice, *Mol. Cell Neurosci.*, 34, 468, 2007.

175. Van Zundert, B., Yoshii, A., and Constantine-Paton, M., Receptor compartmentalization and trafficking at glutamate synapses: a developmental proposal, *Trends Neurosci.*, 27, 428, 2004.

176. Yoshii, A., Sheng, M.H., and Constantine-Paton, M., Eye opening induces a rapid dendritic localization of PSD-95 in central visual neurons, *Proc. Natl. Acad. Sci. USA*, 100, 1334, 2003.

177. Li, Z. and Sheng, M., Some assembly required: development of neuronal synapses, *Nat. Rev. Mol. Cell Biol.*, 4, 833, 2003.

178. Bresler, T. et al., Postsynaptic density assembly is fundamentally different from presynaptic active zone assembly, *J. Neurosci.*, 24, 1507, 2004.

179. Sytnyk, V. et al., Neural cell adhesion molecule promotes accumulation of TGN organelles at sites of neuron-to-neuron contacts, *J. Cell Biol.*, 159, 649, 2002.

180. Polo-Parada, L. et al., Distinct roles of different neural cell adhesion molecule (NCAM) isoforms in synaptic maturation revealed by analysis of NCAM 180 kDa isoform-deficient mice, *J. Neurosci.*, 24, 1852, 2004.

181. Friedman, H. et al., Assembly of new individual excitatory synapses: time course and temporal order of synaptic molecule recruitment, *Neuron*, 27, 57, 2000.

182. Waites, C., Craig, A., and Garner, C., Mechanisms of vertebrate synaptogenesis, *Annu. Rev. Neurosci.*, 28, 251, 2005.

183. Gerrow, K. et al., A preformed complex of postsynaptic proteins is involved in excitatory synapse development, *Neuron*, 49, 547, 2006.

184. Bayer, K.U. et al., Developmental expression of the CaM kinase II isoforms: ubiquitous gamma- and delta-CaM kinase II are the early isoforms and most abundant in the developing nervous system, *Brain Res. Mol. Brain Res.*, 70, 147, 1999.

185. Petralia, R. et al., Selective acquisition of AMPA receptors over postnatal development suggests a molecular basis for silent synapses, *Nat. Neurosci.*, 2, 31, 1999.

186. Rampon, C. et al., Enrichment induces structural changes and recovery from nonspatial memory deficits in CA1 NMDAR1-knockout mice, *Nat. Neurosci.*, 3, 238, 2000.

187. Kennedy, M.B. and Manzerra, P., Telling tails, *Proc. Natl. Acad. Sci. USA*, 98, 12323, 2001.

188. Westphal, R.S. et al., Regulation of NMDARs by an associated phosphatase-kinase signaling complex, *Science,* 285, 93, 1999.
189. Chan, S. and Sucher, N., An NMDAR signaling complex with protein phosphatase 2a, *J. Neurosci.,* 21, 7985, 2001.
190. Dosemeci, A. et al., Glutamate-induced transient modification of the postsynaptic density, *Proc. Natl. Acad. Sci. USA,* 98, 10428, 2001.
191. Shen, K. and Meyer, T., Dynamic control of CaMKII translocation and localization in hippocampal neurons by NMDAR stimulation, *Science,* 284, 162, 1999.
192. Leonard, A.S. et al., Calcium/calmodulin-dependent protein kinase II is associated with the *N*-methyl-D-aspartate receptor, *Proc. Natl. Acad. Sci. USA,* 96, 3239, 1999.
193. Bayer, K.U. et al., Interaction with the NMDAR locks CaMKII in an active conformation, *Nature,* 411, 801, 2001.
194. Lisman, J. and Zhabotinsky, A., A model of synaptic memory: A CaMKII/PP1 switch that potentiates transmission by organizing an AMPA receptor anchoring assembly, *Neuron,* 31, 191, 2001.
195. Andrasfalvy, B.K. and Magee, J.C., Changes in AMPA receptor currents following LTP induction on rat CA1 pyramidal neurones, *J. Physiol.,* 559, 543, 2004.
196. Derkach, V., Barria, A., and Soderling, T.R., Ca^{2+}/calmodulin-kinase II enhances channel conductance of alpha-amino-3-hydroxy-5-methyl-4-isoxazolepropionate type glutamate receptors, *Proc. Natl. Acad. Sci. USA,* 96, 3269, 1999.
197. Gardoni, F. et al., AlphaCaMKII binding to the C-terminal tail of NMDAR subunit NR2A and its modulation by autophosphorylation, *FEBS Lett.,* 456, 394, 1999.
198. Gardoni, F. et al., Protein kinase C activation modulates alpha-calmodulin kinase II binding to NR2A subunit of *N*-methyl-D-aspartate receptor complex, *J. Biol. Chem.,* 276, 7609, 2001.
199. Strack, S. and Colbran, R.J., Autophosphorylation-dependent targeting of calcium/calmodulin-dependent protein kinase II by the NR2B subunit of the *N*-methyl-D-aspartate receptor, *J. Biol. Chem.,* 273, 20689, 1998.
200. Bayer, K.U. et al., Transition from reversible to persistent binding of CaMKII to postsynaptic sites and NR2b, *J. Neurosci.,* 26, 1164, 2006.
201. Gardoni, F. et al., CaMKII-dependent phosphorylation regulates SAP97/NR2A interaction, *J. Biol. Chem.,* 278, 44745, 2003.
202. Mauceri, D. et al., Dual role of CaMKII-dependent SAP97 phosphorylation in mediating trafficking and insertion of NMDAR subunit NR2a, *J. Neurochem.,* 100, 1032, 2007.
203. Fong, D.K. et al., Rapid synaptic remodeling by protein kinase C: reciprocal translocation of NMDARs and calcium/calmodulin-dependent kinase II, *J. Neurosci.,* 22, 2153, 2002.
204. Crump, F.T., Dillman, K.S., and Craig, A.M., cAMP-dependent protein kinase mediates activity-regulated synaptic targeting of NMDARs, *J. Neurosci.,* 21, 5079, 2001.
205. Skeberdis, V.A. et al., Protein kinase A regulates calcium permeability of NMDARs, *Nat. Neurosci.,* 9, 501, 2006.
206. Abel, T. et al., Genetic demonstration of a role for PKA in the late phase of LTP and in hippocampus-based long-term memory, *Cell,* 88, 615, 1997.
207. Colledge, M. et al., Targeting of PKA to glutamate receptors through a MAGUK-AKAP complex, *Neuron,* 27, 107, 2000.
208. Waltereit, R. and Weller, M., Signaling from cAMP/PKA to MAPK and synaptic plasticity, *Mol. Neurobiol.,* 27, 99, 2003.
209. Tingley, W.G. et al., Characterization of protein kinase A and protein kinase C phosphorylation of the *N*-methyl-D-aspartate receptor NR1 subunit using phosphorylation site-specific antibodies, *J. Biol. Chem.,* 272, 5157, 1997.

210. Chen, L. and Huang, L.Y., Protein kinase C reduces Mg^{2+} block of NMDA-receptor channels as a mechanism of modulation, *Nature,* 356, 521, 1992.
211. Gerber, G. et al., Multiple effects of phorbol esters in the rat spinal dorsal horn, *J. Neurosci.,* 9, 3606, 1989.
212. Xiong, Z.G. et al., Regulation of *N*-methyl-D-aspartate receptor function by constitutively active protein kinase C, *Mol. Pharmacol.,* 54, 1055, 1998.
213. Lan, J.Y. et al., Protein kinase C modulates NMDAR trafficking and gating, *Nat. Neurosci.,* 4, 382, 2001.
214. Macdonald, J.F. et al., Convergence of PKC-dependent kinase signal cascades on NMDARs, *Curr. Drug Targets,* 2, 299, 2001.
215. Zheng, X. et al., Protein kinase C potentiation of *N*-methyl-D-aspartate receptor activity is not mediated by phosphorylation of *N*-methyl-D-aspartate receptor subunits, *Proc. Natl. Acad. Sci. USA,* 96, 15262, 1999.
216. Lin, Y. et al., PSD-95 and PKC converge in regulating NMDAR trafficking and gating, *Proc. Natl. Acad. Sci. USA,* 103, 19902, 2006.
217. Ali, D. and Salter, M., NMDAR regulation by Src kinase signalling in excitatory synaptic transmission and plasticity, *Curr. Opin. Neurobiol.,* 11, 336, 2001.
218. Salter, M.W. and Kalia, L.V., Src kinases: A hub for NMDAR regulation, *Nat. Rev. Neurosci.,* 5, 317, 2004.
219. Wang, Y.T. and Salter, M.W., Regulation of NMDARs by tyrosine kinases and phosphatases, *Nature,* 369, 233, 1994.
220. Gingrich, J. et al., Unique domain anchoring of Src to synaptic NMDARs via the mitochondrial protein NADH dehydrogenase subunit 2, *Proc. Natl. Acad. Sci. USA,* 101, 6237, 2004.
221. Tezuka, T. et al., PSD-95 promotes Fyn-mediated tyrosine phosphorylation of the *N*-methyl-D-aspartate receptor subunit NR2A, *Proc. Natl. Acad. Sci. USA,* 96, 435, 1999.
222. Liao, G. et al., The postsynaptic density protein PSD-95 differentially regulates insulin- and Src-mediated current modulation of mouse NMDARs expressed in *Xenopus* oocytes, *J. Neurochem.,* 75, 282, 2000.
223. Seabold, G. et al., Interaction of the tyrosine kinase Pyk2 with the *N*-methyl-D-aspartate receptor complex via the Src homology 3 domains of PSD-95 and SAP102, *J. Biol. Chem.,* 278, 15040, 2003.
224. Yaka, R. et al., NMDAR function is regulated by the inhibitory scaffolding protein, rack1, *Proc. Natl. Acad. Sci. USA,* 99, 5710, 2002.
225. Dunah, A. and Standaert, D., Dopamine D1 receptor-dependent trafficking of striatal NMDA glutamate receptors to the postsynaptic membrane, *J. Neurosci.,* 21, 5546, 2001.
226. Nakazawa, T. et al., Characterization of Fyn-mediated tyrosine phosphorylation sites on GluR epsilon 2 (NR2B) subunit of the *N*-methyl-D-aspartate receptor, *J. Biol. Chem.,* 276, 693, 2001.
227. Wu, H. et al., Fyn-mediated phosphorylation of NR2B TYR-1336 controls calpain-mediated NR2B cleavage in neuron and heterologous systems, *J. Biol. Chem.,* 2007.
228. Lieberman, D.N. and Mody, I., Casein kinase-II regulates NMDA channel function in hippocampal neurons, *Nat. Neurosci.,* 2, 125, 1999.
229. Charriaut-Marlangue, C. et al., Rapid activation of hippocampal casein kinase II during long-term potentiation, *Proc. Natl. Acad. Sci. USA,* 88, 10232, 1991.
230. Li, B.S. et al., Regulation of NMDARs by cyclin-dependent kinase-5, *Proc. Natl. Acad. Sci. USA,* 98, 12742, 2001.
231. Wang, J. et al., Cdk5 activation induces hippocampal CA1 cell death by directly phosphorylating NMDARs, *Nat. Neurosci.,* 6, 1039, 2003.
232. Hawasli, A. et al., Cyclin-dependent kinase 5 governs learning and synaptic plasticity via control of NMDAR degradation, *Nat. Neurosci.,* 10, 880, 2007.

233. Jope, R. and Johnson, G., The glamour and gloom of glycogen synthase kinase-3, *Trends Biochem. Sci.,* 29, 95, 2004.
234. Chen, P. et al., Glycogen synthase kinase 3 regulates *N*-methyl-D-aspartate receptor channel trafficking and function in cortical neurons, *Mol. Pharmacol.,* 72, 40, 2007.
235. Pelkey, K. et al., Tyrosine phosphatase STEP is a tonic brake on induction of long-term potentiation, *Neuron,* 34, 127, 2002.
236. Nguyen, T.H., Liu, J., and Lombroso, P., Striatal enriched phosphatase 61 dephos-phorylates Fyn at phosphotyrosine 420, *J. Biol. Chem.,* 277, 24274, 2002.
237. Braithwaite, S. et al., Regulation of NMDAR trafficking and function by striatal-enriched tyrosine phosphatase (STEP), *Eur. J. Neurosci.,* 23, 2847, 2006.
238. Gu, W. et al., Requirement of PSD-95 for dopamine D1 receptor modulating gluta-mate NR1a/NR2B receptor function, *Acta. Pharmacol. Sin.,* 28, 756, 2007.
239. Snyder, G. et al., A dopamine/D1 receptor/protein kinase A/dopamine- and cAMP-regulated phosphoprotein (Mr 32 kDa)/protein phosphatase-1 pathway regulates dephosphorylation of the NMDAR, *J. Neurosci.,* 18, 10297, 1998.
240. Liu, L. et al., Insulin potentiates *N*-methyl-D-aspartate receptor activity in *Xenopus* oocytes and rat hippocampus, *Neurosci. Lett.,* 192, 5, 1995.
241. Christie, J., Wenthold, R., and Monaghan, D., Insulin causes a transient tyrosine phosphorylation of NR2A and NR2B NMDAR subunits in rat hippocampus, *J. Neu-rochem.,* 72, 1523, 1999.
242. Liao, G. and Leonard, J., Insulin modulation of cloned mouse NMDAR currents in *Xenopus* oocytes, *J. Neurochem.,* 73, 1510, 1999.
243. Kim, E. et al., GKAP, a novel synaptic protein that interacts with the guanylate kinase-like domain of the PSD-95/SAP90 family of channel clustering molecules, *J. Cell Biol.,* 136, 669, 1997.
244. Ehlers, M.D., Activity level controls postsynaptic composition and signaling via the ubiquitin-proteasome system, *Nat. Neurosci.,* 6, 231, 2003.
245. Liao, D. et al., Regulation of morphological postsynaptic silent synapses in develop-ing hippocampal neurons, *Nat. Neurosci.,* 2, 37, 1999.
246. Watt, A. et al., Activity coregulates quantal AMPA and NMDA currents at neocorti-cal synapses, *Neuron,* 26, 659, 2000.
247. El-Husseini, A.E. et al., Dual palmitoylation of PSD-95 mediates its vesiculotubular sorting, postsynaptic targeting, and ion channel clustering, *J. Cell Biol.,* 148, 159, 2000.
248. Huang, K. and El-Husseini, A., Modulation of neuronal protein trafficking and func-tion by palmitoylation, *Curr. Opin. Neurobiol.,* 15, 527, 2005.
249. Greaves, J. and Chamberlain, L., Palmitoylation-dependent protein sorting, *J. Cell Biol.,* 176, 249, 2007.
250. Morabito, M., Sheng, M., and Tsai, L., Cyclin-dependent kinase 5 phosphorylates the N-terminal domain of the postsynaptic density protein PSD-95 in neurons, *J. Neuro-sci.,* 24, 865, 2004.
251. Craven, S., El-Husseini, A.E., and Bredt, D.S., Synaptic targeting of the postsynap-tic density protein PSD-95 mediated by lipid and protein motifs, *Neuron,* 22, 497, 1999.
252. Gardoni, F. et al., Calcium-calmodulin-dependent protein kinase II phosphorylation modulates PSD-95 binding to NMDARs, *Eur. J. Neurosci.,* 24, 2694, 2006.
253. Jaffe, H., Vinade, L., and Dosemeci, A., Identification of novel phosphorylation sites on postsynaptic density proteins, *Biochem. Biophys. Res. Commun.,* 321, 210, 2004.
254. Nada, S. et al., Identification of PSD-93 as a substrate for the Src family tyrosine kinase fyn, *J. Biol. Chem.,* 278, 47610, 2003.
255. Colledge, M. et al., Ubiquitination regulates PSD-95 degradation and AMPA recep-tor surface expression, *Neuron,* 40, 595, 2003.

256. Petralia, R., Wang, Y., and Wenthold, R., The NMDAR subunits NR2A and NR2B show histological and ultrastructural localization patterns similar to those of NR1, *J. Neurosci.*, 14, 6102, 1994.
257. Petralia, R., Yokotani, N., and Wenthold, R., Light and electron microscope distribution of the NMDAR subunit NMDAR1 in the rat nervous system using a selective anti-peptide antibody, *J. Neurosci.*, 14, 667, 1994.
258. Wang, J., Andrews, H., and Thukral, V., Presynaptic glutamate receptors regulate noradrenaline release from isolated nerve terminals, *J. Neurochem.*, 58, 204, 1992.
259. Siegel, S. et al., Regional, cellular, and ultrastructural distribution of *N*-methyl-D-aspartate receptor subunit 1 in monkey hippocampus, *Proc. Natl. Acad. Sci. USA*, 91, 564, 1994.
260. Liu, H. et al., Evidence for presynaptic *N*-methyl-D-aspartate autoreceptors in the spinal cord dorsal horn, *Proc. Natl. Acad. Sci. USA*, 91, 8383, 1994.
261. Liu, H., Mantyh, P.W., and Basbaum, A., NMDA-receptor regulation of substance P release from primary afferent nociceptors, *Nature*, 386, 721, 1997.
262. Giovannucci, D. and Stuenkel, E., An NMDAR on isolated secretory nerve endings, *Brain Res.*, 702, 246, 1995.
263. Bardoni, R. et al., Presynaptic NMDARs modulate glutamate release from primary sensory neurons in rat spinal cord dorsal horn, *J. Neurosci.*, 24, 2774, 2004.
264. Casado, M., Isope, P., and Ascher, P., Involvement of presynaptic *N*-methyl-D-aspartate receptors in cerebellar long-term depression, *Neuron*, 33, 123, 2002.
265. Duguid, I.C. and Smart, T.G., Retrograde activation of presynaptic NMDARs enhances GABA release at cerebellar interneuron-Purkinje cell synapses, *Nat. Neurosci.*, 7, 525, 2004.
266. Lien, C.C. et al., Visual stimuli-induced LTD of GABAergic synapses mediated by presynaptic NMDARs, *Nat. Neurosci.*, 9, 372, 2006.
267. Sjostrom, P.J., Turrigiano, G.G., and Nelson, S.B., Neocortical LTD via coincident activation of presynaptic NMDA and cannabinoid receptors, *Neuron*, 39, 641, 2003.
268. Humeau, Y. et al., Presynaptic induction of heterosynaptic associative plasticity in the mammalian brain, *Nature*, 426, 841, 2003.
269. Wang, J.K. and Thukral, V., Presynaptic NMDARs display physiological characteristics of homomeric complexes of NR1 subunits that contain the exon 5 insert in the n-terminal domain, *J. Neurochem.*, 66, 865, 1996.
270. Glitsch, M. and Marty, A., Presynaptic effects of NMDA in cerebellar Purkinje cells and interneurons, *J. Neurosci.*, 19, 511, 1999.
271. Mameli, M. et al., Neurosteroid-induced plasticity of immature synapses via retrograde modulation of presynaptic NMDARs, *J. Neurosci.*, 25, 2285, 2005.
272. Thompson, C.L. et al., Immunohistochemical localization of *N*-methyl-D-aspartate receptor subunits in the adult murine hippocampal formation: Evidence for a unique role of the NR2D subunit, *Brain Res. Mol. Brain Res.*, 102, 55, 2002.
273. Jourdain, P. et al., Glutamate exocytosis from astrocytes controls synaptic strength, *Nat. Neurosci.*, 10, 331, 2007.
274. Yang, J., Woodhall, G.L., and Jones, R.S., Tonic facilitation of glutamate release by presynaptic NR2B-containing NMDARs is increased in the entorhinal cortex of chronically epileptic rats, *J. Neurosci.*, 26, 406, 2006.
275. Nong, Y., Huang, Y.Q., and Salter, M.W., NMDARs are movin' in, *Curr. Opin. Neurobiol.*, 14, 353, 2004.
276. Perez-Otano, I. and Ehlers, M.D., Learning from NMDAR trafficking: clues to the development and maturation of glutamatergic synapses, *Neurosignals*, 13, 175, 2004.
277. Lau, C. and Zukin, R.S., NMDAR trafficking in synaptic plasticity and neuropsychiatric disorders, *Nat. Rev. Neurosci.*, 8, 413, 2007.

278. Vissel, B. et al., A use-dependent tyrosine dephosphorylation of NMDARs is independent of ion flux, *Nat. Neurosci.,* 4, 587, 2001.

279. Nong, Y. et al., Glycine binding primes NMDAR internalization, *Nature,* 422, 302, 2003.

280. Snyder, E.M. et al., Internalization of ionotropic glutamate receptors in response to mGluR activation, *Nat. Neurosci.,* 4, 1079, 2001.

281. Roche, K.W. et al., Molecular determinants of NMDAR internalization, *Nat. Neurosci.,* 4, 794, 2001.

282. Lavezzari, G. et al., Differential binding of the AP-2 adaptor complex and PSD-95 to the C-terminus of the NMDAR subunit NR2B regulates surface expression, *Neuropharmacology,* 45, 729, 2003.

283. Scott, D.B. et al., Endocytosis and degradative sorting of NMDARs by conserved membrane-proximal signals, *J. Neurosci.,* 24, 7096, 2004.

284. Cottrell, J. et al., CPG2: A brain- and synapse-specific protein that regulates the endocytosis of glutamate receptors, *Neuron,* 44, 677, 2004.

285. Perez-Otano, I. et al., Endocytosis and synaptic removal of NR3A-containing NMDARs by PACSIN1/syndapin1, *Nat. Neurosci.,* 9, 611, 2006.

286. Racz, B. et al., Lateral organization of endocytic machinery in dendritic spines, *Nat. Neurosci.,* 7, 917, 2004.

287. Kirkham, M. and Parton, R.G., Clathrin-independent endocytosis: new insights into caveolae and non-caveolar lipid raft carriers, *Biochim. Biophys. Acta.,* 1746, 349, 2005.

288. Le Roy, C. and Wrana, J., Clathrin- and non-clathrin-mediated endocytic regulation of cell signalling, *Nat. Rev. Mol. Cell Biol.,* 6, 112, 2005.

289. Sigismund, S. et al., Clathrin-independent endocytosis of ubiquitinated cargos, *Proc. Natl. Acad. Sci. USA,* 102, 2760, 2005.

290. Kato, A. et al., Activity-dependent NMDAR degradation mediated by retrotranslocation and ubiquitination, *Proc. Natl. Acad. Sci. USA,* 102, 5600, 2005.

291. Ye, Y. et al., A membrane protein complex mediates retro-translocation from the ER lumen into the cytosol, *Nature,* 429, 841, 2004.

292. Dong, Y., Waxman, E.A., and Lynch, D.R., Interactions of postsynaptic density-95 and the NMDAR 2 subunit control calpain-mediated cleavage of the NMDAR, *J. Neurosci.,* 24, 11035, 2004.

293. Dong, Y. et al., Developmental and cell-selective variations in *N*-methyl-D-aspartate receptor degradation by calpain, *J. Neurochem.,* 99, 206, 2006.

294. Suzuki, T., Lipid rafts at postsynaptic sites: Distribution, function and linkage to postsynaptic density, *Neurosci. Res.,* 44, 1, 2002.

295. Swanwick, C. et al., Interaction of *N*-methyl-D-aspartate (NMDA) receptors with flotillin-1, a lipid raft-associated protein, *Soc. Neurosci. Abs.,* 31, 4, 2006.

296. Allen, J.A., Halverson-Tamboli, R.A., and Rasenick, M.M., Lipid raft microdomains and neurotransmitter signalling, *Nat. Rev. Neurosci.,* 8, 128, 2007.

297. Besshoh, S. et al., Developmental changes in the association of NMDARs with lipid rafts, *J. Neurosci. Res.,* 85, 1876, 2007.

298. Trushina, E. et al., Neurological abnormalities in caveolin-1 knock out mice, *Behav. Brain Res.,* 172, 24, 2006.

299. Trushina, E. et al., Mutant huntingtin inhibits clathrin-independent endocytosis and causes accumulation of cholesterol *in vitro* and *in vivo, Hum. Mol. Genet.,* 15, 3578, 2006.

300. Choi, Y.B. et al., Molecular basis of NMDAR-coupled ion channel modulation by S-nitrosylation, *Nat. Neurosci.,* 3, 15, 2000.

301. Lynch, D. and Guttmann, R., Excitotoxicity: perspectives based on *N*-methyl-D-aspartate receptor subtypes, *J. Pharmacol. Exp. Ther.,* 300, 717, 2002.

302. Tymianski, M. et al., Source specificity of early calcium neurotoxicity in cultured embryonic spinal neurons, *J. Neurosci.*, 13, 2085, 1993.
303. Grossman, S., Rosenberg, L., and Wrathall, J., Temporal-spatial pattern of acute neuronal and glial loss after spinal cord contusion, *Exp. Neurol.*, 168, 273, 2001.
304. Small, D., Mok, S., and Bornstein, J., Alzheimer's disease and amyloid beta toxicity: from top to bottom, *Nat. Rev. Neurosci.*, 2, 595, 2001.
305. Selkoe, D.J., Alzheimer's disease is a synaptic failure, *Science*, 298, 789, 2002.
306. Fan, M. and Raymond, L.A., *N*-methyl-D-aspartate (NMDA) receptor function and excitotoxicity in Huntington's disease, *Prog. Neurobiol.*, 81, 272, 2007.
307. Cheng, C., Fass, D., and Reynolds, I., Emergence of excitotoxicity in cultured forebrain neurons coincides with larger glutamate-stimulated [Ca(2+)](I) increases and NMDAR mRNA levels, *Brain Res.*, 849, 97, 1999.
308. Von Engelhardt, J. et al., Excitotoxicity in vitro by NR2A- and NR2B-containing NMDARs, *Neuropharmacology*, 2007.
309. Al-Hallaq, R.A., Yasuda, R.P., and Wolfe, B.B., Enrichment of *N*-methyl-D-aspartate NR1 splice variants and synaptic proteins in rat postsynaptic densities, *J. Neurochem.*, 77, 110, 2001.
310. Hardingham, G., Fukunaga, Y., and Bading, H., Extrasynaptic NMDARs oppose synaptic NMDARs by triggering CREB shut-off and cell death pathways, *Nat. Neurosci.*, 5, 405, 2002.
311. Soriano, F.X. et al., Preconditioning doses of NMDA promote neuroprotection by enhancing neuronal excitability, *J. Neurosci.*, 26, 4509, 2006.
312. Hardingham, G. and Bading, H., The yin and yang of NMDAR signalling, *Trends Neurosci.*, 26, 81, 2003.
313. Harper, P., The epidemiology of Huntington's disease, *Hum. Genet.*, 89, 365, 1992.
314. Landles, C. and Bates, G.P., Huntingtin and the molecular pathogenesis of Huntington's disease. Fourth in molecular medicine review series, *EMBO Rep.*, 5, 958, 2004.
315. Vonsattel, J. and Difiglia, M., Huntington disease, *J. Neuropathol. Exp. Neurol.*, 57, 369, 1998.
316. Young, A. et al., NMDAR losses in putamen from patients with Huntington's disease, *Science*, 241, 981, 1988.
317. Difiglia, M., Excitotoxic injury of the neostriatum: model for Huntington's disease, *Trends Neurosci.*, 13, 286, 1990.
318. Schwarcz, R. and Kohler, C., Differential vulnerability of central neurons of the rat to quinolinic acid, *Neurosci. Lett.*, 38, 85, 1983.
319. Hantraye, P. et al., A primate model of Huntington's disease: behavioral and anatomical studies of unilateral excitotoxic lesions of the caudate-putamen in the baboon, *Exp. Neurol.*, 108, 91, 1990.
320. Levine, M. et al., Genetic mouse models of Huntington's and Parkinson's diseases: illuminating but imperfect, *Trends Neurosci.*, 27, 691, 2004.
321. Lawrence, A. et al., Executive and mnemonic functions in early Huntington's disease, *Brain*, 119, 1633, 1996.
322. Jason, G.W. et al., Cognitive manifestations of Huntington disease in relation to genetic structure and clinical onset, *Arch. Neurol.*, 54, 1081, 1997.
323. Paulsen, J. et al., Neuropsychiatric aspects of Huntington's disease, *J. Neurol. Neurosurg. Psychiatry*, 71, 310, 2001.
324. Andre, V. et al., Altered cortical glutamate receptor function in the R6/2 model of Huntington's disease, *J. Neurophysiol.*, 95, 2108, 2006.
325. Starling, A. et al., Alterations in *N*-methyl-D-aspartate receptor sensitivity and magnesium blockade occur early in development in the R6/2 mouse model of Huntington's disease, *J. Neurosci. Res.*, 82, 377, 2005.

326. Chen, N. et al., Subtype-specific enhancement of NMDAR currents by mutant huntingtin, *J. Neurochem.*, 72, 1890, 1999.
327. Landwehrmeyer, G. et al., NMDAR subunit mRNA expression by projection neurons and interneurons in rat striatum, *J. Neurosci.*, 15, 5297, 1995.
328. Kuppenbender, K.D. et al., Expression of NMDAR subunit mRNAs in neurochemically identified projection and interneurons in the human striatum, *J. Comp. Neurol.*, 419, 407, 2000.
329. Li, L. et al., Role of NR2B-type NMDARs in selective neurodegeneration in Huntington disease, *Neurobiol. Aging*, 24, 1113, 2003.
330. Zeron, M. et al., Increased sensitivity to *N*-methyl-D-aspartate receptor-mediated excitotoxicity in a mouse model of Huntington's disease, *Neuron*, 33, 849, 2002.
331. Zeron, M. et al., Mutant huntingtin enhances excitotoxic cell death, *Mol. Cell Neurosci.*, 17, 41, 2001.
332. Sun, Y. et al., Polyglutamine-expanded huntingtin promotes sensitization of *N*-methyl-D-aspartate receptors via postsynaptic density 95, *J. Biol. Chem.*, 276, 24713, 2001.
333. Leavitt, B.R. et al., Wild-type huntingtin protects neurons from excitotoxicity, *J. Neurochem.*, 96, 1121, 2006.
334. Luthi-Carter, R. et al., Complex alteration of NMDARs in transgenic Huntington's disease mouse brain: analysis of mRNA and protein expression, plasma membrane association, interacting proteins, and phosphorylation, *Neurobiol. Dis.*, 14, 624, 2003.
335. Jarabek, B., Yasuda, R.P., and Wolfe, B.B., Regulation of proteins affecting NMDAR-induced excitotoxicity in a Huntington's mouse model, *Brain*, 127, 505, 2004.
336. Christopherson, K. et al., PSD-95 assembles a ternary complex with the *N*-methyl-D-aspartic acid receptor and a bivalent neuronal NO synthase PDZ domain, *J. Biol. Chem.*, 274, 27467, 1999.
337. Song, C. et al., Expression of polyglutamine-expanded huntingtin induces tyrosine phosphorylation of *N*-methyl-D-aspartate receptors, *J. Biol. Chem.*, 278, 33364, 2003.
338. Dudman, J.T. et al., Dopamine D1 receptors mediate CREB phosphorylation via phosphorylation of the NMDAR at Ser897-NR1, *J. Neurochem.*, 87, 922, 2003.
339. Obrietan, K. and Hoyt, K., CRE-mediated transcription is increased in Huntington's disease transgenic mice, *J. Neurosci.*, 24, 791, 2004.
340. Lipton, S.A., The molecular basis of memantine action in Alzheimer's disease and other neurologic disorders: low-affinity, uncompetitive antagonism, *Curr. Alzheimer Res.*, 2, 155, 2005.
341. Scarpini, E., Scheltens, P., and Feldman, H., Treatment of Alzheimer's disease: current status and new perspectives, *Lancet Neurol.*, 2, 539, 2003.
342. Davies, C. et al., A quantitative morphometric analysis of the neuronal and synaptic content of the frontal and temporal cortex in patients with Alzheimer's disease, *J. Neurol. Sci.*, 78, 151, 1987.
343. Shrestha, B. et al., Amyloid beta peptide adversely affects spine number and motility in hippocampal neurons, *Mol. Cell Neurosci.*, 33, 274, 2006.
344. Lacor, P.N. et al., Abeta oligomer-induced aberrations in synapse composition, shape, and density provide a molecular basis for loss of connectivity in Alzheimer's disease, *J. Neurosci.*, 27, 796, 2007.
345. Shankar, G. et al., Natural oligomers of the alzheimer amyloid-beta protein induce reversible synapse loss by modulating an NMDA-type glutamate receptor-dependent signaling pathway, *J. Neurosci.*, 27, 2866, 2007.
346. Lambert, M. et al., Diffusible, nonfibrillar ligands derived from A-beta 1-42 are potent central nervous system neurotoxins, *Proc. Natl. Acad. Sci. USA*, 95, 6448, 1998.

347. Cleary, J. et al., Beta-amyloid(1-40) effects on behavior and memory, *Brain Res.*, 682, 69, 1995.
348. Walsh, D.M. and Selkoe, D.J., Deciphering the molecular basis of memory failure in Alzheimer's disease, *Neuron*, 44, 181, 2004.
349. Townsend, M. et al., Orally available compound prevents deficits in memory caused by the alzheimer amyloid-beta oligomers, *Ann. Neurol.*, 60, 668, 2006.
350. Kamenetz, F. et al., APP processing and synaptic function, *Neuron*, 37, 925, 2003.
351. Hynd, M., Scott, H.L., and Dodd, P., Glutamate(NMDA) receptor NR1 subunit mRNA expression in Alzheimer's disease, *J. Neurochem.*, 78, 175, 2001.
352. Hynd, M., Scott, H.L., and Dodd, P., Selective loss of NMDAR NR1 subunit isoforms in Alzheimer's disease, *J. Neurochem.*, 89, 240, 2004.
353. Hynd, M., Scott, H., and Dodd, P., Differential expression of *N*-methyl-D-aspartate receptor NR2 isoforms in Alzheimer's disease, *J. Neurochem.*, 90, 913, 2004.
354. Mishizen-Eberz, A.J. et al., Biochemical and molecular studies of NMDAR subunits NR1/2A/2B in hippocampal subregions throughout progression of Alzheimer's disease pathology, *Neurobiol. Dis.*, 15, 80, 2004.
355. Hynd, M., Scott, H.L., and Dodd, P., Glutamate-mediated excitotoxicity and neurodegeneration in Alzheimer's disease, *Neurochem. Int.*, 45, 583, 2004.
356. Snyder, E. et al., Regulation of NMDAR trafficking by amyloid-beta, *Nat. Neurosci.*, 8, 1051, 2005.
357. Goto, Y. et al., Amyloid beta-peptide preconditioning reduces glutamate-induced neurotoxicity by promoting endocytosis of NMDAR, *Biochem. Biophys. Res. Commun.*, 351, 259, 2006.
358. Hallett, P. et al., Alterations of striatal NMDAR subunits associated with the development of dyskinesia in the MPTP-lesioned primate model of Parkinson's disease, *Neuropharmacology*, 48, 503, 2005.
359. Hallett, P., and Standaert, D., Rationale for and use of NMDAR antagonists in Parkinson's disease, *Pharmacol. Ther.*, 102, 155, 2004.
360. Gardoni, F. and Di Luca, M., New targets for pharmacological intervention in the glutamatergic synapse, *Eur. J. Pharmacol.*, 545, 2, 2006.
361. Nash, J. et al., Antiparkinsonian actions of ifenprodil in the MPTP-lesioned marmoset model of Parkinson's disease, *Exp. Neurol.*, 165, 136, 2000.
362. Loschmann, P. et al., Antiparkinsonian activity of Ro 25-6981, a NR2B subunit specific NMDAR antagonist, in animal models of Parkinson's disease, *Exp. Neurol.*, 187, 86, 2004.
363. Uitti, R.J. et al., Amantadine treatment is an independent predictor of improved survival in Parkinson's disease, *Neurology*, 46, 1551, 1996.
364. Calabresi, P. et al., Electrophysiology of dopamine-denervated striatal neurons. Implications for Parkinson's disease, *Brain*, 116, 433, 1993.
365. Arbuthnott, G., Ingham, C., and Wickens, J., Dopamine and synaptic plasticity in the neostriatum, *J. Anat.*, 196, 587, 2000.
366. Calabresi, P., Centonze, D., and Bernardi, G., Electrophysiology of dopamine in normal and denervated striatal neurons, *Trends Neurosci.*, 23, S57, 2000.
367. Calabresi, P. et al., Levodopa-induced dyskinesia: A pathological form of striatal synaptic plasticity?, *Ann. Neurol.*, 47, S60, 2000.
368. Meoni, P. et al., NMDA NR1 subunit mRNA and glutamate NMDA-sensitive binding are differentially affected in the striatum and pre-frontal cortex of Parkinson's disease patients, *Neuropharmacology*, 38, 625, 1999.
369. Calon, F. et al., Levodopa-induced motor complications are associated with alterations of glutamate receptors in Parkinson's disease, *Neurobiol. Dis.*, 14, 404, 2003.
370. Menegoz, M. et al., Tyrosine phosphorylation of NMDAR in rat striatum: Effects of 6-OH-dopamine lesions, *Neuroreport*, 7, 125, 1995.

371. Dunah, A. et al., Alterations in subunit expression, composition, and phosphorylation of striatal *N*-methyl-D-aspartate glutamate receptors in a rat 6-hydroxydopamine model of parkinson's disease, *Mol. Pharmacol.*, 57, 342, 2000.
372. Picconi, B. et al., Abnormal Ca^{2+}-calmodulin-dependent protein kinase II function mediates synaptic and motor deficits in experimental parkinsonism, *J. Neurosci.*, 24, 5283, 2004.
373. Oh, J. et al., Enhanced tyrosine phosphorylation of striatal NMDAR subunits: Effect of dopaminergic denervation and l-DOPA administration, *Brain Res.*, 813, 150, 1998.
374. Betarbet, R. et al., Differential expression and Ser897 phosphorylation of striatal *N*-methyl-D-aspartate receptor subunit NR1 in animal models of Parkinson's disease, *Exp. Neurol.*, 187, 76, 2004.
375. Oh, J., Vaughan, C., and Chase, T., Effect of dopamine denervation and dopamine agonist administration on serine phosphorylation of striatal NMDAR subunits, *Brain Res.*, 821, 433, 1999.
376. Stahl, S.M., Beyond the dopamine hypothesis to the NMDA glutamate receptor hypofunction hypothesis of schizophrenia, *CNS Spectr.*, 12, 265, 2007.
377. Luby, E. et al., Study of a new schizophrenomimetic drug; sernyl, *AMA Arch. Neurol. Psychiatry*, 81, 363, 1959.
378. Gao, X. et al., Ionotropic glutamate receptors and expression of *N*-methyl-D-aspartate receptor subunits in subregions of human hippocampus: Effects of schizophrenia, *Am. J. Psychiatry*, 157, 1141, 2000.
379. Meador-Woodruff, J.H. and Healy, D.J., Glutamate receptor expression in schizophrenic brain, *Brain Res. Brain Res. Rev.*, 31, 288, 2000.
380. Dracheva, S. et al., N-methyl-D-aspartic acid receptor expression in the dorsolateral prefrontal cortex of elderly patients with schizophrenia, *Am. J. Psychiatry*, 158, 1400, 2001.
381. Mueller, H. and Meador-Woodruff, J.H., NR3A NMDAR subunit mRNA expression in schizophrenia, depression and bipolar disorder, *Schizophr. Res.*, 71, 361, 2004.
382. Mohn, A. et al., Mice with reduced NMDAR expression display behaviors related to schizophrenia, *Cell*, 98, 427, 1999.
383. Kristiansen, L., and Meador-Woodruff, J.H., Abnormal striatal expression of transcripts encoding NMDA interacting PSD proteins in schizophrenia, bipolar disorder and major depression, *Schizophr. Res.*, 78, 87, 2005.
384. Large, C., Do NMDAR antagonist models of schizophrenia predict the clinical efficacy of antipsychotic drugs? *J. Psychopharmacol.*, 21, 283, 2007.
385. Stefansson, H. et al., Neuregulin 1 and susceptibility to schizophrenia, *Am. J. Hum. Genet.*, 71, 877, 2002.
386. Gu, Z. et al., Regulation of NMDARs by neuregulin signaling in prefrontal cortex, *J. Neurosci.*, 25, 4974, 2005.
387. Kwon, O. et al., Neuregulin-1 reverses long-term potentiation at CA1 hippocampal synapses, *J. Neurosci.*, 25, 9378, 2005.
388. Bjarnadottir, M. et al., Neuregulin1 (NRG1) signaling through fyn modulates NMDAR phosphorylation: differential synaptic function in NRG1+/− knockouts compared with wild-type mice, *J. Neurosci.*, 27, 4519, 2007.
389. Zarate, C.A. et al., Modulators of the glutamatergic system: Implications for the development of improved therapeutics in mood disorders, *Psychopharmacol. Bull.*, 36, 35, 2002.
390. Zarate, C.J. et al., A randomized trial of an *N*-methyl-D-aspartate antagonist in treatment-resistant major depression, *Arch. Gen. Psychiatry*, 63, 856, 2006.
391. Boyce-Rustay, J.M. and Holmes, A., Genetic inactivation of the NMDAR NR2A subunit has anxiolytic- and antidepressant-like effects in mice, *Neuropsychopharmacology*, 31, 2405, 2006.

392. Kitamura, T., Mishina, M., and Sugiyama, H., Enhancement of neurogenesis by running wheel exercises is suppressed in mice lacking NMDAR epsilon 1 subunit, *Neurosci. Res.,* 47, 55, 2003.
393. Becker, S. and Wojtowicz, J., A model of hippocampal neurogenesis in memory and mood disorders, *Trends Cogn. Sci.,* 11, 70, 2007.
394. Basselin, M. et al., Chronic lithium chloride administration attenuates brain NMDAR-initiated signaling via arachidonic acid in unanesthetized rats, *Neuropsychopharmacology,* 31, 1659, 2006.
395. Phiel, C. and Klein, P., Molecular targets of lithium action, *Annu. Rev. Pharmacol. Toxicol.,* 41, 789, 2001.
396. McClung, C., Circadian genes, rhythms and the biology of mood disorders, *Pharmacol. Ther.,* 114, 222, 2007.
397. Chen, C. et al., Altered NMDAR trafficking contributes to sleep deprivation-induced hippocampal synaptic and cognitive impairments, *Biochem. Biophys. Res. Commun.,* 340, 435, 2006.
398. McDermott, C. et al., Sleep deprivation-induced alterations in excitatory synaptic transmission in the CA1 region of the rat hippocampus, *J. Physiol.,* 570, 553, 2006.
399. Kopp, C., Longordo, F., and Luthi, A., Experience-dependent changes in NMDAR composition at mature central synapses, *Neuropharmacology,* 2007.
400. Woo, T., Walsh, J.P., and Benes, F.M., Density of glutamic acid decarboxylase 67 messenger RNA-containing neurons that express the *N*-methyl-D-aspartate receptor subunit NR2A in the anterior cingulate cortex in schizophrenia and bipolar disorder, *Arch. Gen. Psychiatry,* 61, 649, 2004.
401. McCullum-Smith, R. et al., Decreased NR1, NR2A, and SAP102 transcript expression in the hippocampus in bipolar disorder, *Brain Res.,* 1127, 108, 2007.
402. Degiorgio, L. et al., A subset of lupus anti-DNA antibodies cross-reacts with the NR2 glutamate receptor in systemic lupus erythematosus, *Nat. Med.,* 7, 1189, 2001.
403. Kowal, C. et al., Human lupus autoantibodies against NMDARs mediate cognitive impairment, *Proc. Natl. Acad. Sci. USA,* 103, 19854, 2006.
404. Bleakman, D., Alt, A., and Nisenbaum, E., Glutamate receptors and pain, *Semin. Cell Dev. Biol.,* 17, 592, 2006.
405. Wang, C. and Shuaib, A., NMDA/NR2B selective antagonists in the treatment of ischemic brain injury, *Curr. Drug Targets CNS Neurol. Disord.,* 4, 143, 2005.
406. Eisen, A. and Weber, M., Treatment of amyotrophic lateral sclerosis, *Drugs Aging,* 14, 173, 1999.
407. Kaul, M. and Lipton, S., Signaling pathways to neuronal damage and apoptosis in human immunodeficiency virus type 1-associated dementia: chemokine receptors, excitotoxicity, and beyond, *J. Neurovirol.,* 10 Suppl 1, 97, 2004.

9 NMDA Receptor-Mediated Calcium Transients in Dendritic Spines

Brenda L. Bloodgood and Bernardo L. Sabatini

CONTENTS

9.1 INTRODUCTION

In pyramidal neurons of the hippocampus and cortex, NMDA-type glutamate receptors (NMDARs) are the predominant sources of synaptically evoked calcium (Ca) signals[1-7] (Figure 9.1A through C). Ca influx through NMDARs regulates diverse processes including kinase and phosphatase activity, protein trafficking, structural and functional synaptic plasticity, cell growth, cell survival, and apoptosis.[8-11] Which of these many Ca-dependent processes are triggered when NMDARs open depend on the context of receptor activation and the magnitude, kinetics, timing, and spatial spread of the resulting Ca transients. This chapter reviews the features of NMDARs that determine Ca influx through the receptors and discusses how the context of NMDAR activation shapes synaptically evoked Ca transients.

9.2 SUBUNIT DEPENDENCE OF NMDA RECEPTOR-MEDIATED CALCIUM INFLUX

NMDARs are heteromeric tetramers typically composed of NR1 subunits and NR2 or NR3 subunits.[12,13] Each subunit has multiple isoforms and in some cases multiple splice variants.[14,15] This structural diversity is functionally relevant; the specific

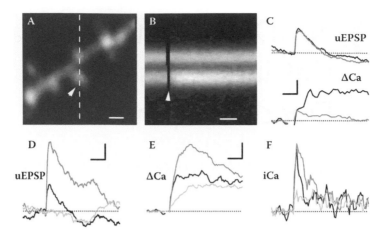

FIGURE 9.1 (See color insert following page 212). NMDAR-dependent calcium influx into active spines is modulated by AMPARs and SK channels. (A) Spiny dendrite from a mouse CA1 pyramidal neuron filled through a somatic whole-cell recording electrode with the red Ca-insensitive fluorophore Alexa-594 and the green Ca-sensitive fluorophore Fluo-5F. (B) Line scans through the dendrite and spine as indicated by the dashed line in (A) Arrow heads in (A) and (B) indicate locations and timings of 500 μsec laser pulse used to trigger 2-photon mediated photolysis of MNI glutamate. Uncaged glutamate binds to and opens NMDARs, resulting in Ca influx into the spine head seen as an increase in green fluorescence. The accompanying uncaging-evoked excitatory postsynaptic potential (uEPSP) is not shown. (C) Examples of uEPSPs (top) and spine head Ca transients (bottom) generated by glutamate uncaging at a single spine in control conditions (black line) or in the presence of the NMDAR antagonists CPP and MK-801 (red). (D) uEPSPs from individual spines in control conditions (black line) in the presence of the SK channel antagonist apamin (red) and the AMPAR antagonist NBQX (green). (E) Fluorescence transients measured in the spine head in response to the uEPSPs shown in (D) The amplitude of the fluorescence transient was directly proportional to spine head Ca. (F) Time course of the calculated spine head calcium currents in the three conditions. Ca currents were estimated by deconvolving the spine head Ca transients shown in (E) with the impulse response function of spine head Ca handling (not shown). Scale bars: (A) 1 μm; (B) 25 msec; (C) 0.5 mV, 5% $\Delta G/G_{sat}$, 25 msec; (D) 0.5 mV, 25 msec; (E) 5% $\Delta G/G_{sat}$, 25 msec.

subunit composition of a receptor along with the timing and magnitude of local membrane potential fluctuations determines the duration and magnitude of Ca current through NMDARs.

Each of the four NR2 subunits (NR2A through D) has a glutamate binding site.[16–18] However, the affinity for glutamate differs among the isoforms such that NR2A has the lowest affinity, NR2D the highest, and NR2B and NR2C have intermediate affinities. Generally, high binding affinity indicates a low dissociation rate of glutamate from the receptor and more prolonged NMDAR opening following glutamate binding. Thus, differential glutamate affinity may explain some of the variability in NMDAR deactivation kinetics. Receptors containing the NR2A subunit generate currents that decay rapidly (t ~120 msec) in comparison to those containing NR2B and NR2C (t ~400 msec) or NR2D (t ~5 sec).[19–21]

Similarly, receptor affinity for Magnesium (Mg) also varies with the NR2 subunit such that NR2A and NR2B are more susceptible to block by extracellular Mg and show greater voltage dependence than NR2C and NR2D.[19,22,23] Finally, single-channel conductance is subunit-dependent such that NR2A- and NR2B-containing receptors conduct nearly twice as much current as NR2C- and NR2D-containing receptors.[20,23–25] For these reasons, activation of NR2A-containing receptors generates relatively large and fast currents. In comparison, current influx through NR2B-containing receptors is also large but lasts far longer. NR2C- and NR2D-containing receptors generate the smallest and longest lasting currents. For similar reasons, receptors containing different NR2 subunits generate Ca transients with different time courses.

The influence of NMDAR subunit composition on Ca signaling suggests that activation of receptors composed of distinct subunit combinations may trigger different biological pathways. This may partially explain the wide range of physiological outcomes associated with NMDAR signaling. Moreover, regulation of receptor subunit composition may provide a cell or even an individual synapse with an efficient mechanism for determining which Ca-dependent signaling cascades are engaged.

Recent studies employing a wide range of techniques suggest this may be the case. Immunogold electron microscopy suggests that NR2A- and NR2B-containing receptors are often segregated so that most spines express NR2A or NR2B but not both.[26] Stimulation of single postsynaptic terminals using two-photon uncaging of glutamate has shown that the contributions of NR2A- and NR2B-containing receptors to NMDAR-dependent currents and evoked Ca transients vary widely from spine to spine.[27] Furthermore, antagonism of NR2B-containing receptors with ifenprodil reduced the intraspine variability and the amplitude of NMDAR-mediated calcium transients, indicating that NR2A- and NR2B-containing receptors are present in spines and that NR2B-containing receptors flux more calcium.

A developmental switch from NR2B- to NR2A-containing NMDARs occurs in many brain areas.[28,29] However, a recent study indicates that the subunit composition is also rapidly regulated in response to plasticity inducing stimuli.[30] Thus, induction of long-term potentiation at CA3 to CA1 synapses in hippocampus of young rats is accompanied by a switch from NR2B- to NR2A-containing synaptic NMDARs. This switch accelerates the decay kinetics of NMDAR-mediated synaptic currents and, although not measured directly in the cited study, should also alter the time course of NMDAR-dependent Ca influx and Ca transients in spines.

Differential coupling to downstream signaling systems[31,32] may allow opening of NR2A- versus NR2B-containing receptors to have different functional implications for plasticity induction[33–35] (but see references 36–38). Therefore, developmental and activity-dependent changes in NMDAR subunit composition, through regulation of synaptically evoked Ca influx, may constitute a form of metaplasticity that regulates the induction of activity-dependent synaptic plasticity.

9.3 PHOSPHORYLATION-DEPENDENT REGULATION OF NMDA RECEPTOR-DEPENDENT CALCIUM ENTRY

Regulation by phosphorylation provides a rapid means to alter the Ca permeability of NMDARs. PKA activity enhances the Ca permeability of both NR2A- and NR2B-containing receptors.[39] Furthermore, NMDAR Ca signaling is controlled by a negative feedback loop by which repetitive activation of NR2B-containing

NMDARs activates a serine–threonine phosphatase that decreases Ca permeability of NMDARs.[31] These data suggest that the Ca permeability of NMDARs may be regulated by an AKAP protein complex[40] associated with the NR2B subunit, although this has not been explicitly demonstrated.

9.4 VOLTAGE-DEPENDENT REGULATION OF SYNAPTICALLY EVOKED CALCIUM INFLUX

As discussed above, the amount of Ca entering via open NMDARs is governed by many intrinsic features of the receptors including Ca permeability, glutamate affinity, and Mg affinity. When a synapse is stimulated, NMDAR-dependent Ca transients will also be shaped by extrinsic factors such as the context of receptor activation. For example, Ca current through a receptor is greatly regulated by membrane potential. Since the membrane potential is controlled by a wide array of ion channels, the activities of many channels have the capability to shape Ca influx through NMDARs. Furthermore, the concentration of Ca reached in postsynaptic terminals is determined by the Ca buffering capacity, Ca extrusion rate, and diffusional isolation of the terminals. This section and the following one consider the impacts of these extrinsic receptor factors on NMDAR dependent Ca transients.

The most powerful and rapidly adjustable factor that regulates NMDAR-dependent Ca flux is membrane potential. Changes in membrane potential alter the driving force for Ca entry and the degree of Mg block of the receptor.[41] The vastly asymmetric distribution of Ca across membranes results in a high Ca reversal potential (~125 mV assuming Ca concentration is 2 mM outside and 100 nM inside the cell). Because of this large driving force, a 20 mV depolarization from rest will reduce the driving force for Ca entry by roughly 10 to 15%. This depressive effect is modest compared to the large enhancement of Ca entry caused by partial relief of Mg block. As demonstrated in classic studies, the affinity of Mg for the NMDAR is decreased nearly 10-fold by a 20 mV depolarization in the subthreshold range.[42] Thus, depolarization from −70 to −50 mV, despite decreasing the driving force for Ca entry, increases current flow through NMDARs and the magnitude of evoked Ca transients.

The effects of voltage-dependent Mg block on synaptically evoked Ca transients can be seen in several contexts. *In vivo*, many neurons show large fluctuations in the resting membrane potential that, when mimicked by *in vitro* whole cell recordings, exert large effects on NMDAR-dependent Ca influx. For example, in striatal medium spiny neurons (MSNs) the amplitude of NMDAR-mediated Ca transients in active dendritic spines increases nearly four fold with depolarization from −80 to −50 mV.[43] This effect may contribute to the dependence of the induction of long-term synaptic plasticity on resting membrane potential in these cells.[44–46] As expected for NMDAR-mediated signals, similar effects of resting or holding potential on synaptically evoked Ca transients have been described in other cell types.[1,3,4,47]

Transient changes in spine membrane potential occur during back-propagating action potentials (bAPs) that, in many cell types, can travel from the soma into the proximal dendrite and dendritic spines. Back-propagation of an action potential into the spine transiently relieves the Mg block of NMDARs. The rapid kinetics of Mg block[48] provides a brief enhancement of Ca influx through NMDARs that lasts approximately as long as the bAP and can be seen as a nonlinear increase in Ca entry into active spines.[5,6,43,49–51]

Several lines of evidence indicate that, even in the absence of bAPs, the membrane of an active spine or a stretch of dendrite with multiple active synapses can experience large depolarizations that shape NMDAR-dependent Ca influx. These dynamic effects modulate Ca influx during the synaptic potential. In both hippocampal CA1 and lateral amygdala pyramidal neurons, blockade of SK-type Ca-activated K channels (SKs) modulates NMDAR-mediated synaptic currents in a Mg-dependent manner.[52,53] These studies suggested that SKs either repolarize or hyperpolarize the membrane near the active synapse and rapidly alter the Mg block of synaptically activated NMDARs.

This signaling cascade has been examined in more detail in spines of hippocampal CA1 pyramidal neurons[2] (Figure 9.2). In these cells, the blockade of SK channels with apamin nearly doubled the amplitude of NMDAR-mediated Ca transients in active spines. Opening of SKs in the spine is triggered by the entry of Ca through SNX-482 sensitive voltage-sensitive Ca channels (VSCCs, presumably $Ca_V2.3$). Since these are high, voltage activated VSCCs, this suggests that the spine must be

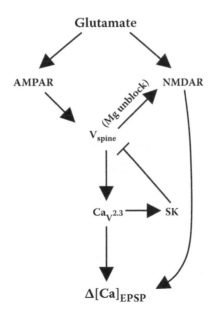

FIGURE 9.2 Model of regulation of spine head Ca transients by glutamate receptors and ion channels. Glutamate release activates AMPARs and NMDARs in the spine head. Opening of AMPARs depolarizes the spine head, enhancing current flow through NMDARs by relief of Mg block. AMPAR-mediated depolarization of the spine also activates voltage-gated Ca channels located on spine membranes. Ca influx through $Ca_V2.3$ channels specifically activates SK channels that repolarize the spine and terminate NMDAR signaling.

depolarized many tens of millivolts to reach the threshold for activation of the channels. Furthermore, since the opening of other VSCCs that are known to be present in the spine does not activate SKs, SK channels must lie in the Ca microdomain of $Ca_V2.3$ VSCCs, possibly due to a physical association of the two ion channels.

In addition to shaping single synapse responses, interactions of VSCCs and NMDARs determine synaptic responses and nonlinearities during near-synchronous stimulation of groups of synapses. Rapid activation of ~20 synapses on an individual segment of a radial oblique dendrite of a CA1 pyramidal neuron generated a Ca spike in the dendrite that was detectable in the soma as a rapid, all-or-none rising phase to the compound EPSP.[54] Interactions of NMDARs, VSCCs, and voltage-sensitive Na channels in these dendrites boosted the somatic potential and the dendritic Ca influx, presumably through increased relief of Mg block of synaptically activated NMDARs. Such locally confined dendritic spikes may play a role in the induction of associative plasticity at distal synapses.[55]

In cortical layer 5 pyramidal neurons, activation of clustered synapses on basal dendrites produced an NMDAR-mediated spike detectable as an all-or-nothing ~5 to 10 mV depolarization at the soma and Ca transient in the dendrite.[56] The proposed mechanism for this spike is that synaptic depolarization relieves Mg block of the NMDARs, and this increases inward current flux through the receptors, further depolarizing the dendrite and further relieving Mg block. This positive feedback loop is enhanced by VSCCs and voltage-sensitive Na channel opening but neither of these channels is strictly necessary for spike initiation.

Finally, in striatal medium spiny neurons (MSNs), clustered activation of synapses on a short stretch of dendrite (5 synapses in ~10 microns) also boosted synaptic potentials and Ca transients in an NMDAR- and VSCC-dependent manner.[57] However, no dendrite spike was elicited and graded increases in the amplitude and duration of the EPSP were seen. Ca influx into the active spine was enhanced, presumably due to increased relief of Mg block during the enhanced potential.

9.5 POSSIBLE EFFECTS OF COMPARTMENTALIZED ELECTRICAL SIGNALING ON NMDA RECEPTORS

The mechanisms described above nonlinearly boost synaptic signals by modulating the Mg block of NMDARs. They either require or are enhanced by the activation of VSCCs. The depolarization of 20 to 40 mV required for opening high-voltage-activated VSCCs such as $Ca_V2.3$ suggests that the submillivolt unitary EPSPs measured at the soma represent highly filtered versions of much larger depolarizations in the spine. Since synaptic depolarization is principally mediated by current flow through AMPARs, these results predict that blocking AMPARs should produce a significant impact on synaptically evoked Ca transients in spines. The reported effects of AMPAR blockade on synaptic Ca transients range broadly from small reductions to complete blockades.[1,51,58]

Measuring the amplitude of Ca transients may miss large effects of AMPARs on NMDAR-dependent Ca influx. It is likely that the potential in the spine during synaptic activation rises quickly and, due to the activation of repolarizing currents such as SK channels, falls quickly. For these reasons, the depolarization in the spine is likely brief and blocking AMPARs may only affect NMDAR-dependent Ca influx in a short period immediately after synaptic stimulation. Since Ca indicators slow Ca clearance and report the integrated Ca influx through NMDARs over hundreds of milliseconds,[4,27,59,60] the effects of AMPARs on Ca influx in active spines may be largely obscured.

To fully reveal the effects of spine depolarization of synaptically evoked Ca transients, quantitative models of the action of Ca buffers[61–65] on Ca handling may be used to calculate the Ca current into the spine. This can be accomplished by deconvolving the Ca transient with the impulse response of the spine to a brief, small increase in Ca.[4,59] In regimes of linear Ca handling, this approach reveals the time course of the Ca current into the spine. In preliminary studies using this approach, the Ca current that underlies synaptically evoked Ca transients in spines of mouse CA1 pyramidal neurons is comprised of fast and slow components (Figure 9.1D through F; authors' unpublished data).

The amplitude and time course of the fast component are regulated by the actions of ion channels in the spine such as SKs and AMPARs (Figure 9.1D through F), whereas the slow component depends only on the number of open NMDARs. The fast phase represents Ca influx through VSCCs and NMDARs and is modulated by the amplitude and duration of synaptic depolarization in the spine. The slow component outlasts the EPSP and reflects Ca influx through NMDARs after the spine returns to its resting potential. Regulation of these two distinguishable phases of Ca influx may have important functional consequences for activation of downstream Ca-dependent processes such as synaptic plasticity, although this has yet to be demonstrated.

9.6 IMPACT OF SPINE MORPHOLOGY ON CALCIUM SIGNALING IN DENDRITIC SPINES

The morphology of the spine may impact Ca transient in the spine head in two ways. First, since the surface-to-volume ratio of a sphere is inversely proportional to the radius, changes in the size of the spine head may impact both the amplitude and kinetics of Ca transients. If Ca channels are present on the spine head at constant density or number, the amplitudes of Ca transients should be smaller in larger spines than in smaller spines. However, this simple relationship is not found experimentally and the sizes of synaptically evoked Ca transients and spine volumes are only poorly correlated.[27] Similarly, at a constant density of Ca transporters and pumps, the clearance of Ca should be slower in larger spines. If Ca diffusion across the spine neck plays an important role in clearing Ca from the head, spines with longer and thinner necks or larger heads should clear Ca more slowly than spines with shorter and thicker necks or smaller heads, leading to longer lasting and larger Ca transients.[27,66–68]

The role of Ca diffusion across the spine neck in clearing Ca from the spine head is contentious. Several studies of Ca handling in the spine at room temperature led to the conclusion that Ca diffusion across the neck is a significant mechanism of Ca clearance and that differences in spine morphology directly account for interspine variability in the amplitude and kinetics of Ca transients.[60,66,67,69] In contrast, studies performed at near physiological temperatures revealed that variability in spine neck morphology does not significantly impact synaptically evoked Ca transients in dendritic spines and that most Ca clearance occurs via pumping or transport from the cytoplasm.[4,27,59]

9.7 CONCLUSION

NMDARs contribute most synaptically evoked Ca influx into dendritic spines. Ca influx through NMDARs depends on many factors intrinsic to receptors such as subunit composition and phosphorylation state. Ca influx through receptors and the properties of synaptically evoked Ca accumulations in the spine are also regulated by many extrinsic receptor factors. These include relatively stable parameters such as the number of ion channels and Ca pumps and the buffer capacity and morphology of the spine. In addition, changes in the resting potentials of neurons and rapid changes in membrane potential at synapses during synaptic potential strongly influence the amplitude and kinetics of Ca influx through NMDARs. These factors act together to determine the amplitude and kinetics of synaptically evoked Ca transients in dendritic spines and, in ways that are not yet clear, determine the coupling of NMDAR opening to the activation of downstream Ca-dependent processes.

REFERENCES

1. Kovalchuk, Y. et al., NMDAR-mediated subthreshold Ca(2+) signals in spines of hippocampal neurons, *J. Neurosci.*, 20, 1791, 2000.
2. Bloodgood, B.L. and Sabatini, B.L., Nonlinear regulation of unitary synaptic signals by CaV(2.3) voltage-sensitive calcium channels located in dendritic spines, *Neuron*, 53, 249, 2007.
3. Yuste, R. et al., Mechanisms of calcium influx into hippocampal spines: heterogeneity among spines, coincidence detection by NMDARs, and optical quantal analysis, *J. Neurosci.*, 19, 1976, 1999.
4. Sabatini, B.L., Oertner, T.G., and Svoboda, K., The life cycle of Ca(2+) ions in dendritic spines, *Neuron*, 33, 439, 2002.
5. Koester, H.J. and Sakmann, B., Calcium dynamics in single spines during coincident pre- and postsynaptic activity depend on relative timing of back-propagating action potentials and subthreshold excitatory postsynaptic potentials, *Proc. Natl. Acad. Sci. USA*, 95, 9596, 1998.
6. Nevian, T. and Sakmann, B., Spine Ca2+ signaling in spike-timing-dependent plasticity, *J. Neurosci.*, 26, 11001, 2006.
7. Schiller, J., Schiller, Y., and Clapham, D.E., NMDARs amplify calcium influx into dendritic spines during associative pre- and postsynaptic activation, *Nat. Neurosci.*, 1, 114, 1998.
8. Tada, T. and Sheng, M., Molecular mechanisms of dendritic spine morphogenesis, *Curr. Opin. Neurobiol.*, 16, 95, 2006.
9. Kennedy, M.B. et al., Integration of biochemical signalling in spines, *Nat. Rev. Neurosci.*, 6, 423, 2005.
10. Kennedy, M.J. and Ehlers, M.D., Organelles and trafficking machinery for postsynaptic plasticity, *Annu. Rev. Neurosci.*, 2006.
11. Alvarez, V.A. and Sabatini, B.L., Anatomical and physiological plasticity of dendritic spines, *Annu. Rev. Neurosci.*, 2007.

12. Monyer, H. et al., Heteromeric NMDARs: molecular and functional distinction of subtypes, *Science*, 256, 1217, 1992.
13. Buller, A.L. et al., The molecular basis of NMDAR subtypes: native receptor diversity is predicted by subunit composition, *J. Neurosci.*, 14, 5471, 1994.
14. Sugihara, H. et al., Structures and properties of seven isoforms of the NMDAR generated by alternative splicing, *Biochem. Biophys. Res. Commun.*, 185, 826, 1992.
15. Nakanishi, N., Axel, R., and Shneider, N.A., Alternative splicing generates functionally distinct N-methyl-D-aspartate receptors, *Proc. Natl. Acad. Sci. USA*, 89, 8552, 1992.
16. Anson, L.C. et al., Identification of amino acid residues of the NR2A subunit that control glutamate potency in recombinant NR1/NR2A NMDARs, *J. Neurosci.*, 18, 581, 1998.
17. Anson, L.C. et al., Single-channel analysis of an NMDAR possessing a mutation in the region of the glutamate binding site, *J. Physiol. (Lond.)*, 527, 225, 2000.
18. Laube, B., Kuhse, J., and Betz, H., Evidence for a tetrameric structure of recombinant NMDARs, *J. Neurosci.*, 18, 2954, 1998.
19. Monyer, H. et al., Developmental and regional expression in the rat brain and functional properties of four NMDARs, *Neuron*, 12, 529, 1994.
20. Wyllie, D.J., Behe, P., and Colquhoun, D., Single-channel activations and concentration jumps: comparison of recombinant NR1a/NR2A and NR1a/NR2D NMDARs, *J. Physiol.*, 510 (Pt 1), 1, 1998.
21. Vicini, S., Functional and pharmacological differences between recombinant N-methyl-D-aspartate receptors, *J. Neurophysiol.*, 79, 555, 1998.
22. Momiyama, A., Feldmeyer, D., and Cull-Candy, S.G., Identification of a native low-conductance NMDA channel with reduced sensitivity to Mg2+ in rat central neurones, *J. Physiol.*, 494 (Pt 2), 479, 1996.
23. Wyllie, D.J. et al., Single-channel currents from recombinant NMDA NR1a/NR2D receptors expressed in *Xenopus* oocytes, *Proc. Biol. Sci.*, 263, 1079, 1996.
24. Cheffings, C.M. and Colquhoun, D., Single channel analysis of a novel NMDA channel from *Xenopus* oocytes expressing recombinant NR1a, NR2A and NR2D subunits, *J. Physiol.*, 526 (Pt 3), 481, 2000.
25. Cull-Candy, S., Brickley, S., and Farrant, M., NMDAR subunits: diversity, development and disease, *Curr. Opin. Neurobiol.*, 11, 327, 2001.
26. He, Y., Janssen, W.G., and Morrison, J.H., Synaptic coexistence of AMPA and NMDARs in the rat hippocampus: a postembedding immunogold study, *J. Neurosci. Res.*, 54, 444, 1998.
27. Sobczyk, A., Scheuss, V., and Svoboda, K., NMDAR subunit-dependent [Ca2+] signaling in individual hippocampal dendritic spines, *J. Neurosci.*, 25, 6037, 2005.
28. Monyer, H. et al., Developmental and regional expression in the rat brain and functional properties of four NMDARs, *Neuron*, 12, 529, 1994.
29. Sheng, M. et al., Changing subunit composition of heteromeric NMDARs during development of rat cortex, *Nature*, 368, 144, 1994.
30. Bellone, C. and Nicoll, R.A., Rapid bidirectional switching of synaptic NMDARs, *Neuron*, 55, 779, 2007.
31. Sobczyk, A. and Svoboda, K., Activity-dependent plasticity of the NMDA-receptor fractional Ca2+ current, *Neuron*, 53, 17, 2007.
32. Barria, A. and Malinow, R., Subunit-specific NMDAR trafficking to synapses, *Neuron*, 35, 345, 2002.
33. Massey, P.V. et al., Differential roles of NR2A- and NR2B-containing NMDARs in cortical long-term potentiation and long-term depression, *J. Neurosci.*, 24, 7821, 2004.
34. Liu, L. et al., Role of NMDAR subtypes in governing the direction of hippocampal synaptic plasticity, *Science*, 304, 1021, 2004.

35. Edwards, J. et al., Identification of loci associated with putative recurrence genes in transitional cell carcinoma of the urinary bladder, *J. Pathol.*, 196, 380, 2002.
36. Barria, A. et al., Regulatory phosphorylation of AMPA-type glutamate receptors by CaM-KII during long-term potentiation, *Science*, 276, 2042, 1997.
37. Morishita, W. et al., Activation of NR2B-containing NMDARs is not required for NMDAR-dependent long-term depression, *Neuropharmacology*, 52, 71, 2007.
38. Berberich, S. et al., Lack of NMDAR subtype selectivity for hippocampal long-term potentiation, *J. Neurosci.*, 25, 6907, 2005.
39. Skeberdis, V.A. et al., Protein kinase A regulates calcium permeability of NMDARs, Nature Neurosci 9, 501, 2006.
40. Dell'Acqua, M.L. et al., Regulation of neuronal PKA signaling through AKAP targeting dynamics, *Eur. J. Cell. Biol.*, 85, 627, 2006.
41. Burnashev, N. et al., Fractional calcium currents through recombinant GluR channels of the NMDA, AMPA and kainate receptor subtypes, *J. Physiol.*, 485 (Pt 2), 403, 1995.
42. Jahr, C.E. and Stevens, C.F., Voltage dependence of NMDA-activated macroscopic conductances predicted by single-channel kinetics, *J. Neurosci.*, 10, 3178, 1990.
43. Carter, A.G. and Sabatini, B.L., State-dependent calcium signaling in dendritic spines of striatal medium spiny neurons, *Neuron*, 44, 483, 2004.
44. Kreitzer, A.C. and Malenka, R.C., Dopamine modulation of state-dependent endocannabinoid release and long-term depression in the striatum, *J. Neurosci.*, 25, 10537, 2005.
45. Calabresi, P. et al., Long-term potentiation in the striatum is unmasked by removing the voltage-dependent magnesium block of NMDAR channels, *Eur. J. Neurosci.*, 4, 929, 1992.
46. Adermark, L. and Lovinger, D.M., Combined activation of L-type Ca2+ channels and synaptic transmission is sufficient to induce striatal long-term depression, *J. Neurosci.*, 27, 6781, 2007.
47. Egger, V., Svoboda, K., and Mainen, Z.F., Dendrodendritic synaptic signals in olfactory bulb granule cells: local spine boost and global low-threshold spike, *J. Neurosci.*, 25, 3521, 2005.
48. Lester, R.A. et al., Channel kinetics determine the time course of NMDAR-mediated synaptic currents, *Nature*, 346, 565, 1990.
49. Yuste, R. and Denk, W., Dendritic spines as basic functional units of neuronal integration, *Nature*, 375, 682, 1995.
50. Magee, J.C. and Johnston, D., A synaptically controlled, associative signal for Hebbian plasticity in hippocampal neurons, *Science*, 275, 209, 1997.
51. Nevian, T. and Sakmann, B., Single spine Ca2+ signals evoked by coincident EPSPs and backpropagating action potentials in spiny stellate cells of layer 4 in the juvenile rat somatosensory barrel cortex, *J. Neurosci.*, 24, 1689, 2004.
52. Faber, E.S., Delaney, A.J., and Sah, P., SK channels regulate excitatory synaptic transmission and plasticity in the lateral amygdala, *Nat. Neurosci.*, 8, 635, 2005.
53. Ngo-Anh, T.J. et al., SK channels and NMDARs form a Ca2+-mediated feedback loop in dendritic spines, *Nat. Neurosci.*, 8, 642, 2005.
54. Losonczy, A. and Magee, J.C., Integrative properties of radial oblique dendrites in hippocampal CA1 pyramidal neurons, *Neuron*, 50, 291, 2006.
55. Golding, N.L., Staff, N.P., and Spruston, N., Dendritic spikes as a mechanism for cooperative long-term potentiation, *Nature*, 418, 326, 2002.
56. Schiller, J. et al., NMDA spikes in basal dendrites of cortical pyramidal neurons, *Nature*, 404, 285, 2000.
57. Carter, A.G., Soler-Llavina, G.J., and Sabatini, B.L., Timing and location of synaptic inputs determine modes of subthreshold integration in striatal medium spiny neurons, *J. Neurosci.*, 27, 8967, 2007.

58. Emptage, N., Bliss, T.V., and Fine, A., Single synaptic events evoke NMDAR-mediated release of calcium from internal stores in hippocampal dendritic spines, *Neuron*, 22, 115, 1999.
59. Scheuss, V. et al., Nonlinear [Ca2+] signaling in dendrites and spines caused by activity-dependent depression of Ca2+ extrusion, *J. Neurosci.*, 26, 8183, 2006.
60. Majewska, A. et al., Mechanisms of calcium decay kinetics in hippocampal spines: role of spine calcium pumps and calcium diffusion through the spine neck in biochemical compartmentalization, *J. Neurosci.*, 20, 1722, 2000.
61. Neher, E. and Augustine, G.J., Calcium gradients and buffers in bovine chromaffin cells, *J. Physiol.*, 450, 273, 1992.
62. Helmchen, F., Imoto, K., and Sakmann, B., Ca2+ buffering and action potential-evoked Ca2+ signaling in dendrites of pyramidal neurons, *Biophys. J.*, 70, 1069, 1996.
63. Sabatini, B.L., Maravall, M., and Svoboda, K., Ca(2+) signaling in dendritic spines, *Curr. Opin. Neurobiol.*, 11, 349, 2001.
64. Maravall, M. et al., Estimating intracellular calcium concentrations and buffering without wave length ratioing, *Biophys. J.*, 78, 2655, 2000.
65. Sabatini, B.L. and Regehr, W.G., Optical measurement of presynaptic calcium currents, *Biophys. J.*, 74, 1549, 1998.
66. Majewska, A., Tashiro, A., and Yuste, R., Regulation of spine calcium ddynamics by rapid spine motility, *J. Neurosci.*, 20, 8262, 2000.
67. Noguchi, J. et al., Spine neck geometry determines NMDAR-dependent Ca2+ signaling in dendrites, *Neuron*, 46, 609, 2005.
68. Svoboda, K., Tank, D.W., and Denk, W., Direct measurement of coupling between dendritic spines and shafts, *Science*, 272, 716, 1996.
69. Korkotian, E., Holcman, D., and Segal, M., Dynamic regulation of spine-dendrite coupling in cultured hippocampal neurons, *Eur. J. Neurosci.*, 20, 2649, 2004.

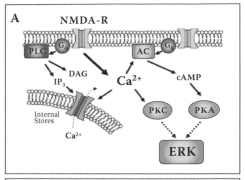

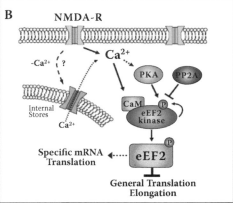

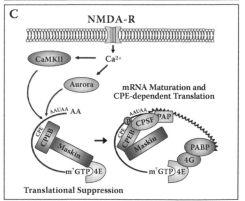

FIGURE 6.1 (A) NMDAR activation is linked to numerous intracellular signaling pathways. N-methyl-D-aspartate (NMDA) receptor (NMDAR) activation is the major source of activity-dependent calcium (Ca^{2+}) entry into the neuron. In addition, NMDAR activation may promote the generation of other second messengers [cAMP; diacylglyercol (DAG); and inositol-3,4,5-trisphosphate (IP_3)] through NMDAR association with membrane bound G-protein (Gs, Go) signaling to adenylate cyclase (AC) and calcium activated phospholipase (PLC). These second messengers may promote the activation of signal kinases such as cAMP-dependent kinase (PKA) and protein kinase C (PKC). Increased intracellular Ca^{2+} and IP_3 can trigger additional calcium influx via stimulation of calcium-release channels regulating calcium release from internal stores. (B) Signaling pathways activated by NMDARs involved in translational elongation. Activation of NMDARs results in extracellular calcium (Ca^{2+}) entry that eventually activates cAMP-dependent protein kinase (PKA). PKA-dependent phosphorylation and calcium-bound calmodulin promote the activity of eEF2 kinase (also known as CaMKIII). eEF2 kinase exhibits autophosphorylation activity that allows it to remain active after upstream signaling ceases. eEF2 kinase phosphorylates eEF2 (p-eEF2), which suppresses translational elongation. This has the effect of repressing general protein synthesis but can also produced enhanced translation of some mRNAs (i.e., 5′ TOP mRNA) that under normal conditions are translated with low efficiency. Protein phosphatase 2A (PP2A) can dephosphorylate eEF2 kinase to reduce its activity and de-repress translation elongation. NMDAR activation may also lead to eEF2 kinase activation via a mechanism independent of extracellular Ca^{2+} entry. (C) NMDAR regulation of mRNA maturation through CPEB. Activation of NMDARs leads to extracellular Ca^{2+} influx that activates Ca^{2+}/calmodulin-dependent protein kinase II (CaMKII) and Aurora kinase. Specific mRNAs contain 3′ untranslated region sequences (UUUUUAU) called CPEs. Immature mRNAs with CPEs are bound by cytoplasmic polyadenylation element-binding protein (CPEB) which binds maskin that also associates with eIF4E (4E) bound to the m^7GTP cap of the mRNA transcript. Translation of the transcript is inhibited by short poly(A) tails and the sequestration of 4E. Transcript de-repression is achieved via the phosphorylation of CPEB (p-CPEB) by CaMKII and Aurora. This promotes increased interaction between CPEB and polyadenylation specificity factor (CPSF). CPEB–CPSF association results in the recruitment of poly(A) polymerase (PAP) to lengthen the poly(A) tail of the immature mRNA transcript. Poly(A)-binding protein (PABP) then binds to the extended poly(A) tail and in turn interacts with eIF4G (4G). 4G displaces maskin binding with 4E, which permits the bound transcript to be translated. m^7GTP = 7-methyl GTP. AAUAA = polyadenylation signal.

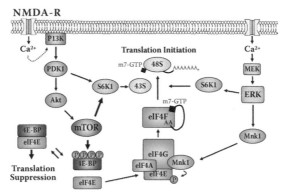

FIGURE 6.2 Signaling pathways activated by NMDARs and involved in translational initiation. NMDAR activation in turn activates the phosphatidylinositol 3-kinase (PI3K) and the mitogen-activated protein kinase (MEK)/extracellular signal-regulated kinase (ERK) signaling pathways. NMDAR activation produces sequential activation of PI3K, phosphoinositide-dependent kinase 1 or 2 (PDK1), protein kinase-B (Akt), and mammalian target of rapamycin (mTOR). mTOR activation leads to activation of S6 kinase 1 (S6K1) and phosphorylation (P) of 4E-binding proteins (4E-BPs). This phosphorylation causes disassociation of 4E-BPs from initiation factor 4E (eIF4E). Released eIF4E interacts with initiation factor 4G (eIF4G) and forms the active eIF4F (eIF4E-eIF4A-eIF4G) complex. eIF4F promotes mRNA binding to the 43S pre-initiation complex to form the 48S pre-initiation complex. ERK-dependent phosphorylation of both MAPK-interacting serine/threonine kinase 1 (Mnk1) that can phosphorylate eIF4E and S6K1 that can phosphorylate ribosomal protein S6 is correlated with enhanced translation initiation. m7G = 7-methyl-GTP. AAAAAAA$_n$ = poly(A) tail.

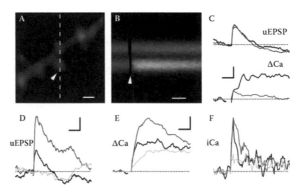

FIGURE 9.1 NMDAR-dependent calcium influx into active spines is modulated by AMPARs and SK channels. (A) Spiny dendrite from a mouse CA1 pyramidal neuron filled through a somatic whole-cell recording electrode with the red Ca-insensitive fluorophore Alexa-594 and the green Ca-sensitive fluorophore Fluo-5F. (B) Line scans through the dendrite and spine as indicated by the dashed line in (A) Arrow heads in (A) and (B) indicate locations and timings of 500 µsec laser pulse used to trigger 2-photon mediated photolysis of MNI glutamate. Uncaged glutamate binds to and opens NMDARs, resulting in Ca influx into the spine head seen as an increase in green fluorescence. The accompanying uncaging-evoked excitatory postsynaptic potential (uEPSP) is not shown. (C) Examples of uEPSPs (top) and spine head Ca transients (bottom) generated by glutamate uncaging at a single spine in control conditions (black line) or in the presence of the NMDAR antagonists CPP and MK-801 (red). (D) uEPSPs from individual spines in control conditions (black line) in the presence of the SK channel antagonist apamin (red) and the AMPAR antagonist NBQX (green). (E) Fluorescence transients measured in the spine head in response to the uEPSPs shown in (D) The amplitude of the fluorescence transient was directly proportional to spine head Ca. (F) Time course of the calculated spine head calcium currents in the three conditions. Ca currents were estimated by deconvolving the spine head Ca transients shown in (E) with the impulse response function of spine head Ca handling (not shown). Scale bars: (A) 1 µm; (B) 25 msec; (C) 0.5 mV, 5% ΔG/G$_{sat}$, 25 msec; (D) 0.5 mV, 25 msec; (E) 5% ΔG/G$_{sat}$, 25 msec.

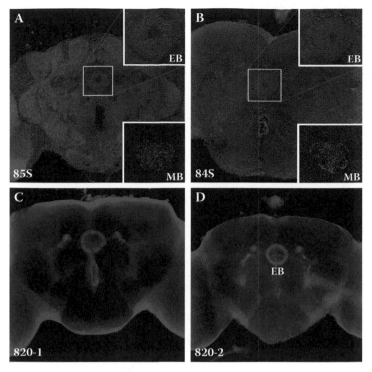

FIGURE 10.2 Expression of dNR1 and dNR2 proteins in adult brain. (A) Confocal imaging of dNR1 immunostaining in whole-mount adult brain with α-85S, a specific polyclonal anti-dNR1 antibody. All neurons appear to show weak expression of dNR1. The immunopositive signals were detected in the ellipsoid body (EB, upper inset) or in the calyx of the mushroom body (MB, lower inset). Many immunopositive signals appear as synapse-like puncta (insets), indicating synaptic localization of dNR1. (B) Immunolabeling of dNR2 proteins with α-84S, a specific polyclonal anti-dNR2 antibody. Similar to α-85S, all neurons show weak expression of dNR2, and many synapse-like puncta are found in the ellipsoid body (EB, upper inset) or in the calyx of the mushroom body (lower inset). (C) and (D) Confocal imaging of dNR2 immunostaining in whole-mount adult brain with α-820-1 (C) and α-820-2 (D), two independent polyclonal anti-dNR2 antibodies. Immunostaining reveals similar widespread expression of dNR2 proteins. Interestingly, strong expression is detected in the R4m neurons of the EB, suggesting that dNR2 may be preferentially expressed in the EB. (*Source:* Adapted from Wu, C.L. et al., *Nat. Neurosci.*, 10, 1578, 2007. With permission.)

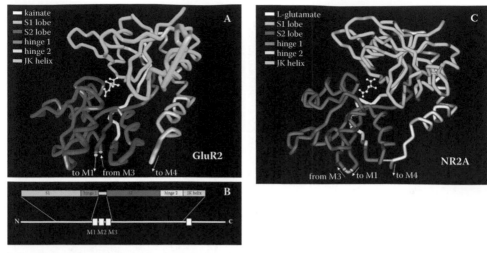

FIGURE 13.4 Crystal structures of GluR2 and NR2A ligand binding domains. X-ray crystallographic structures are shown for the isolated ligand binding domains (LBDs) of the AMPAR GluR2 in complex with kainate (A) and the NMDAR NR2A subunit in complex with L-glutamate (C). Only the peptide backbone is shown as a C-α trace. (B) Relationship of LBDs and linear amino acid sequence.

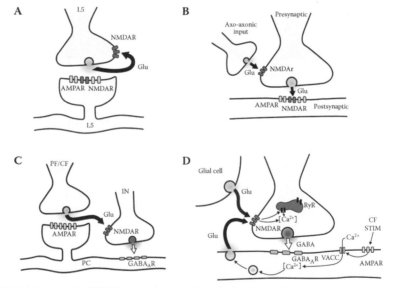

FIGURE 14.1 Presynaptic NMDAR activation by released glutamate. (A) Presynaptic autoreceptor activation. High frequency afferent stimulation (100 Hz) onto a layer 5 (L5) neocortical neuron enables presynaptically released glutamate (Glu) to activate presynaptic NMDA autoreceptors (NMDAR). (B) Axo-axonic NMDAR activation. Direct afferent input onto axon terminals enables released glutamate to activate presynaptic NMDARs, thereby regulating glutamate release onto AMPA receptors (AMPAR) and NMDARs on postsynaptic neurons. (C) Spillover-dependent NMDAR activation. High frequency stimulation of cerebellar climbing (CF) or parallel fibers (PF) results in glutamate pooling and saturation of juxtaposed excitatory amino acid transporters. Synaptic glutamate can diffuse out of the synapse (spillover) to activate presynaptic NMDARs on inhibitory interneurons. This leads to increased GABA release and postsynaptic GABA$_A$ receptor (GABA$_A$R) activation. (D) Retrograde/paracrine release-dependent NMDAR activation. At the interneuron (IN)–Purkinje cell (PC) synapse, postsynaptic depolarization via CF stimulation and AMPAR activation enables Ca^{2+} influx via voltage-gated Ca^{2+} channels (VGCC) to induce the retrograde release of vesicular glutamate from the PC. Enhanced GABA release results from Ca^{2+} induced Ca^{2+} release from ryanodine-sensitive stores (RyR). Alternatively, activation of the perforant path in the hippocampus leads to the paracrine release of glutamate from adjacent glial cells. Either source of released glutamate activates presynaptic NMDARs to enhance synaptic efficacy.

10 NMDA Receptors in *Drosophila*

Shouzhen Xia and Ann-Shyn Chiang

CONTENTS

10.1 INTRODUCTION

NMDA receptors (NMDARs), a subtype of ionotropic glutamate receptors, mediate the vast majority of excitatory neurotransmission in the central brains of vertebrates. NMDARs form heteromeric complexes usually comprised of a principal NR1 subunit and various NR2 subunits.[1,2] The NMDAR channel is highly permeable to Ca^{2+} and Na^+, and its opening requires simultaneous binding of glutamate and postsynaptic membrane depolarization.[1,3,4] Once activated, the NMDAR channel allows calcium influx into postsynaptic cells, where calcium triggers a cascade of biochemical events resulting in synaptic changes. NMDARs play diverse roles in normal central nervous system activity and development including regulation of synaptic development and function, and refinement of synaptic connections with experience and synaptic plasticity. NMDARs have also been widely investigated as targets for pharmacological management of seizures, pain, and a variety of neurological disorders including Schizophrenia, Parkinson's, Alzheimer's, and Huntington's diseases.[2,5–14]

Various studies in invertebrates suggest the existence of functional NMDA-like receptors and their requirement for synaptic and behavioral plasticity.[15–37] This chapter highlights the recent characterization of *Drosophila* NMDARs,[25,29–33] with emphasis on their physiological role during memory processing after Pavlovian olfactory conditioning—a well-defined and widely used elemental learning paradigm.[38]

10.2 NMDA RECEPTOR HOMOLOGUES IN *DROSOPHILA*

Three mammalian families of NMDAR subtypes have been identified: NR1, NR2, and NR3 subunits. Eight functional isoforms of the NR1 subunit are generated by alternative splicing of a single NR1 gene, while four distinct NR2 (A through D) subunits and two NR3 (A and B) subunits are encoded by six different genes.[2,6,39,40] The consensus is that most native NMDARs function as heteromeric tetramers composed of two NR1 subunits and two NR2 subunits.[2] The NR1 subunit is the essential constituent of NMDARs, expressed ubiquitously in the central nervous system.[41] The NR2 subunits regulate the biophysical and pharmacological properties of the NMDAR channel, including its high affinity for glutamate, modulation by glycine, Mg^{2+} block, and channel kinetics.[2,6,39,40] The NR3 subunits that appear not to be essential components of most native NMDARs may coassemble with the NR1 and NR2 complexes and thus regulate channel function.[6,39]

The situation is much simpler in *Drosophila* where homologues of the NR1 and NR2 subunits have been characterized. dNR1, composed of 15 exons, appears to be the only gene encoding the fly homologue of the NR1 subunit.[30,33] Although two different transcripts are generated by alternative splicing of the noncoding exon 1, they differ only in the 5′ untranslated region and contain the same coding sequence.[33] Therefore, dNR1 encodes a single NR1 subunit, different from the rodent NR1 gene that encodes multiple NR1 isoforms.[2,6,39] dNR2 may also be the only gene encoding the *Drosophila* homologue of the NR2 subunit.[33] Consistent with the fact that most NR2 genes are subject to alternative splicing in vertebrates,[2,6,39,40] dNR2 also undergoes alternative splicing, generating eight different transcripts that may encode three different protein isoforms.[33]

The major structural features of NMDARs are well conserved in both dNR1 and dNR2,[30,33] including three hydrophobic transmembrane regions (TM1, TM3, and TM4), one hydrophobic pore-forming segment (TM2) in the carboxyl terminal half,[2] and two ligand binding domains (S1 and S2) with high homology to bacterial amino acid–binding proteins.[42,43] Also conserved are the major determinants for ligand binding including amino acid residues in dNR1 (F430, Y432, D491, F494, V699, S702, D747, and F769) for coagonist glycine binding[42,44–46] and those in dNR2 (E511, K591, S618, R625, T792, T798, and V841) for glutamate binding.[47,48] Finally, many of the binding determinants for the noncompetitive or competitive antagonists are conserved, including the critical amino acids in dNR1 (W626, N631, and A660) for binding of dizocilpine (MK-801) and phencyclidine[49] and those in dNR2 (K591, S618, T798, and V841) for binding of D-2-amino-5-phosphonopentanoate (AP5) and 3-((R)-2-carboxypiperazin-4-yl)-propyl-1-phosphonic acid (R-CPP).[47,48]

Fly and rodent NMDARs exhibit several interesting differences.[33] Three asparagine residues present in the channel-forming TM2 domains of NMDA subunits control Ca^{2+} permeability and voltage-dependent Mg^{2+} block.[2,50,51] One such asparagine residue is conserved in dNR1 (N631), but the other two are not conserved in dNR2, suggesting that Mg^{2+} block may be relatively weak for *Drosophila* NMDARs. Fly NMDARs may mainly interact with PDZ domain-containing proteins through dNR1 but not dNR2, which is usually the case in mammals.[52,53] PDZ domains are found on the basis of sequence repeats in PSD-95, Dlg, and ZO-1 proteins.[54]

They can bind the carboxyl terminal sequences of proteins through a consensus sequence present in many glutamate receptors.[55] The type I PDZ binding motif (X–S/T–X–V) is not present in dNR2 although it is well conserved in all mammalian NR2 homologues.[52,53] Interestingly, dNR1 has a putative type II PDZ domain-binding motif (X–Ψ–X–Ψ in which Ψ is a hydrophobic amino acid) at its C terminus (and a potential type I PDZ-binding motif preceding this type II motif), suggesting that fly NMDARs may interact with PDZ domain-containing proteins via dNR1 subunits. Finally, although the entire size and domain structures of dNR2 show high homology to its vertebrate counterpart, its active pharmacological and physiological sites only moderately mimic its mammalian counterparts.[33]

10.3 FUNCTIONAL EXPRESSION OF FLY NMDA RECEPTORS

An initial attempt to express cloned dNR1 cDNA in *Xenopus* oocytes failed to generate a reliable NMDA-selective response.[30] Xia et al.[33] thus re-cloned dNR1 and showed that it alone could produce weak but significant NMDA-dependent responses in oocytes (Figure 10.1A). This weak response appears to support the notion that dNR1 alone can form functional NMDARs in oocytes. Notably, dNR1 has a RSS (retention signal sequence) motif at its C terminus, similar to its mammalian homologues.[33] The motif regulates the insertion of NMDARs in cell membranes by retaining the NR1 subunit in the endoplasmic reticulum (ER) when not assembled in functional receptors.[56,57]

dNR1 thus may be largely kept at the ER rather than inserted in membrane surfaces, making it possible to generate a weak response even if it can form a functional channel. Similarly, mammalian functional NMDARs may be formed by expression of NR1 alone in oocytes, exhibiting many of the properties of native NMDARs.[41,58] Nevertheless, one should be cautious in concluding that NR1 alone can form a homomeric functional channel. *Xenopus* oocytes express endogenous XenU1, a glutamate receptor subunit that can assemble with mammalian NR1 to form functional NMDARs.[59] This appears to reinforce the notion that NR1 must assemble with one or more NR2 subunits to form functional channels. However, NMDARs formed by NR1 in oocytes in fact do not contain the XenU1 subunit,[60] prompting further investigation why NR1 alone can form functional channels in oocytes but not in mammalian cell lines.[41,58]

Coexpression of dNR1 and dNR2 in *Xenopus* oocytes generates much stronger NMDA-selective responses (Figure 10.1A), consistent with the formation of highly potent NMDAR channels when the NR1 subunit is coexpressed with NR2.[2,3,61] Combined expression of dNR1 and dNR2 also exhibits several physiological features (Figure 10.1A and B) that distinguish NMDARs from other ionotropic glutamate receptors, including selective activation by NMDA and L-asparate[3,62] and modulation by glycine as the coagonist for glutamate.[63]

The NMDA-selective response, however, is not sensitive to Mg^{2+} blockade in oocytes.[33] This observation highlights certain facts. First, replacement of the asparagine residue in the channel-forming TM2 domain of the NR2 subunit disrupts Mg^{2+} block for mammalian NMDARs.[51,64] This crucial asparagine residue is replaced by glutamine in the dNR2 subunit, suggesting that Mg^{2+} block may be relatively weak

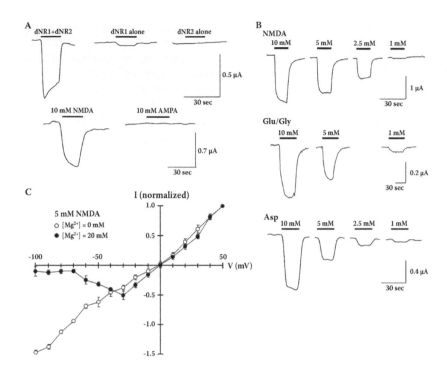

FIGURE 10.1 Coexpression of dNR1 and dNR2-2 yields a functional NMDAR. A: NMDA response in *Xenopus* oocytes expressing both dNR1 and dNR2-2. Oocytes injected with dNR1 and dNR2-2 cRNAs exhibited inward currents upon application of NMDA (10 *mM*) but not upon application of AMPA (10 *mM*; bottom). Oocytes expressing dNR1 alone showed modest inward currents upon application of 10 *mM* NMDA, while the oocytes expressing dNR2-2 alone showed no significant NMDA-selective responses (top). This suggests that dNR1 and dNR2 subunits function as heterodimers to form a functional NMDAR channel. B: NMDA, glutamate in combination with glycine, and L-asparate activate fly NMDARs in a concentration-dependent manner. Besides NMDA (top), coexpression of dNR1 and dNR2-2 may be activated by glutamate in the presence of glycine as coagonist (Glu/Gly, middle) and by L-asparate (Asp, bottom). Current responses were observed in a dosage-dependent manner. C: Voltage dependence of NMDARs in *Drosophila* S2 cells. Coexpression of dNR1 and dNR2-2 yielded a voltage-dependent effect on conductance (mean ± SEM, same for all following figures) at physiological concentrations of Mg^{2+} (20 *mM*). Conductance is linear in the absence of external Mg^{2+} (n = 8). (*Source:* Reproduced from Xia, S. et al., *Curr. Biol.,* 15, 603, 2005. With permission.)

for *Drosophila* NMDARs. TM1 and TM4 domains are also important for Mg^{2+} block,[65] but both domains are poorly conserved in dNR2.[33]

Finally, proper external ionic conditions for oocytes and insect cells are dramatically different. The appropriate Mg^{2+} concentration for fly muscle cells, for instance, is about 10 times higher than that for oocytes,[66] suggesting that fly NMDARs may have evolved to be less sensitive to Mg^{2+}. Nevertheless, *Drosophila* NMDARs may still be regulated by Mg^{2+} block *in vivo*. In support, MK-801, a compound requiring binding to the asparagine residue in the NR1 subunit to execute its antagonist effect,[49] has been shown to abolish NMDAR-dependent locomotor rhythm in fly larvae.[29] It also suppresses NMDA-mediated juvenile hormone biosynthesis in cockroaches.[25]

Consistently, coexpression of dNR1 and dNR2 in *Drosophila* S2 cells reveals voltage-dependent conductance blocked by external Mg^{2+} (Figure 10.1C). Therefore, the electrophysiological profile of coexpressing dNR1 and dNR2 in oocytes or S2 cells reveals most of the distinguishing characteristics of mammalian NMDARs, including the unique requirement for coagonist and voltage-dependent Mg^{2+} block.[33] This suggests that *Drosophila* likely has functional NMDARs consisting of two subunits, dNR1 and dNR2.

10.4 EXPRESSION OF FLY NMDA RECEPTORS IN ADULT BRAIN

The expression of dNR1 and dNR2 in adult brains has been extensively studied with multiple antibodies.[32,33] Both proteins are widely expressed throughout the entire brain, including all neuropils that consist of neural processes and dendritic regions (Figure 10.2). In the central brain, all neurons show weak expression of dNR1 and dNR2 (Figure 10.2A and B). In both cases, immunopositive signals are detected in the calyx of the mushroom body (MB, lower insets) and in the ellipsoid body (EB, upper insets), a substructure of the central complex. Both proteins are also detected throughout the optical lobes (not shown). Interestingly, many immunopositive signals are clustered as synapse-like puncta (Figure 10.2A and B, insets) and distribute along neural fibers.[33] This observation indicates that dNR1 and dNR2 may be localized to synapses, consistent with their contribution to associative learning and memory formation (see below).

The dNR2 protein may be preferentially expressed in the ellipsoid body, as indicated by strong immunopositive signals in its R4m large-field neurons from two of the anti-dNR2 antibodies (Figure 10.2C and D). This is particularly interesting because the ellipsoid body is a substructure of the central complex, one prominent neuropil located in the insect central brain that forms intricate connections to a variety of brain centers and proposed to mediate communication between the two hemispheres and many behavioral outputs.[67,68] The fan-shaped body, a closely related substructure of the central complex, may house a short-term memory trace for visual learning[69] and regulate long-term memory (LTM) formation after courtship conditioning.[70] Wu et al.[32] observed that fly NMDARs function in the ellipsoid body to regulate LTM consolidation after olfactory conditioning. However, it is unclear whether dNR1 is also preferentially expressed in the ellipsoid body.

Fly NMDARs appear to be only weakly expressed in the mushroom body, as all six (two polyclonal anti-dNR1, one monoclonal, and three polyclonal anti-dNR2) antibodies do not strongly label the structure.[32] The calyx, dendritic arborization of intrinsic neurons of the mushroom body, receives efferent inputs from several regions, including projection neurons from antennal lobes.[71] The axons of these neurons project rostrally as densely packed and stalk-like structures called pedunculi to the anterior face of the brain where they split and give rise to the dorsally projecting α and α′ lobes and the medially projecting β, β′, and γ lobes.[72] Output neurons from the MB project to many parts of the central brain.[73]

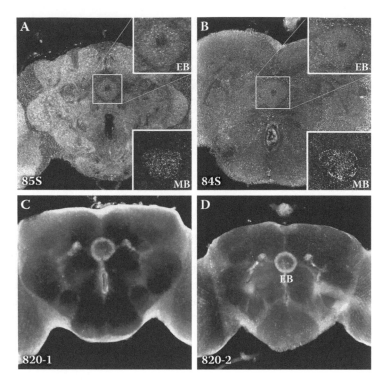

FIGURE 10.2 (See color insert following page 212.) Expression of dNR1 and dNR2 proteins in adult brain. (A) Confocal imaging of dNR1 immunostaining in whole-mount adult brain with α-85S, a specific polyclonal anti-dNR1 antibody. All neurons appear to show weak expression of dNR1. The immunopositive signals were detected in the ellipsoid body (EB, upper inset) or in the calyx of the mushroom body (MB, lower inset). Many immunopositive signals appear as synapse-like puncta (insets), indicating synaptic localization of dNR1. (B) Immunolabeling of dNR2 proteins with α-84S, a specific polyclonal anti-dNR2 antibody. Similar to α-85S, all neurons show weak expression of dNR2, and many synapse-like puncta are found in the ellipsoid body (EB, upper inset) or in the calyx of the mushroom body (lower inset). (C) and (D) Confocal imaging of dNR2 immunostaining in whole-mount adult brain with α-820-1 (C) and α-820-2 (D), two independent polyclonal anti-dNR2 antibodies. Immunostaining reveals similar widespread expression of dNR2 proteins. Interestingly, strong expression is detected in the R4m neurons of the EB, suggesting that dNR2 may be preferentially expressed in the EB. (*Source:* Adapted from Wu, C.L. et al., *Nat. Neurosci.*, 10, 1578, 2007. With permission.)

The mushroom body, one of the most prominent and well-characterized neuro-pillar structures in the insect central brain, has long been shown to mediate associative learning and early memory processing.[74,75] The weak expression of dNR1 and dNR2 in the mushroom body is intriguing, as targeted dsRNA-mediated knockdown of either protein disrupts the formation of middle-term memory,[2] an earlier phase of memory processing that depends on the normal function of the mushroom body.[76] Also intriguing are the strong immunopositive signals in scattered cell bodies and parts of their fibers that were detected with two of these antibodies but not with the remaining four.[32,33] Finally, as noted above, the ellipsoid body was preferentially labeled with only two of the polyclonal anti-dNR2 antibodies.[32,33]

10.5 NMDA RECEPTOR-DEPENDENT LEARNING AND LONG-TERM MEMORY CONSOLIDATION

Accumulating evidence over the past two decades has established that NMDARs and their downstream signaling pathways play a crucial role in the regulation of synaptic and behavioral plasticity by mediating long-lasting changes in synapse strength [long-term depression (LTD) and long-term potentiation (LTP)] in mammalian and human brains.[9,14,77–85] Opening of the NMDAR channel requires simultaneous binding of presynaptically released neurotransmitters and postsynaptic membrane depolarization as it is subjected to a unique voltage-dependent Mg^{2+} block.[1,3,4] This suggests that NMDARs may serve as "Hebbian coincidence detectors" underlying associative learning.[81,86–90] Molecular and physiological characterization of functional NMDARs in *Drosophila*[33] makes it possible to extend these finding to invertebrates.[91–93]

Using the Pavlovian olfactory conditioning paradigm involving well-defined odors as conditioned stimuli (CSs) and footshocks as unconditioned stimuli (US),[38] Xia et al. demonstrated that NMDARs are required acutely for associative learning and subsequent LTM consolidation in *Drosophila*.[33] By limiting rapid inducible knockdown of dNR1 with a specific anti-dNR1 message in adults, olfactory learning is transiently disrupted (Figure 10.3A), suggesting that NMDARs play an acute, physiological role in associative learning. This observation rules out a potential developmental explanation for adult learning deficits, something that past genetic studies did not achieve.[91,93]

The physiological requirement for NMDARs during olfactory learning strengthens the idea that these receptors play a central role in synaptic and behavioral plasticity, potentially by acting as coincidence detectors.[81,86–90] Extended (massed or spaced) training can overcome such an acute requirement for dNR1.[33] Nevertheless, the acute adult-specific knockdown of dNR1 appears to abolish LTM, which is specifically induced by extended spaced training (Figure 10.3B). Considering that this acute effect is specific to LTM consolidation but not retrieval (see below), this observation suggests that NMDARs are acutely required for LTM consolidation—an interesting idea supported by past genetic but not pharmacological studies.[94,95]

The relevant results also support the idea that NMDARs are involved with LTM consolidation and storage but not retrieval (Figure 10.3). The acute knockdown of dNR1, induced 15 hr before training, disrupted initial learning after one-session training, suggesting that NMDARs are involved in early encoding of olfactory memory (Figure 10.3A). Such a disruptive effect on learning was reversed when knockdown of dNR1 was induced 36 hr before training, suggesting that the dNR1 protein returns to its preinduction level (Figure 10.3A, right panels).

LTM tested 24 hours after spaced training was specifically abolished by the acute knockdown of dNR1 induced 15 hr before training (Figure 10.3B). The induction was delivered 15 hr before spaced training that lasted about 3 hr,[96] and then LTM

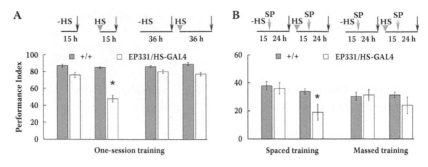

FIGURE 10.3 Disruption of olfactory learning and LTM consolidation by acute induction of anti-dNR1 mRNA. EP331 flies contain EP elements inserted downstream of and in an opposite orientation to the transcription start site of dNR1. The EP element yields a specific anti-dNR1 transcript[33] in the presence of GAL4, a yeast transcription factor that binds to its target sequence, UAS (upstream activation sequence, constructed in the EP element[150]), then activates transcription of the downstream gene,[151] HS-GAL4 flies contain a GAL4 transgene under the control of a heat shock promoter. A: Learning in transheterozygous EP331, hs-GAL4/+, + (EP331/P26) flies was significantly reduced after heat shock (HS, arrowhead; 15 hr recovery time after heat shock; * P < 0.001) and was mildly lower in the absence of heat shock (–HS), compared to wild-type controls (+/+; two left panels). When tested 36 hr after heat shock, learning returned to pre-heat shock level (two right panels), suggesting that heat shock-specific disruption of learning is transient. B: Transheterozygous EP331, hs-GAL4/+, + (EP331/P26) flies were subjected to spaced (two left panels) or massed (two right panels) training (gray arrow)[33] 15 hr after heat shock. One-day memory after spaced training was significantly disrupted (* P < 0.05) and was normal after massed training. One-day memory after spaced training in EP331, hs-GAL4/+, + flies, in fact, was reduced 47% to a level similar to normal one-day memory after massed training. Typically, one-day memory after spaced training consisted of 50% LTM and 50% ARM (anesthesia-resistant memory). LTM was specifically disrupted in transgenic flies inducibly overexpressing CREB repressor; one-day memory after massed training contained only ARM.[96] These results suggest that ARM is normal and LTM is completely abolished in EP331/+, hs-GAL4/+ flies after acute disruption of dNR1. (*Source:* Adapted from Xia, S. et al., *Curr. Biol.,* 15, 603, 2005. With permission.)

was tested 24 hr after training. Therefore, LTM was tested 42 hr (15 +24 + 3) after induction of the antisense dNR1 transcript, a time when the dNR1 protein returned to preinduction level, suggesting that LTM consolidation but not retrieval was abolished. Spaced training can overcome the learning defect (present after one training session) after the acute knockdown of dNR1 induced 15 hr before training,[33] again ruling out the possibility that NMDARs are involved in memory retrieval. Therefore, *Drosophila* NMDARs are acutely and selectively required for early encoding and consolidation of olfactory memory but not retrieval,[33] consistent with studies of mammals.[83,94,97–104]

Specific abolition of LTM consolidation by acute knockdown of dNR1 (Figure 10.3B) is similar to that produced by induced expression of a CREB-repressor transgene and indicates a specific disruption of cycloheximide-dependent LTM.[105] Fly NMDARs are required for CREB-dependent LTM formation, consistent with mammalian experiments revealing NMDAR-dependent activation of CREB during LTP and LTM in both amygdala and hippocampus.[106–109] The cAMP/PKA/CREB signaling pathway plays an important role in diverse processes from barrel formation and hippocampal LTP to learning and memory in invertebrates and vertebrates[48,110–119; but see.120,121] Interestingly, two types of functionally distinct NMDAR signaling complexes have been identified: synaptic and extra-synaptic.[122] Synaptic NMDARs can cause sustained CREB phosphorylation and CRE-mediated gene expression. Extra-synaptic NMDARs suppress CREB activity. It seems possible that synaptic NMDAR complexes regulate memory consolidation by controlling nuclear signaling to CREB.

10.6 LOCALIZATION OF NMDA RECEPTOR-DEPENDENT MEMORIES

Memory formation after olfactory conditioning proceeds through several temporal phases, all of which have been proposed to be predominantly processed in the mushroom body.[74,75] Surprisingly, dsRNA-mediated silencing of dNR1 or dNR2 in the mushroom body disrupts middle-term memory, an earlier memory phase proposed to be processed upstream of LTM,[96,123,124] without affecting LTM consolidation.[32] Because the mushroom body is required for LTM retrieval,[32,124] this observation suggests that it may be involved with LTM processing via an NMDAR-independent pathway, and NMDAR-dependent middle-term memory in the structure may not be necessary for LTM consolidation.

Interestingly, aging-dependent memory impairment is regulated by the amnesiac peptide [encoding the fly homologue of PACAP (pituitary adenylate cyclase-activating polypeptide)[125,126]] and DC0 (encoding a catalytic subunit of cAMP-dependent protein kinase known as PKA[127]), and is specific to middle-term memory in *Drosophila*.[128–130] Considering that both PACAP and PKA are known to mediate normal NMDAR function through phosphorylation,[131,132] the appearance of NMDAR-dependent middle-term memory in the mushroom body raises the interesting possibility that amnesiac peptide and PKA may regulate aging-dependent memory impairment by phosphorylation of fly NMDARs in the mushroom body.

When targeted specifically to the R4m neurons of the ellipsoid body, where dNR2 (and presumably dNR1) may be preferentially expressed (Figure 10.2), dsRNA-mediated silencing of dNR2 (or dNR1[32]) specifically abolishes protein synthesis-dependent LTM (Figure 10.4A and B), suggesting that the ellipsoid body plays a critical role during LTM processing. The involvement of NMDARs during LTM processing is physiological rather than developmental, because induction of the dsRNA transgene is limited to adults (Figure 10.4A and B).

The abolition is specific to LTM consolidation and storage but not retrieval (Figure 10.4C through E). This requirement for NMDARs is also specific for a memory phase (LTM only) and brain region (ellipsoid body, but not mushroom body), as initial learning and early memories are normal when NMDARs are silenced in the ellipsoid body, and LTM forms normally when these receptors are silenced in the mushroom body.[32] Therefore, functional NMDARs contribute specifically to the consolidation and storage, but not to the retrieval of protein synthesis-dependent LTM in the ellipsoid body.

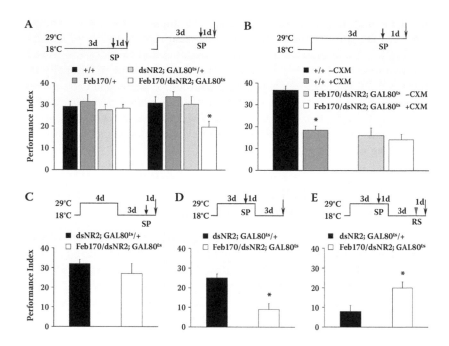

FIGURE 10.4 Inducible knockdown of dNR2 specifically blocks consolidation and storage but not retrieval of protein synthesis-dependent LTM. UAS-dsNR2 is a dsRNA-based transgene that can silence expression of dNR2 gene in the presence of GAL4.[32] dsRNA triggers RNAi interference, an evolutionarily conserved process of sequence-specific posttranscriptional gene silencing.[152,153] Feb170 is a GAL4 driver that targets gene expression preferentially in R4m neurons of the ellipsoid body.[32,154] Tub-GAL80[ts] flies contain a temperature-sensitive GAL80[ts] transgene, expressed ubiquitously from the tubulin 1α promoter.[155] At a permissive temperature (18°C), GAL80[ts] suppresses GAL4-controlled UAS transgene expression; at a restrictive temperature (29°C), GAL80[ts] releases the suppression so that the transgene can be expressed. A: Adult-specific knockdown of dNR2 using the tub-GAL80[ts] repressor of GAL4-mediated UAS-dsNR2 expression specifically abolished LTM at the restrictive temperature. One-day memory after spaced training was significantly disrupted in Feb170/+; UAS-dsNR2/+; GAL80[ts]/+ (Feb170/dsNR2; GAL80[ts]) flies when shifted from 18°C to 29°C for 3 days (right; * P < 0.05) before spaced training (SP, gray arrow) and testing (black arrow) at 29°C, but was normal when kept at 18°C (left). B: Adult-specific knockdown of dNR2 abolished protein synthesis-dependent LTM. One-day memory after spaced training decreased 50% in cycloheximide (CXM)-fed wild-type flies (+/+ +CXM; left panel), but was not disrupted further in CXM-fed Feb170/+; UAS-dsNR2/+; GAL80[ts]/+ (Feb170/dsNR2; GAL80[ts] +CXM) flies, kept for 3 days at 29°C before spaced training and testing. C: One-day memory after spaced training was normal in Feb170/+; UAS-dsNR2/+; tub-GAL80[ts]/+ (Feb170/dsNR2; GAL80[ts]) flies after 4 days at 29°C, shifted to 18°C for 3 days, followed by training and testing at 18°C, indicating a recovery of NMDAR function at permissive temperature. D: Four-day memory after spaced training was significantly impaired in the same flies when incubated at 29°C for 3 days, subjected to spaced training, maintained at 29°C for 1 day, then returned to 18°C for 3 days before testing, indicating that knockdown of NMDARs for an additional day after training is sufficient to block LTM consolidation and storage. E: A reversal training protocol was designed to distinguish consolidation and storage defects from retrieval failures. The flies were subjected to a second phase of reversal-spaced (RS, arrow head) training 4 days after

These data identify for the first time a brain region outside the mushroom body for LTM consolidation and storage. The specific involvement of the R4m subtype large-field neurons in the ellipsoid body during LTM consolidation and storage along with the requirement of LTM formation for neuronal activity from the mushroom body and correlation with the appearance of an asymmetrical body near the central complex,[124,133,134] supports a broader neuroanatomical circuitry involving both brain structures that subserves olfactory memory consolidation in *Drosophila*.

The acute and specific requirement for NMDARs in the ellipsoid body for LTM consolidation and storage along with the occurrence of associative learning within or upstream of the mushroom body[74,75,135–138] raises the provocative hypothesis that the acquired olfactory experience may be transferred from the mushroom body to the ellipsoid body for LTM consolidation, in agreement with observations from other species.[83,139,140] Consistent with this hypothesis, blocking the synaptic output from the mushroom body but not from the ellipsoid body during and within the first 6 hr after training abolished the later consolidation of LTM (Figure 10.5A and B). A second consistent finding is that the synaptic output of the ellipsoid body is specifically required for LTM retrieval but not for acquisition and consolidation (Figure 10.5B and C), suggesting that NMDARs function in the ellipsoid body to support LTM consolidation and storage, while neuronal activity from the structure regulates NMDAR-independent LTM retrieval. These results reveal a distributed brain system subserving olfactory memory formation and the existence of a system-level memory consolidation in *Drosophila* that was previously only demonstrated in mammals.[83,141–143]

10.7 NMDA RECEPTORS IN OTHER INVERTEBRATES

NMDARs have been shown to exist in several other invertebrate species including *Caenorhabditis elegans* (nematodes),[19,21] *Aplysia californica*, *Lymnaea stagnalis*, and *Sepioteuthis sepioidea* (mollusks),[16,22,23,28,34–37] *Hirudo medicinalis* (annelids),[24] *Procambarus clarkia* and *Chasmagnathus* (crustaceans),[15,18] and *Apis mellifera* and *Diploptera punctata* (insects).[25–27] The characterization of those invertebrate NMDARs was achieved through electrophysiological and/or pharmacological analyses, mostly in the context of cellular recording,[15–18,21,24,35–37] behavior,[21–23,26] and recent molecular cloning of the NR1 homologues.[21,27,28]

FIGURE 10.4 *(Continued)* the first spaced training; the original CS– became the CS+ and vice versa. One-day memory was quantified relative to the second CS+ during reversal training. In the absence of memory formation after the first spaced training session, reversal memory would be expected to approach a performance index of 35. If memory formation after the first spaced training was normal, it would counteract memory formation after reversal training, thereby producing a performance index near zero. One-day memory after RS was higher in the same flies than in controls when incubated at 29°C for 3 days, subjected to spaced training, maintained at 29°C for 1 day, shifted to 18°C for 3 days, subjected to reversal training, then tested at 18°C 1 day later, indicating normal retrieval. Therefore, functional NMDARs contribute specifically to consolidation and storage, but not to retrieval of a protein synthesis-dependent LTM in the ellipsoid body. (*Source*: Adapted from Wu, C.L. et al., *Nat. Neurosci.*, 10, 1578, 2007. With permission.)

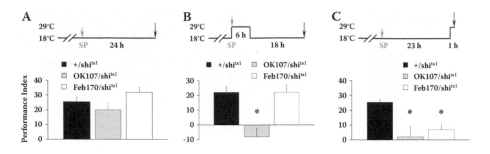

FIGURE 10.5 Transference of memory from MB to EB during LTM consolidation. The UAS-shi[ts1] transgene was shown to block neuronal transmission in a temperature-dependent, dominant-negative fashion.[156] OK107 is a GAL4 driver that targets gene expression in most mushroom body neurons.[32,157] Wild-type (+/+), Feb170 or OK107 males were crossed to UAS-shi[ts1] females. All progeny were raised at 18°C to minimize potential "leaky" effects of UAS-shi[ts1] on development. A: Flies were kept at permissive temperature (18°C) throughout experiment. One-day memory after spaced training (SP, gray arrow) did not differ among UAS-shi[ts1]/+ (+/shi[ts1]), UAS-shi[ts1]/+; OK107/+ (OK107/shi[ts1]) or Feb170/+; UAS-shi[ts1]/+ (Feb170/shi[ts1]) flies. B: Flies were shifted to restrictive temperature (29°C), subjected to spaced training immediately thereafter, maintained at 29°C for another 6 hr, shifted back to 18°C for 18 hr and tested at 18°C. One-day memory after spaced training was abolished in UAS-shi[ts1]/+; OK107/+ (OK107/shi[ts1]) flies but not in Feb170/+; UAS-shi[ts1]/+ (Feb170/shi[ts1]) flies, compared to +/shi[ts1] controls, suggesting that neural activity from MB (but not from the EB) is required during memory consolidation. C: Flies were subjected to spaced training at 18°C, kept at 18°C for 23 more hr, shifted to 29°C, then tested 1 hr later. One-day memory after spaced training was impaired in UAS-shi[ts1]/+; OK107/+ (OK107/shi[ts1]) and in Feb170/+; UAS-shi[ts1]/+ (Feb170/shi[ts1]) flies, compared to +/shi[ts1] controls, indicating that output of neural activity from both MB and EB is required during retrieval of LTM. (*Source:* Reproduced from Wu, C.L. et al., *Nat. Neurosci.,* 10, 1578, 2007. With permission.)

NMDARs in invertebrates seem to share major structural hallmarks and some biophysical and pharmacological characteristics with their vertebrate counterparts. Besides dNR1 and dNR2,[30,33] the NR1 homologues have been fully cloned in *Caenorhabditis elegans,*[21] *Lymnaea stagnalis,*[28] *Aplysia californica,*[28] and *Apis mellifera.*[27] The partial sequence of ~400 amino acids (including the predicted transmembrane segments, pore-forming regions, and ligand-binding regions) was also cloned for the NR2 homologue (NMR-2) in *Caenorhabditis elegans.*[19] All seven fully or partially cloned receptor subunits contain all the signature features of NMDARs including three hydrophobic transmembrane segments (TM1, TM3, and TM4), one hydrophobic pore-forming region (TM2), two ligand-binding domains (S1 and S2) with high homology to bacterial amino acid–binding proteins,[42,43] and the highly conserved SYTANLAAF amino acid sequence in TM3.[19,21,27,28,33]

Most of the amino acids for glycine binding in the NR1 homologues[21,27,28,33] and glutamate binding in the NR2 homologues are well conserved.[19,33] In addition all five fully cloned NR1 homologues contain one or more putative PDZ binding motifs,[21,27,28,33] allowing them to interact with PDZ domain-containing proteins and thus form huge signaling complexes.[52–55,144] Consistently, the basic biophysical characteristics of NMDARs appear to be conserved in most if not all of these invertebrate species including selective activation by NMDA, modulation via glycine as the coagonist for glutamate, calcium permeability and even relatively slow kinetics, as supported by electrophysiological experiments.[15–18,21,24,25,29,33]

Invertebrate and mammalian NMDARs exhibit certain pharmacological similarities. MK-801, for example, has been shown to block NMDA-selective responses or NMDA-dependent processes in all the invertebrate models tested.[21–23,25,26,29] It binds to the same asparagine residue in the channel-forming TM2 of the NR1 subunit that also controls the calcium permeability and voltage-dependent Mg^{2+} block.[2,50,51] This residue is conserved in all the invertebrate NR1 homologues cloned to date.[21,27,28,33]

Mammalian NMDARs emerged as major targets for studying synaptic and behavioral plasticity since their discovery in the 1970s.[2,7–9,14] Many pharmacological studies of invertebrate NMDARs focused on their contributions to long-term synaptic plasticity and memory formation.[22–24,26,35–37] In particular NMDAR-dependent LTP of the *Aplysia* sensorimotor synapse mediates associative learning of the withdrawal reflexes, leading to a hypothesis that classical conditioning in *Aplysia* is partially mediated by Hebbian-type LTP due to the hypothetical activation of NMDARs located at postsynaptic neurons.[35–37] The hypothesis has been further elaborated, assuming that the critical role of NMDARs is paralleled during long-term synaptic plasticity both in *Aplysia* and mammals.[145,146] Since the molecular identities of the NR1 homologues (AcNR1-1 and AcNR1-2) were identified,[28] this hypothesis may be further explored.

Nevertheless, invertebrate and vertebrate NMDARs reveal substantial differences. In particular, the two asparagine residues controlling calcium permeability and Mg^{2+} block[2,50,51] are not conserved in both dNR2 and NMR-2, leading to a less sensitive Mg^{2+} block in *Drosophila*[33] and possible absence of Mg^{2+} block in *Caenorhabditis elegans*.[21] Similarly, Mg^{2+} block is relatively weak[15,16,25] and even absent[17] in other invertebrate species, suggesting that NMDARs may have evolved to be increasingly sensitive to Mg^{2+} blockade. Also, the active pharmacological and physiological binding sites appear less conserved in invertebrate NR2 homologues.[19,33] Some antagonists including AP5 and R-CPP are less effective and showed no block effects in some experiments.[17,21,29]

10.8 SUMMARY

Molecular and physiological characterizations of cloned dNR1 and dNR2 reveal functional NMDARs in *Drosophila* that consist of dNR1 and dNR2 subunits. Coexpression of dNR1 and dNR2 in *Xenopus* oocytes or *Drosophila* S2 cells produces an electrophysiological profile exhibiting most of the distinguishing properties specific to mammalian NMDARs including selective activation by NMDA and L-aspartate, modulation by glycine as a coagonist for glutamate and voltage- and Mg^{2+}-dependent conductance.

Genetic analyses of the dNR1 gene reveal an acute and physiological role for NMDARs in associative learning and subsequent LTM consolidation. This extends genetic findings in vertebrates to invertebrates. Many intracellular signaling proteins are known to be physically associated with vertebrate NMDARs.[144,147] Obvious

Drosophila homologues can be identified for most of these proteins and many have been shown to be important for associative learning and memory formation.[112,148,149] It will be important identify more of the biochemical signaling pathway from NMDAR to CREB during LTM formation and the functional genomics of NMDAR-dependent memory consolidation.

Subsequent identification of the ellipsoid body for LTM consolidation and storage supports a much broader and more complex neuronal circuitry subserving memory consolidation in *Drosophila*. Distinct components of this extensive neuronal circuitry seem to be independently involved with different temporal stages of memory consolidation, with the mushroom body responsible for acquisition and earlier memory processing, while the ellipsoid body specifically controls LTM consolidation and storage. This discovery implies a "transference" of memory from one anatomic location (mushroom body) to another (ellipsoid body) as consolidation progresses.

The conservation of functional NMDARs and their involvement during behavioral plasticity in invertebrates further demonstrate that a unified mechanism may underlie associative learning and memory across species. Because behavioral plasticity is tightly associated with synaptic plasticity, we speculate that similar cellular mechanisms of NMDAR-mediated long-term changes including LTP and LTD may also exist in the insect brain. We expect that *Drosophila* genetics will likely continue to discover additional genes and signaling pathways important for these forms of plasticity.

ACKNOWLEDGMENTS

We thank Dr. Tim Tully for valuable comments and discussion and Drs. Josh Dubnau and Glenn Turner for critical reading of the manuscript. S.X. is a senior research scientist in Dr. Tully's laboratory at Cold Spring Harbor. This work was supported by grants to Dr. Tully from Dart Neurosciences, LLC, and to A.S.C. from the National Science Council and the Ministry of Education of Taiwan.

REFERENCES

1. McBain, C.J. and Mayer, M.L., N-methyl-D-aspartic acid receptor structure and function, *Physiol. Rev.*, 74, 723, 1994.
2. Dingledine, R. et al., The glutamate receptor ion channels, *Pharmacol. Rev.*, 51, 7, 1999.
3. Monaghan, D.T., Bridges, R.J., and Cotman, C.W., The excitatory amino acid receptors: their classes, pharmacology, and distinct properties in the function of the central nervous system, *Annu. Rev. Pharmacol. Toxicol.*, 29, 365, 1989.
4. Nowak, L. et al., Magnesium gates glutamate-activated channels in mouse central neurones, *Nature*, 307, 462, 1984.
5. Contestabile, A., Roles of NMDAR activity and nitric oxide production in brain development, *Brain Res. Brain Res. Rev.*, 32, 476, 2000.
6. Cull-Candy, S., Brickley, S., and Farrant, M., NMDAR subunits: diversity, development and disease, *Curr. Opin. Neurobiol.*, 11, 327, 2001.
7. Kullmann, D.M., Asztely, F., and Walker, M.C., The role of mammalian ionotropic receptors in synaptic plasticity: LTP, LTD and epilepsy, *Cell. Mol. Life Sci.*, 57, 1551, 2000.

8. Platenik, J., Kuramoto, N., and Yoneda, Y., Molecular mechanisms associated with long-term consolidation of the NMDA signals, *Life Sci.*, 67, 335, 2000.
9. Riedel, G., Platt, B., and Micheau, J., Glutamate receptor function in learning and memory, *Behav. Brain Res.*, 140, 1, 2003.
10. Chohan, M.O. and Iqbal, K., From tau to toxicity: emerging roles of NMDAR in Alzheimer's disease, *J. Alzheimer's Dis.*, 10, 81, 2006.
11. Fan, M.M. and Raymond, L.A., N-methyl-D-aspartate (NMDA) receptor function and excitotoxicity in Huntington's disease, *Prog. Neurobiol.*, 81, 272, 2007.
12. Hallett, P.J. and Standaert, D.G., Rationale for and use of NMDAR antagonists in Parkinson's disease, *Pharmacol. Ther.*, 102, 155, 2004.
13. Lau, C.G. and Zukin, R.S., NMDAR trafficking in synaptic plasticity and neuropsychiatric disorders, *Nat. Rev. Neurosci.*, 8, 413, 2007.
14. Newcomer, J.W. and Krystal, J.H., NMDAR regulation of memory and behavior in humans, *Hippocampus*, 11, 529, 2001.
15. Pfeiffer-Linn, C., and Glantz, R.M., An arthropod NMDAR, *Synapse*, 9, 35, 1991.
16. Dale, N. and Kandel, E.R., L-glutamate may be the fast excitatory transmitter of *Aplysia* sensory neurons, *Proc. Natl. Acad. Sci. USA*, 90, 7163, 1993.
17. Moroz, L.L., Gyori, J., and Salanki, J., NMDA-like receptors in the CNS of molluscs, *Neuroreport*, 4, 201, 1993.
18. Feinstein, N. et al., Functional and immunocytochemical identification of glutamate autoreceptors of an NMDA type in crayfish neuromuscular junction, *J. Neurophysiol.*, 80, 2893, 1998.
19. Brockie, P.J. et al., Differential expression of glutamate receptor subunits in the nervous system of *Caenorhabditis elegans* and their regulation by the homeodomain protein UNC-42, *J. Neurosci.*, 21, 1510, 2001.
20. Brockie, P.J. and Maricq, A.V., Ionotropic glutamate receptors in *Caenorhabditis elegans*, *Neurosignals*, 12, 108, 2003.
21. Brockie, P.J. et al., The *C. elegans* glutamate receptor subunit NMR-1 is required for slow NMDA-activated currents that regulate reversal frequency during locomotion, *Neuron*, 31, 617, 2001.
22. Pedreira, M.E. et al., Reactivation and reconsolidation of long-term memory in the crab *Chasmagnathus*: protein synthesis requirement and mediation by NMDA-type glutamatergic receptors, *J. Neurosci.*, 22, 8305, 2002.
23. Troncoso, J. and Maldonado, H., Two related forms of memory in the crab *Chasmagnathus* are differentially affected by NMDAR antagonists, *Pharmacol. Biochem. Behav.*, 72, 251, 2002.
24. Burrell, B.D. and Sahley, C.L., Multiple forms of long-term potentiation and long-term depression converge on a single interneuron in the leech CNS, *J. Neurosci.*, 24, 4011, 2004.
25. Chiang, A.S. et al., Insect NMDARs mediate juvenile hormone biosynthesis, *Proc. Natl. Acad. Sci. USA*, 99, 37, 2002.
26. Si, A., Helliwell, P., and Maleszka, R., Effects of NMDAR antagonists on olfactory learning and memory in the honeybee (*Apis mellifera*), *Pharmacol. Biochem. Behav.*, 77, 191, 2004.
27. Zannat, M.T. et al., Identification and localisation of the NR1 sub-unit homologue of the NMDA glutamate receptor in the honeybee brain, *Neurosci. Lett.*, 398, 274, 2006.
28. Ha, T.J. et al., Molecular characterization of NMDA-like receptors in *Aplysia* and *Lymnaea*: relevance to memory mechanisms, *Biol. Bull.*, 210, 255, 2006.
29. Cattaert, D. and Birman, S., Blockade of the central generator of locomotor rhythm by noncompetitive NMDAR antagonists in *Drosophila* larvae, *J. Neurobiol.*, 48, 58, 2001.

30. Ultsch, A. et al., Glutamate receptors of *Drosophila melanogaster*. Primary structure of a putative NMDAR protein expressed in the head of the adult fly, *FEBS Lett.*, 324, 171, 1993.
31. Volkner, M. et al., Novel CNS glutamate receptor subunit genes of *Drosophila melanogaster*, *J. Neurochem.*, 75, 1791, 2000.
32. Wu, C.-L. et al., Specific requirement of NMDARs for long-term memory consolidation in *Drosophila* ellipsoid body, *Nat. Neurosci.*, 10, 1578, 2007.
33. Xia, S. et al., NMDARs mediate olfactory learning and memory in *Drosophila*, *Curr. Biol.*, 15, 603, 2005.
34. Evans, P.D. et al., N-methyl-D-aspartate (NMDA) and non-NMDA (metabotropic) type glutamate receptors modulate the membrane potential of the Schwann cell of the squid giant nerve fibre, *J. Exp. Biol.*, 173, 229, 1992.
35. Antonov, I. et al., Activity-dependent presynaptic facilitation and Hebbian LTP are both required and interact during classical conditioning in *Aplysia*, *Neuron*, 37, 135, 2003.
36. Lin, X.Y. and Glanzman, D.L., Hebbian induction of long-term potentiation of *Aplysia* sensorimotor synapses: partial requirement for activation of an NMDA-related receptor, *Proc. Biol. Sci.*, 255, 215, 1994.
37. Murphy, G.G. and Glanzman, D.L., Mediation of classical conditioning in *Aplysia californica* by long-term potentiation of sensorimotor synapses, *Science*, 278, 467, 1997.
38. Tully, T. and Quinn, W.G., Classical conditioning and retention in normal and mutant *Drosophila melanogaster*, *J. Comp. Physiol. [A]*, 157, 263, 1985.
39. Cull-Candy, S.G. and Leszkiewicz, D.N., Role of distinct NMDAR subtypes at central synapses, *Sci. STKE*, re16, 2004.
40. Yamakura, T. and Shimoji, K., Subunit- and site-specific pharmacology of the NMDAR channel, *Prog. Neurobiol.*, 59, 279, 1999.
41. Moriyoshi, K. et al., Molecular cloning and characterization of the rat NMDAR, *Nature*, 354, 31, 1991.
42. Kuryatov, A. et al., Mutational analysis of the glycine-binding site of the NMDAR: structural similarity with bacterial amino acid-binding proteins, *Neuron*, 12, 1291, 1994.
43. Stern-Bach, Y. et al., Agonist selectivity of glutamate receptors is specified by two domains structurally related to bacterial amino acid-binding proteins, *Neuron*, 13, 1345, 1994.
44. Wafford, K.A. et al., Identification of amino acids in the N-methyl-D-aspartate receptor NR1 subunit that contribute to the glycine binding site, *Mol. Pharmacol.*, 47, 374, 1995.
45. Hirai, H. et al., The glycine binding site of the N-methyl-D-aspartate receptor subunit NR1: identification of novel determinants of co-agonist potentiation in the extracellular M3-M4 loop region, *Proc. Natl. Acad. Sci. USA*, 93, 6031, 1996.
46. Williams, K. et al., An acidic amino acid in the N-methyl-D-aspartate receptor that is important for spermine stimulation, *Mol. Pharmacol.*, 48, 1087, 1995.
47. Laube, B. et al., Molecular determinants of agonist discrimination by NMDAR subunits: analysis of the glutamate binding site on the NR2B subunit, *Neuron*, 18, 493, 1997.
48. Anson, L.C. et al., Identification of amino acid residues of the NR2A subunit that control glutamate potency in recombinant NR1/NR2A NMDARs, *J. Neurosci.*, 18, 581, 1998.
49. Ferrer-Montiel, A.V., Sun, W., and Montal, M., Molecular design of the N-methyl-D-aspartate receptor binding site for phencyclidine and dizolcipine, *Proc. Natl. Acad. Sci. USA*, 92, 8021, 1995.

50. Wollmuth, L.P., Kuner, T., and Sakmann, B., Adjacent asparagines in the NR2-subunit of the NMDAR channel control the voltage-dependent block by extracellular Mg²⁺, *J. Physiol.*, 506, 13, 1998.

51. Burnashev, N. et al., Control by asparagine residues of calcium permeability and magnesium blockade in the NMDAR, *Science*, 257, 1415, 1992.

52. Sheng, M. and Sala, C., PDZ domains and the organization of supramolecular complexes, *Annu. Rev. Neurosci.*, 24, 1, 2001.

53. Nourry, C., Grant, S.G., and Borg, J.P., PDZ domain proteins: plug and play!, *Sci. STKE*, RE7, 2003.

54. Kennedy, M.B., Origin of PDZ (DHR, GLGF) domains, *Trends Biochem. Sci.*, 20, 350, 1995.

55. Kornau, H.C., Seeburg, P.H., and Kennedy, M.B., Interaction of ion channels and receptors with PDZ domain proteins, *Curr. Opin. Neurobiol.*, 7, 368, 1997.

56. Scott, D.B. et al., An NMDAR ER retention signal regulated by phosphorylation and alternative splicing, *J. Neurosci.*, 21, 3063, 2001.

57. Standley, S. et al., PDZ domain suppression of an ER retention signal in NMDAR NR1 splice variants, *Neuron*, 28, 887, 2000.

58. Hollmann, M. et al., Zinc potentiates agonist-induced currents at certain splice variants of the NMDAR, *Neuron*, 10, 943, 1993.

59. Soloviev, M.M. and Barnard, E.A., *Xenopus oocytes* express a unitary glutamate receptor endogenously, *J. Mol. Biol.*, 273, 14, 1997.

60. Green, T. et al., NMDARs formed by NR1 in Xenopus laevis oocytes do not contain the endogenous subunit XenU1, *Mol. Pharmacol.*, 61, 326, 2002.

61. Mori, H. and Mishina, M., Structure and function of the NMDAR channel, *Neuropharmacol.*, 34, 1219, 1995.

62. Patneau, D.K. and Mayer, M.L., Structure-activity relationships for amino acid transmitter candidates acting at N-methyl-D-aspartate and quisqualate receptors, *J. Neurosci.*, 10, 2385, 1990.

63. Kleckner, N.W. and Dingledine, R., Requirement for glycine in activation of NMDA-receptors expressed in *Xenopus oocytes*, *Science*, 241, 835, 1988.

64. Mori, H. et al., Identification by mutagenesis of a Mg²⁺-block site of the NMDAR channel, *Nature*, 358, 673, 1992.

65. Kuner, T. and Schoepfer, R., Multiple structural elements determine subunit specificity of Mg²⁺ block in NMDAR channels, *J. Neurosci.*, 16, 3549, 1996.

66. Stewart, B.A. et al., Improved stability of *Drosophila* larval neuromuscular preparations in haemolymph-like physiological solutions, *J. Comp. Physiol. [A]*, 175, 179, 1994.

67. Strauss, R. and Heisenberg, M., A higher control center of locomotor behavior in the *Drosophila* brain, *J. Neurosci.*, 13, 1852, 1993.

68. Hanesch, U., Fischback, K.-F., and Heisenberg, M., Neuronal architecture of the central complex in *Drosophila melanogaster*, *Cell Tissue Res.*, 257, 343, 1989.

69. Liu, G. et al., Distinct memory traces for two visual features in the *Drosophila* brain, *Nature*, 439, 551, 2006.

70. Sakai, T. et al., A clock gene, *period*, plays a key role in long-term memory formation in *Drosophila Proc. Natl. Acad. Sci. USA*, 101, 16058, 2004.

71. Stocker, R.F., The organization of the chemosensory system in *Drosophila melanogaster*: a review, *Cell Tissue Res.*, 275, 3, 1994.

72. Jefferis, G.S. et al., Development of neuronal connectivity in *Drosophila* antennal lobes and mushroom bodies, *Curr. Opin. Neurobiol.*, 12, 80, 2002.

73. Strausfeld, N.J., *Atlas of an Insect Brain*, Springer, Heidelberg, 1976.

74. Gerber, B., Tanimoto, H., and Heisenberg, M., An engram found? Evaluating the evidence from fruit flies, *Curr. Opin. Neurobiol.*, 14, 737, 2004.

75. Davis, R.L., Olfactory memory formation in *Drosophila*: from molecular to systems neuroscience, *Annu. Rev. Neurosci.*, 28, 275, 2005.

76. Krashes, M.J. et al., Sequential use of mushroom body neuron subsets during *drosophila* odor memory processing, *Neuron*, 53, 103, 2007.

77. Morris, R.G., Davis, S., and Butcher, S.P., Hippocampal synaptic plasticity and NMDARs: a role in information storage? In *Long-Term Potentiation: A Debate of Current Issues*, Baudry, M. and Davis, J., eds. MIT Press, Cambridge, 1991, p. 267.

78. Bliss, T.V. and Collingridge, G.L., A synaptic model of memory: long-term potentiation in the hippocampus, *Nature*, 361, 31, 1993.

79. Malenka, R.C. and Nicoll, R.A., Long-term potentiation: a decade of progress?, *Science*, 285, 1870, 1999.

80. Martin, S.J., Grimwood, P.D., and Morris, R.G., Synaptic plasticity and memory: an evaluation of the hypothesis, *Annu. Rev. Neurosci.*, 23, 649, 2000.

81. Tsien, J.Z., Linking Hebb's coincidence-detection to memory formation, *Curr. Opin. Neurobiol.*, 10, 266, 2000.

82. Wittenberg, G.M. and Tsien, J.Z., An emerging molecular and cellular framework for memory processing by the hippocampus, *Trends Neurosci.*, 25, 501, 2002.

83. Nakazawa, K. et al., NMDARs, place cells and hippocampal spatial memory, *Nat. Rev. Neurosci.*, 5, 361, 2004.

84. Wang, H., Hu, Y., and Tsien, J.Z., Molecular and systems mechanisms of memory consolidation and storage, *Prog. Neurobiol.*, 79, 123, 2006.

85. Wittenberg, G.M., Sullivan, M.R., and Tsien, J.Z., Synaptic reentry reinforcement based network model for long-term memory consolidation, *Hippocampus*, 12, 637, 2002.

86. Brown, T.H., Kairiss, E.W., and Keenan, C.L., Hebbian synapses: biophysical mechanisms and algorithms, *Annu. Rev. Neurosci.*, 13, 475, 1990.

87. Collingridge, G.L., Kehl, S.J., and McLennan, H., Excitatory amino acids in synaptic transmission in the Schaffer collateral-commissural pathway of the rat hippocampus, *J. Physiol.*, 334, 33, 1983.

88. Hebb, D.O., *The Organization of Behavior*, Wiley, New York, 1949.

89. Morris, R.G. et al., Elements of a neurobiological theory of the hippocampus: the role of activity-dependent synaptic plasticity in memory, *Philos. Trans. R. Soc. Lond. B Biol. Sci.*, 358, 773, 2003.

90. Tonegawa, S., Nakazawa, K., and Wilson, M.A., Genetic neuroscience of mammalian learning and memory, *Philos. Trans. R. Soc. Lond. B. Biol. Sci.*, 358, 787, 2003.

91. Sakimura, K. et al., Reduced hippocampal LTP and spatial learning in mice lacking NMDAR epsilon 1 subunit, *Nature*, 373, 151, 1995.

92. Tang, Y.P. et al., Genetic enhancement of learning and memory in mice, *Nature*, 401, 63, 1999.

93. Tsien, J.Z., Huerta, P.T., and Tonegawa, S., The essential role of hippocampal CA1 NMDAR-dependent synaptic plasticity in spatial memory, *Cell*, 87, 1327, 1996.

94. Shimizu, E. et al., NMDAR-dependent synaptic reinforcement as a crucial process for memory consolidation, *Science*, 290, 1170, 2000.

95. Day, M. and Morris, R.G., Memory consolidation and NMDARs: discrepancy between genetic and pharmacological approaches, *Science*, 293, 755, 2001.

96. Tully, T. et al., Genetic dissection of consolidated memory in *Drosophila*, *Cell*, 79, 35, 1994.

97. Bast, T., da Silva, B.M., and Morris, R.G., Distinct contributions of hippocampal NMDA and AMPA receptors to encoding and retrieval of one-trial place memory, *J. Neurosci.*, 25, 5845, 2005.

98. Kim, J.J. et al., Selective impairment of long-term but not short-term conditional fear by the N-methyl-D-aspartate antagonist APV, *Behav. Neurosci.*, 106, 591, 1992.

99. Day, M., Langston, R., and Morris, R.G., Glutamate-receptor-mediated encoding and retrieval of paired-associate learning, *Nature*, 424, 205, 2003.

100. Takehara-Nishiuchi, K., Kawahara, S., and Kirino, Y., NMDAR-dependent processes in the medial prefrontal cortex are important for acquisition and the early stage of consolidation during trace, but not delay eyeblink conditioning, *Learn. Mem.*, 12, 606, 2005.

101. Cui, Z. et al., Requirement of NMDAR reactivation for consolidation and storage of nondeclarative taste memory revealed by inducible NR1 knockout, *Eur. J. Neurosci.*, 22, 755, 2005.

102. Winters, B.D. and Bussey, T.J., Glutamate receptors in perirhinal cortex mediate encoding, retrieval, and consolidation of object recognition memory, *J. Neurosci.*, 25, 4243, 2005.

103. Nakazawa, K. et al., Requirement for hippocampal CA3 NMDARs in associative memory recall, *Science*, 297, 211, 2002.

104. Robbins, T.W. and Murphy, E.R., Behavioural pharmacology: 40+ years of progress, with a focus on glutamate receptors and cognition, *Trends Pharmacol. Sci.*, 27, 141, 2006.

105. Yin, J.C. et al., Induction of a dominant negative CREB transgene specifically blocks long-term memory in *Drosophila*, *Cell*, 79, 49, 1994.

106. Cammarota, M. et al., Learning-associated activation of nuclear MAPK, CREB and Elk-1, along with Fos production, in the rat hippocampus after a one-trial avoidance learning: abolition by NMDAR blockade, *Brain Res. Mol. Brain. Res.*, 76, 36, 2000.

107. Poser, S. and Storm, D.R., Role of Ca^{2+}-stimulated adenylyl cyclases in LTP and memory formation, *Int. J. Dev. Neurosci.*, 19, 387, 2001.

108. Schulz, S. et al., Direct evidence for biphasic cAMP responsive element-binding protein phosphorylation during long-term potentiation in the rat dentate gyrus *in vivo*, *J. Neurosci.*, 19, 5683, 1999.

109. Walker, D.L. and Davis, M., Involvement of NMDARs within the amygdala in short- versus long-term memory for fear conditioning as assessed with fear-potentiated startle, *Behav. Neurosci.*, 114, 1019, 2000.

110. Silva, A.J. et al., CREB and memory, *Annu. Rev. Neurosci.*, 21, 127, 1998.

111. Brandon, E.P., Idzerda, R.L., and McKnight, G.S., PKA isoforms, neural pathways, and behaviour: making the connection, *Curr. Opin. Neurobiol.*, 7, 397, 1997.

112. Dubnau, J. and Tully, T., Gene discovery in *Drosophila*: new insights for learning and memory, *Annu. Rev. Neurosci.*, 21, 407, 1998.

113. Abdel-Majid, R.M. et al., Loss of adenylyl cyclase I activity disrupts patterning of mouse somatosensory cortex, *Nat. Genet.*, 19, 289, 1998.

114. Mayford, M. and Kandel, E.R., Genetic approaches to memory storage, *Trends Genet.*, 15, 463, 1999.

115. Frankland, P.W. et al., Consolidation of CS and US representations in associative fear conditioning, *Hippocampus*, 14, 557, 2004.

116. Pittenger, C. et al., Reversible inhibition of CREB/ATF transcription factors in region CA1 of the dorsal hippocampus disrupts hippocampus-dependent spatial memory, *Neuron*, 34, 447, 2002.

117. Falls, W.A. et al., Fear-potentiated startle, but not prepulse inhibition of startle, is impaired in CREBα-/- mutant mice, *Behav. Neurosci.*, 114, 998, 2000.

118. Kogan, J.H. et al., Spaced training induces normal long-term memory in CREB mutant mice, *Curr. Biol.*, 7, 1, 1997.

119. Bourtchuladze, R. et al., Deficient long-term memory in mice with a targeted mutation of the cAMP-responsive element-binding protein, *Cell*, 79, 59, 1994.

120. Leroy, E. et al., The ubiquitin pathway in Parkinson's disease, *Nature*, 395, 451, 1998.

121. Rammes, G. et al., Synaptic plasticity in the basolateral amygdala in transgenic mice expressing dominant-negative cAMP response element-binding protein (CREB) in forebrain, *Eur. J. Neurosci.*, 12, 2534, 2000.

122. Hardingham, G.E., Fukunaga, Y., and Bading, H., Extrasynaptic NMDARs oppose synaptic NMDARs by triggering CREB shut-off and cell death pathways, *Nat. Neurosci.*, 5, 404, 2002.

123. DeZazzo, J. and Tully, T., Dissection of memory formation: from behavioral pharmacology to molecular genetics, *Trends Neurosci.*, 18, 212, 1995.

124. Isabel, G., Pascual, A., and Preat, T., Exclusive consolidated memory phases in *Drosophila*, *Science*, 304, 1024, 2004.

125. Feany, M.B. and Quinn, W.G., A neuropeptide gene defined by the *Drosophila* memory mutant amnesiac, *Science*, 268, 869, 1995.

126. Moore, M.S. et al., Ethanol intoxication in *Drosophila*: Genetic and pharmacological evidence for regulation by the cAMP signaling pathway, *Cell*, 93, 997, 1998.

127. Kalderon, D. and Rubin, G.M., Isolation and characterization of *Drosophila* cAMP-dependent protein kinase genes, *Genes Dev.*, 2, 1539, 1988.

128. Saitoe, M. et al., *Drosophila* as a novel animal model for studying the genetics of age-related memory impairment, *Rev. Neurosci.*, 16, 137, 2005.

129. Tamura, T. et al., Aging specifically impairs *amnesiac*-dependent memory in *Drosophila*, *Neuron*, 40, 1003, 2003.

130. Yamazaki, D. et al., The *Drosophila DCO* mutation suppresses age-related memory impairment without affecting lifespan, *Nat. Neurosci.*, 10, 478, 2007.

131. Tingley, W.G. et al., Characterization of protein kinase A and protein kinase C phosphorylation of the N-methyl-D-aspartate receptor NR1 subunit using phosphorylation site-specific antibodies, *J. Biol. Chem.*, 272, 5157, 1997.

132. Yaka, R. et al., Pituitary adenylate cyclase-activating polypeptide (PACAP(1-38)) enhances N-methyl-D-aspartate receptor function and brain-derived neurotrophic factor expression via RACK1, *J. Biol. Chem.*, 278, 9630, 2003.

133. Pascual, A. et al., Neuroanatomy: brain asymmetry and long-term memory, *Nature*, 427, 605, 2004.

134. Yu, D. et al., *Drosophila* DPM neurons form a delayed and branch-specific memory trace after olfactory classical conditioning, *Cell*, 123, 945, 2005.

135. Dubnau, J. et al., Disruption of neurotransmission in *Drosophila* mushroom body blocks retrieval but not acquisition of memory, *Nature*, 411, 476, 2001.

136. Connolly, J.B. et al., Associative learning disrupted by impaired Gs signaling in *Drosophila* mushroom bodies, *Science*, 274, 2104, 1996.

137. de Belle, J.S. and Heisenberg, M., Associative odor learning in *Drosophila* abolished by chemical ablation of mushroom bodies, *Science*, 263, 692, 1994.

138. Xia, S. and Tully, T., Segregation of odor identity and intensity during odor discrimination in *Drosophila* mushroom body, *PLoS Biol.*, 5, e264, 2007.

139. Margulies, C., Tully, T., and Dubnau, J., Deconstructing memory in *Drosophila*, *Curr. Biol.*, 15, R700, 2005.

140. Dash, P.K., Hebert, A.E., and Runyan, J.D., A unified theory for systems and cellular memory consolidation, *Brain Res. Brain Res. Rev.*, 45, 30, 2004.

141. Dudai, Y., The neurobiology of consolidations, or, how stable is the engram?, *Annu. Rev. Psychol.*, 55, 51, 2004.

142. Frankland, P.W. and Bontempi, B., The organization of recent and remote memories, *Nat. Rev. Neurosci.*, 6, 119, 2005.

143. Wiltgen, B.J. et al., New circuits for old memories: the role of the neocortex in consolidation, *Neuron*, 44, 101, 2004.

144. Husi, H. et al., Proteomic analysis of NMDAR-adhesion protein signaling complexes, *Nat. Neurosci.*, 3, 661, 2000.

145. Bailey, C.H. et al., Is heterosynaptic modulation essential for stabilizing Hebbian plasticity and memory?, *Nat. Rev. Neurosci.*, 1, 11, 2000.
146. Roberts, A.C. and Glanzman, D.L., Learning in *Aplysia*: looking at synaptic plasticity from both sides, *Trends Neurosci.*, 26, 662, 2003.
147. Sheng, M. and Pak, D.T., Ligand-gated ion channel interactions with cytoskeletal and signaling proteins, *Annu. Rev. Physiol.*, 62, 755, 2000.
148. Guo, H.F. et al., A neurofibromatosis-1-regulated pathway is required for learning in *Drosophila*, *Nature*, 403, 895, 2000.
149. Drier, E.A. et al., Memory enhancement and formation by atypical PKM activity in *Drosophila melanogaster*, *Nat. Neurosci.*, 5, 316, 2002.
150. Rorth, P., A modular misexpression screen in *Drosophila* detecting tissue-specific phenotypes, *Proc. Natl. Acad. Sci. USA*, 93, 12418, 1996.
151. Brand, A.H. and Perrimon, N., Targeted gene expression as a means of altering cell fates and generating dominant phenotypes, *Development*, 118, 401, 1993.
152. Fire, A. et al., Potent and specific genetic interference by double-stranded RNA in *Caenorhabditis elegans*, *Nature*, 391, 806, 1998.
153. Hannon, G.J., RNA interference, *Nature*, 418, 244, 2002.
154. Siegmund, T. and Korge, G., Innervation of the ring gland of *Drosophila melanogaster*, *J. Comp. Neurol.*, 431, 481, 2001.
155. McGuire, S.E. et al., Spatiotemporal rescue of memory dysfunction in *Drosophila*, *Science*, 302, 1765, 2003.
156. Kitamoto, T., Conditional modification of behavior in *Drosophila* by targeted expression of a temperature-sensitive shibire allele in defined neurons, *J. Neurobiol.*, 47, 81, 2001.
157. Lee, T., Lee, A., and Luo, L., Development of the *Drosophila* mushroom bodies: sequential generation of three distinct types of neurons from a neuroblast, *Development*, 126, 4065, 1999.

11 Extracellular Modulation of NMDA Receptors

Keith Williams

CONTENTS

11.1 INTRODUCTION

A number of natural and synthetic molecules including polyamines, protons, Zn^{2+}, steroids, redox reagents, ifenprodil, and ethanol are modulators of NMDA receptors (NMDARs), acting at extracellular sites on the receptors to increase or decrease macroscopic currents and Ca^{2+} flux through NMDA channels. Some of these modulators, for example Zn^{2+}, polyamines, and protons, are endogenous molecules that may exert important regulatory effects on NMDARs under physiological and/or pathological conditions. Others, for example ifenprodil, are synthetic molecules that serve as experimental tools to study the properties of NMDARs and receptor subtypes and may provide lead compounds for the development of therapeutically useful NMDAR antagonists.

This chapter provides an overview of some of these modulators, with a focus on polyamines, ifenprodil, and Zn^{2+} and on recent studies detailing their mechanisms of action, interactions, and proposed sites of action on NMDARs. Much of the early work in this area was reviewed previously[1-11] and is not dealt with in detail in this chapter. Similarly, the list of citations at the end of this chapter is not exhaustive; rather, it draws on recent and older publications to present an overview of how the field has developed and where it stands at present (mid 2007).

11.2 POLYAMINES

Effects of the endogenous polyamines, spermidine and spermine (Figure 11.1) on NMDARs were first observed in ligand binding assays in which the polyamines were found to increase binding of the use-dependent open-channel blockers [³H]MK-801 and [³H]TCP[12–15]. It was proposed that polyamines bind to unique sites on NMDARs distinct from the agonist binding sites to potentiate receptor activity and thus potentiate binding of [³H]MK-801.[12,13] Subsequent work demonstrated effects of polyamines on the function of NMDARs studied electrophysiologically using both native and recombinant receptors.[16–27] In studies of macroscopic NMDA currents recorded from neurons and oocytes or mammalian cells expressing recombinant receptors,

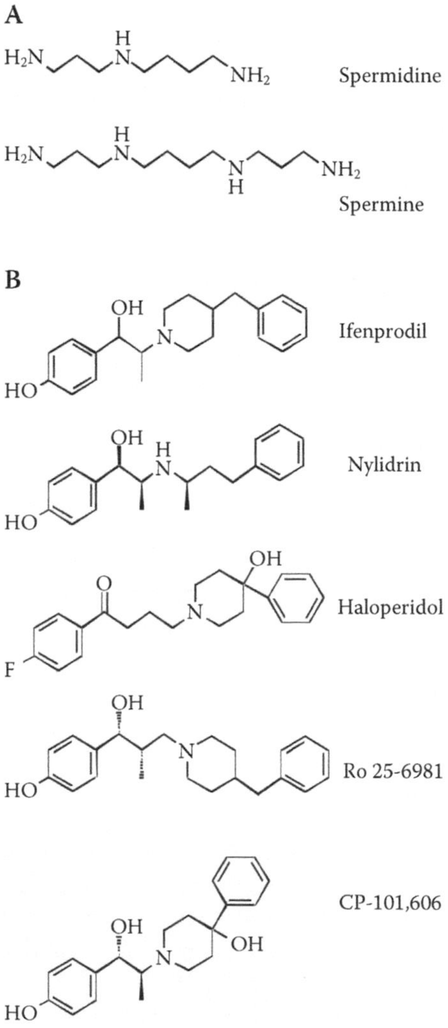

FIGURE 11.1 Structures of the endogenous polyamines, spermidine and spermine (A), and of ifenprodil and related antagonists that selectively inhibit NR1/NR2B receptors (B).

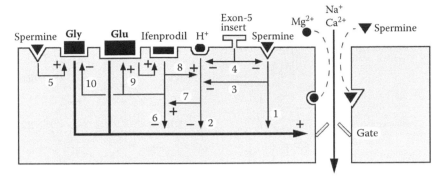

FIGURE 11.2 Model of the NMDAR to illustrate some of the effects and interactions of the various binding and modulatory sites. The receptor is gated by coagonists glutamate (Glu) and glycine (Gly), and currents are potentiated or inhibited by spermine, ifenprodil, and protons. Not all known interactions are shown. The effects of some modulators (e.g., Mg^{2+}) are simplified; certain modulators (e.g., Zn^{2+}, steroids, redox reagents) are not illustrated. Numbers refer to mechanisms or interactions discussed in the text.

four effects of spermine were described and the effects could be studied in relative isolation by adjusting experimental conditions such as the membrane potential and agonist concentration.

One effect of spermine is to increase the size of whole-cell currents evoked by saturating concentrations of glutamate and glycine, so-called glycine-independent stimulation—somewhat of a misnomer since it requires glycine. The key point is that stimulation is seen in the presence of saturating concentrations of glycine (mechanism 1, Figure 11.2).[17–20,22,24–27] At least part of this effect involves a decrease in desensitization of NMDARs in the presence of spermine.[19] Spermine can also alter deactivation of NMDARs.[28]

Protons inhibit NMDARs, with a tonic inhibition of 40 to 50% at physiologic pH (mechanism 2, Figure 11.2), and there is evidence that spermine stimulation involves a relief of tonic proton inhibition (mechanism 3, Figure 11.2).[29,30] This stimulation is subunit-dependent and is seen at NR1/NR2B receptors but not at binary NR1/NR2 receptors containing NR2A, NR2C, or NR2D.[26,27,31] Furthermore, the effects of spermine and protons are both reduced in NR1/NR2B receptors containing splice variants of the NR1 subunit that include the exon-5 insert.[18,24,30] This 21-amino acid insert is located in the extracellular amino terminal domain (ATD) or regulatory (R) domain of the NR1 subunit (Figure 11.3A). The exon-5 insert, containing six basic (Lys or Arg) residues,[32,33] may act as a spermine-like moiety, changing the conformation of the R domain and thus its interactions with other domains, or the insert may interfere with the binding of spermine or the modulatory effects of protons (mechanism 4, Figure 11.2).

If spermine stimulation involves a relief of tonic proton inhibition (mechanism 3, Figure 11.2), there is a conundrum because spermine can alter deactivation and desensitization of NMDARs including recombinant NR1/NR2B receptors[19,29] whereas protons do not affect deactivation or desensitization of these receptors.[34] Relief of proton inhibition may represent only part of the mechanism of spermine stimulation at these receptors and changes in deactivation and desensitization involving proton-insensitive gating mechanisms may also be involved.

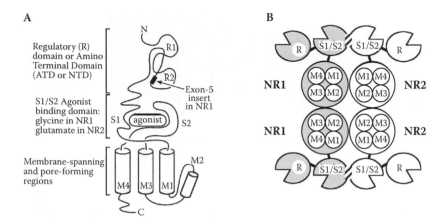

FIGURE 11.3 Domain-based structures of NMDAR subunits. A: Schematic of an NMDAR subunit with two large extracellular domains—the regulatory R domain (comprised of R1 and R2) and the S1/S2 agonist binding domain—and a membrane-spanning/pore-forming core domain comprised of M1-M4. The R domain is also referred to as the amino terminal domain (ATD or NTD) or the LIVBP-like domain because of its homology with bacterial leucine–isoleucine–valine binding protein (LIVBP).[53–55,70,113,115] In some other glutamate receptor subtypes, the R domain has been referred to as the X domain.[11,145] The NR1 subunit gene is subject to alternative splicing of its mRNA, including the insertion or deletion in the R domain of a 21-amino acid peptide encoded by exon-5. NR1 subunits have short intracellular C terminal domains. NR2 subunits have larger intracellular C terminal domains important for trafficking, attachment to intracellular scaffolding proteins, and regulation by intracellular factors (not illustrated). B: Intact receptors are proposed to be tetramers composed of dimers of heterodimers (NR1–NR2 dimers) based on studies of tandem subunits[96] and x-ray crystallographic studies of the S1/S2 domains of NR1 and NR2A. These domains form heterodimers with an interface between NR1 and NR2A.[62] The dimer interface is indicated by hatching between the S1/S2 domains of NR1 and NR2. Whether the R domains of NR1 and NR2 subunits interact in a heteromeric or a homomeric fashion in intact receptors or even if they interact at all is unknown although evidence indicates that the R domain of an AMPAR GluR1 subunit can form dimers and that four R domains are located in close proximity in intact GluR1 receptors.[68,70] Similarly, in intact GluR receptors, the R domains appear to exist as dimers.[66]

A second effect of spermine is so-called glycine-dependent stimulation seen in the presence of subsaturating concentrations of glycine (mechanism 5, Figure 11.2).[16,22,26] This effect appears to be due to an increase in the affinity of the receptor for glycine (mechanism 5, Figure 11.2) and is distinct from glycine-independent stimulation (mechanism 1) because it is seen at both NR1/NR2A and NR1/NR2B receptors (as opposed to only NR1/NR2B), it is not affected by the presence of the exon-5 insert in NR1, and it is not sensitive to extracellular pH.[7] Thus, it may involve a second, distinct spermine binding site (Figure 11.2).

A third effect of spermine is to reduce the sensitivity to glutamate (or other glutamate site agonists) at NR1/NR2B receptors (not illustrated), presumably by reducing the affinity of the receptor for glutamate.[25] The result is that spermine stimulation is smaller at subsaturating concentrations of glutamate than at saturating concentrations. This effect is seen at NR1/NR2B receptors but not at NR1/NR2A,

NR1/NR2C, or NR1/NR2D receptors and is likely mediated via the spermine binding site responsible for glycine-independent stimulation (mechanism 1).[25,31]

The fourth effect of spermine is voltage-dependent channel block due to binding of spermine within the ion channel pore at a site near to or overlapping the binding sites for extracellular Mg^{2+} (Figure 11.2).[21,22,35-37] Voltage-dependent block by spermine and by related natural and synthetic polyamines has been reported for different types of cation channels including some subtypes of AMPA receptors, kainate receptors, and inward rectifier K^+ channels that are blocked by *intracellular* spermine.[7,38-46] Spermine is a weak blocker of NMDA channels compared to the other types of cation channels.

Whether endogenous polyamines are involved in modulation of NMDARs *in vivo* is still unknown. The effects of intracellular polyamines on AMPA channels and inward rectifier K^+ channels where polyamines are responsible for rectification, are well-established.[7,38,47,48] The role, if any, of extracellular polyamines in the nervous system is unclear. Selective polyamine transport systems have been identified in neurons and glia[49] and depolarization-evoked release of polyamines has been described.[1,7,49] Since NR1/NR2B receptors are most sensitive to polyamine stimulation and NR2B is the predominant subunit expressed in embryonic and neonatal forebrain,[50-52] polyamines have pronounced effects during the development of the nervous system—a time when polyamine levels in the brain are higher than in the adult nervous system—and may conceivably play a role in NMDA-dependent plasticity during development.

11.3 STRUCTURES OF NMDA RECEPTOR SUBUNITS AND RELEVANCE TO THE POLYAMINE SITE

Mutations at a number of residues in the R domain of the NR1 subunit can reduce spermine stimulation and reduce sensitivity to protons; these residues may form part of a spermine binding site in the R domain of NR1.[53] The R domains of glutamate receptor subunits are proposed to be bilobar, similar in structure to the S1/S2 domains, based on homology with bacterial amino acid–binding proteins such as the leucine–isoleucine–valine binding protein (LIVBP) as well as weak homology with the amino terminal domain of the metabotropic glutamate receptor-1.[53-57]

X-ray crystallographic studies show that the S1/S2 domains of several glutamate receptors including NMDARs do indeed have such bilobed or clamshell structures.[58-65] The S1/S2 domains (and presumably entire subunits) of NR1/NR2A receptors assemble as dimers of heterodimers, i.e., as two sets of NR1/NR2A dimers[62] (Figure 11.3B). Although no high resolution structural data are available for a glutamate receptor R domain, evidence from single-particle electron microscopy supports the idea that the R domains can exist as dimers (presumably corresponding to the S1/S2 dimers to which they are attached) and, interestingly, that their positions or orientations can change during receptor activation or desensitization.[66]

Evidence from biochemical studies also suggests that R domains can assemble as dimers but, at least in the case of NR1/NR2A receptors, as homodimers (i.e., NR1–NR1 and NR2A–NR2A) or that entire subunits can assemble as homodimers with the R domain being important for assembly.[67-69] At first glance, this appears to contradict data from x-ray crystallographic studies of NR1/NR2A receptors in which the isolated S1/S2 domains are assembled as heterodimers (i.e., NR1–NR2A; Figure 11.3B).[62] It is possible that homodimers of NR subunits are formed during the initial assembly and trafficking of the subunits, followed by later formation of heterodimers and assembly of the final, intact four-subunit receptor, or that the formations of some types of dimers are artifacts of the experimental conditions. Another possibility is that heterodimers are formed initially involving contacts between S1/S2 domains of native NR1 and NR2 subunits and that homodimerization of the R domains is involved in subsequent assembly of tetrameric NMDARs. This would differ from the assembly of AMPAr channels in which the R domains are important for the formation of initial heterodimers and contacts between the S1/S2 domains (together with other determinants) are important for the subsequent assembly of heterodimers into tetramers.[57,70] In this context, it is notable that the fundamental gating mechanisms of NMDARs and AMPA receptors are different. NMDARs undergo concerted channel opening, requiring the simultaneous binding of two molecules of glutamate and two molecules of glycine. AMPARs undergo subunit-specific gating in which binding of glutamate to one subunit causes partial opening, binding to two subunits causes further opening, and so on, with full opening requiring binding of glutamate to all four subunits.[71–75] Based on the fundamental differences in gating mechanisms, it is not unreasonable to suppose that differences also exist in the rules that guide subunit assembly and quaternary structures of NMDARs and AMPA receptors.

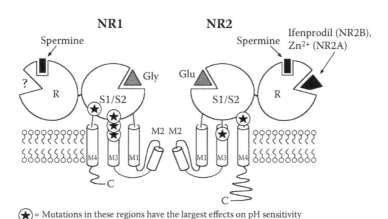

★ = Mutations in these regions have the largest effects on pH sensitivity

FIGURE 11.4 Putative locations of spermine, ifenprodil, and Zn^{2+} binding sites and domain-based organization of NR1 and NR2 subunits. In NR1/NR2A and NR1/NR2B receptors, spermine may bind to the R domain of NR1 and to a second site in the R domain of NR2. Ifenprodil may bind within the cleft of the R domain in NR2B. Zn^{2+} may bind to a homologous site in the R domain of NR2A. Glutamate binds in the cleft of the S1/S2 domain of the NR2 subunit and glycine binds to a homologous site in the NR1 subunit. Residues at which mutations produce the largest shifts in proton sensitivity have been identified in M3, in the M3–S2 linker, in S2 just distal to that linker and just proximal to the S2–M4 linker, and in the S2–M4 linker as shown by circled stars.

Based on homology modeling of the structure of the R domain, the putative spermine binding site appears to lie outside the central cleft of the NR1 R domain.[53] The identity of the endogenous ligand (if any) that binds within the cleft is unknown (see Figure 11.4). Evidence also indicates a discrete spermine binding site on the NR2 subunit, at least with regard to NR2A and NR2B[53]. This is shown in Figure 11.4.

It is possible that the spermine binding site on NR1 is responsible for glycine-independent stimulation and a second site on NR2A or NR2B underlies glycine-dependent stimulation, but no direct evidence supports this hypothesis. Nonetheless, it would be consistent with the effects of the exon-5 insert (mechanism 4, Figure 11.2) present in the R domains of some slice variants of NR1 (Figure 11.3A) that reduces glycine-independent spermine stimulation but not glycine-dependent stimulation at NR1/NR2B receptors. Similarly, mutations in the R domain of NR1 that affect glycine-dependent stimulation do not affect glycine-independent stimulation by spermine.[76]

11.4 IFENPRODIL

Ifenprodil (Figure 11.1) was originally developed as a vasodilator, based on its activity as an $\alpha 1$ adrenergic antagonist.[77,78] Studies in the 1980s and early 1990s demonstrated that it was also a noncompetitive NMDAR antagonist with a novel profile and mechanism of action.[79–82] Widespread interest in the properties of ifenprodil as a tool to study NMDARs and as a potential lead compound for novel therapeutic agents followed the discovery that it is a highly selective antagonist for receptors containing the NR2B subunit.[83,84] Ifenprodil was considered an atypical antagonist[83] because it was noncompetitive and subtype-selective but did not act as an open-channel blocker like MK-801 and phencyclidine. Ifenprodil is sometimes called an 'allosteric antagonist' (just as spermine and other modulators are referred to as allosteric modulators), but in the absence of a detailed mechanistic understanding of how ifenprodil binds and what it does after it binds to NMDARs, the allosteric designation is probably best avoided.

Other compounds with structures similar to ifenprodil, including haloperidol and nylidrin (Figure 11.1), were subsequently found to be selective for NR1/NR2B receptors[85,86] and additional NR2B-selective antagonists were developed based on the structure of ifenprodil. These compounds include CP-101,606, Ro 8-4304, and Ro 25-6981 (Figure 11.1) and are presumed to share the same binding site as ifenprodil on NMDARs.[87–89] In addition to their subtype selectivity, ifenprodil and related antagonists exhibit a novel form of use-dependency that may contribute to a favorable *in vivo* profile if these compounds are eventually used in clinical settings.[89,90]

11.5 STRUCTURES OF NMDA RECEPTOR SUBUNITS AND RELEVANCE TO THE IFENPRODIL SITE

Ifenprodil and related compounds such as Ro 25-6981 are several hundred-to several thousand-fold more potent at NR1/NR2B receptors than at NR1/NR2 receptors containing NR2A, NR2C, or NR2D. For example, the IC_{50} of ifenprodil was reported to be 0.3 μM at NR1/NR2B receptors and 146 μM at NR1/NR2A receptors—a selectivity of about 500-fold.[83] Several explanations may account for

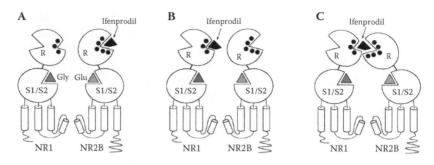

FIGURE 11.5 Where is the ifenprodil binding site on NR1/NR2B receptors? The schematics show the domain-based organization of the NR1 and NR2B subunits. Mutations within the R domains of NR1 and NR2B (small circles) alter sensitivity to ifenprodil, which could bind to the R domain of NR2B (A), the R domain of NR1 (B), or both R domains (C).

the selectivity of ifenprodil for heteromeric NR1/NR2B receptors. The ifenprodil binding site may be located on the NR2B subunit (Figure 11.5A) and be absent or have a much lower affinity on NR2A, NR2C, and NR2D. Another possibility is that the ifenprodil binding site is located on the NR1 subunit and its properties or transduction mechanisms are influenced by NR2 subunits, requiring NR2B for high affinity inhibition (Figure 11.5B). A third possibility is that the binding site involves regions in both NR1 and NR2B (Figure 11.5C). The weight of the available evidence suggests that ifenprodil binds to the NR2B subunit as illustrated in Figure 11.5A, possibly within the cleft of the bilobed R domain (analogous to binding of glutamate or glycine in the S1/S2 domains). The evidence comes largely from site-directed mutagenesis studies in which mutations in the R domain of NR2B reduced ifenprodil inhibition of recombinant NR1/NR2B receptors studied electrophysiologically.[55] The same mutations also reduced the ability of ifenprodil to protect against proteolytic degradation of an isolated, purified soluble NR2B R domain.[55] In other studies, an isolated soluble NR2B R domain was found to bind ifenprodil with high affinity (Kd, 0.13 μM) based on shifts in circular dichroism.[91]

Interestingly, residues at which mutations in NR2B produced the most pronounced effects on sensitivity to ifenprodil or ifenprodil-like ligands are in positions analogous to residues in NR2A that appear to form the high affinity Zn^{2+} binding site.[54-56,92] Surprisingly, many of these residues are actually identical in NR2A and NR2B, despite the marked differences in the chemical natures of ifenprodil and Zn^{2+}. Equally surprising, in light of the large degree of selectivity of ifenprodil for NR1/NR2B over NR1/NR2A, is that many residues in the proposed ifenprodil binding site of NR2B are identical to their corresponding residues in NR2A, at least on the basis of alignment of linear amino acid sequences.[55] This may suggest that only one or two key residues influence selectivity for ifenprodil or that the overall structure or folding of the R domain in NR2B is somewhat different from that of the R domain of NR2A despite their close sequence similarity in regions that affect sensitivity to ifenprodil. By homology modeling, these residues were proposed to lie within the cleft between the two lobes of the R domain.[54,55]

Although the evidence favors a high affinity ifenprodil binding site on the NR2B subunit (Figure 11.5A), it has also been shown that mutations in the R domain of NR1 can exert profound and selective effects on ifenprodil inhibition at NR1/NR2B receptors.[53] Based on homology modeling, these mutations are in a region of the NR1 R domain that is likely outside the cleft (Figure 11.5). It is possible that at least part of the ifenprodil binding site in intact NR1/NR2B receptors is formed by regions in the R domain of NR1 (Figure 11.5C). Alternatively, some mutations in NR1-R may disrupt the properties of the NR2B R domain, particularly if these mutations are at an interface between the NR1 and NR2B subunits, perhaps at an interface between adjacent R domains. It is also notable that ifenprodil inhibits apparent homomeric NR1 receptors expressed in *Xenopus* oocytes with a potency similar to that at NR1/NR2B receptors[84] consistent with the idea that a high affinity ifenprodil binding site can be formed by the NR1 subunit alone, although the question whether these recombinant NR1 receptors are truly homomeric or involve the inclusion of endogenous *Xenopus* NR2-like subunits is still unresolved.[93]

Like AMPA receptors, NMDARs are thought to be tetramers,[71,94,95] likely assembled as dimers of dimers,[62,96] although there is also evidence consistent with a pentameric rather than a tetrameric subunit structure for NMDARs.[97,98] Assuming a tetrameric structure (a similar argument holds if the receptor were a pentamer), then each receptor must have two identical and presumably equivalent ifenprodil binding sites just as it has two glutamate binding sites and two glycine binding sites. In the case of glutamate and glycine, the agonist concentration–response curves are steep, with Hill coefficients close to 2.0, suggesting cooperativity of binding of the agonists. In the case of ifenprodil, the Hill coefficient for inhibition of NR1/NR2B receptors is close to 1.0,[55,99] suggesting, perhaps surprisingly, that there is no positive or negative cooperativity between the two ifenprodil binding sites.

In triheteromeric receptors engineered to contain one NR2A subunit and one NR2B subunit together with two NR1 subunits, ifenprodil was found to inhibit responses with high affinity (via the NR2B subunit) but the maximum degree of inhibition was greatly reduced.[92] A similar profile was seen for high affinity inhibition of these receptors by Zn^{2+}; high affinity inhibition mediated via the NR2A subunit (see below) was still present, but the degree of inhibition was greatly reduced. This may suggest that the R domains can influence channel gating in an independent rather than a concerted manner[92]—in contrast to the concerted gating of the channel by agonist binding to the S1/S2 domains.[73–75]

Initial studies suggested that ifenprodil acts as an antagonist at the stimulatory polyamine site,[100] but subsequent work has shown this to not be the case,[81,83] and ifenprodil is thought to bind to a distinct site on the NMDAR (Figures 11.2 and 11.4). This is consistent with results from mutagenesis studies in which different residues were found to influence sensitivity to ifenprodil and spermine.[53]

Ifenprodil inhibits currents activated by glutamate and glycine (mechanism 6, Figure 11.2), and this effect is not voltage-dependent but is dependent on agonist concentration. Nonetheless, there are documented interactions between spermine and ifenprodil. Spermine can reduce the affinity of the receptor for ifenprodil and, conversely, ifenprodil can reduce the affinity for spermine.[101] Spermine can affect proton inhibition (mechanism 3, Figure 11.2) and protons can, in turn, affect ifenprodil inhibition (mechanism 7, Figure 11.2) and vice versa (mechanism 8, Figure 11.2). Thus, spermine may indirectly alter sensitivity to ifenprodil by changing proton sensitivity of the receptor.

The use-dependent properties of ifenprodil arise because the affinity for ifenprodil is increased by glutamate binding and vice versa (mechanism 9, Figure 11.2).[89,90] In addition, ifenprodil reduces the affinity for glycine at NR1/NR2B receptors (mechanism 10, Figure 11.2), which presumably contributes to its inhibitory effects at these receptors.[82,83,102]

In addition to its interactions with spermine, inhibition by ifenprodil is also dependent on extracellular pH, and it was proposed that the mechanism of action of ifenprodil is to potentiate tonic proton inhibition.[103] Inhibition of NR1/NR2B receptors by ifenprodil is increased at acidic pH and reduced at alkaline pH, i.e., protons increase the apparent affinity for ifenprodil (mechanism 7, Figure 11.2).[99] At the same time, ifenprodil (and similar NR2B-selective antagonists) can enhance the inhibitory effects of protons at NR1/NR2B receptors, and this may be the major mechanism that underlies ifenprodil inhibition (mechanism 8, Figure 11.2).[103] The presence or absence of the exon-5 insert or the addition of spermine produces only modest effects on sensitivity to ifenprodil and related compounds. These effects are likely indirect due to changes in proton sensitivity (mechanisms 3, 4, 7, and 8, Figure 11.2).[30,99,103]

11.6 ZINC IONS

Zinc ions are present in high concentrations in some areas of the nervous system and may modulate synaptic transmission.[104,105] Their effects on NMDARs were first described in studies of native receptors on isolated neurons in which Zn^{2+} was found to be a potent inhibitor of NMDA currents.[106–108] That work indicated that Zn^{2+} likely had two effects at native NMDARs (at that time receptor subunits had not yet been cloned and NMDAR subtypes were uncharacterized): a relatively high affinity inhibition that was not voltage-dependent, presumed to be mediated at a unique extracellular site on the receptor, and a low affinity, voltage-dependent block, possibly mediated at the Mg^{2+} binding site or at a nearby site within the channel pore.[106–108] Subsequent studies characterized effects of Zn^{2+} on recombinant NMDARs. Pronounced differences in subunit-dependent sensitivity to Zn^{2+} were reported, and recent studies have identified potential Zn^{2+} binding sites in the R domains of NR2 subunits and shed some light on the mechanism of action of Zn^{2+}.

Initial studies of recombinant NMDARs found marked differences in the sensitivity to Zn^{2+} of NR1/NR2A and NR1/NR2B receptors. At NR1/NR2B receptors, Zn^{2+} inhibition was not voltage-dependent and was monophasic with an IC_{50} of 0.5 to 9 μM.[109–111] In contrast, inhibition at NR1/NR2A receptors was biphasic, with a high affinity component (IC_{50} 5 to 80 nM) that was not voltage-dependent and a low affinity component (IC_{50} 26 to 79 μM) that was voltage-dependent.[109–111] This low affinity, voltage-dependent component likely represents a weak open-channel block by high concentrations of Zn^{2+}. Another discovery was that residual or contaminating traces of Zn^{2+} in experimental solutions can produce marked inhibitory effects at NR1/NR2A receptors, and that solutions should be buffered with a Zn^{2+} chelator such as tricine (certainly for low concentrations of Zn^{2+}) to accurately determine the concentration-dependence of Zn^{2+} inhibition at these receptors.[111,112] At NR1/NR2C and NR1/NR2D receptors, Zn^{2+} was subsequently found to have an even lower affinity than at NR1/NR2B receptors.[113]

Putting aside the low affinity, voltage-dependent block of NMDA channels by Zn^{2+}, the difference uncovered in early studies of recombinant receptors was a very high affinity inhibition at NR1/NR2A receptors versus a lower affinity inhibition at NR1/NR2B receptors. Another difference was that inhibition at NR1/NR2B receptors was monophasic and complete (Zn^{2+} produced a complete inhibition of NMDA currents), whereas the high affinity inhibition at NR1/NR2A receptors was incomplete. Zn^{2+} inhibited responses by only 40 to 70%.[109,112,114]

This is reminiscent of the effects of ifenprodil at NR1/NR2B receptors, where the maximum inhibition is about 80 to 90% at physiologic pH.[83] Interactions occur between high affinity Zn^{2+} inhibition of NR1/NR2A receptors and proton inhibition at these same receptors, and it was proposed that the mechanism of Zn^{2+} inhibition involves an increase in tonic proton inhibition at NR1/NR2A receptors, analogous to the proposed mechanism of ifenprodil inhibition at NR1/NR2B receptors (equivalent to mechanism 8, Figure 11.2).[112,115,116] In the proposed model of Zn^{2+} inhibition of NR1/NR2A receptors, binding of glutamate to the S1/S2 domain led to increased affinity for Zn^{2+} in the R domain and binding of Zn^{2+} led to a conformational change that enhanced binding of protons to proton-sensitive gating elements leading, in turn, to a reduction in channel open probability.[114,116]

11.7 STRUCTURES OF NMDA RECEPTOR SUBUNITS AND RELEVANCE TO THE Zn^{2+} SITE

Swapping the R domains between NR2A and NR2B led to reciprocal changes in sensitivity to Zn^{2+} and ifenprodil.[54,55] Thus, NR2B subunits containing R domains of NR2A showed very high sensitivity to Zn^{2+} (similar to the native NR2A subunit) whereas NR2A subunits containing R domains of NR2B showed high sensitivity to ifenprodil (similar to the native NR2B subunit) and a reduced sensitivity to Zn^{2+}.[54,55] Mutations at a number of positions in the R domain of the NR2A subunit produced marked and in some cases specific effects on inhibition by Zn^{2+}.[54,112,115,117]

Using homology modeling based on LIVBP and related proteins, the residues that affect Zn^{2+} sensitivity were proposed to lie within the cleft of the R domain of NR2A[54] (Figure 11.6). Some of these residues are conserved between NR2A and NR2B, and may form part of the high affinity Zn^{2+} binding site in NR2A and the ifenprodil binding site in NR2B.[54,55] However, there are a few other residues that are identical or similar in NR2A and NR2B at which mutations affect Zn^{2+} sensitivity in NR2A but not ifenprodil sensitivity in NR2B and vice versa.[54,55]

As with most mutagenesis studies, the results may mean that the residues (their side chains or peptide bond backbones) (1) interact directly with Zn^{2+} and ifenprodil, (2) are key components of the backbone of a binding pocket, or (3) interact with water molecules within the binding site that in turn make contacts with Zn^{2+} and ifenprodil. Structural studies will be required to determine whether Zn^{2+} and ifenprodil really bind within the R domain clefts and what role particular residues play in those domains.

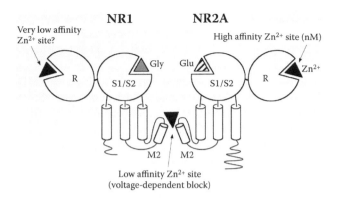

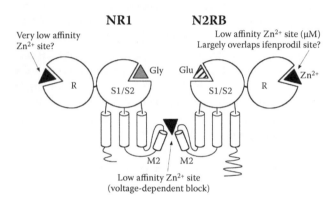

FIGURE 11.6 Putative locations of Zn^{2+} binding sites in NR1/NR2A and NR1/NR2B receptors and domain-based organization of these receptors, with a high affinity Zn^{2+} binding site in the cleft of the R domain of NR2A and a lower affinity site in the homologous region of the R domain in NR2B. Zn^{2+} can also produce a voltage-dependent block by binding to a site within the channel pore. In the absence of the R domain of the NR2 subunit voltage-dependent block by Zn^{2+}, a residual, very low affinity Zn^{2+} inhibition is seen that may involve a Zn^{2+} binding site on the R domain of the NR1 subunit.

As with ifenprodil, if the Zn^{2+} binding site lies within the cleft of the R domain in the NR2 subunit, then each tetrameric receptor must have two presumably identical and equivalent Zn^{2+} binding sites. Again, as with ifenprodil acting at NR1/NR2B receptors, the Hill coefficient for inhibition of NR1/NR2A receptors by Zn^{2+} is close to unity,[109,113,114,117] suggesting a lack of cooperativity between the two Zn^{2+} binding sites. As discussed above with regard to ifenprodil, triheteromeric NR1/NR2A/NR2B receptors still had a very high affinity for Zn^{2+} acting at the NR2A subunit, but the degree of inhibition by Zn^{2+} was greatly reduced.[92]

Although there is good evidence for localization of the high affinity Zn^{2+} binding site on the NR2A subunit (Figure 11.6), the NR1 subunit can also influence Zn^{2+} sensitivity of NR1/NR2A receptors. Thus, the exon-5 insert in the R domain of the NR1 subunit (Figure 11.3A) has been shown to alter inhibition by Zn^{2+},[118] as have mutations at residues in the extracellular region preceding M1 and in the M3-M4 loop region, both of which contribute to the S1/S2 domain in the NR1 subunit.[118,119] However, the effects of these mutations on sensitivity to Zn^{2+} may be indirect and arise from changes in pH sensitivity and/or changes in the redox state of the receptor protein, both of which can alter Zn^{2+} inhibition in NR1/NR2A receptors.[118,119]

Results of studies using truncated NR2 subunits in which the R domain was removed and the remainder of the subunit was intact, and studies of point mutations in the R domain of NR2B suggest that the low affinity Zn^{2+} binding site in NR2B is located in the R domain and is structurally homologous to the high affinity Zn^{2+} binding site in NR2A (Figure 11.6), and that this site in NR2B shares many structural determinants with the ifenprodil binding site in NR2B and the high affinity Zn^{2+} binding site in NR2A.[113] Absent an R domain in the NR2 subunit, NR1/NR2A and NR1/NR2B receptors are still inhibited by Zn^{2+}, albeit with very low affinity similar to NR1/NR2C and NR1/NR2D receptors.[113] This suggests the presence of yet another, very low affinity Zn^{2+} binding site that must be located somewhere other than the R domain of the NR2 subunit.

One possibility is that the very low affinity site is located on the R domain of the NR1 subunit (Figure 11.6). It is notable that some splice variants of NR1, when expressed as homomeric NR1 receptors, are inhibited by Zn^{2+} whereas others are potentiated by Zn^{2+}. The presence or absence of the exon-5 insert located in the R domain of NR1 influences sensitivity to Zn^{2+}.[120] However, this interpretation is complicated by the interactions between Zn^{2+} inhibition and proton inhibition, and by the fact that proton inhibition is, itself, influenced by the absence or presence of the exon-5 insert.[30,118]

11.8 PROTONS AND EXTRACELLULAR pH

Protons and thus extracellular pH exert profound effects on NMDARs over the normal physiologic pH range. The effects of protons were discussed in the preceding sections because of their interactions with modulation by polyamines, ifenprodil, and Zn^{2+}. An investigation and understanding of proton modulation of NMDARs developed concurrent with studies of those other modulators, and proton inhibition may represent a common denominator linking the effects of the other modulators and a common end point through which those modulators exert their effects on activation of NMDA channels (mechanism 2, Figure 11.2). Thus, spermine stimulation may at least in part involve a relief of tonic proton inhibition (mechanism 3, Figure 11.2) whereas inhibition by ifenprodil and Zn^{2+} may involve an increase in tonic proton inhibition (mechanism 8, Figure 11.2).

As with many other modulatory effects characterized at NMDARs, effects of extracellular pH were first documented in studies of native NMDARs expressed on isolated neurons.[121–124] Protons inhibited NMDA responses with an IC_{50} around pH 7.0 to 7.3, indicating that the receptors were tonically inhibited by about 50% at physiologic pH. Proton inhibition is not voltage-dependent and presumably does not involve block of the ion channel pore.[121–124] The effects of protons involve a decrease in channel opening frequency but not changes in unitary conductance or dwell time.[34,121,123,124] At recombinant NR1/NR2B receptors, proton inhibition can occur independent of agonist binding, and it was suggested that protonated receptors are shifted into a state from which they cannot open.[34]

It was assumed that the effects of protons were due to protonation of one or more ionizable residues on the extracellular surface of the NMDAR.[121–124] Studies of recombinant receptors have begun to uncover the molecular bases for proton inhibition, but the site and mechanism of proton inhibition remain unclear and do not appear to involve only one (or even several) ionizable residues on the NR1 or NR2 subunit.

Proton sensitivity of NMDARs is influenced by the exon-5 insert in the NR1 subunit. The presence of the insert reduces proton inhibition.[30] The mechanism underlying this effect is not understood, but the insert may function similar to spermine to alter the conformation of the R domain that influences proton-sensitive gating of the intact receptor. Proton sensitivity is also influenced by NR2 subunits; receptors containing NR2C are much less sensitive to protons than receptors containing NR2A or NR2B.[125] However, most mechanistic studies of pH sensitivity focused on NR1/NR2A and NR1/NR2B receptors.

Mutations in various regions of the NR1 subunit were reported to affect proton sensitivity in NR1/NR2 receptors. These include mutations in the R domain,[53,76,115,118] the S2 portion of the S1/S2 domain formed by the loop between M3 and M4,[29,125,126] the linker between M3 and S2,[125] and at the critical Asn residue in the M2 loop that controls Mg^{2+} block and Ca^{2+} permeability of the channel.[36] Some of these residues, particularly those in the R domain and within the channel pore, likely have indirect effects on pH sensitivity rather than being ionizable residues that directly form part of a 'proton sensor' on the NMDAR. Residues in NR1 at which mutations have the largest effects on proton sensitivity are clustered in several regions that may be important for channel gating—at the top of M3, in the M3–S2 linker, in S2 near that linker, and in S2 near the S2–M4 linker (Figure 11.4).[125]

Results of other studies showed that the M3 and M3–S2 linker regions are associated with gating; movement of the M3 domain may be critical for channel gating.[74,127,130] Residues that appear to be important for proton sensitivity have been identified in similar regions of the NR2A subunit—in the M3 domain and in the S2–M4 linker (Figure 11.4).[125] Structural differences in the S2–M4 linker regions of NR2C versus NR2A or NR2B appear to account for the reduced proton sensitivity of NR1/NR2C receptors compared to NR1/NR2A or NR1/NR2B receptors.[125] Whether any of these residues form part of a proton sensor per se, or whether the mutations have indirect effects on proton sensitivity are still unknown.

11.9 WHITHER NEXT?

In the early 1990s, the cloning of cDNAs encoding NMDAR subunits and the subsequent widespread availability of these clones ushered in the era of molecular studies of NMDARs.[32,33,131–136] This activity followed molecular studies of AMPA receptors and kainate receptors begun in 1989 with the cloning of the GluR1 AMPA receptor subunit and the subsequent cloning of cDNAs encoding a family of related GluR subunits.[137,138]

Subunits of NMDARs, AMPARs, and kainate receptors were recognized as members of the same superfamily of genes, likely with common ancestors and similar tertiary and quaternary structures. Subsequent work from many laboratories focused on the effects for agonists, antagonists, and intra- and extracellular modulators of NMDARs at the molecular level, often combined with studies of the properties and proposed structures of the receptor subunits and their interactions with other intracellular or membrane-bound proteins. Some of that work is reviewed in this chapter and elsewhere in this book. However, even when coupled with homology modeling and powerful algorithms for structure prediction, the results of mutagenesis and similar studies do not provide definitive information about the locations of binding sites for modulators or their mechanisms of action and interactions. These studies provide important clues and allow the formulation of testable hypotheses and descriptive models such as those illustrated in Figures 11.3 through 11.6.

Another major step forward in molecular studies of glutamate receptors came in 1998 with the determination of a high-resolution x-ray crystallographic structure of the S1/S2 domain of a GluR2 AMPA receptor subunit[59], following the demonstration three years earlier, that it was possible to isolate and purify a GluR subunit S1/S2 fusion protein that retained the glutamate binding site with appropriate pharmacological characteristics.[139] The 1998 work[59] confirmed that the S1/S2 domain is composed of two major lobes; that it is structurally related to bacterial amino acid binding proteins such as the glutamine binding protein (QBP); that glutamate (or the agonist kainate) binds within the cleft between the two lobes; and that glutamate receptor subunits can thus be considered to have modular architectures, perhaps arising minimally during evolution from a combination of two or three genes encoding an ion channel and one or more amino acid–binding proteins. The structures of a number of other GluR subunit S1/S2 domains have also been reported, in some cases with an agonist, partial agonist, or antagonist bound within the domain cleft, leading to detailed models of the binding and activation of these receptors.[58,59,64,65,140–144] Recently, structures for the S1/S2 domains of the NMDAR NR1 and NR2A subunits have been reported, providing crucial new insights into activation, regulation, and desensitization of NMDARs.[60,62,63,75]

Three extracellular modulators discussed in this chapter—spermine, ifenprodil, and Zn^{2+}—are thought to bind to the R domains of NMDAR subunits. High resolution structural studies of the R domains, both in isolation and together with the S1/S2 domains and ultimately together with the pore-forming regions, will be required to elucidate definitively the sites and mechanisms of action of these modulators. For example, if ifenprodil really does bind within the cleft of the R domain in NR2B, does it promote closure of the cleft, analogous to glutamate binding within the cleft of the S1/S2 domain of GluR1 and glycine binding within the cleft of the S1/S2 domain of NR1?[58,60] If so, how does that structural change in the R domain ultimately translate into a reduction of current through the channel and/or to a change in proton inhibition mediated elsewhere in the receptor protein? Or does ifenprodil stabilize an open conformation of the R domain in NR2B? If so, how does that affect receptor activity? Is there an interface between the R domains of the NR1 and NR2 subunits and, if so, are there interactions between these domains? The exact same questions can be asked of Zn^{2+} at the NR2A and NR2B subunits if it binds within the clefts of the R domains on the subunits (Figure 11.6). Similarly, if spermine binds elsewhere on the R domain (Figure 11.4), what is its site and mechanisms of action?

What is the spatial relationship of the R domain and the S1/S2 domain, and does this change during receptor activation or desensitization? A powerful experimental approach, single particle electron microscopy, has begun to provide tentative and fascinating information about the structure, position, and role of R domains in AMPA receptor GluR subunits.[66] Clearly, the models and hypotheses outlined in Figures 11.3 through 11.6 beg many questions for experimentation when techniques such as x-ray crystallography and single particle electron microscopy can be applied to isolated R domains and larger structural components of NMDAR subunits (e.g., R–S1/S2; NR1-R–NR1-R; NR1-R–NR2-R; etc.) and to fully assembled NMDARs together with the biochemical and functional approaches already being applied to the study of these receptors.

REFERENCES

1. Williams, K. et al., Minireview: modulation of the NMDAR by polyamines, *Life Sci.*, 48, 469, 1991.
2. Nakanishi, S., Molecular diversity of glutamate receptors and implications for brain function, *Science*, 258, 597, 1992.
3. McBain, C.J. and Mayer, M.L., *N*-Methyl-D-aspartic acid receptor structure and function, *Physiol. Rev.*, 74, 723, 1994.
4. Mori, H. and Mishina, M., Structure and function of the NMDAR channel, *Neuropharmacology*, 34, 1219, 1995.
5. Zukin, R.S. and Bennett, M.V., Alternatively spliced isoforms of the NMDAR1 receptor subunit, *Trends Neurosci.*, 18, 306, 1995.
6. Sucher, N.J. et al., NMDARs: from genes to channels, *Trends Pharmacol. Sci.*, 17, 348, 1996.
7. Williams, K., Interactions of polyamines with ion channels, *Biochem. J.*, 325, 289, 1997.
8. Williams, K., Modulation and block of ion channels: a new biology of polyamines, *Cell Signal.*, 9, 1, 1997.
9. Dingledine, R. et al., The glutamate receptor ion channels, *Pharmacol. Rev.*, 51, 7, 1999.
10. Williams, K., Ifenprodil, a novel NMDAR antagonist: site and mechanism of action, *Current Drug Targets*, 2, 285, 2001.
11. Herin, G.A. and Aizenman, E., Amino terminal domain regulation of NMDAR function, *Eur. J. Pharmacol.*, 500, 101, 2004.
12. Ransom, R.W. and Stec, N.L., Cooperative modulation of [^{3}H]MK-801 binding to the *N*-methyl-D-aspartate receptor-ion channel complex by L-glutamate, glycine, and polyamines, *J. Neurochem.*, 51, 830, 1988.
13. Williams, K., Romano, C., and Molinoff, P.B., Effects of polyamines on the binding of [^{3}H]MK-801 to the *N*-methyl-D-aspartate receptor: pharmacological evidence for the existence of a polyamine recognition site, *Mol. Pharmacol.*, 36, 575, 1989.
14. Reynolds, I.J., Arcaine is a competitive antagonist of the polyamine site on the NMDAR, *Eur. J. Pharmacol.*, 177, 215, 1990.
15. Sacaan, A.I. and Johnson, K.M., Characterization of the stimulatory and inhibitory effects of polyamines on [^{3}H]*N*-(1-[thienyl]cyclohexyl)piperidine binding to the *N*-methyl-D-aspartate receptor ionophore complex, *Mol. Pharmacol.*, 37, 572, 1990.
16. McGurk, J.F., Bennett, M.V.L., and Zukin, R.S., Polyamines potentiate responses of *N*-methyl-D-aspartate receptors expressed in *Xenopus* oocytes, *Proc. Natl. Acad. Sci. USA*, 87, 9971, 1990.

17. Williams, K. et al., Characterization of polyamines having agonist, antagonist, and inverse agonist effects at the polyamine recognition site of the NMDAR, *Neuron*, 5, 199, 1990.
18. Durand, G.M. et al., Cloning of an apparent splice variant of the rat *N*-methyl-D-aspartate receptor NMDAR1 with altered sensitivity to polyamines and activators of protein kinase C, *Proc. Natl. Acad. Sci. USA*, 89, 9359, 1992.
19. Lerma, J., Spermine regulates N-methyl-D-aspartate receptor desensitization, *Neuron*, 8, 343, 1992.
20. Rock, D.M. and Macdonald, R.L., The polyamine spermine has multiple actions on N-methyl-D-aspartate receptor single-channel currents in cultured cortical neurons, *Mol. Pharmacol.*, 41, 83, 1992.
21. Rock, D.M. and Macdonald, R.L., Spermine and related polyamines produce a voltage-dependent reduction of *N*-methyl-D-aspartate receptor single-channel conductance, *Mol. Pharmacol.*, 42, 157, 1992.
22. Benveniste, M. and Mayer, M.L., Multiple effects of spermine on N-methyl-D-aspartic acid receptor responses of rat cultured hippocampal neurones, *J. Physiol.*, 464, 131, 1993.
23. Araneda, R.C., Zukin, R.S., and Bennett, M.V., Effects of polyamines on NMDA-induced currents in rat hippocampal neurons: a whole-cell and single-channel study, *Neurosci. Lett.*, 152, 107, 1993.
24. Durand, G.M., Bennett, M.V.L., and Zukin, R.S., Splice variants of the *N*-methyl-D-aspartate receptor NR1 identify domains involved in regulation by polyamines and protein kinase C, *Proc. Natl. Acad. Sci. USA*, 90, 6731, 1993.
25. Williams, K., Mechanisms influencing stimulatory effects of spermine at recombinant *N*-methyl-D-aspartate receptors, *Mol. Pharmacol.*, 46, 161, 1994.
26. Williams, K. et al., Sensitivity of the *N*-methyl-D-aspartate receptor to polyamines is controlled by NR2 subunits, *Mol. Pharmacol.*, 45, 803, 1994.
27. Zhang, L. et al., Spermine potentiation of recombinant *N*-methyl-D-aspartate receptors is affected by subunit composition, *Proc. Natl. Acad. Sci. USA*, 91, 10883, 1994.
28. Rumbaugh, G. et al., Exon 5 and spermine regulate deactivation of NMDAR subtypes, *J. Neurophysiol.*, 83, 1300, 2000.
29. Sullivan, J.M. et al., Identification of two cysteine residues that are required for redox modulation of the NMDA subtype of glutamate receptor, *Neuron*, 13, 929, 1994.
30. Traynelis, S.F., Hartley, M., and Heinemann, S.F., Control of proton sensitivity of the NMDAR by RNA splicing and polyamines, *Science*, 268, 873, 1995.
31. Williams, K., Pharmacological properties of recombinant *N*-methyl-D-aspartate (NMDA) receptors containing the e4 (NR2D) subunit, *Neurosci. Lett.*, 184, 181, 1995.
32. Moriyoshi, K. et al., Molecular cloning and characterization of the rat NMDAR, *Nature*, 354, 31, 1991.
33. Sugihara, H. et al., Structures and properties of seven isoforms of the NMDAR generated by alternative splicing, *Biochem. Biophys. Res. Comm.*, 185, 826, 1992.
34. Banke, T.G., Dravid, S.M., and Traynelis, S.F., Protons trap NR1/NR2B NMDARs in a nonconducting state, *J. Neurosci.*, 25, 42, 2005.
35. Araneda, R. et al., Spermine and arcaine block and permeate *N*-methyl-D-aspartate receptor channels, *Biophys. J.*, 76, 2899, 1999.
36. Kashiwagi, K. et al., Block and modulation of *N*-methyl-D-aspartate receptors by polyamines and protons: role of amino acid residues in the transmembrane and pore-forming regions of NR1 and NR2 subunits, *Mol. Pharmacol.*, 52, 701, 1997.
37. Zheng, X. et al., Mutation of structural determinants lining the *N*-methyl-D-aspartate receptor channel differentially affects phencyclidine block and spermine potentiation and block, *Neuroscience*, 93, 125, 1999.

38. Lopatin, A.N., Makhina, E.N., and Nichols, C.G., Potassium channel block by cytoplasmic polyamines as the mechanism of intrinsic rectification, *Nature*, 372, 366, 1994.
39. Ficker, E. et al., Spermine and spermidine as gating molecules for inward rectifier K$^+$ channels, *Science*, 266, 1068, 1994.
40. Lopatin, A.N., Makhina, E.N., and Nichols, C.G., The mechanism of inward rectification of potassium channels. Long-pore plugging by cytoplasmic polyamines, *J. Gen. Physiol.*, 106, 923, 1995.
41. Fakler, B. et al., Strong voltage-dependent inward rectification of inward rectifier K$^+$ channels is caused by intracellular spermine, *Cell*, 80, 149, 1995.
42. Koh, D.S., Burnashev, N., and Jonas, P., Block of native Ca^{2+}-permeable AMPA receptors in rat brain by intracellular polyamines generates double rectification, *J. Physiol.*, 486, 305, 1995.
43. Kamboj, S.K., Swanson, G.T., and Cull-Candy, S.G., Intracellular spermine confers rectification on rat calcium-permeable AMPA and kainate receptors, *J. Physiol.*, 486, 297, 1995.
44. Isa, T. et al., Spermine mediates inward rectification of Ca^{2+}-permeable AMPA receptor channels, *Neuroreport*, 6, 2045, 1995.
45. Donevan, S.D. and Rogawski, M.A., Intracellular polyamines mediate inward rectification of Ca^{2+}-permeable alpha-amino-3-hydroxy-5-methyl-4-isoxazolepropionic acid receptors, *Proc. Natl. Acad. Sci. USA*, 92, 9298, 1995.
46. Bowie, D. and Mayer, M.L., Inward rectification of both AMPA and kainate subtype glutamate receptors generated by polyamine-mediated ion channel block, *Neuron*, 15, 453, 1995.
47. Shyng, S.L. et al., Depletion of intracellular polyamines relieves inward rectification of potassium channels, *Proc. Natl. Acad. Sci. USA*, 93, 12014, 1996.
48. Bowie, D., Lange, G.D., and Mayer, M.L., Activity-dependent modulation of glutamate receptors by polyamines, *J. Neurosci.*, 18, 8175, 1998.
49. Masuko, T. et al., Polyamine transport, accumulation and release in brain, *J. Neurochem.*, 84, 610, 2003.
50. Monyer, H. et al., Developmental and regional expression in the rat brain and functional properties of four NMDARs, *Neuron*, 12, 529, 1994.
51. Sheng, M. et al., Changing subunit composition of heteromeric NMDARs during development of rat cortex, *Nature*, 368, 144, 1994.
52. Zhong, J. et al., Expression of mRNAs encoding subunits of the NMDAR in developing rat brain, *J. Neurochem.*, 64, 531, 1995.
53. Masuko, T. et al., A regulatory domain (R1–R2) in the amino terminus of the *N*-methyl-D-aspartate receptor: effects of spermine, protons, and ifenprodil, and structural similarity to bacterial leucine/isoleucine/valine binding protein, *Mol. Pharmacol.*, 55, 957, 1999.
54. Paoletti, P. et al., Molecular organization of a zinc binding N-terminal modulatory domain in a NMDAR subunit, *Neuron*, 28, 911, 2000.
55. Perin-Dureau, F. et al., Mapping the binding site of the neuroprotectant ifenprodil on NMDARs, *J. Neurosci.*, 22, 5955, 2002.
56. Malherbe, P. et al., Identification of critical residues in the amino terminal domain of the human NR2B subunit involved in the RO 25-6981 binding pocket, *J. Pharmacol. Exp. Ther.*, 307, 897, 2003.
57. Ayalon, G. et al., Two regions in the N-terminal domain of ionotropic glutamate receptor 3 form the subunit oligomerization interfaces that control subtype-specific receptor assembly, *J. Biol. Chem.*, 280, 15053, 2005.
58. Armstrong, N. and Gouaux, E., Mechanisms for activation and antagonism of an AMPA-sensitive glutamate receptor: crystal structures of the GluR2 ligand binding core, *Neuron*, 28, 165, 2000.

59. Armstrong, N. et al., Structure of a glutamate-receptor ligand-binding core in complex with kainate, *Nature*, 395, 913, 1998.
60. Furukawa, H. and Gouaux, E., Mechanisms of activation, inhibition and specificity: crystal structures of the NMDAR NR1 ligand-binding core, *EMBO J.*, 22, 2873, 2003.
61. Gouaux, E., Structure and function of AMPA receptors, *J. Physiol.*, 554, 249, 2004.
62. Furukawa, H. et al., Subunit arrangement and function in NMDARs, *Nature*, 438, 185, 2005.
63. Inanobe, A., Furukawa, H., and Gouaux, E., Mechanism of partial agonist action at the NR1 subunit of NMDARs, *Neuron*, 47, 71, 2005.
64. Mayer, M.L., Crystal structures of the GluR5 and GluR6 ligand binding cores: molecular mechanisms underlying kainate receptor selectivity, *Neuron*, 45, 539, 2005.
65. Nanao, M.H. et al., Structure of the kainate receptor subunit GluR6 agonist-binding domain complexed with domoic acid, *Proc. Natl. Acad. Sci. USA*, 102, 1708, 2005.
66. Nakagawa, T. et al., Structure and different conformational states of native AMPA receptor complexes, *Nature*, 433, 545, 2005.
67. Papadakis, M., Hawkins, L.M., and Stephenson, F.A., Appropriate NR1-NR1 disulfide-linked homodimer formation is requisite for efficient expression of functional, cell surface N-methyl-D-aspartate NR1/NR2 receptors, *J. Biol. Chem.*, 279, 14703, 2004.
68. Matsuda, S., Kamiya, Y., and Yuzaki, M., Roles of the N-terminal domain on the function and quaternary structure of the ionotropic glutamate receptor, *J. Biol. Chem.*, 280, 20021, 2005.
69. Qiu, S. et al., Subunit assembly of N-methyl-D-aspartate receptors analyzed by fluorescence resonance energy transfer, *J. Biol. Chem.*, 280, 24923, 2005.
70. Greger, I.H., Ziff, E.B., and Penn, A.C., Molecular determinants of AMPA receptor subunit assembly, *Trends Neurosci.*, 30, 407, 2007.
71. Rosenmund, C., Stern-Bach, Y., and Stevens, C.F., The tetrameric structure of a glutamate receptor channel, *Science*, 280, 1596, 1998.
72. Smith, T.C. and Howe, J.R., Concentration-dependent substate behavior of native AMPA receptors, *Nat. Neurosci.*, 3, 992, 2000.
73. Banke, T.G. and Traynelis, S.F., Activation of NR1/NR2B NMDARs, *Nat. Neurosci.*, 6, 144, 2003.
74. Wollmuth, L.P. and Sobolevsky, A.I., Structure and gating of the glutamate receptor ion channel, *Trends Neurosci.*, 27, 321, 2004.
75. Kristensen, A.S. et al., Glutamate receptors: variation in structure-function coupling, *Trends Pharmacol. Sci.*, 27, 65, 2006.
76. Williams, K. et al., An acidic amino acid in the *N*-methyl-D-aspartate receptor that is important for spermine stimulation, *Mol. Pharmacol.*, 48, 1087, 1995.
77. Carron, C., Jullien, A., and Bucher, B., Synthesis and pharmacological properties of a series of 2-piperidino alkanol derivatives, *Arzneimittelforschung*, 21, 1992, 1971.
78. Adeagbo, A.S., Vascular relaxation by ifenprodil in the isolated perfused rat mesenteric artery, *J. Cardiovasc. Pharmacol.*, 6, 1142, 1984.
79. Carter, C. et al., Ifenprodil and SL 82.0715 as cerebral anti-ischemic agents. II. Evidence for N-methyl-D-aspartate receptor antagonist properties, *J. Pharmacol. Exp. Ther.*, 247, 1222, 1988.
80. Carter, C., Rivy, J.-P., and Scatton, B., Ifenprodil and SL 82.0715 are antagonists at the polyamine site of the N-methyl-D-aspartate (NMDA) receptor, *Eur. J. Pharmacol.*, 164, 611, 1989.
81. Reynolds, I.J. and Miller, R.J., Ifenprodil is a novel type of *N*-methyl-D-aspartate receptor antagonist: interaction with polyamines, *Mol. Pharmacol.*, 36, 758, 1989.

82. Legendre, P. and Westbrook, G.L., Ifenprodil blocks *N*-methyl-D-aspartate receptors by a two-component mechanism, *Mol. Pharmacol.*, 40, 289, 1991.

83. Williams, K., Ifenprodil discriminates subtypes of the *N*-methyl-D-aspartate receptor: selectivity and mechanisms at recombinant heteromeric receptors, *Mol. Pharmacol.*, 44, 851, 1993.

84. Williams, K. et al., Developmental switch in the expression of NMDARs occurs in vivo and in vitro, *Neuron*, 10, 267, 1993.

85. Ilyin, V.I. et al., Subtype-selective inhibition of *N*-methyl-D-aspartate receptors by haloperidol, *Mol. Pharmacol.*, 50, 1541, 1996.

86. Whittemore, E.R. et al., Subtype-selective antagonism of NMDARs by nylidrin, *Eur. J. Pharmacol.*, 337, 197, 1997.

87. Fischer, G. et al., Ro 25-6981, a highly potent and selective blocker of N-methyl-D-aspartate receptors containing the NR2B subunit. Characterization in vitro, *J. Pharmacol. Exp. Ther.*, 283, 1285, 1997.

88. Menniti, F. et al., CP-101,606, a potent neuroprotectant selective for forebrain neurons, *Eur. J. Pharmacol.*, 331, 117, 1997.

89. Kew, J.N.C., Trube, G., and Kemp, J.A., State-dependent NMDAR antagonism by Ro 8-4304, a novel NR2B selective, non-competitive, voltage-independent antagonist, *Br. J. Pharmacol.*, 123, 463, 1998.

90. Kew, J.N.C., Trube, G., and Kemp, J.A., A novel mechanism of activity-dependent NMDAR antagonism describes the effect of ifenprodil in rat cultured cortical neurones, *J. Physiol.*, 497, 761, 1996.

91. Wong, E. et al., Expression and characterization of soluble amino-terminal domain of NR2B subunit of N-methyl-D-aspartate receptor, *Protein Sci.*, 14, 2275, 2005.

92. Hatton, C.J. and Paoletti, P., Modulation of triheteromeric NMDARs by N-terminal domain ligands, *Neuron*, 46, 261, 2005.

93. Green, T. et al., NMDARs formed by NR1 in Xenopus laevis oocytes do not contain the endogenous subunit XenU1, *Mol. Pharmacol.*, 61, 326, 2002.

94. Laube, B., Kuhse, J., and Betz, H., Evidence for a tetrameric structure of recombinant NMDARs, *J. Neurosci.*, 18, 2954, 1998.

95. Mano, I. and Teichberg, V.I., A tetrameric subunit stoichiometry for a glutamate receptor-channel complex, *Neuroreport*, 9, 327, 1998.

96. Schorge, S. and Colquhoun, D., Studies of NMDAR function and stoichiometry with truncated and tandem subunits, *J. Neurosci.*, 23, 1151, 2003.

97. Premkumar, L.S. and Auerbach, A., Stoichiometry of recombinant *N*-methyl-D-aspartate receptor channels inferred from single-channel current patterns, *J. Gen. Physiol.*, 110, 485, 1997.

98. Hawkins, L.M., Chazot, P.L., and Stephenson, F.A., Biochemical evidence for the co-association of three *N*-methyl-D-aspartate (NMDA) R2 subunits in recombinant NMDARs, *J. Biol. Chem.*, 274, 27211, 1999.

99. Pahk, A.J. and Williams, K., Influence of extracellular pH on inhibition by ifenprodil at *N*-methyl-D-aspartate receptors in *Xenopus* oocytes, *Neurosci. Lett.*, 225, 29, 1997.

100. Carter, C.J. et al., Ifenprodil and SL 82.0715 as cerebral antiischemic agents. III. Evidence for antagonistic effects at the polyamine modulatory site within the N-methyl-D-aspartate receptor complex, *J. Pharmacol. Exp. Ther.*, 253, 475, 1990.

101. Kew, J.N.C. and Kemp, J.A., An allosteric interaction between the NMDAR polyamine and ifenprodil sites in rat cultured cortical neurones, *J. Physiol.*, 512, 17, 1998.

102. Ransom, R.W., Polyamine and ifenprodil interactions with the NMDAR's glycine site, *Eur. J. Pharmacol.*, 208, 67, 1991.

103. Mott, D.D. et al., Phenylethanolamines inhibit NMDARs by enhancing proton inhibition, *Nat. Neurosci.*, 1, 659, 1998.
104. Xie, X. and Smart, T.G., A physiological role for endogenous zinc in rat hippocampal synaptic transmission, *Nature*, 349, 521, 1991.
105. Harrison, N.L. and Gibbons, S.J., Zn^{2+}: an endogenous modulator of ligand- and voltage-gated ion channels, *Neuropharmacology*, 33, 935, 1994.
106. Peters, S., Koh, J., and Choi, D.W., Zinc selectively blocks the action of N-methyl-D-aspartate on cortical neurons, *Science*, 236, 589, 1987.
107. Westbrook, G.L. and Mayer, M.L., Micromolar concentrations of Zn^{2+} antagonize NMDA and GABA responses of hippocampal neurons, *Nature*, 328, 640, 1987.
108. Christine, C.W. and Choi, D.W., Effect of zinc on NMDAR-mediated channel currents in cortical neurons, *J. Neurosci.*, 10, 108, 1990.
109. Williams, K., Separating dual effects of zinc at recombinant *N*-methyl-D-aspartate receptors, *Neurosci. Lett.*, 215, 9, 1996.
110. Chen, N., Moshaver, A., and Raymond, L.A., Differential sensitivity of recombinant N-methyl-D-aspartate receptor subtypes to zinc inhibition, *Mol. Pharmacol.*, 51, 1015, 1997.
111. Paoletti, P., Ascher, P., and Neyton, J., High-affinity zinc inhibition of NMDA NR1-NR2A receptors, *J. Neurosci.*, 17, 5711, 1997.
112. Choi, Y.-B. and Lipton, S.A., Identification and mechanism of action of two histidine residues underlying high-affinity Zn^{2+} inhibition of the NMDAR, *Neuron*, 23, 171, 1999.
113. Rachline, J. et al., The micromolar zinc-binding domain on the NMDAR subunit NR2B, *J. Neurosci.*, 25, 308, 2005.
114. Zheng, F. et al., Allosteric interaction between the amino terminal domain and the ligand binding domain of NR2A, *Nat. Neurosci.*, 4, 894, 2001.
115. Low, C.-M. et al., Molecular determinants of coordinated proton and zinc inhibition of *N*-methyl-D-aspartate NR1/NR2A receptors, *Proc. Natl. Acad. Sci. USA*, 97, 11062, 2000.
116. Erreger, K. and Traynelis, S.F., Allosteric interaction between zinc and glutamate binding domains on NR2A causes desensitization of NMDARs, *J. Physiol.*, 569, 381, 2005.
117. Fayyazuddin, A. et al., Four residues of the extracellular N-terminal domain of the NR2A subunit control high-affinity Zn^{2+} binding to NMDARs, *Neuron*, 25, 683, 2000.
118. Traynelis, S.F. et al., Control of voltage-independent zinc inhibition of NMDARs by the NR1 subunit, *J. Neurosci.*, 18, 6163, 1998.
119. Choi, Y., Chen, H.V., and Lipton, S.A., Three pairs of cysteine residues mediate both redox and zn^{2+} modulation of the NMDAR, *J. Neurosci.*, 21, 392, 2001.
120. Hollmann, M. et al., Zinc potentiates agonist-induced currents at certain splice variants of the NMDAR, *Neuron*, 10, 943, 1993.
121. Tang, C.M., Dichter, M., and Morad, M., Modulation of the *N*-methyl-D-aspartate channel by extracellular H+, *Proc. Natl. Acad. Sci. USA*, 87, 6445, 1990.
122. Traynelis, S.F. and Cull-Candy, S.G., Proton inhibition of *N*-methyl-D-aspartate receptors in cerebellar neurones, *Nature*, 245, 247, 1990.
123. Vyklicky, L.J., Vlachova, V., and Krusek, J., The effect of external pH changes on responses to excitatory amino acids in mouse hippocampal neurones, *J. Physiol.*, 430, 497, 1990.
124. Traynelis, S.F. and Cull-Candy, S.G., Pharmacological properties and H+ sensitivity of excitatory amino acid receptor channels in rat cerebellar granule neurones, *J. Physiol.*, 433, 727, 1991.

125. Low, C. et al., Molecular determinants of proton sensitive NMDAR gating, *Mol. Pharmacol.*, 63, 1212, 2003.
126. Kashiwagi, K. et al., An aspartate residue in the extracellular loop of the *N*-methyl-D-aspartate receptor controls sensitivity to spermine and protons, *Mol. Pharmacol.*, 49, 1131, 1996.
127. Kohda, K., Wang, Y., and Yuzaki, M., Mutation of a glutamate receptor motif reveals its role in gating and delta2 receptor channel properties., *Nat. Neurosci.*, 3, 315, 2000.
128. Jones, K.S., VanDongen, H.M., and VanDongen, A.M., The NMDAR M3 segment is a conserved transduction element coupling ligand binding to channel opening, *J. Neurosci.*, 22, 2044, 2002.
129. Sobolevsky, A.I., Beck, C., and Wollmuth, L.P., Molecular rearrangements of the extracellular vestibule in NMDAR channels during gating, *Neuron*, 33, 75, 2002.
130. Yuan, H. et al., Conserved structural and functional control of N-methyl-D-aspartate receptor gating by transmembrane domain M3, *J. Biol. Chem.*, 280, 29708, 2005.
131. Ikeda, K. et al., Cloning and expression of the e4 subunit of the NMDAR channel, *FEBS Lett.*, 313, 34, 1992.
132. Kutsuwada, T. et al., Molecular diversity of the NMDAR channel, *Nature*, 358, 36, 1992.
133. Meguro, H. et al., Functional characterization of a heteromeric NMDAR channel expressed from cloned cDNAs, *Nature*, 357, 70, 1992.
134. Yamazaki, M. et al., Cloning, expression and modulation of a mouse NMDAR subunit, *FEBS Lett.*, 300, 39, 1992.
135. Monyer, H. et al., Heteromeric NMDARs: molecular and functional distinction of subtypes, *Science*, 256, 1217, 1992.
136. Ishii, T. et al., Molecular characterization of the family of the *N*-methyl-D-aspartate receptor subunits, *J. Biol. Chem.*, 268, 2836, 1993.
137. Hollmann, M. et al., Cloning by functional expression of a member of the glutamate receptor family, *Nature*, 342, 643, 1989.
138. Hollmann, M. and Heinemann, S., Cloned glutamate receptors, *Annu. Rev. Neurosci.*, 17, 31, 1994.
139. Kuusinen, A., Arvola, M., and Keinänen, K., Molecular dissection of the agonist binding site of an AMPA receptor, *EMBO J.*, 14, 6327, 1995.
140. Mayer, M.L., Olson, R., and Gouaux, E., Mechanisms for ligand binding to GluR0 ion channels: crystal structures of the glutamate and serine complexes and a closed apo state, *J. Mol. Biol.*, 311, 815, 2001.
141. Madden, D.R., The structure and function of glutamate receptor ion channels, *Nature Rev. Neurosci.*, 3, 91, 2002.
142. Sun, Y. et al., Mechanism of glutamate receptor desensitization, *Nature*, 417, 245, 2002.
143. Jin, R. et al., Structural basis for partial agonist action at ionotropic glutamate receptors, *Nat. Neurosci.*, 6, 803, 2003.
144. Mayer, M.L. et al., Crystal structures of the kainate receptor GluR5 ligand binding core dimer with novel GluR5-selective antagonists, *J. Neurosci.*, 26, 2852, 2006.
145. Pasternack, A. et al., a-amino-3-hydroxy-5-methyl-4-isoxazolepropionic acid (AMPA) receptor channels lacking the N-terminal domain, *J. Biol. Chem.*, 277, 49662, 2002.

12 Pharmacology of NMDA Receptors

Daniel T. Monaghan and David E. Jane

CONTENTS

12.1   INTRODUCTION

The discovery of NMDA receptors (NMDARs) was made possible by the synthesis and study of NMDA (Figure 12.1) and various NMDAR antagonists by Jeff Watkins and colleagues.[1] These compounds, most notably (R)-α-aminoadipate $((R)$-α-AA) and (R)-2-amino-5-phosphonopentanoate (Figure 12.2), were shown to block neuronal responses to applied NMDA, but not to block responses to kainate or quisqualate.[2,3] As a result, NMDARs were shown to represent a distinct subpopulation of excitatory amino acid receptors.

Over the next several years, these and other NMDAR antagonists led to the discovery that NMDARs play key roles in synaptic transmission, synaptic plasticity, learning and memory, neuronal development, excitotoxicity, stroke, seizures, and many other physiological and pathological processes. These studies generated great excitement about the potential use of NMDAR antagonists to treat neuropathological and neurodegenerative diseases. However, with the exception of the

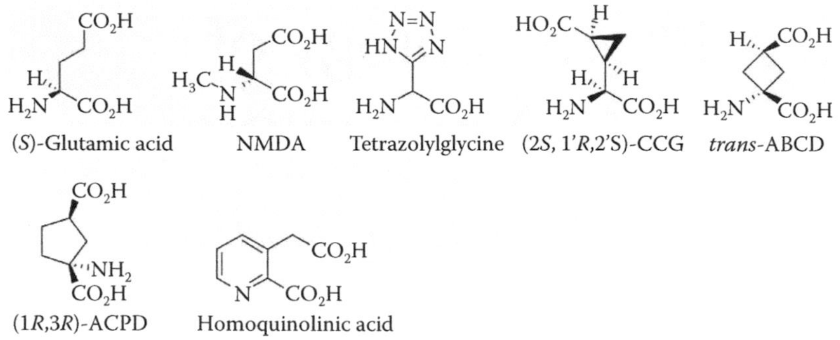

FIGURE 12.1 Structures of NMDAR agonists interacting with the glutamate binding site on the NR2 subunit.

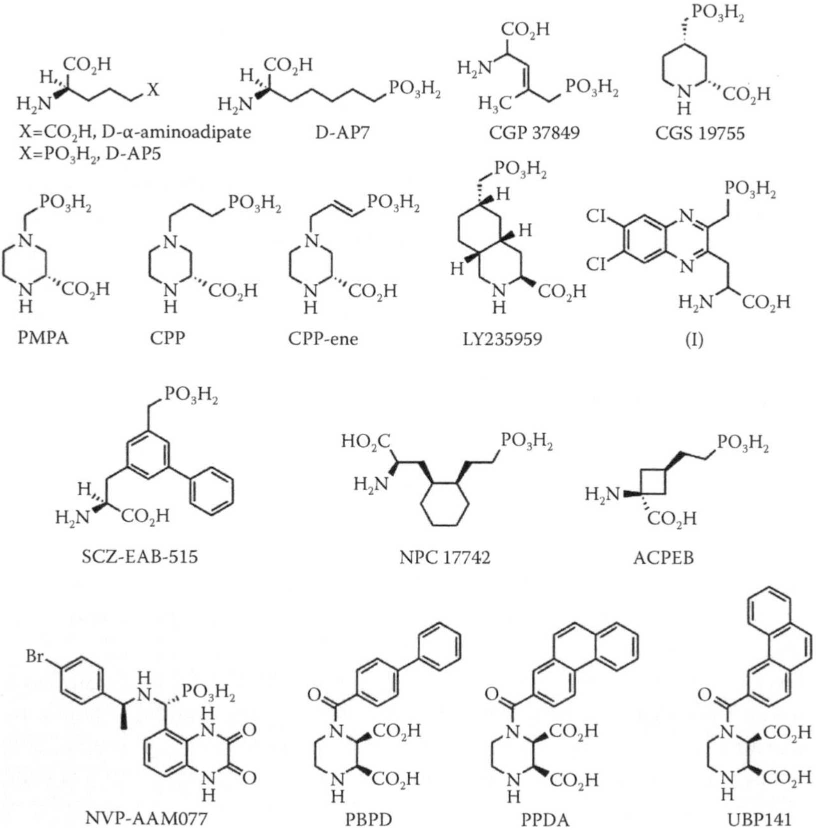

FIGURE 12.2 Structures of NMDAR antagonists interacting with the glutamate binding site on the NR2 subunit.

use of memantine for Alzheimer's disease, the development of NMDAR-targeted therapeutics has been disappointing. Several agents failed in clinical trials due to adverse effects and/or a lack of clinical efficacy. Despite this disappointment, NMDAR therapeutics continue to exhibit significant potential. Of the multiple drug binding sites on the various NMDAR subunits, many potential types of NMDAR antagonists exist, and some of these reveal distinct patterns of selectivity. This chapter will summarize the current understanding of the various sites of drug action on the NMDAR complex.

NMDARs are heteromeric complexes composed of four subunits derived from three related families: NR1, NR2, and NR3 subunits.[4–6] The well-characterized glutamate- and glycine-responsive NMDAR requires both NR1 and NR2 subunits. The NR1 subunit contains a glycine binding site,[7,8] while the homologous domain on the NR2 subunit contains the (S)-glutamate binding site.[9,10] Multiple lines of evidence suggest that a single NMDAR complex contains two NR1 subunits and two NR2 subunits.[11] The NR3 subunit can complex with NR1 subunits to form a glycine-responsive excitatory receptor that does not require L-glutamate.[12]

The NR1 subunit gene consists of 22 exons; exons 5, 21, and 22 can be alternatively spliced to produce eight distinct NR1 isoforms.[13,14] As discussed below, exon 5 of NR1 inserts a 21-amino acid sequence in the N-terminal extracellular domain that significantly alters receptor responses to pH and polyamines such as spermine.[15] The other two alternative splice cassettes are at the intracellular C terminus and do not affect NMDAR pharmacological properties.[14] The three NMDAR families (NR1, NR2, and NR3) display 27 to 31% identity to each other. Within the NR2 family, NR2A and NR2B are more closely related to each other (57%) than to NR2C or NR2D (43 to 47%), which are closely related to each other (54%). Thus, with respect to the NR1/NR2 NMDAR complex, the pharmacological heterogeneity is primarily determined by the NR2 subunit and exon 5 of the NR1 subunit.

NMDAR pharmacology has its basis in the domain structure of the NMDAR subunits. Each subunit is composed of an extracellular amino terminal, four hydrophobic segments (M1 through M4), and an intracellular carboxy terminal.[5,6] Each subunit contains two regions that have homology to bacterial amino acid–binding proteins. The first 350 amino acid residues contain the amino terminal domain (ATD) that has homology to the bacterial amino acid–binding protein known as LIVBP (leucine–isoleucine–valine binding protein).[16,17] This region is thought to be an allosteric regulatory domain that binds zinc in NR2A and polyamines in NR2B.[18–20]

The second structure with homology to bacterial amino acid–binding proteins is the glutamate–glycine binding domain formed by the pairing of two discrete segments, S1 and S2. S1 is a sequence of 120 amino acids located between the ATD and the first transmembrane domain (M1). The S2 segment is found on the extracellular loop between the third and fourth hydrophobic domains (M3 and M4). Together, S1 and S2 form a bilobed structure with structural homology to the bacterial leucine–arginine–ornithine binding protein (LAOBP).[21] The (S)-glutamate and glycine binding sites are found in the cavity between the two lobes of the S1/S2 structure in NR2 and NR1 subunits, respectively.

The ion permeating channel represents an additional drug binding site, a binding site for NMDAR channel blockers such as PCP, MK-801, and memantine

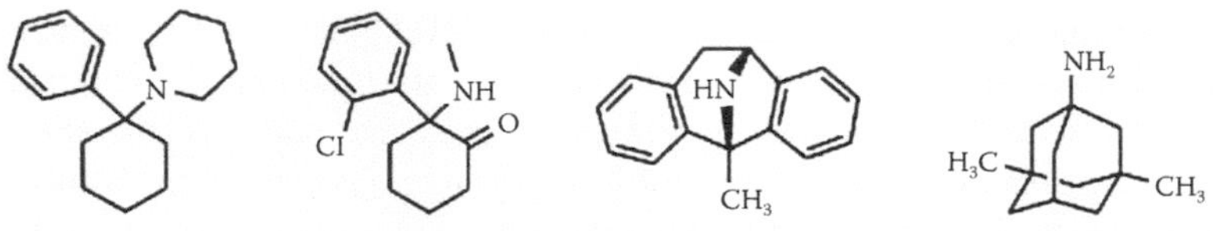

Phencyclidine   Ketamine   MK-801   Memantine

FIGURE 12.3 Structures of antagonists that bind to a site inside the channel of NMDAR complex.

(Figure 12.3). The channel structure is structurally related to potassium channels wherein one hydrophobic segment forms a P loop within the membrane and this segment is flanked by transmembrane domains.[22,23] The P loop contributes to the selectivity filter of the channel. Near the tip of this loop is a critical asparagine residue that is important for the binding of several channel blockers. The other transmembrane domains contribute to the pore lining in the extracellular facing half of the membrane and thus can contribute to channel blocker binding.

12.2 PHARMACOLOGY OF THE NR2 GLUTAMATE BINDING SITE

12.2.1 AGONISTS

Early structure–activity studies established that an ideal structure for activating NMDARs (and for activating EAA receptors in general) is represented by (S)-glutamate.[1] Excitatory activity requires one positive and two negative charge centers. The positive charge center (e.g., NH_3^+) should be positioned α to a carboxyl group. For optimal agonist action, the two negative charge groups (preferably both carboxylic acids) should be separated by four carbon–carbon bond lengths, and the α carbon should be in the S configuration. These findings are consistent with the three-point attachment pharmacophore model proposed by Curtis and Watkins[24] and recently confirmed by the publication of the X-ray crystal structure of glutamate bound to the ligand binding core of NR2A.[25] The ω acid group can also be a sulfonate or a tetrazole. In the latter case, the carbon chain should be shorter, (as in the very potent tetrazol-5-glycine NMDAR agonist[26] (Figure 12.1).

NMDA is several-fold weaker as an agonist than (S)-glutamate. However, NMDA has a low affinity for the plasma membrane transporters and thus can appear more potent than glutamate in some physiological assays. It is perhaps surprising that such a simple structure as NMDA is so selective; in the micromolar range, NMDA displays no activity at other glutamate receptors. The critical difference between NMDARs and the non-NMDA ionotropic glutamate receptors that allow NMDA to bind in the NR2 subunit binding pocket is an aspartate residue (D731 in NR2A) that is a glutamate residue in the AMPA and kainate receptors. This residue binds the agonist's amino group and by being one methylene group shorter in the NR2 subunit, allows space for the N-methyl group of NMDA.[25]

By incorporating ring systems into the glutamate structure, rigid glutamate analogues that are potent NMDAR agonists have been developed. They mimic the active, partially folded, conformation of (S)-glutamate and include homoquinolinate,[27]

($2S,1'R,2'S$) 2-(carboxycyclopropyl)glycine (L-CCG-IV),[28,29] ($1R,3R$) 1-aminocyclo-pentane-1,3-dicarboxylic acid (ACPD),[30] and 1-aminocyclobutane-1,3-dicarboxylic acid (ACBD).[31–33] See Figure 12.1 for structures. With resolution of the NR2A crystal structure with (S)-glutamate bound, the precise features that underlie high affinity (S)-glutamate binding are now known.[25]

12.2.2 Antagonists

The first NMDAR antagonists were variations of the (S)-glutamate structure. For example, by extending the glutamate backbone by one carbon, antagonist activity was observed for (RS)-a-AA.[34] Antagonist activity arose from the (R) isomer.[2,35] (R)-α-AA (Figure 12.2) was found to inhibit NMDA-evoked depolarizations while having little effect upon kainate- or quisqualate-evoked responses.[1,2,36] Hence, NMDA was shown to activate a receptor that is distinct from those activated by kainate or quisqualate. Even greater antagonist potency was found by replacing the ω carboxy group of (R)-α-AA with a phosphonate group, resulting in (R)-2-amino-5-phosphonopentanoate ((R)-AP5 or D-AP5, Figure 12.2)[37–39], also known as D-2-amino-5-phosphonovalerate (D-APV). For both (R)-α-AA and (R)-AP5, extending the chain length by adding a –CH_2 group diminished affinity, yet adding two carbons to the chain restored potency ((R)-α-aminosuberate and (R)-2-amino-7-phosphonoheptanoate (Figure 12.2), respectively).

As found for agonists, glutamate binding site antagonists display at least three charge centers, one positive and two negative.[33] The two negative charge centers are generally provided by a carboxyl group that is α to an amino group and by a distal acid group that is frequently a phosphonate group. The positive charge center can be provided by a primary or secondary amine. The distal phosphonate group may provide two charge–charge interactions with a receptor since phosphonates provide significantly greater affinity than a corresponding carboxylate or sulfonate.[40] The ω phosphonate group of NMDAR antagonists can sometimes be replaced by a tetrazole,[41] but this modification reduces potency. The chiral carbon attached to both the carboxyl and amino groups generally should be in the R configuration.

Further increases in antagonist potency can be achieved by constraining the AP5/AP7 chain in various ring structures and by adding specific groups (bulky hydrophobic groups, methyl groups, or double bonds) to this backbone. Several potent and selective NMDAR antagonists are generated by incorporating the AP5 or AP7 backbone into a piperidine or piperazine ring (see Figure 12.2 for structures). Hence, 4-phosphonomethyl-2-piperidine carboxylic acid (CGS19755)[42] is a potent AP5 analogue where the amino group is part of a piperidine ring, and 4-(3-phosphonopropyl) piperazine-2-carboxylic acid (CPP)[43,44] is a potent AP7 analogue incorporated into a piperazine ring (Figure 12.2). A further increase in potency results when a double bond is introduced into the carbon chain of D-CPP to make D-CPPene [(R,E)-4-(3-phosphonoprop-2-enyl) piperazine-2-carboxylic acid].[45]

A variety of other ring structures and additional groups have also been shown to increase the antagonist potency of the basic AP5/AP7 structure. The addition of a cyclohexane ring (NPC 17742),[46] biphenyl group (EAB 515),[47] methyl group plus a double bond (CGP 37849),[48] and quinoxaline ring[49] all yield compounds of increased affinity for NMDARs. Unlike the parent compound, the ethyl ester of CGP 37849,

CGP 39551 displayed oral bioavailability as an anticonvulsant, presumably acting as a prodrug form of CGP 37849.[48]

A photoaffinity probe has been developed based on the structure of CGP 39653.[50] NMDAR antagonists with benzene rings include a variety of phenylglycine and phenylalanine derivatives with a wide range of potencies.[33] The incorporation of the unsaturated bicyclic decahydroisoquinoline ring or a partially unsaturated tetrahydroisoquinoline ring into the AP7 backbone produced a wide variety of NMDAR antagonists of varying activities.[51] The phosphono derivative LY 274614 was the most potent. Interestingly, some of these compounds display distinctive NMDAR subtype selectivities.[52,53] A number of radioligands have been developed, e.g., [³H]AP5,[54,55] [³H]CGS19755,[56] [³H]CPP,[57] and [³H]CGP 39653[58] (K_D value 7 nM). The latter is potent enough to be used in a filtration binding assay, facilitating compound throughput. These ligands, however, are limited to the labelling of NR2A- or NR2A- and NR2B-containing NMDARs. In contrast, (S)-[³H]glutamate can label all four NR2 subunits.[59]

The general rules listed above for NMDAR antagonist activity have few exceptions. One example is the preference for six-bond lengths between the acidic groups to achieve optimal activity. The insertion of a chlorinated quinoxaline ring[49] into the (R)-AP6 structure results in α-amino-6,7-dichloro-3-(phosphonomethyl)-2-quinoxalinepropanoic acid (I in Figure 12.2), a highly potent NMDAR antagonist. Likewise, the addition of a cyclobutane ring into D-AP6 yields two 1-aminocyclobutanecarboxylic acid derivatives (ACPED in Figure 12.2) that are antagonists.[60]

While for most potent NMDAR antagonists the R configuration at the α carbon has greater activity than the corresponding S isomer, some S isomer antagonists are more potent, for example, the EAB515-related antagonists in which a biphenyl (or triphenyl) group is incorporated into the AP7 chain. The S isomer displays higher affinity than the R isomer.[61] Likewise, the bicyclic decahydroisoquinoline LY-235959 (Figure 12.2) has greater activity associated with the S isomer.[51]

Pharmacophore modeling studies describe the optimal antagonist structure as having 5.1 to 6.6 Å between the two negative charge centers.[62–64] This conforms to the straight chain, piperazine, and piperidine phosphonate antagonists such as (R)-AP5, (R)-CPP, and CGS19755. However, in the biphenyl/phenanthrene antagonists, (2R*,3S*)-1-(4-phenylbenzoyl)piperazine-2,3-dicarboxylic acid (PBPD) and (2R*,3S*)-1-(phenanthrenyl-2-carbonyl)piperazine-2,3-dicarboxylic acid (PPDA), the distance between the two carboxyl carbons is 3.4 Å (see Figure 12.2). The structure has two carboxylic acids separated by only three carbon–carbon bonds and an additional carbonyl group four bond lengths away from the amino carbon. Site-directed mutagenesis results support molecular modeling studies indicating that a histidine residue in the active site interacts with the distal carboxyl group in PPDA but does not interact with the phosphonate group in CGS19755.[65] Thus, the pharmacophores for PPDA and CGS19755 are not identical.

12.2.3 NR2 Subunit Selectivity of Glutamate Binding Site Ligands

NR2 subunits provide the greatest potential for pharmacologically distinguishing different types of NMDARs. This subunit family is generated by four distinct genes, each coding for a slightly different glutamate binding site and different ATD regulatory sites.[66–70] They also contribute similar (but not identical) channel lining structures. In contrast, the NR1 subunits are generated by only one gene that produces identical glycine binding sites and identical channel-lining residues.[71] The exon 5 extracellular alternative splice site introduces a modified ATD region. Since NR2 subunits also confer distinct physiological and biochemical properties to NMDARs, the selective blockade of differing NR2 subunit types should yield compounds with distinct therapeutic and adverse effect profiles.

An important consideration for subunit-specific antagonists is to define their actions in a heteromeric receptor complex. Functional NMDARs are thought to consist of two NR1 subunits and two NR2 subunits,[11,72] although some tetrameric NMDAR complexes may contain NR3 subunits.[73] Coimmunoprecipitation studies indicate that multiple types of NR1 subunits and NR2 subunits may be coassembled into the same receptor complex.[74–76] Physiological studies indicate that both glutamate- and glycine-binding sites must be occupied to achieve channel activation.[77] Thus, an NMDAR with both NR2A and NR2B subunits may be highly sensitive to a selective NR2A glutamate-binding site antagonist and an NR2B glutamate-binding site antagonist. Agents acting at the ATD regulate activity via domain–domain interactions; hence their actions in a heteromeric assembly may depend upon the specific complex.

Another possibility is that the subunit in the heteromeric assembly may alter the pharmacological specificities of adjacent subunits. For example, the glycine-site antagonist CGP 61594 displays nearly a 10-fold higher affinity in a complex containing NR2B subunits than those containing NR2A subunits.[78] The adjacent NR2 subunit alters the pharmacological specificity of the NR1 subunit. Similar examples can be found for kainate receptor complexes. If NR2 subunits can likewise alter the pharmacological specificity of an adjacent NR2 subunit, NMDARs may possess even greater pharmacological diversity. To date, however, studies of native NMDARs expressed in rat brains identified only four pharmacologically distinct populations of glutamate recognition sites.[52,79,80] The anatomical distribution and pharmacological profile of these four pharmacologically distinct sites correspond well to the four NR2 subunits in the brain.[81–84]

No glutamate-binding site antagonists display high degrees of NR2-subunit selectivity. In a survey of more than 75 compounds at native NMDARs,[85] most displayed similar weak selectivity patterns corresponding to the highest affinity at NR2A with progressively lower affinities at NR2B, NR2C, and NR2D. This is the typical pattern observed for antagonists such as (R)-AP5, (R)-CPP, and CGS-19755.[86] Of the compounds examined, only large, multiring antagonists (biphenyl compounds EAB515 and PBPD and the bicyclic decahydroisoquinoline LY233536) displayed varied selectivity patterns confirmed via recombinant receptors.[53] Each exhibited reduced relative affinity for recombinant NR2A-containing receptors; EAB515 and PBPD had higher affinities for NR2B- and NR2D-containing receptors; and LY233536 had higher affinity for NR2B- and NR2C-containing receptors. LY233536 displayed approximately 10-fold selectivity for NR2B- over NR2A-containing receptors at both recombinant[53] and native NMDARs.[82] Nevertheless, each of these compounds displayed low levels of selectivity that limit their utility.

In characterizing a series of derivatives of PBPD, a higher affinity compound PPDA (Figure 12.2) displayed a small improvement in selectivity for NR2C- and NR2D-containing NMDARs.[87,88] PPDA has been successfully used to demonstrate that long-term potentiation and long-term depression are mediated by pharmacologically distinct NMDARs[89] and that NMDAR-mediated synaptic responses in adult hippocampal CA3-CA1 synapses have two pharmacologically distinct components.[90] This agent has been improved via a closely related compound known as UBP141 (Figure 12.2). It should be useful for distinguishing NR2B and NR2D subunit-containing NMDARs because it displays a several-fold higher affinity for NR1/NR2D receptors than for NR1/NR2B receptors and intermediate affinity for NR1/NR2A.

Another large, quinoxaline-2,3-dione based antagonist with unusual sub-unit-selectivity is the widely-used NR2A-selective antagonist NVP-AAM077 (Figure 12.2).[91] It displays a 100-fold selectivity for human NR2A-containing NMDARs compared to NR2B-containing receptors. At rodent NMDARs, however, the degree of selectivity is about 10-fold.[87,92–94] NVP-AAM077 also has high affinity for NR2C subunits and lower affinity for NR2D-containing receptors[87] and thus is modestly selective for NR2A and NR2C subunits.

A major challenge in developing agents to distinguish NR2 subunits is the highly conserved aspect of the glutamate-binding pocket.[65] Of the amino acid residues that line the binding pocket, only a few are variable between NR2 subunits and all are at a distance from the central glutamate binding core. Modest differences also exist in the selectivity of small antagonists such as (R)-CPP and (RS)-4-(phosphonomethyl)-piperazine-2-carboxylic acid (PMPA, Figure 12.2).[87] While (R)-CPP displays a 50-fold higher affinity for NR2A than for NR2D subunits, the two-carbon shorter analogue PMPA shows only a five-fold difference in affinity. Hence, the NR2 subunits appear to have structural differences in the binding pocket. Recent modeling studies suggest that the position of helix F in the S2 domain of NR2A is slightly different in NR2D.[95] This places a small groove in the NR2D subunit that can accommodate the methyl group of the agonist (2S,4R)-4-methylglutamate and thus contributes to the 46-fold higher affinity displayed by NR2D subunits for this compound.

Most agonists studied to date exhibit the reverse selectivity patterns of most small antagonists; agonists tend to have high affinities for NR2D > NR2C > NR2B > NR2A subunits. In large surveys of compounds at native NMDARs[85] and at recombinant receptors,[95] homoquinolinate stands out as having higher affinity for NR2A- and NR2B-containing NMDARs.

12.3 PHARMACOLOGY OF GLYCINE BINDING SITE ON NR1

12.3.1 AGONISTS

Glycine binds to the S1S2 site on the NR1 subunit and is a necessary coagonist for activation of NMDARs.[96,97] Initially it was thought that endogenous levels of extracellular glycine were enough to saturate the glycine binding site; however, later studies suggest that this is not the case and it may be possible to develop positive modulators of NMDAR function via interaction with the glycine binding site.[98] Amino acids such as (R)-alanine and (R)-serine (Figure 12.4) display high affinities

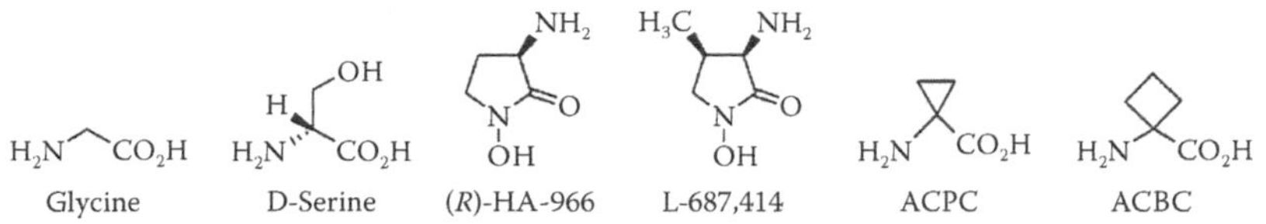

FIGURE 12.4 Structures of agonists and partial agonists that interact with the glycine binding site on NR1.

for the glycine site and behave as full agonists.[99] Conformationally constrained analogues of glycine such as ACPC, a cyclopropyl analogue,[100,101] and ACBC, a cyclobutane analogue,[102] are partial agonists with different degrees of efficacy. At lower doses, they show antischizophrenic properties in animal models but this effect is reversed at higher doses when they act like antagonists.[103] Other partial agonists include HA-966 (Figure 12.4), one of the first compounds identified as an NMDAR antagonist,[36] and L-687,414.[104]

Interestingly, the cocrystal structures of the NR1 ligand binding core with the partial agonists ACPC and ACBC show the same degrees of domain closure as found in the complex with the full glycine agonist.[105] Thus the mechanism by which partial agonism occurs for the NR1 subunit is distinct from that of the related GluR2 AMPA receptor in which partial opening of the binding domains results from partial agonist binding; full agonists stabilize the closed form and antagonists the open form.[106,107]

12.3.2 Antagonists

The development of antagonists acting at a glycine binding site associated with an NMDAR and the therapeutic potential of such compounds were reviewed[99] and the first full antagonist found to bind to the glycine site was kynurenic acid (Figure 12.5).[108,109] It was nonselective and antagonized a range of glutamate receptors. The AMPA/kainate receptor antagonists designated CNQX and DNQX (Figure 12.5)[110] also act as weak NMDAR antagonists.[111] These lead compounds were used as templates to develop more potent antagonists via structure–activity relationship studies.

Structural modification of kynurenic acid led to a series of potent antagonists such as 5,7-dichlorokynurenic acid (5,7-DCKA),[112] L-683,344,[112] L-689,560,[113] L-701,324,[114] GV150526A,[115] and GV196771A[116] (see Figure 12.5). Analogues of CNQX such as ACEA-1021[117] (Figure 12.5) were described as potent and selective glycine site antagonists, but quinoxalinedione derivatives suffered from poor water solubility. A SAR study of the quinoxaline-2,3-dione structure provided α-phosphoalanine-substituted compounds with >500-fold selectivity for the glycine site (compared to AMPA receptors), enhanced water solubility, and excellent *in vivo* anticonvulsant activity.[118] Pharmacophore models for the NMDAR glycine site[99,119,120] have been superseded by X-ray crystal structures of antagonists bound to the ligand binding core of NR1.[121]

Glycine site antagonists have improved therapeutic ratios (retain anticonvulsant, neuroprotective, and analgesic properties and exhibit reduced psychotomimetic effects) in comparison to conventional orthosteric antagonists.[99] However, the brain bioavailability of these compounds is questionable (high affinity plasma protein binding is the main problem)[122]. None of these compounds have achieved clinical

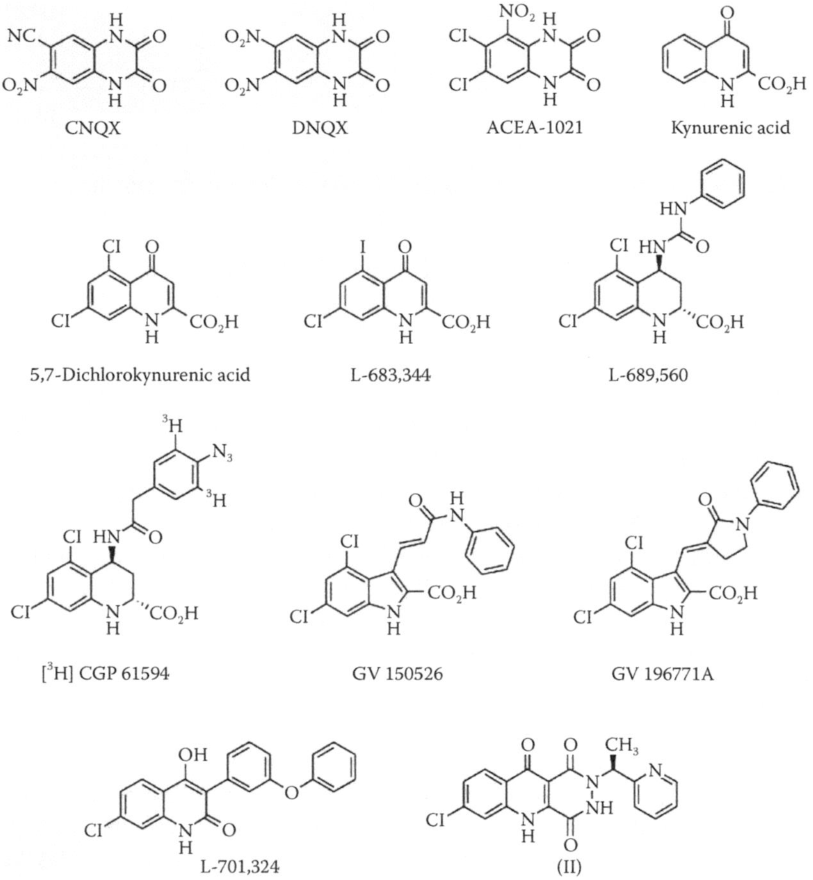

FIGURE 12.5 Structures of antagonists that interact with the glycine binding site on NR1.

use to treat stroke or epilepsy. Recently, a range of antagonists based on the quinoline nucleus (II in Figure 12.5) have been developed and dosed orally displayed good aqueous solubility and excellent bioavailability based on plasma concentration and activity in an *in vivo* model of neuropathic pain.[123]

A photoaffinity label, [³H]CGP 61594 (Figure 12.5) has been developed for the NMDAR glycine site.[124] An early report indicated that CGP 61594 displayed higher affinity for the NR1/NR2B receptor subtype over NMDARs containing NR2A, NR2C, or NR2D subunits.[78] The dependency of the affinity of agonists for the glycine site of the NR1 subunit on the type of NR2 subunit in the tetrameric complex has been reported.[70,125]

12.4 PHARMACOLOGY OF GLYCINE BINDING SITE ON NR3

The NR3A and NR3B subunits reveal only a 24 to 29% sequence homology with NR1 and NR2. When NR3A or NR3B subunits are coexpressed with NR1 and NR2, they act as negative modulators, reducing single-channel conductance and

Ca^{2+} permeability.[73,126] However, when NR1 and NR3A are coexpressed in *Xenopus* oocytes, the excitatory glycine receptors formed are Ca^{2+} impermeable.[12] Whether these NR1/NR3 excitatory glycine receptors exist in neurons remains controversial. Studies using the ligand binding cores of NR1 and NR3A revealed that glycine has a 650-fold higher affinity for NR3A compared to NR1.[127] Reports suggest that in NR1/NR3 receptors glycine binds to the NR3 subunit leading to ion channel opening while glycine binding to NR1 leads to inhibition due to rapid desensitization.[128,129] This is in contrast to the NR1/NR2 subunit combination in which glycine binding to NR1 potentiates NMDAR function. The reduced current through triheteromeric NR1/NR2/NR3 receptors may arise from inhibition via glycine binding to the NR1 subunit in the NR1/NR3 dimer (assuming the tetramer consists of a dimer of dimers).

Isolated ligand binding cores were used to investigate the pharmacology of NR3A. Interestingly glutamate can bind to NR3A with very low affinity but would not bind to NR3A at physiologically relevant concentrations.[127] The rank order of affinity for NR1 based on testing of partial agonists was ACPC > ACBC > cycloleucine. The rank order for NR3 was ACBC > ACPC > cycloleucine. Indeed, ACBC (Figure 12.5) showed 65-fold higher affinity for NR3 compared to NR1.[127] A number of NR1 glycine site antagonists were tested and the quinoxalinedione analogue CNQX (Figure 12.5) was found to have low micromolar affinity for NR3A and ~2.5-fold higher affinity for NR3A versus NR1. Importantly, a number of antagonists with nanomolar affinities for NR1 had only low affinity for NR3A (5,7-DCKA and L-689,560, Figure 12.5), suggesting that the binding site of NR3A is different from that of NR1. It should therefore be possible to develop selective NR3A antagonists. Homology models of NR3A and NR3B provided insights into differences in the pharmacology of NR1 and NR3.[127,130] The binding site cavity in NR3 is likely to be larger than that in NR1 because two amino acids (V689 and W731) in the NR1 ligand binding core are replaced by alanine and methionine residues, respectively. The ACPC and ACBC partial agonists (Figure 12.5) make van der Waals contacts with V689 in NR1.[105] The replacement of this residue by an alanine residue in NR3 along with the W731M switch may explain the differences in affinities of these two agonists for NR1 compared to NR3.[127] In addition, the W731M switch in NR3 may at least partially explain why the 5,7-DCKA NR1 antagonist has low affinity for NR3; W731 makes an important contact with the 5-chloro substituent of 5,7-DCKA in NR1.

Little is known about the functions of NR3A subunits in the CNS, although increased dendritic spine formation in early postnatal cerebrocortical neurons of NR3$^{-/-}$ mice has been reported.[73] Recent studies revealed that oligodendrocytes express NR3A subunit-containing NMDARs.[131–133] The NMDARs appear to be key players in glutamate-mediated damage of oligodendrocytes and show potential as new therapeutic targets to prevent white matter damage in a range of conditions. The precise subunit composition of these oligodendroglial NMDARs is unknown.

12.5 ALLOSTERIC MODULATORY SITES ON NMDA Receptors

12.5.1 POLYAMINES

Studies of native and recombinant NMDARs revealed three effects of polyamines on NMDAR activity: (1) glycine-dependent stimulation characterized by an increase in glycine affinity for its binding site, (2) glycine-independent stimulation characterized by increases in the maximal amplitudes of NMDAR responses at saturating concentrations of glycine, and (3) voltage-dependent inhibition. In the absence of glutamate and glycine, polyamines have no effect on NMDAR activity. However, they increase glycine affinity[134–137] and thus increase NMDAR responses at subsaturating glycine concentrations by increasing glycine association.

Under saturating glycine conditions, polyamines still potentiate NMDAR responses (glycine-independent potentiation). In addition, at negative potentials, polyamines reduce channel conductance by partial channel block. Consistent with early studies,[138] these polyamine effects are noncompetitive with glutamate, glycine, and channel blockers, suggesting distinct binding sites for polyamines.[139,140]

Polyamine responses are dependent upon specific NR1 and NR2 subunits. Glycine-independent stimulation by spermine in recombinant receptors expressed in *Xenopus* oocytes is inhibited by the N-terminal insert of the NR1 subunit coded by exon 5.[15,141,142] The E342 residue in the amino terminus of the NR1 subunit is necessary for glycine-independent spermine stimulation[143] but has no effect upon polyamine glycine-dependent potentiation or voltage-dependent channel block. Mutations at equivalent positions in NR2A and NR2B subunits had no effect on spermine stimulation.

The NR2 subunit also contributes to both the stimulatory and inhibitory effects of polyamines at NMDARs.[144–146] Polyamines cause glycine-independent stimulation and decrease the affinity for glutamate site agonists at NR1a/NR2B receptors but not at NR1a/NR2A, NR1a/NR2C, or NR1a/NR2D receptors. However, glycine-dependent stimulation and voltage-dependent inhibition are seen at both NR1a/NR2A and NR1a/NR2B receptors. These data suggest the existence of at least three distinct polyamine binding sites on NMDARs.

12.5.2 IFENPRODIL AND RELATED NR2B-SELECTIVE COMPOUNDS

A large number of pharmacological agents bind and inhibit NMDAR activity specifically at NR2B-containing receptors. The prototype is ifenprodil (Figure 12.6), a phenylethanolamine that binds at a site distinct from the glutamate- and glycine-binding sites.[147,148] Ifenprodil exhibits greater than a 100-fold selectivity for NR2B over NR2A containing receptors[149] and very low affinity at NR2C- and NR2D-containing receptors.[145] The ifenprodil binding site appears to be located on the ATD region and involves amino acid residues distinct from (and possibly partially overlapping) residues that contribute to polyamine binding.[150] The NR1 insert (exon 5), which alters polyamine modulation of NMDARs had no effect on ifenprodil inhibition of NMDAR activity. This suggests that the glycine-independent polyamine binding sites on NMDARs are separate from those of ifenprodil binding sites.[151]

A variety of other compounds show NR2B selectivity, including haloperidol,[152] CP-101,606,[153] and Ro 25-6981[154] (Figure 12.6). These compounds display the highest degree of subtype selectivity among the different classes of NMDAR antagonists. They have been useful for defining the actions of NR2B-containing receptors in the brain.

Structure–activity analysis of ifenprodil-like compounds has been explored extensively and multiple series of compounds have been optimized for selective high affinity

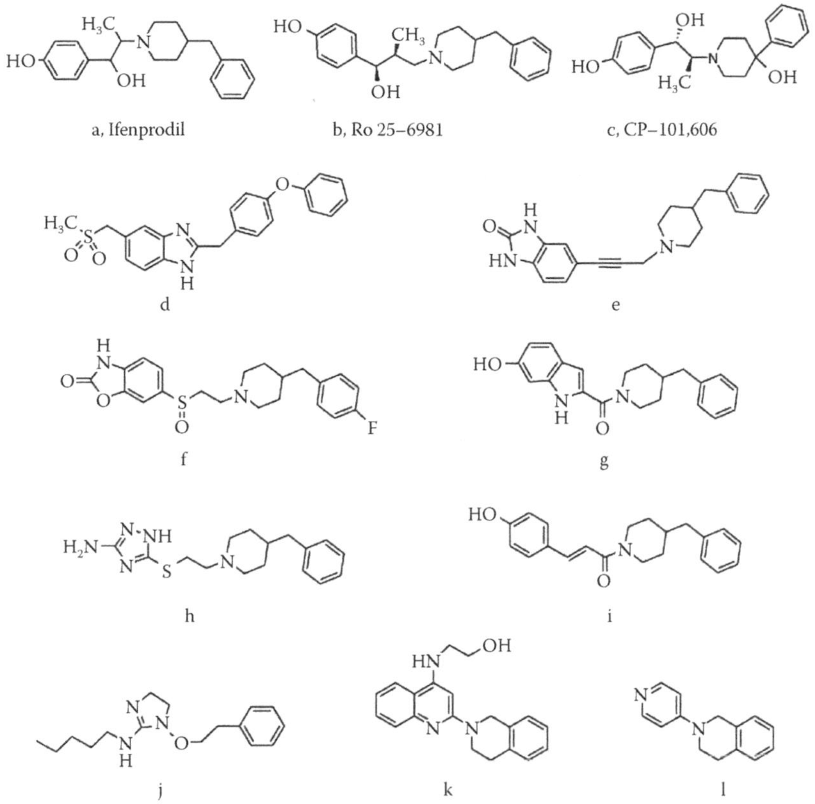

FIGURE 12.6 Examples of polyamine site antagonists.

binding. One challenge already overcome is the α-1 adrenergic receptor antagonist activity and/or human ether a go-go (hERG) potassium channel blocking activity (which may lead to cardiac arrhythmias) of many ifenprodil-like agents.[155] Another success was identifying agents that are metabolically stable and active *in vivo*. Several lead compounds are now providing interesting preclinical data regarding the role of NR2B subunits in neuropathic pain and excitotoxicity.

The general pharmacophore structure, as represented by ifenprodil (Figure 12.6, compound a), has two aromatic rings separated by a linker with a basic nitrogen in the center of the linker. Commonly, each ifenprodil-like compound has a 4-benzyl-piperidine group that provides one aromatic ring and the basic nitrogen. This moiety is then linked to a second aromatic ring system that optimally has a hydrogen bond donor. Thus, the potency of ifenprodil is reduced by removal of its phenol hydroxy group. This general structure is similar to those of the well-characterized NR2B antagonists, Ro-25,6981[154] and CP-101,606[153] (Figure 12.6, compounds b and c).

Optimization of different initial lead compounds indicates that removal of an aromatic ring or basic nitrogen can be tolerated if combined with other changes. The phenol ring can be replaced by a number of heterocyclics such as a benzimidazole,[156] benzimidazolone,[157] benzoxazole-2(3H)-one,[158] indole-2-carboxamides,[159] and aminotriazole,[160] especially if they contain an H-bond donor (Figure 12.6, compounds d through h). Likewise, the linker between 4-benzylpiperidine and phenol can be replaced by number of structures. Significantly, a basic nitrogen in the linker is not essential. A nonbasic nitrogen correlated with reduced hERG and α-1 NE activity (Figure 12.6, compound i).[161] In a series of dihydroimidazoline derivatives (Figure 12.6, compound j), replacement of a terminal aromatic group by an aliphatic chain was also tolerated, resulting in high affinity NR2B-selective antagonists.[162] A series of 4-aminoquinolines (Figure 12.6, compound k)[163] and 4-(3,4-dihydro-1H-isoquinolin-2yl)-pyridines (Figure 12.6, compound l) diverge from the original ifenprodil structure. They retain at least an aromatic ring at each end with a nitrogen in the center.

12.6   ZINC

Zinc displays subunit-specific actions at recombinant NMDARs. It displays a voltage-dependent inhibition of NMDAR responses in heteromeric NR1/NR2A and NR1/NR2B receptors. At lower concentrations, it shows a voltage-independent inhibition of NR1/NR2A receptors.[164,165] The NR2A selectivity accounts for observations that the addition of heavy metal chelators to buffer solutions significantly potentiates NR1a/NR2A but not NR1a/NR2B receptor responses. This result may be due to chelation of contaminant traces of heavy metals in solutions that tonically inhibit NR1a/NR2A NMDAR responses. Two effects of zinc were also seen in cultured murine cortical neurons.[166] At low concentrations (3 μM), it produced a voltage-independent reduction in channel open probability. At higher concentrations (10 to 100 μM), it produced a voltage-dependent reduction in single channel amplitude associated with an increase in channel noise, suggesting a fast channel block. Since zinc is co-released with glutamate from pre-synaptic terminals, zinc modulation of NMDARs may be physiologically relevant.[167,168]

Molecular modeling experiments paired with site-directed mutagenesis indicate that the ATD region forms a bilobed structure with an apparent binding cavity in the center, much like that found for the glycine or glutamate binding S1/S2 domain.[169] In NR2A, specific histidine residues are necessary for zinc inhibition. Interestingly, these sites line both sides of the binding cleft in the ATD structure. This suggests that zinc binding may induce domain closure and this is transmitted to the S1/S2 domain as an inhibitory signal. The observation that zinc binding alters the trypsin sensitivity of purified ATD protein supports this model. The implications of the model are significant for potential drug development.

12.7   UNCOMPETITIVE ANTAGONISTS (CHANNEL BLOCKERS)

In the mammalian CNS, Mg^{2+} ions block NMDAR channels at resting membrane potentials.[170] This block is voltage-dependent. At depolarized membrane potentials, the channel block is relieved and ion flux occurs.[171,172] Nonhomologous asparagine residues on NR1 and NR2 subunits produce a constriction in NMDAR ion channels, allowing Ca^{2+} but not Mg^{2+} ions to enter.[173] The low affinity binding site for Mg^{2+} ions is deep within the channel and NMDAR complexes containing NR2A or NR2B subunits have a higher affinity for Mg^{2+} than those containing NR2C or NR2D.[66]

A number of compounds block NMDAR channels by a use-dependent (channels must be opened via binding of glycine and glutamate to their respective binding sites for access to and dissociation from the binding site) and voltage-dependent mechanism.[174,175] These compounds include the dissociative anaesthetics, phencyclidine (PCP) and ketamine.[176] Site-specific mutagenesis revealed that an asparagine residue (N598) deep within the pore lining M2 segment of an NMDAR is important for channel blocking.[177] Since the mechanism of these channel blockers is use-dependent, the suggestion was made to use them to treat ischemia in which neurons degenerate due to excessive Ca^{2+} entry through NMDARs. This led to the development of selective high affinity NMDAR channel blockers such as MK-801, which is used widely as an experimental tool.[178,179] The kinetic action of channel blocking and unblocking exhibited by MK-801 depends on the NR2 subunit composition of the NMDAR complex. Slower channel blocking kinetics were observed for NR2C-containing receptors compared to those containing NR2A or NR2B.[180] This is consistent with the shorter open times of NR2C-containing receptors.

High affinity channel blockers such as PCP and MK-801 induced psychotomimetic-like effects in animals. This result coupled with adverse effects such as ataxia, memory and learning impairment, and neuronal vacuolization has prevented development of high affinity channel blockers for clinical use.[181,182] The propensity of these compounds to produce adverse side effects has been linked to their slow kinetics of dissociation from their binding site in the NMDAR channel. Indeed, the slow dissociation rate of MK-801 allows it to be trapped inside the channel.

High affinity channel blockers such as PCP mimic the symptoms of schizophrenia and have served as animal models of this disorder. Low affinity channel blockers such as memantine exhibit fast on-and-off kinetics and reduced tendencies to produce adverse reactions such as psychotomimetic effects.[181] Memantine is now in clinical use under the trade names Ebixa, Axura, and Namenda for treatment of cognitive deficits in moderate to severe Alzheimer's disease. Although it is a channel blocker, memantine exhibits three- to five-fold greater potency for NR2C- versus NR2A-containing NMDARs but the relevance of this modest subunit selectivity to the improved therapeutic profile has not been established.

12.8 CONCLUDING REMARKS

The pharmacology of NMDAR complexes is highly diverse due mainly to the complexity of the subunit composition of NMDARs. Despite many years of sustained effort in developing drugs that interact selectively with NMDAR complexes, only memantine, a low affinity channel blocker, has made it into the clinic. However, recent advances in solving the X-ray crystal structures of ligand binding cores of NR1 and NR2 subunits have made possible the development of selective agonists and antagonists for individual NR2 subunits.

In addition, advances in our understanding of the pharmacology and function of the NR3 subunit are likely to lead to the development of selective antagonists for this subunit. The combination of subunit-selective pharmacological tools for NMDARs and molecular biological methods will provide significant information about the functions of NMDARs and the roles played by the individual subunits in the CNS. In addition, these advances are likely to herald new possibilities for treating a range of CNS disorders in which NMDARs play a role.

REFERENCES

1. Watkins, J.C., Pharmacology of excitatory amino acid transmitters, *Adv. Biochem. Psychopharmacol.*, 29, 205, 1981.
2. Biscoe, T.J. et al., D-alpha-aminoadipate as a selective antagonist of amino acid-induced and synaptic excitation of mammalian spinal neurones, *Nature*, 270, 743, 1977.
3. Davies, J. and Watkins, J.C., Actions of D and L forms of 2-amino-5-phosphonovalerate and 2-amino-4-phosphonobutyrate in the cat spinal cord, *Brain Res.*, 235, 378, 1982.
4. Nakanishi, S., Molecular diversity of glutamate receptors and implications for brain function, *Science*, 258, 597, 1992.
5. Mori, H. and Mishina, M., Structure and function of the NMDAR channel, *Neuropharmacology*, 34, 1219, 1995.
6. Seeburg, P.H. et al., The NMDAR channel: molecular design of a coincidence detector, *Recent Prog. Horm. Res.*, 50, 19, 1995.
7. Kuryatov, A. et al., Mutational analysis of the glycine-binding site of the NMDAR: structural similarity with bacterial amino acid-binding proteins, *Neuron*, 12, 1291, 1994.
8. Hirai, H. et al., The glycine binding site of the N-methyl-D-aspartate receptor subunit NR1: identification of novel determinants of co-agonist potentiation in the extracellular M3-M4 loop region, *Proc. Natl. Acad. Sci. USA*, 93, 6031, 1996.
9. Anson, L.C. et al., Identification of amino acid residues of the NR2A subunit that control glutamate potency in recombinant NR1/NR2A NMDARs, *J. Neurosci.*, 18, 581, 1998.
10. Laube, B. et al., Molecular determinants of agonist discrimination by NMDAR subunits: analysis of the glutamate binding site on the NR2B subunit, *Neuron*, 18, 493, 1997.
11. Laube, B., Kuhse, J., and Betz, H., Evidence for a tetrameric structure of recombinant NMDARs, *J. Neurosci.*, 18, 2954, 1998.
12. Chatterton, J.E. et al., Excitatory glycine receptors containing the NR3 family of NMDAR subunits, *Nature*, 415, 793, 2002.
13. Sugihara, H. et al., Structures and properties of seven isoforms of the NMDAR generated by alternative splicing, *Biochem. Biophys. Res. Commun.*, 185, 826, 1992.
14. Hollmann, M. et al., Zinc potentiates agonist-induced currents at certain splice variants of the NMDAR, *Neuron*, 10, 943, 1993.
15. Traynelis, S.F., Hartley, M., and Heinemann, S.F., Control of proton sensitivity of the NMDAR by RNA splicing and polyamines, *Science*, 268, 873, 1995.
16. Dingledine, R., Myers, S.J., and Nicholas, R.A., Molecular biology of mammalian amino acid receptors, *FASEB J.*, 4, 2636, 1990.
17. O'Hara, P.J. et al., The ligand-binding domain in metabotropic glutamate receptors is related to bacterial periplasmic binding proteins, *Neuron*, 11, 41, 1993.
18. Masuko, T. et al., A regulatory domain (R1-R2) in the amino terminus of the N-methyl-D-aspartate receptor: effects of spermine, protons, and ifenprodil, and structural similarity to bacterial leucine/isoleucine/valine binding protein, *Mol. Pharmacol.*, 55, 957, 1999.

19. Fayyazuddin, A. et al., Four residues of the extracellular N-terminal domain of the NR2A subunit control high-affinity Zn^{2+} binding to NMDARs, *Neuron*, 25, 683, 2000.

20. Rachline, J. et al., The micromolar zinc-binding domain on the NMDAR subunit NR2B, *J. Neurosci.*, 25, 308, 2005.

21. Stern-Bach, Y. et al., Agonist selectivity of glutamate receptors is specified by two domains structurally related to bacterial amino acid-binding proteins, *Neuron*, 13, 1345, 1994.

22. Kuner, T., Seeburg, P.H., and Guy, H.R., A common architecture for K+ channels and ionotropic glutamate receptors?, *Trends Neurosci.*, 26, 27, 2003.

23. Zhorov, B.S. and Tikhonov, D.B., Potassium, sodium, calcium and glutamate-gated channels: pore architecture and ligand action, *J. Neurochem.*, 88, 782, 2004.

24. Curtis, D.R. and Watkins, J.C., The excitation and depression of spinal neurones by structurally related amino acids, *J. Neurochem.*, 6, 117, 1960.

25. Furukawa, H. et al., Subunit arrangement and function in NMDARs, *Nature*, 438, 185, 2005.

26. Lunn, M.L. et al., Three-dimensional structure of the ligand-binding core of GluR2 in complex with the agonist (S)-ATPA: implications for receptor subunit selectivity, *J. Med. Chem.* 46, 872, 2003.

27. Brown, J.C., et al., [3H]homoquinolinate binds to a subpopulation of NMDARs and to a novel binding site, *J. Neurochem.*, 71, 1464, 1998.

28. Shinozaki, H. et al., A conformationally restricted analogue of L-glutamate, the (2S,3R,4S) isomer of L-alpha-(carboxycyclopropyl)glycine, activates the NMDA-type receptor more markedly than NMDA in the isolated rat spinal cord, *Brain Res.*, 480, 355, 1989.

29. Kawai, M. et al., 2-(Carboxycyclopropyl)glycines: binding, neurotoxicity and induction of intracellular free Ca^{2+} increase, *Eur. J. Pharmacol.*, 211, 195, 1992.

30. Sunter, D.C., Edgar, G.E., Pook, P.C.-K., Howard, J.A.K., Udvarhelyi, P.M., and Watkins, J.C., Actions of the four isomers of 1-aminocyclopentane-1,3-dicarboxylate (ACPD) in the hemisected spinal cord of the neonatal rat, *Br. J. Pharmacol.*, 104, 377P, 1991.

31. Allan, R.D. et al., Synthesis and activity of a potent N-methyl-D-aspartic acid agonist, trans-1-aminocyclobutane-1,3-dicarboxylic acid, and related phosphonic and carboxylic acids, *J. Med. Chem.*, 33, 2905, 1990.

32. Lanthorn, T.H. et al., Cis-2,4-methanoglutamate is a potent and selective N-methyl-D-aspartate receptor agonist, *Eur. J. Pharmacol.*, 182, 397, 1990.

33. Jane, D.E., Olverman, H.J., and Watkins, J.C., *Agonists and Competitive Antagonists: Structure–Activity and Molecular Modelling Studies*, Watkins, J.C., Ed., Oxford University Press, Oxford, 1994, p. 31.

34. Hall, J.G., McLennan, H., and Wheal, H.V., The actions of certain amino acids as neuronal excitants *J. Physiol.*, 272, 52P, 1977.

35. Biscoe, T.J. et al., D-alpha-aminoadipate, alpha, epsilon-diominopimelic acid and HA-966 as antagonists of amino acid-induced and synaptic excitation of mammalian spinal neurones *in vivo*, *Brain. Res.*, 148, 543, 1978.

36. Evans, R.H., Francis, A.A., and Watkins, J.C., Mg2+-like selective antagonism of excitatory amino acid-induced responses by alpha, epsilon-diaminopimelic acid, D-alpha-aminoadipate and HA-966 in isolated spinal cord of frog and immature rat, *Brain Res.*, 148, 536, 1978.

37. Davies, J. et al., Differential activation and blockade of excitatory amino acid receptors in the mammalian and amphibian central nervous systems, *Comp. Biochem. Physiol. C.*, 72, 211, 1982.

38. Davies, J. et al., 2-Amino-5-phosphonovalerate (2APV), a potent and selective antagonist of amino acid-induced and synaptic excitation, *Neurosci. Lett.*, 21, 77, 1981.
39. Evans, R.H. et al., The effects of a series of omega-phosphonic alpha-carboxylic amino acids on electrically evoked and excitant amino acid-induced responses in isolated spinal cord preparations, *Br. J. Pharmacol.*, 75, 65, 1982.
40. Olverman, H.J. et al., Structure/activity relations of N-methyl-D-aspartate receptor ligands as studied by their inhibition of [3H]D-2-amino-5-phosphonopentanoic acid binding in rat brain membranes, *Neuroscience*, 26, 17, 1988.
41. Ornstein, P.L. et al., 4-(Tetrazolylalkyl)piperidine-2-carboxylic acids: potent and selective N-methyl-D-aspartic acid receptor antagonists with a short duration of action, *J. Med. Chem.*, 34, 90, 1991.
42. Lehmann, J. et al., CGS 19755, a selective and competitive N-methyl-D-aspartate-type excitatory amino acid receptor antagonist, *J. Pharmacol. Exp. Ther.*, 246, 65, 1988.
43. Davies, J. et al., CPP, a new potent and selective NMDA antagonist: depression of central neuron responses, affinity for [3H]D-AP5 binding sites on brain membranes and anticonvulsant activity, *Brain Res.*, 382, 169, 1986.
44. Harris, E.W. et al., Action of 3-[(+/−)-2-carboxypiperazin-4-yl]-propyl-1-phosphonic acid (CPP): a new and highly potent antagonist of N-methyl-D-aspartate receptors in the hippocampus, *Brain Res.*, 382, 174, 1986.
45. Lowe, D.A. et al., The pharmacology of SDZ EAA 494, a competitive NMDA antagonist, *Neurochem. Int.*, 25, 583, 1994.
46. Ferkany, J.W. et al., Pharmacological profile of NPC 17742 [2R,4R,5S-(2-amino-4,5-(1, 2-cyclohexyl)-7-phosphonoheptanoic acid)], a potent, selective and competitive N-methyl-D-aspartate receptor antagonist, *J. Pharmacol. Exp. Ther.*, 264, 256, 1993.
47. Urwyler, S. et al., Biphenyl-derivatives of 2-amino-7-phosphono-heptanoic acid, a novel class of potent competitive N-methyl-D-aspartate receptor antagonists II. Pharmacological characterization *in vivo*, *Neuropharmacology*, 35, 655, 1996.
48. Fagg, G.E. et al., CGP 37849 and CGP 39551: novel and potent competitive N-methyl-D-aspartate receptor antagonists with oral activity, *Br. J. Pharmacol.*, 99, 791, 1990.
49. Baudy, R.B. et al., Potent quinoxaline-spaced phosphono α-amino acids of the AP-6 type as competitive NMDA antagonists: synthesis and biological evaluation, *J. Med. Chem.*, 36, 331, 1993.
50. Heckendorn, R. et al., Synthesis and binding properties of 2-amino-5-phosphono-3-pentenoic acid photoaffinity ligands as probes for the glutamate recognition site of the NMDAR, *J. Med. Chem.*, 36, 3721, 1993.
51. Ornstein, P.L. et al., 6-substituted decahydroisoquinoline-3-carboxylic acids as potent and selective conformationally constrained NMDAR antagonists, *J. Med. Chem.*, 35, 3547, 1992.
52. Beaton, J.A., Stemsrud, K., and Monaghan, D.T., Identification of a novel N-methyl-D-aspartate receptor population in the rat medial thalamus, *J. Neurochem.*, 59, 754, 1992.
53. Buller, A.L. and Monaghan, D.T., Pharmacological heterogeneity of NMDARs: characterization of NR1a/NR2D heteromers expressed in *Xenopus* oocytes, *Eur. J. Pharmacol.*, 320, 87, 1997.
54. Olverman, H.J., Jones, A.W., and Watkins, J.C., [3H]D-2-amino-5-phosphonopentanoate as a ligand for N-methyl-D-aspartate receptors in the mammalian central nervous system, *Neuroscience*, 26, 1, 1988.
55. Olverman, H.J., Jones, A.W., and Watkins, J.C., L-glutamate has higher affinity than other amino acids for [3H]-D-AP5 binding sites in rat brain membranes, *Nature*, 307, 460, 1984.

56. Murphy, D.E. et al., Characterization of the binding of [3H]-CGS 19755: a novel N-methyl-D-aspartate antagonist with nanomolar affinity in rat brain, *Br. J. Pharmacol.*, 95, 932, 1988.
57. Olverman, H.J. et al., [3H]CPP, a new competitive ligand for NMDARs, *Eur. J. Pharmacol.*, 131, 161, 1986.
58. Sills, M.A. et al., [3H]CGP 39653: a new N-methyl-D-aspartate antagonist radioligand with low nanomolar affinity in rat brain, *Eur. J. Pharmacol.*, 192, 19, 1991.
59. Monaghan, D.T., Andaloro, V.J., and Skifter, D.A., Molecular determinants of NMDAR pharmacological diversity, *Prog. Brain Research*, 1998.
60. Gaoni, Y. et al., Synthesis, NMDAR antagonist activity, and anticonvulsant action of 1-aminocyclobutanecarboxylic acid derivatives, *J. Med. Chem.*, 37, 4288, 1994.
61. Müller, W. et al., Synthesis and N-methyl-D-aspartate (NMDA) antagonist properties of the enantiomers of a-amino-5-(phosphonomethyl)[1,1'-biphenyl]-3-propranoic acid: use of a new chiral glycine derivative, *Helv. Chim. Acta.*, 75, 855, 1992.
62. Hutchison, A.J. et al., 4-(Phosphonoalkyl)- and 4-(phosphonoalkenyl)-2-piperidinecarboxylic acids: synthesis, activity at N-methyl-D-aspartic acid receptors, and anticonvulsant activity, *J. Med. Chem.*, 32, 2171, 1989.
63. Dorville, A. et al., Preferred antagonist binding state of the NMDAR: synthesis, pharmacology, and computer modeling of (phosphonomethyl)phenylalanine derivatives, *J. Med. Chem.*, 35, 2551, 1992.
64. Ortwine, D.F. et al., Generation of N-methyl-D-aspartate agonist and competitive antagonist pharmacophore models: design and synthesis of phosphonoalkyl-substituted tetrahydroisoquinolines as novel antagonists, *J. Med. Chem.*, 35, 1345, 1992.
65. Kinarsky, L. et al., Identification of subunit- and antagonist-specific amino acid residues in the N-Methyl-D aspartate receptor glutamate-binding pocket, *J. Pharmacol. Exp. Ther.*, 313, 1066, 2005.
66. Monyer, H. et al., Developmental and regional expression in the rat brain and functional properties of four NMDARs, *Neuron*, 12, 529, 1994.
67. Monyer, H. et al., Heteromeric NMDARs: molecular and functional distinction of subtypes, *Science*, 256, 1217, 1992.
68. Ishii, T. et al., Molecular characterization of the family of the N-methyl-D-aspartate receptor subunits, *J. Biol. Chem.*, 268, 2836, 1993.
69. Ikeda, K. et al., Cloning and expression of the epsilon 4 subunit of the NMDAR channel, *FEBS Lett.*, 313, 34, 1992.
70. Kutsuwada, T. et al., Molecular diversity of the NMDAR channel, *Nature*, 358, 36, 1992.
71. Nakanishi, N., Axel, R., and Shneider, N.A., Alternative splicing generates functionally distinct N-methyl-D-aspartate receptors, *Proc. Natl. Acad. Sci. USA*, 89, 8552, 1992.
72. Behe, P. et al., Determination of NMDA NR1 subunit copy number in recombinant NMDARs, *Proc. R. Soc. Lond. B. Biol. Sci.*, 262, 205, 1995.
73. Das, S. et al., Increased NMDA current and spine density in mice lacking the NMDAR subunit NR3A, *Nature*, 393, 377, 1998.
74. Sheng, M. et al., Changing subunit composition of heteromeric NMDARs during development of rat cortex, *Nature*, 368, 144, 1994.
75. Chazot, P.L. and Stephenson, F.A., Molecular dissection of native mammalian forebrain NMDARs containing the NR1 C2 exon: direct demonstration of NMDARs comprising NR1, NR2A, and NR2B subunits within the same complex, *J. Neurochem.*, 69, 2138, 1997.
76. Dunah, A.W. et al., Subunit composition of N-methyl-D-aspartate receptors in the central nervous system that contain the NR2D subunit, *Mol. Pharmacol.*, 53, 429, 1998.

77. Benveniste, M. and Mayer, M.L., Kinetic analysis of antagonist action at N-methyl-D-aspartic acid receptors. Two binding sites each for glutamate and glycine, *Biophys. J.*, 59, 560, 1991.
78. Honer, M. et al., Differentiation of glycine antagonist sites of N-methyl-D-aspartate receptor subtypes. Preferential interaction of CGP 61594 with NR1/2B receptors, *J. Biol. Chem.*, 273, 11158, 1998.
79. Monaghan, D.T. et al., Two classes of N-methyl-D-aspartate recognition sites: differential distribution and differential regulation by glycine, *Proc. Natl. Acad. Sci. USA*, 85, 9836, 1988.
80. Monaghan, D.T. and Beaton, J.A., Quinolinate differentiates between forebrain and cerebellar NMDARs, *Eur. J. Pharmacol.*, 194, 123, 1991.
81. Buller, A.L. et al., The molecular basis of NMDAR subtypes: native receptor diversity is predicted by subunit composition, *J. Neurosci.*, 14, 5471, 1994.
82. Christie, J.M., Jane, D.E., and Monaghan, D.T., Native N-methyl-D-aspartate receptors containing NR2A and NR2B subunits have pharmacologically distinct competitive antagonist binding sites, *J. Pharmacol. Exp. Ther.*, 292, 1169, 2000.
83. Laurie, D.J. and Seeburg, P.H., Ligand affinities at recombinant N-methyl-D-aspartate receptors depend on subunit composition, *Eur. J. Pharmacol.*, 268, 335, 1994.
84. Watanabe, M. et al., Distinct spatio-temporal distributions of the NMDAR channel subunit mRNAs in the brain, *Ann. NY Acad. Sci.*, 707, 463, 1993.
85. Andaloro, V.J. et al., Pharmacology of NMDAR subtypes, *Soc. Neurosci. Abstr.*, 604, 1996.
86. Feng, B. et al., The effect of competitive antagonist chain length on NMDAR subunit selectivity, *Neuropharmacology*, 48, 354, 2005.
87. Feng, B. et al., Structure-activity analysis of a novel NR2C/NR2D-preferring NMDAR antagonist: 1-(phenanthrene-2-carbonyl) piperazine-2,3-dicarboxylic acid, *Br. J. Pharmacol.*, 141, 508, 2004.
88. Morley, R.M. et al., Synthesis and pharmacology of N1-substituted piperazine-2,3-dicarboxylic acid derivatives acting as NMDAR antagonists, *J. Med. Chem.*, 48, 2627, 2005.
89. Hrabetova, S. et al., Distinct NMDAR subpopulations contribute to long-term potentiation and long- term depression induction, *J. Neurosci. (Online)*, 20, RC81, 2000.
90. Lozovaya, N.A. et al., Extrasynaptic NR2B and NR2D subunits of NMDARs shape superslow afterburst EPSC in rat hippocampus, *J. Physiol.*, 558, 451, 2004.
91. Auberson, Y.P. et al., 5-Phosphonomethylquinoxalinediones as competitive NMDAR antagonists with a preference for the human 1A/2A, rather than 1A/2B receptor composition, *Bioorg. Med. Chem. Lett.*, 12, 1099, 2002.
92. Massey, P.V. et al., Differential roles of NR2A and NR2B-containing NMDARs in cortical long-term potentiation and long-term depression, *J. Neurosci.*, 24, 7821, 2004.
93. Berberich, S. et al., Lack of NMDAR subtype selectivity for hippocampal long-term potentiation, *J. Neurosci.*, 25, 6907, 2005.
94. Frizelle, P.A., Chen, P.E., and Wyllie, D.J., Equilibrium constants for NVP-AAM077 acting at recombinant NR1/NR2A and NR1/NR2B NMDARs: implications for studies of synaptic transmission, *Mol. Pharmacol.*, 2006.
95. Erreger, K. et al., Subunit-specific agonist activity at NR2A, NR2B, NR2C, and NR2D containing N-methyl-D-aspartate glutamate receptors, *Mol. Pharmacol.*, 2007.
96. Johnson, J.W. and Ascher, P., Glycine potentiates the NMDA response in cultured mouse brain neurons, *Nature*, 325, 529, 1987.
97. Kleckner, N.W. and Dingledine, R., Requirement for glycine in activation of NMDA-receptors expressed in Xenopus oocytes, *Science*, 241, 835, 1988.

98. Danysz, W. and Parsons, A.C., Glycine and N-methyl-D-aspartate receptors: physiological significance and possible therapeutic applications, *Pharmacol. Rev.*, 50, 597, 1998.

99. Leeson, P.D. and Iversen, L.L., The glycine site on the NMDAR: structure-activity relationships and therapeutic potential, *J. Med. Chem.*, 37, 4053, 1994.

100. Watson, G.B. and Lanthorn, T.H., Pharmacological characteristics of cyclic homologues of glycine at the N-methyl-D-aspartate receptor-associated glycine site, *Neuropharmacology*, 29, 727, 1990.

101. Marvizon, J.C., Lewin, A.H., and Skolnick, P., 1-Aminocyclopropane carboxylic acid: a potent and selective ligand for the glycine modulatory site of the N-methyl-D-aspartate receptor complex, *J. Neurochem.*, 52, 992, 1989.

102. Hood, W.F. et al., 1-Aminocyclobutane-1-carboxylate (ACBC): a specific antagonist of the N- methyl-D-aspartate receptor coupled glycine receptor, *Eur. J. Pharmacol.*, 161, 281, 1989.

103. Javitt, D.C., Ionotropic glutamate receptors as therapeutic targets, in *Schizophrenia*, F.P. Graham Publishing, Johnson City, TN, 2002, p. 151.

104. Leeson, P.D. et al., Derivatives of 1-hydroxy-3-aminopyrrolidin-2-one (HA-966). Partial agonists at the glycine site of the NMDAR, *Bioorg. Med. Chem. Lett.*, 3, 71, 1993.

105. Inanobe, A., Furukawa, H., and Gouaux, E., Mechanism of partial agonist action at the NR1 subunit of NMDARs, *Neuron*, 47, 71, 2005.

106. Hogner, A. et al., Competitive antagonism of AMPA receptors by ligands of different classes: crystal structure of ATPO bound to the GluR2 ligand-binding core, in comparison with DNQX, *J. Med. Chem.* 46, 214, 2003.

107. Armstrong, N. and Gouaux, E., Mechanisms for activation and antagonism of an AMPA-sensitive glutamate receptor: crystal structures of the GluR2 ligand binding core, *Neuron*, 28, 165, 2000.

108. Watson, G.B. et al., Kynurenate antagonizes actions of N-methyl-D-aspartate through a glycine sensitive receptor, *Neurosci. Res. Commun.*, 2, 169, 1988.

109. Kessler, M. et al., A glycine site associated with N-methyl-D-aspartic acid receptors: characterization and identification of a new class of antagonists, *J. Neurochem.*, 52, 1319, 1989.

110. Honore, T. et al., Quinoxalinediones: potent competitive non-NMDA glutamate receptor antagonists, *Science*, 241, 701, 1988.

111. Birch, P.J., Grossman, C.J., and Hayes, A.G., 6,7-Dinitro-quinoxaline-2,3-dion and 6-nitro,7-cyano-quinoxaline-2,3-dion antagonise responses to NMDA in the rat spinal cord via an action at the strychnine-insensitive glycine receptor, *Eur. J. Pharmacol.*, 156, 177, 1988.

112. Leeson, P.D. et al., Kynurenic acid derivatives. Structure-activity relationships for excitatory amino acid antagonism and identification of potent and selective antagonists at the glycine site on the N-methyl-D-aspartate receptor, *J. Med. Chem.*, 34, 1243, 1991.

113. Leeson, P.D. et al., 4-Amido-2-carboxytetrahydroquinolines. Structure–activity relationships for antagonism at the glycine site of the NMDAR, *J. Med. Chem.*, 35, 1954, 1992.

114. Kulagowski, J.J. et al., 3′-(Arylmethyl)- and 3′-(aryloxy)-3-phenyl-4-hydroxyquinolin-2(1H)-ones: orally active antagonists of the glycine site on the NMDAR, *J. Med. Chem.*, 37, 1402, 1994.

115. Di Fabio, R. et al., Substituted indole-2-carboxylates as *in vivo* potent antagonists acting as the strychnine-insensitive glycine binding site, *J. Med. Chem.*, 40, 841, 1997.

116. Carignani, C. et al., NMDAR subunit characterization of the glycine site antagonist GV196771A and its action on the spinal cord wind-up, *Naun. Schmied. Arch. Pharmacol.*, 358, P1119, 1998.

117. Cai, S.X. et al., Structure-activity relationships of alkyl- and alkoxy-substituted 1,4-dihydroquinoxaline-2,3-diones: potent and systemically active antagonists for the glycine site of the NMDAR, *J. Med. Chem.*, 40, 730, 1997.

118. Auberson, Y.P. et al., N-phosphonoalkyl-5-aminomethylquinoxaline-2,3-diones: *in vivo* active AMPA and NMDA(glycine) antagonists, *Bioorg. Med. Chem. Lett.*, 9, 249, 1999.

119. Nikam, S.S. et al., Design and synthesis of novel quinoxaline-2,3-dione AMPA/GlyN receptor antagonists: amino acid derivatives, *J. Med. Chem.*, 42, 2266, 1999.

120. Bigge, C.F. et al., Synthesis of 1,4,7,8,9,10-hexahydro-9-methyl-6-nitropyrido[3,4-f]-quinoxaline-2,3-dione and related quinoxalinediones: characterization of alpha-amino-3-hydroxy-5-methyl-4-isoxazolepropionic acid (and N-methyl-D-aspartate) receptor and anticonvulsant activity, *J. Med. Chem.*, 38, 3720, 1995.

121. Furukawa, H. and Gouaux, E., Mechanisms of activation, inhibition and specificity: crystal structures of the NMDAR NR1 ligand-binding core, *EMBO J.*, 22, 2873, 2003.

122. Kew, J.N. and Kemp, J.A., Ionotropic and metabotropic glutamate receptor structure and pharmacology, *Psychopharmacology*, 179, 4, 2005.

123. Bare, T.M. et al., Pyridazinoquinolinetriones as NMDA glycine site antagonists with oral antinociceptive activity in a model of neuropathic pain, *J. Med. Chem.*, 50, 3113, 2007.

124. Benke, D. et al., [³H]CGP 61594, the first photoaffinity ligand for the glycine site of NMDARs, *Neuropharmacology*, 38, 233, 1999.

125. Buller, A.L. et al., Glycine modulates ethanol inhibition of heteromeric N-methyl-D-aspartate receptors expressed in *Xenopus* oocytes, *Mol. Pharmacol.*, 48, 717, 1995.

126. Sasaki, Y.F. et al., Characterization and comparison of the NR3A subunit of the NMDAR in recombinant systems and primary cortical neurons, *J. Neurophysiol.*, 87, 2052, 2002.

127. Yao, Y. and Mayer, M.L., Characterization of a soluble ligand binding domain of the NMDAR regulatory subunit NR3A, *J. Neurosci.*, 26, 4559, 2006.

128. Madry, C. et al., Principal role of NR3 subunits in NR1/NR3 excitatory glycine receptor function, *Biochem. Biophys. Res. Commun.*, 354, 102, 2007.

129. Awobuluyi, M. et al., Subunit-specific roles of glycine-binding domains in activation of NR1/NR3 N-methyl-D-aspartate receptors, *Mol. Pharmacol,.* 71, 112, 2007.

130. Nilsson, A. et al., Characterisation of the human NMDAR subunit NR3A glycine binding site, *Neuropharmacol.*, 52, 1151, 2007.

131. Karadottir, R. et al., NMDARs are expressed in oligodendrocytes and activated in ischaemia, *Nature*, 438, 1162, 2005.

132. Micu, I. et al., NMDARs mediate calcium accumulation in myelin during chemical ischaemia, *Nature*, 439, 988, 2006.

133. Salter, M.G. and Fern, R., NMDARs are expressed in developing oligodendrocyte processes and mediate injury, *Nature*, 438, 1167, 2005.

134. Sacaan, A.I. and Johnson, K.M., Spermine enhances binding to the glycine site associated with the N-methyl-D-aspartate receptor complex, *Mol. Pharmacol.*, 36, 836, 1989.

135. McGurk, J.F., Bennett, M.V., and Zukin, R.S., Polyamines potentiate responses of N-methyl-D-aspartate receptors expressed in *Xenopus* oocytes, *Proc. Natl. Acad. Sci. USA*, 87, 9971, 1990.

136. Ransom, R.W. and Deschenes, N.L., Polyamines regulate glycine interaction with the N-methyl-D-aspartate receptor, *Synapse*, 5, 294, 1990.

137. Benveniste, M. and Mayer, M.L., Multiple effects of spermine on N-methyl-D-aspartic acid receptor responses of rat cultured hippocampal neurones, *J. Physiol.*, 464, 131, 1993.

138. Ransom, R.W. and Stec, N.L., Cooperative modulation of [^{3}H]MK-801 binding to the N-methyl-D-aspartate receptor-ion channel complex by L-glutamate, glycine, and polyamines, *J. Neurochem.*, 51, 830, 1988.
139. McBain, C.J. and Mayer, M.L., N-methyl-D-aspartic acid receptor structure and function, *Physiol. Rev.*, 74, 723, 1994.
140. Williams, K., Interactions of polyamines with ion channels, *Biochem. J.*, 325, 289, 1997.
141. Durand, G.M., Bennett, M.V., and Zukin, R.S., Splice variants of the N-methyl-D-aspartate receptor NR1 identify domains involved in regulation by polyamines and protein kinase C, *Proc. Natl. Acad. Sci. USA*, 90, 6731, 1993.
142. Zheng, X. et al., Mutagenesis rescues spermine and Zn2+ potentiation of recombinant NMDARs, *Neuron*, 12, 811, 1994.
143. Williams, K. et al., An acidic amino acid in the N-methyl-D-aspartate receptor that is important for spermine stimulation, *Mol. Pharmacol.*, 48, 1087, 1995.
144. Williams, K. et al., Sensitivity of the N-methyl-D-aspartate receptor to polyamines is controlled by NR2 subunits, *Mol. Pharmacol.*, 45, 803, 1994.
145. Williams, K., Pharmacological properties of recombinant N-methyl-D-aspartate (NMDA) receptors containing the epsilon 4 (NR2D) subunit, *Neurosci. Lett.*, 184, 181, 1995.
146. Zhang, L. et al., Spermine potentiation of recombinant N-methyl-D-aspartate receptors is affected by subunit composition, *Proc. Natl. Acad. Sci. USA*, 91, 10883, 1994.
147. Carter, C., Rivy, J.P., and Scatton, B., Ifenprodil and SL 82.0715 are antagonists at the polyamine site of the N-methyl-D-aspartate (NMDA) receptor, *Eur. J. Pharmacol.*, 164, 611, 1989.
148. Legendre, P. and Westbrook, G.L., Ifenprodil blocks N-methyl-D-aspartate receptors by a two-component mechanism, *Mol. Pharmacol.*, 40, 289, 1991.
149. Williams, K., Ifenprodil discriminates subtypes of the N-methyl-D-aspartate receptor: selectivity and mechanisms at recombinant heteromeric receptors, *Mol. Pharmacol.*, 44, 851, 1993.
150. Galli, A. and Mori, F., Acetylcholinesterase inhibition and protection by dizocilpine (MK-801) enantiomers, *J. Pharm. Pharmacol.*, 48, 71, 1996.
151. Gallagher, M.J. et al., Interactions between ifenprodil and the NR2B subunit of the N-methyl-D-aspartate receptor, *J. Biol. Chem.*, 271, 9603, 1996.
152. Gallagher, M.J., Huang, H., and Lynch, D.R., Modulation of the N-methyl-D-aspartate receptor by haloperidol: NR2B-specific interactions, *J. Neurochem.*, 70, 2120, 1998.
153. Chenard, B.L. et al., (1S,2S)-1-(4-hydroxyphenyl)-2-(4-hydroxy-4-phenylpiperidino)-1-propanol: a potent new neuroprotectant which blocks N-methyl-D-aspartate responses, *J. Med. Chem.*, 38, 3138, 1995.
154. Fischer, G. et al., Ro 25-6981, a highly potent and selective blocker of N-methyl-D-aspartate receptors containing the NR2B subunit. Characterization *in vitro*, *J. Pharmacol. Exp. Ther.*, 283, 1285, 1997.
155. Layton, M.E., Kelly, M.J., and Rodzinak, K.J., Recent advances in the development of NR2B subtype-selective NMDAR antagonists, *Curr. Top. Med. Chem.*, 6, 697, 2006.
156. McCauley, J.A. et al., NR2B-selective N-methyl-D-aspartate antagonists: synthesis and evaluation of 5-substituted benzimidazoles, *J. Med. Chem.*, 47, 2089, 2004.
157. Wright, J.L. et al., Subtype-selective N-methyl-D-aspartate receptor antagonists: synthesis and biological evaluation of 1-(heteroarylalkynyl)-4-benzylpiperidines, *J. Med. Chem.*, 43, 3408, 2000.

158. Whittemore, E.R., Illyin, V.I., and Woodward, R.M., Electrophysiological characterization of CI-1041 on cloned and native NMDARs, *Soc. Neurosci. Abstr.*, 26, 527, 2000.
159. Borza, I. et al., Indole-2-carboxamides as novel NR2B selective NMDAR antagonists, *Bioorg. Med. Chem. Lett.*, 13, 3859, 2003.
160. Gregory, T.F. et al., Parallel synthesis of a series of subtype-selective NMDAR antagonists, *Bioorg. Med. Chem. Lett.*, 10, 527, 2000.
161. Tamiz, A.P. et al., Structure-activity relationship of N-(phenylalkyl)cinnamides as novel NR2B subtype-selective NMDAR antagonists, *J. Med. Chem.*, 42, 3412, 1999.
162. Alanine, A. et al., 1-Benzyloxy-4,5-dihydro-1H-imidazol-2-yl-amines, a novel class of NR1/2B subtype selective NMDAR antagonists, *Bioorg. Med. Chem. Lett.*, 13, 3155, 2003.
163. Pinard, E. et al., 4-Aminoquinolines as a novel class of NR1/2B subtype selective NMDAR antagonists, *Bioorg. Med. Chem. Lett.*, 12, 2615, 2002.
164. Paoletti, P., Ascher, P., and Neyton, J., High-affinity zinc inhibition of NMDA NR1-NR2A receptors, *J. Neurosci.*, 17, 5711, 1997.
165. Chen, N., Moshaver, A., and Raymond, L.A., Differential sensitivity of recombinant N-methyl-D-aspartate receptor subtypes to zinc inhibition, *Mol. Pharmacol.*, 51, 1015, 1997.
166. Christine, C.W. and Choi, D.W., Effect of zinc on NMDAR-mediated channel currents in cortical neurons, *J. Neurosci.*, 10, 108, 1990.
167. Aniksztejn, L., Charton, G., and Ben-Ari, Y., Selective release of endogenous zinc from the hippocampal mossy fibers *in situ*, *Brain. Res.*, 404, 58, 1987.
168. Assaf, S.Y. and Chung, S.H., Release of endogenous Zn2+ from brain tissue during activity, *Nature*, 308, 734, 1984.
169. Paoletti, P. et al., Molecular organization of a zinc binding N-terminal modulatory domain in a NMDAR subunit, *Neuron*, 28, 911, 2000.
170. Evans, R.H., Francis, A.A., and Watkins, J.C., Selective antagonism by Mg2+ of amino acid-induced depolarization of spinal neurones, *Experientia*, 33, 489, 1977.
171. Nowak, L. et al., Magnesium gates glutamate-activated channels in mouse central neurones, *Nature*, 307, 462, 1984.
172. Mayer, M.L., Westbrook, G.L., and Guthrie, P.B., Voltage-dependent block by Mg^{2+} of NMDA responses in spinal cord neurones, *Nature*, 309, 261, 1984.
173. Wollmuth, L.P. and Sobolevsky, A.I., Structure and gating of the glutamate receptor ion channel, *Trends Neurosci.*, 27, 321, 2004.
174. Lodge, D. and Johnson, K.M., Noncompetitive excitatory amino acid receptor antagonists, *Trends Pharmacol. Sci.*, 11, 81, 1990.
175. Huettner, J.E. and Bean, B.P., Block of N-methyl-D-aspartate-activated current by the anticonvulsant MK-801: selective binding to open channels, *Proc. Natl. Acad. Sci. USA*, 85, 1307, 1988.
176. Anis, N.A. et al., The dissociative anaesthetics, ketamine and phencyclidine, selectively reduce excitation of central mammalian neurones by N-methyl-aspartate, *Br. J. Pharmacol.*, 79, 565, 1983.
177. Sakurada, K., Masu, M., and Nakanishi, S., Alteration of Ca2+ permeability and sensitivity to Mg^{2+} and channel blockers by a single amino acid substitution in the N-methyl-D-aspartate receptor, *J. Biol. Chem.*, 268, 410, 1993.
178. Wong, E.H. et al., The anticonvulsant MK-801 is a potent N-methyl-D-aspartate antagonist, *Proc. Natl. Acad. Sci. USA*, 83, 7104, 1986.
179. Gill, R., Foster, A.C., and Woodruff, G.N., Systemic administration of MK-801 protects against ischemia-induced hippocampal neurodegeneration in the gerbil, *J. Neurosci.*, 7, 3343, 1987.

180. Monaghan, D.T. and Larson, H., NR1 and NR2 subunit contributions to N-methyl-D-aspartate receptor channel blocker pharmacology, *J. Pharmacol. Exp. Ther.*, 280, 614, 1997.
181. Parsons, C.G., Danysz, W., and Quack, G., Memantine is a clinically well tolerated N-methyl-D-aspartate (NMDA) receptor antagonist: review of preclinical data, *Neuropharmacol.*, 38, 735, 1999.
182. Iversen, L.L. and Kemp, J.A., Non-competitive NMDA antagonists as drugs, in *The NMDAR*, Oxford Press, Oxford, 1994.

13 Activation Mechanisms of the NMDA Receptor

Marie L. Blanke and Antonius M.J. VanDongen

CONTENTS

13.1 INTRODUCTION

NMDA receptors (NMDARs) are glutamate-gated cation channels with high calcium permeability that play important roles in many aspects of the biology of higher organisms. They are critical for the development of the central nervous system (CNS), generation of rhythms for breathing and locomotion, and the processes underlying learning, memory, and neuroplasticity. Consequently, abnormal expression levels and altered NMDAR function have been implicated in numerous neurological disorders and pathological conditions. NMDAR hypofunction can result in cognitive defects, whereas overstimulation causes excitotoxicity and subsequent neurodegeneration. Therefore, NMDARs are important therapeutic targets for many CNS disorders[1–8] including stroke, hypoxia, ischemia, head trauma, Huntington's, Parkinson's, and Alzheimer's diseases, epilepsy, neuropathic pain, alcoholism, schizophrenia, and mood disorders. To date, drugs targeting NMDARs have had only limited success clinically due to poor efficacy and unacceptable side effects, including hallucinations, catatonia, ataxia, nightmares, and memory deficits.

A detailed understanding of the mechanisms underlying agonist-induced receptor activation would facilitate development of more selective drugs that target specific NMDAR subtypes and alter their function to a well-defined extent. This chapter will investigate the physiological roles NMDARs play in the mammalian nervous system and the molecular and structural basis of NMDAR activation. One of the main questions that will be addressed is how agonist binding results in opening of the NMDAR ion channel. Although the mechanism coupling ligand binding to channel opening remains incompletely understood for NMDARs, we propose that this process suggests promising approaches to drug design.

13.2 FUNCTION OF NMDA RECEPTORS

NMDARs belong to a class of ionotropic glutamate receptors (iGluRs) that also includes the AMPA receptors (AMPARs) and kainate receptors.[9] The names of these subclasses derived from the selective synthetic agonists that can be used to distinguish them (see Chapter 12 for detailed description of NMDAR pharmacology). This pharmacological distinction is mirrored by distinct neurophysiological roles for each of the iGluR subtypes.

13.2.1 SYNAPTIC FUNCTION

Excitatory synaptic transmission in the vertebrate brain relies on the release of L-glutamate from presynaptic terminals that diffuses across the synaptic cleft and binds to postsynaptic AMPARs and NMDARs. Activation of AMPARs is fast and transient, causing brief depolarizations that last no longer than a few milliseconds. NMDARs are not critical for this basal synaptic transmission, but instead they regulate functional and structural plasticity of individual synapses, dendrites, and neurons by allowing activation of specific calcium-dependent signaling cascades. Several unique properties of NMDARs prevent their activation by L-glutamate released during a single synaptic event.

First, NMDARs activate significantly slower than AMPARs and kainate receptors. Glutamate released from a presynaptic terminal following arrival of an action potential is removed efficiently from the synaptic cleft by the actions of glutamate transporters located in the presynaptic terminal and nearby astrocytes.[10,11] Consequently, glutamate is available for receptor binding only briefly during low frequency synaptic transmission. Because NMDARs have relatively a high affinity for glutamate, the millisecond-long neurotransmitter pulses should be able to partially (and slowly) activate NMDARs. However, individual excitatory synaptic inputs received during baseline activity do not result in calcium (Ca^{2+}) influx because of a second NMDAR property: its pronounced voltage dependence.

At resting membrane potentials, external magnesium (Mg^{2+}) ions enter the NMDAR pore, but unlike the permeant Ca^{2+} ions, they bind tightly and prevent further ion permeation.[12,13] Mg^{2+} ions are present at millimolar concentrations in the external milieu of neurons, while intracellular Mg^{2+} concentrations are in the micromolar range, resulting in a net inward driving force for Mg^{2+} ions at negative membrane potentials. A depolarization of sufficient amplitude and duration is required to dislodge and repel the Mg^{2+} ions from the pore, thereby allowing the flow of permeant ions. As a result, the NMDAR acts as a molecular coincidence detector[14]: efficient activation and ion permeation through the NMDAR requires both a sufficiently strong depolarization and synaptic release of glutamate. This dual input requirement, together with their slow activation and deactivation kinetics allows NMDARs to integrate and decode incoming synaptic activity. The high Ca^{2+} permeability of NMDARs enables them to transduce specific synaptic input patterns into long-lasting alterations in synaptic strength.

13.2.2 LONG-TERM POTENTIATION AND DEPRESSION

The strong depolarization required to remove Mg block from synaptically localized NMDARs can be achieved in several ways. High frequency synaptic inputs may allow the excitatory postsynaptic potentials (EPSPs) generated by AMPAR activation to accumulate and build over time. This phenomenon underlies the paradigm of long-term potentiation (LTP) discovered in 1973 by Bliss and Lomo[15] in which a short burst of high frequency synaptic input (15 Hz for 15 sec or 100 Hz for 3 sec) results in strengthening of excitatory synapses for a prolonged period (hours to days).

The role of NMDARs in LTP induction in the hippocampal CA1 area is well documented.[16] LTP is NMDAR-dependent in many other regions of the brain,[17] although NMDAR-independent LTP has also been observed.[18,19] The synaptic strengthening observed during LTP has been attributed to two major mechanisms: (1) phosphorylation of AMPARs, resulting in an increased open probability and (2) enhanced trafficking of AMPARs to the postsynaptic membrane (see Chapter 8 for details).

Long-term depression (LTD), the counterpart of LTP, may be experimentally induced by prolonged low frequency (0.5 to 3 Hz) stimulation of excitatory synapses.[20] Induction of LTD is also NMDAR-dependent in the hippocampal CA1 region and, like LTP induction, requires Ca^{2+} influx through NMDARs.[21] LTD can also be induced by other mechanisms including stimulation of metabotropic glutamate receptors.[22] The mechanism by which NMDA-dependent LTD reduces synaptic strength is to reverse the effects of LTP: dephosphorylation of AMPARs, thus reducing their open probability[23] and removal of AMPARs from the synaptic plasma membrane by endocytosis (see Chapter 8). In these examples of homosynaptic LTP and LTD, the synapse that receives the low/high frequency input is weakened/strengthened. However, synaptic strength can also be altered in either direction if a single synaptic input is coupled with a postsynaptic depolarization, resulting in heterosynaptic LTP, which has been proposed as a model for associative memory. Postsynaptic depolarizations can occur by various mechanisms. In many cases, neuronal dendrites play critical roles in the generation and processing of these plasticity-inducing signals.

13.2.3 DENDRITIC FUNCTION

Dendrites exhibit active conductances, mediated by voltage-gated Na and Ca channels, as well as the NMDAR itself, which allow them to generate a back-propagating action potential (bAP).[24,25] Dendritic bAPs have a longer duration than axonal spikes and therefore permit the removal of Mg^{2+} block from NMDARs. Any glutamate release occurring during a bAP-induced depolarization may therefore result in Ca^{2+} influx through activated NMDARs and subsequent alterations in synapse strength.[26,27] Because bAPs take time to propagate down distal dendrites, efficient activation of NMDARs by this mechanism requires precise timing of synaptic input relative to bAP generation.

The direction and magnitude of the resulting alteration in synaptic strength depend critically on the temporal relationship between the two processes.[28] The precise timing dependence of presynaptic firing relative to the firing of the postsynaptic neuron, has given rise to a model of long-term synaptic plasticity called spike timing-dependent plasticity (STDP).[29-35] This promising model extends the LTP–LTD paradigm by proposing that the coupling strength between neurons depends on the degree of correlation of their spiking activities.[36]

Thus, STDP implements an important and long ignored aspect of the most influential model for synaptic strengthening formulated in 1949 by Donald Hebb.[37] In Hebb's words (relevant wording in bold): "When an axon of cell A is near enough to excite a cell B and **repeatedly or persistently takes part in firing it**, some growth process or metabolic change takes place in one or both cells such that A's efficiency, as one of the cells firing B, is increased." Hebb's insistence that the presynaptic neuron causes the postsynaptic neuron to fire an action potential appears visionary.

The voltage dependence and ion selectivity of NMDARs provide them with regenerative properties that allow the generation of dendritic action potentials (NMDA spikes) that do not originate from bAPs, but from highly synchronized excitatory synaptic inputs to a region of the dendrite[38-42] (see Section 9.4). The ability to generate NMDA spikes endows dendrites with interesting nonlinear computational capabilities, the impact of which is only now beginning to become clear.[43-45]

13.2.4 PRIVILEGED CA²⁺ IONS

The long-lasting effects on synaptic efficacy resulting from activation of NMDARs depend on Ca^{2+} influx into the postsynaptic compartment (see Chapter 9 for a review of NMDAR-mediated Ca signaling in dendritic spines). It is intriguing that both synaptic strengthening and weakening are mediated by Ca^{2+} influx through NMDARs. Several hypotheses have been proposed to explain this phenomenon. The magnitude of the rise in Ca^{2+} concentration may be an important determinant. The direction of change in synaptic strength may also be determined by the temporal properties of the internal Ca concentration changes.[46]

Alternatively, there may be a spatial difference in calcium rises associated with by LTP and LTD. NR2B receptors have been proposed to be localized extrasynaptically[47] (although this is not a universal finding[48]). Such spatial separation of NR2A and NR2B receptors would allow Ca^{2+} ions to activate distinct signaling complexes.[49,50] An earlier proposal that the NR2 subtype determines the direction of synaptic change, with activation of NR2A- and NR2B-containing NMDARs resulting in LTP and LTD, respectively,[51,52] has not held up in later experiments.[53,54] More work is needed to define the molecular mechanism by which spatiotemporal differences in Ca concentrations resulting from activation of synaptic and extrasynaptic

NMDARs of varying subtype compositions lead to bidirectional, long-term changes in synaptic efficacy.

Ca^{2+} ions entering the neuron through the NMDAR are privileged, because they are able to act locally on large signal transduction complexes associated with synaptic NMDARs, which consist of calcium-dependent enzymes, second messengers, protein kinases and phosphatases, scaffolding proteins, cytoskeletal elements, GTP binding proteins and their regulators, and adhesion molecules.[55–57] NMDAR-mediated calcium influx activates these complexes, generating signal transduction cascades that produce long-lasting changes in synaptic function and structure.[57–63]

The ability of an NMDAR to integrate and decode synaptic inputs and generate or amplify dendritic spikes depends on its kinetics of activation, deactivation, and desensitization. The amount of Ca influx is also largely determined by kinetics, in particular the rate of deactivation. Subunit composition is an important determinant for NMDAR (de-)activation kinetics and intracellular calcium dynamics, as discussed in Chapter 9. The amount of Ca influx resulting from NMDAR activation is further modulated by other external physiological factors including pH, zinc ions, and polyamines that are discussed in Chapters 11 and 12. The next section considers the roles of the two coagonists, glutamate and glycine, in activation of NMDARs.

13.2.5 Coagonists Glutamate and Glycine

NMDARs are unique among ligand-gated ion channels in that their activation requires binding of two coagonists, glycine and L-glutamate.[64–66] Glycine is sometimes cited in the literature as an NMDAR modulator—to set it apart from the agonist L-glutamate, but as explained below, their binding sites are structurally similar and seem to play equivalent roles in receptor activation. Physiologically, however, glycine and glutamate have distinct functions. While L-glutamate is released from specific presynaptic terminals, low concentrations of ambient glycine present at the synapse are thought to be sufficient to allow receptor activation. Interestingly, a recent paper suggests that D-serine released by astrocytes is the endogenous glycine site agonist in certain brain regions, allowing glial cells to actively control synaptic metaplasticity.[67]

Because glycine plays a more modulatory role *in vivo*,[68,64] while glutamate is the 'active', released neurotransmitter, the glycine and glutamate binding sites on the NMDAR represent two distinct therapeutic targets. Particular efforts have been devoted to the development of partial agonists for the glycine site[69–76] that may act as negative modulators of NMDAR function by competitively displacing the full agonist glycine from its binding site. The pharmacology of NMDARs is described in detail in Chapter 12.

An interesting but poorly studied property of glycine is its ability to partially activate NMDARs in the absence of glutamate. This action has been observed in NMDARs containing NR1 and NR2A subunits expressed in *Xenopus laevis* oocytes, where $10\,\mu M$ glycine alone activated the receptor by a few percent.[77] Due to the inhibition of this effect by the competitive glutamate-site antagonist APV, it has been suggested that glycine may act as a partial agonist at the glutamate binding site.[77] It is not clear whether ambient glycine in the CNS also exhibits this property, which would allow a small amount of Ca influx through NMDARs independent of synaptic input during large depolarizations that remove Mg^{2+} block. Also, it is not known whether D-serine, which can be released from astrocytes, has the same ability as glycine to partially activate NMDARs in the absence of glutamate.

The importance of ambient glycine binding to NMDARs for cognitive function is underscored by experiments with transgenic mice carrying mutant alleles of the NR1 subunit.[78] Mice that express NR1 subunits with lowered glycine affinity display a spectrum of cognitive and learning defects including nonhabituating hyperactivity, increased stereotyped behavior, disruptions of nest building activity, and poor performance in the Morris water maze[78], a measure of cued learning. The behavioral phenotypes of these glycine-insensitive mutant mice resemble some of the positive and negative symptoms displayed by schizophrenia patients, consistent with NMDAR hypofunction as one of the leading hypotheses for schizophrenia.[79–82] Section 13.3.7 discusses the (limited) successes of glycine site partial agonists in the clinic. The next two sections describe studies of the behavior of individual NMDAR channels.

13.2.5 SINGLE CHANNEL BEHAVIOR

Ion channels are unique among proteins in that their behavior can be studied at the level of a single molecule. This important experimental paradigm was made possible by the advent of the patch clamp technique pioneered by Neher and Sakmann in the mid 1970s.[83] In single channel recordings, individual ion channels are seen to alternate stochastically between two states, open and closed. No measurable ion flux is present in the closed state whereas the open state is characterized by a constant, channel-specific conductance. Activation of an ion channel results from an increase in the probability of being in the open state, not a change in its single channel conductance.

Analysis of single channel behavior provides detailed information about mechanisms of action that cannot be extracted from macroscopic currents measured in whole cell recordings.[84] Early single channel studies suggested that Mg^{2+} ions inhibited NMDARs via an open channel block mechanism, as indicated by the presence of short closings within otherwise stable open periods.[12] The allosteric modulator spermine exerts multiple effects on single channel behavior, increasing opening frequency at low concentrations, while decreasing single channel conductance and mean open time at higher concentrations, suggesting the existence of two binding sites.[85] Single channel analysis of ethanol inhibition of NMDARs (see Chapter 4) indicated that ethanol exerts its effect through an allosteric mechanism by reducing agonist efficacy.[86]

Single channel behavior of NMDARs from hippocampal CA1 neurons was studied using very low glutamate concentrations to improve temporal resolution of individual glutamate binding events.[87] Openings resulting from individual receptor activations showed surprising complexity: they consist of a long cluster of bursts of openings. Furthermore, the NMDARs appeared to have different gating modes, occasionally entering periods of very high open probability.[87] These results demonstrated that the slow deactivation kinetics of the NMDAR result from intrinsic gating properties.

Single channel analysis can also provide more insight in how NMDARs function at the synapse. In response to a brief pulse of glutamate, mimicking synaptic release, NMDARs activate slowly over hundreds of milliseconds and continue activating long after all glutamate has been removed from the synaptic cleft, thereby briefly "memorizing" the occurrence of a synaptic input. Single channel analysis of NR1 and NR2A receptors indicates that after a brief pulse of glutamate, receptors enter a high affinity closed state from which either channel opening or agonist unbinding occurs with approximately equal probability.[88] A single synaptic event is therefore expected to only partially activate NMDARs. Consequently, a closely spaced second pulse of agonist is able to further increase the open probability, endowing the NMDAR with an ability to decode synaptic input frequency (Figure 13.1).

Comparison of the single channel behavior of NR2A- and NR2B-containing NMDARs revealed several kinetic differences in these receptor subtypes. NR2A receptors respond faster to brief synaptic-like pulses of glutamate and reach higher open probabilities.[89] It has been proposed that these differences in channel gating kinetics result in preferential opening of NR2A-containing receptors during high frequency synaptic inputs that stimulate LTP. Conversely, NR2B receptors are associated with lower frequency inputs that cause LTD. However, the NR2A/LTP and NR2B/LTD connections were discredited by recent experiments.[54]

13.2.6 SUBCONDUCTANCE LEVELS, PERMEATION, AND GATING

One of the more intriguing aspects of NMDAR single channel behavior is that it is not strictly binary. Channels occasionally fail to open or close fully and instead visit intermediate conductance levels. Such subconductances levels (sublevels) have been observed in both native[90] and recombinant glutamate receptors.[91–93] Typically, these sublevels are visited during transitions between closed and fully open states, not as isolated opening events. NR2A- and NR2B-containing receptors display similar short-lived sublevels whose conductance is approximately 80% of the main open state. In contrast, NR2C- and NR2D-containing receptors briefly visit a 50% sublevel and a fully open state with approximately equal probability.

Subconductance levels have been observed in virtually every type of ion channel, although the number of levels, stability, and abundance vary widely. Sublevels have been well-characterized in K channels, which are evolutionarily related to NMDARs.[94] Our laboratory has described sublevels observed during open–closed transitions in the voltage-gated K channel $K_v2.1$ (drk1),[95] and we have proposed a model that attributes sublevels to heteromeric pore conformations visited when one, two, or three of the four subunits move to the open conformation (Figure 13.2). Because channel opening appears to be a strongly cooperative process, the heteromeric pore conformations are predicted to be highly unstable and the associated sublevels may be very short-lived.

One prediction of the subunit–sublevel hypothesis is that sublevels should be more abundant when a receptor or channel is incompletely activated, which was experimentally confirmed for $K_v2.1$.[95] In a more stringent test of this hypothesis, a $K_v2.1$ tandem dimer combining two K channel subunits with different activation thresholds was created. Single channel behavior of this tandem dimer at potentials

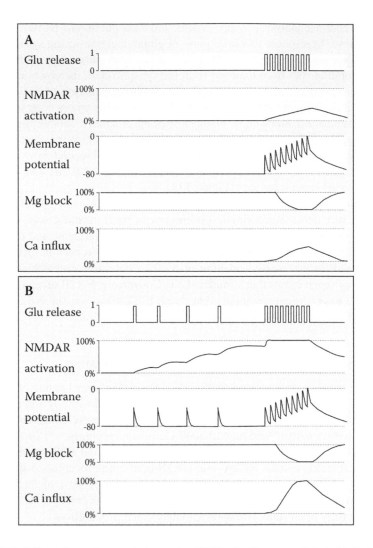

FIGURE 13.1 Decoding of synaptic inputs by NMDARs. Because of their high glutamate affinity and relatively slow activation and deactivation kinetics, NMDARs can "memorize" recent low frequency synaptic inputs that alone cannot elicit ion permeation through the NMDAR because of the persistent Mg block at negative membrane potentials. Panels A and B illustrate how the amount of Ca influx elicited by the same high-frequency input (tetanus) depends on the recent history of synaptic activity preceding the tetanus.

between the two thresholds was dominated by two sublevels, whose kinetics and voltage dependence indicated that they resulted from the activation of one and two subunits.[96] These results support the subunit–subconductance hypothesis, which implies that gating and permeation are strictly coupled. In this model, the selectivity filter is directly responsible for opening and closing the channel.[97,98] A specific mechanism has been proposed for how the selectivity filter may function as the channel gate, in which the filter alternates between two conformations with high and low affinity for

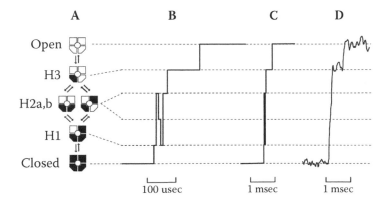

FIGURE 13.2 Subunit-subconductance model for NMDAR gating. The permeation pathway of the NMDAR is formed by two NR1 and two NR2 subunits (A). The conformation of each subunit alters between a "closed" conformation that does not support permeation (filled symbols) and an "open" conformation that does (open symbols). When the channel moves from the closed to the open state, each subunit must undergo a conformational change from closed to open. Unless the movements in the four subunits are strictly synchronized (due to an extreme form of positive cooperativity), heteromeric pore conformations (H1 through H3) will be visited in which one, two, or three subunits are in the open state. It was proposed that these heteromeric pore conformations can give rise to subconductance levels,[95–97] predicting a complex structure for all open, close transitions (B). Because the heteromeric pore conformations are transition states expected to be short-lived, most of them may go undetected in single channel recordings because of the limited bandwidth of the measurements (C, D). NMDARs containing NR2A or NR2B subunits display an 80% subconductance level visited during transitions between the closed and fully open state. The subunit–subconductance model would explain this by assuming that one of the heteromeric pore conformations is relatively stable. Based on the relatively large size of the subconductance level, the relatively stable state would likely be H3.

the permeant ion.[98] Experimental support for an affinity-switching selectivity filter has recently been shown by NMR experiments using the KcsA K channel.[99]

Single channel analysis of AMPARs and NMDARs also provided evidence for a strong coupling between permeation and gating, as well as support for a link between sublevels and partial receptor activation. Two sublevels associated with partially activated GluR3 AMPARs were proposed to result from the activation of two and three subunits.[100] Mutations in the selectivity filter were shown to stabilize sublevels both in K channels[101,103] and NMDARs.[102,104] The ion selectivity of the sublevels in these mutant NMDARs is different from that of the fully open state,[104] as was reported for the *Shaker* K channel,[101] thereby providing a direct causal linkage between subconductance gating and the selectivity filter.

Under bi-ionic (Na^+/Cs^+) conditions, gating of these mutant NMDARs becomes strongly asymmetric, with sublevels visited either during openings (external Cs^+) or during closings (external Na^+),[104] providing additional evidence for a strong coupling of gating and permeation. Finally, partial agonists were shown to determine the open probability of subconductance levels in the GluR2 AMPAR,[105] and a model was suggested which is identical to the subunit–subconductance model (Figure 13.2) proposed by Chapman and VanDongen in 1997.[97,95] Section 13.3.7 addresses the structural basis of partial agonism in more detail. The next section discusses the structures of NMDARs and their mechanisms of activation.

13.3 STRUCTURE AND ACTIVATION

Functional NMDARs generally form as heterotetramers of two glycine-binding NR1 subunits and two glutamate-binding NR2 subunits assembled around a central permeation pathway. The inhibitory NR3 subunit, which also binds glycine, can substitute for one NR2 subunit or replace both to form a glycine-activated receptor, although the result would not technically be considered an NMDAR. Experiments in which NMDAR subunits were covalently linked as tandem dimers suggested that the heteromeric tetramer assembles according to an NR1–NR1–NR2–NR2 arrangement.[106] Assembly of the receptor complex is thought to proceed via a "dimer-of-dimers" mechanism. Whether the initial assembly step involves homomeric or heteromeric dimers[107,108] remains controversial. Furthermore, the NR1 and NR2 subunits exhibit significant sequence homology with each other and other iGluRs, and are therefore expected to adopt similar overall domain structures.

13.3.1 DOMAIN STRUCTURE OF SUBUNITS

Ionotropic glutamate receptor subunits are organized into four discrete functional domains (Figure 13.3): an extracellular N-terminal domain (NTD), a ligand binding domain (LBD), a pore forming transmembrane region, and an intracellular C-terminal domain. Hydrophobicity analysis performed after expression cloning of the first iGluR (AMPAR GluR1)[109] identified four hydrophobic segments (M1 through M4), the first three of which are closely spaced and separated by a long linker from the fourth segment.

The three-plus-one hydrophobicity profile is very similar to previously characterized ligand gated ion channels that include the nicotinic acetylcholine receptor (nAChR) family, $GABA_A$, glycine, and 5-HT3 receptors. It was therefore initially assumed that iGluRs would have the same membrane topology as these receptors, in which each hydrophobic domain is thought to form an α helix that crosses the membrane once.[109,110] The N-termini of all ligand-gated ion channels contain signal peptides, ensuring their extracellular localization. Therefore, in the membrane topology model for the nAChR super-family, the linker between the third and fourth transmembrane segment is predicted to be cytoplasmic.

The similarities of the nAChR and iGluR hydrophobicity profiles were later proven to be "red herrings." A functional N-glycosylation site was identified in the M3–M4 linker of the GluR6 kainate receptor.[111,112] Since the enzymes responsible for sugar modification of asparagine residues reside in the lumen of the Golgi, N-glycosylation is a reliable marker of extracellular localization. Initially, a new topology model was proposed with an additional transmembrane segment in the middle of the M3–M4 linker.[111,112] However, functional N-glycosylation was also observed in the M3–M4 linker of kainate receptors from goldfish brain, and modification was not affected by deletion of the M2 segment, prompting the authors to conclude that M2 does not cross the membrane and the M3–M4 linker is extracellular.[113]

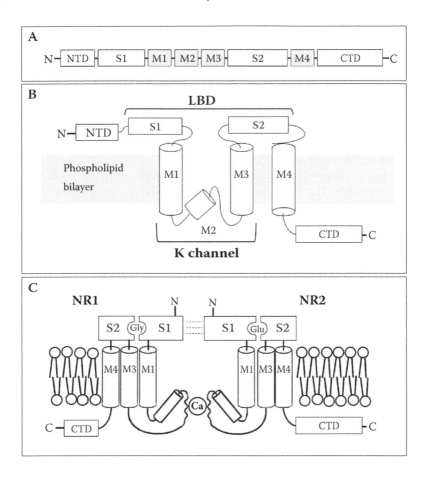

FIGURE 13.3 Domain structure of the NMDAR. A: Modular nature of linear amino acid sequences of NMDAR subunits: four hydrophobic domains (M1 through M4), two ligand binding domains (S1 and S2), and amino and carboxy terminal domains (NTD, CTD). B: Membrane topology of an individual NMDAR subunit. The amino terminus contains a signal peptide, placing the NTD in the extracellular space. M1, M3, and M4 are predicted to cross the membrane as helices. The M2 segment is predicted to form a cytoplasmic reentrant hairpin loop that connects M1 and M3. The ligand binding domains S1 and S2 are both extracellular. The CTD localizes to the cytoplasm. C: Two-dimensional membrane folding model of glycine binding NR1 and glutamate binding NR2 subunits. The ligand binding domains form a heteromeric dimer whose interface is formed by the D1 lobe.

Similar results were obtained from N-glycosylation studies of the GluR1 AMPAR and NR1 NMDAR subunits, confirming that the M2 segment does not cross the membrane.[94,114] Mutagenesis experiments on the NMDAR NR1 subunit identified residues in the M3–M4 linker critical for binding of glycine site ligands, indicating it must be extracellular.[115] An identical conclusion was reached by exchanging domains between the GluR3 AMPAR and GluR6 kainate receptor.[116] These studies also observed that two glutamate receptor domains display homology to bacterial periplasmic amino acid–binding proteins for which X-ray crystal structures already existed. This homology was first noted by Nakanishi et al.[117] and extended by O'Hara et al.[118] The first domain (S1) immediately precedes M1, while the second (S2) comprises most of the M3–M4 linker[115,116] (Figure 13.3A).

An extracellular location of the M3–M4 segment implies that the first three hydrophobic domains must span the membrane an even number of times. This discrepancy was explained by the significant amino acid sequence homology discovered between the M2 segments of iGluRs and the pore forming P regions of K channels[94], which fold as a reentrant hairpin loop. Figure 13.3 illustrates the current model for the membrane topology of NMDAR subunits.[94,114]

The extracellular domains come together to form a ligand binding domain (LBD) consisting of the S1 segment preceding M1 and the S2 segment sandwiched between M3 and M4. The NTD is located N terminal to S1 and exhibits homology with metabotropic glutamate receptor (mGluR) binding domains, suggesting that it forms a ligand binding structure distinct from the agonist binding S1–S2 domain. In the NR2 subunits, the NTD has been proposed to bind zinc ions (NR2A) or polyamines (NR2B), both of which modulate NMDAR function (see Chapters 11 and 12). The ligand for the NTD of the NR1 subunit, if one exists, is currently not known.

13.3.2 STRUCTURE OF THE LIGAND BINDING DOMAIN

The modular nature of glutamate receptors prompted experiments to isolate the ligand binding module as a separate soluble protein. Early experiments with the GluR4 AMPAR demonstrated that the isolated LBD consisting of the S1 and S2 domains connected by a short peptide linker retained the pharmacology of the full-length receptor, binding antagonists and agonists with normal affinity.[119] This approach enabled the crystallization of the GluR2 AMPAR LBD in complex with kainate,[120] illustrated in Figure 13.4A.

The bilobate structure consists of two globular domains or lobes (D1 and D2), connected by a flexible hinge, and bears a striking similarity to the bacterial periplasmic amino acid–binding proteins. The S1 segment forms lobe 1 (purple) and hinge 1 (blue) that connect to the first transmembrane segment M1 in the intact receptor. S2 forms lobe 2 (green) and hinge 2 (yellow), which crosses back to lobe 1. Two helices (J and K, gray) follow hinge 2, running across the backs of the two lobes and connecting to the M4 segment. The kainate agonist is sandwiched between the two domains and forms hydrogen bonds with both lobes, thereby stabilizing the closed cleft conformation.

GluR2 structures were obtained in the absence of ligand (apo) and in complex with an antagonist (DNQX), the partial agonist kainate, and full agonists AMPA and glutamate.[121] Separation of the two domains was significantly increased in the apo state, and all ligands tested produced some amount of cleft closure. The degree of domain closure increased as follows: apo < DNQX < kainate < AMPA = glutamate. This result led to the suggestion that agonist-induced domain closure activates the channel and that degree of domain closure determines the extent of activation. Although the competitive DNQX antagonist still produces a small amount of domain closure relative to the apo state, it apparently is not sufficient to activate the channel. Section 13.3.7 discusses the structural basis for partial agonism in more detail.

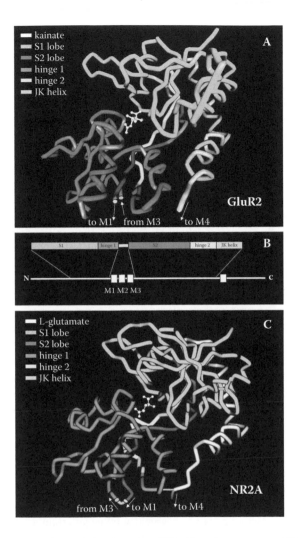

FIGURE 13.4 (See color insert following page 212.) Crystal structures of GluR2 and NR2A ligand binding domains. X-ray crystallographic structures are shown for the isolated ligand binding domains (LBDs) of the AMPAR GluR2 in complex with kainate (A) and the NMDAR NR2A subunit in complex with L-glutamate (C). Only the peptide backbone is shown as a C-α trace. (B) Relationship of LBDs and linear amino acid sequence.

Crystallographic structures for the S1–S2 LBD were obtained for additional members of the iGluR family including the GluR5 and GluR6 kainate receptors,[122] GluR0,[123] a prokaryotic glutamate-gated K channel, and the NMDAR subunits NR1[124] and NR2A.[125] All LBDs have the same basic structure, although certain conformational differences exist between the lobes and hinge regions. Figure 13.4C shows the LBD structure of the NMDAR NR2A subunit in complex with glutamate; this can be compared with the structure of GluR2 in Figure 13.4A. The availability of high-resolution crystal structures greatly facilitates the design of specific point mutations and the interpretations of resulting phenotypes.

13.3.3 Affinity and Efficacy

The interaction of a ligand with its receptor is characterized by two fundamental pharmacodynamic properties: affinity and efficacy. Affinity measures how tightly a ligand binds to a receptor and is characterized by the equilibrium dissociation constant, K_D. Efficacy measures how effectively a ligand, once bound, activates the receptor. Antagonists have no efficacy. The efficacies of partial agonists are lower than those of full agonists binding to the same site. The processes of agonist binding and receptor activation are strongly coupled, with the former initiating the latter.

This coupling complicates the interpretation of phenotypes caused by mutations. A point mutation that causes a shift in the concentration–response curve can do so without affecting binding, because the midpoint of the curve (EC_{50}) is dependent on both the binding equilibrium (affinity) and the activation process (efficacy).[126] Examples are provided by the previously published mutations in the pore of the nicotinic acetylcholine receptor, many of which altered acetylcholine sensitivity by affecting open state stability.[127,128] Another complication is that a change in efficacy caused by a mutation does not necessarily result in a measurable change in maximum response. When efficacy is very high, even substantial changes in efficacy exert minimal effects on maximum response. Moreover, when a change in maximum response *is* observed for a mutation, it is often difficult to exclude altered protein folding efficiency, protein stability, and inefficient expression as possible explanations.

To solve this dilemma, we developed an approach that allows the evaluation of affinity and efficacy roles for individual amino acid positions in a receptor. First, the position to be investigated is mutated to cysteine, which ideally should have little effect on receptor pharmacology. Second, concentration–response curves are collected for a full agonist and a partial agonist before and after covalent modification of the introduced cysteine.[129] Due to their small size, MTS (methanethiosulfonate) compounds are very useful for this purpose. These experiments yield values for the EC_{50} and intrinsic activity (α), a measure of relative efficacy, before and after covalent modification of the cysteine.

The *in situ* mutagenesis produced by MTS modification does not alter the population of receptors studied because it typically takes less than a minute to complete; therefore, any change in maximum response must result from an alteration in efficacy. Using a partial agonist guarantees that any changes in efficacy caused by MTS modification will reflect a change in intrinsic activity. Based on the values of EC_{50} and α before and after MTS modification, it is possible to quantitate changes in affinity and efficacy caused by the modification.[129]

Using this approach, Kalbaugh *et al.* assigned affinity and efficacy roles to positions in the LBDs of the NR1 and NR2A NMDAR subunits. *In situ* mutation of residues in direct contact with bound ligand affected both efficacy and affinity, while positions that stabilize the closed cleft conformation without a direct ligand interaction contributed only to efficacy.[129] These results provide a molecular basis for the tight coupling of agonist binding and receptor activation. The same residues that mediate stabilization of the closed cleft conformation by the bound ligand are also critical for ligand binding to the apo state, resulting in an agonist-bound open cleft conformation.

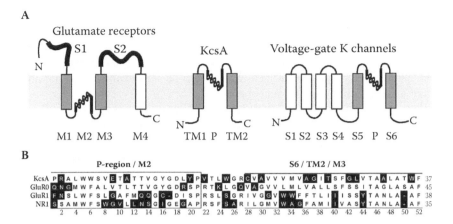

FIGURE 13.5 Relationship of glutamate receptors and K channels. A: Comparison of membrane topologies of ionotropic glutamate receptors, the prokaryotic K channel KcsA, and voltage-gated K channels. A common pore forming motif (M1–M2–M3, TM1-P-TM2, S5-P–S6) is shown in gray. The P regions in glutamate receptors are inverted, resulting in cytoplasmic localization of the reentrant hairpin loop M2. B: Amino acid homology of pore forming regions of KcsA, prokaryotic glutamate-gated K channel GluR0, AMPAR subunit GluR1, and NMDAR receptor subunit NR1. Nonconserved amino acids are shown as white symbols on a black background. In GluR0, 45 of 51 amino acids are conserved; it represents a missing link in the evolution of glutamate receptors and K channels from a common ancestor.

Crystallization of the isolated and soluble S1 and S2 LBDs for several iGluRs has significantly increased our understanding of the molecular bases of ligand binding and receptor activation mechanisms. The structure and function of the membrane domain (M1 through M4) will be discussed next.

13.3.4 STRUCTURE OF THE MEMBRANE DOMAIN: K CHANNEL AS MODEL

High-resolution structures are available for many LBDs of iGluRs but intact receptors, which are large integral membrane proteins, have thus far resisted crystallization. However, the M1–M3 segment appears to be structurally and functionally related to K channels, for several of which X-ray structures do exist.[130–132] Wo and Oswald (1994) originally hinted at a structural relationship between K channels and iGluRs by suggesting that the kainate receptor M2 segment folds back into the membrane similar to the P regions of K channels[113] (Figure 13.5A). N-glycosylation experiments in the M3–M4 linker by Wood et al. (1995) were prompted by a high degree of amino acid homology between NMDAR M2 segments and K channel P regions.[94]

Interestingly, the K channel GYGD signature sequence that forms part of the selectivity filter is replaced in NR1 with GIGE (Figure 13.5B). However, transplantation of the selectivity filter sequence TVGYG from K channels into iGluRs failed to transfer K selectivity to these channels, although functional channels were obtained and many pore properties were altered.[133] If the transplanted TVGYG regions fold similarly in iGluRs and K channels, as suggested by the functionality of the chimeras, it is likely that permeating ions are coordinated by backbone carbonyls in both channel families. Because iGluRs do not distinguish between Na^+ and K^+ ions and some members conduct divalent Ca^{2+} ions, the atomic distance between backbone carbonyls forming the oxygen cage should be significantly more flexible and dynamic in these nonselective cation channels.

The amino acid homology in the pore-forming regions prompted us to suggest that K channels and iGluRs have a common evolutionary ancestor.[94] This is supported by the identification of a "missing link," the prokaryotic GluR0 glutamate receptor that has a pore-forming domain with high homology to K channel P regions and exhibits K^+ selectivity.[134] Using the GluR0 amino acid sequence as a guide, the homology of K channels and iGluRs can be extended to the transmembrane domain following the pore region: M3 in iGluRs and S6 or TM2 in K channels[135] (Figure 13.5B). In K channels, the S6 and TM2 segments undergo conformational changes during channel activation.[136–138] Evidence suggests that the iGluR M3 segment may play a similarly important role during receptor activation.

13.3.5 Role of the M3 Domain

The M3 segment contains a 9-amino acid sequence (SYTANLAAF) that is highly conserved among glutamate receptors; many cysteine substitutions in this region display state-dependent accessibility.[135,139] Altered residue accessibility often indicates a conformational change,[140] suggesting that the SYTANLAAF region moves in response to receptor activation; thus, M3 has been proposed to function as a transduction element, coupling ligand-binding to channel gating.[135] A recent study suggests that M3 is the only transmembrane domain contributing to the deepest portion of the pore, supporting a prominent role in gating.[141]

The functional importance of the SYTANLAAF motif was originally identified in the GluRδ2 receptor, an orphan receptor with homology to the iGluR family, but no known glutamatergic agonist. A single-point mutation, A8T, was found to cause an inherited neurological defect in mice.[142] Known as *lurcher*, the phenotype, characterized by ataxia and neurodegeneration, is caused by constitutive activation of δ2 receptors, which produces excitotoxicity and apoptosis of cerebellar Purkinje cells. Introduction of the *lurcher* mutation in GluR1 and GluR6 produced some constitutive activity, increased agonist potency, slower deactivation, and conversion of an antagonist into an agonist.[143–145] The same substitution in NR1 and NR2A did not produce constitutive activity, but exhibited very slow deactivation; interestingly, increased glutamate potency was observed only in the NR1 mutant.[143]

Similar to the *lurcher* phenotype, constitutive activity and/or current potentiation were observed upon thiol modification of substituted cysteines at several positions within the SYTANLAAF region.[135,139,146–148] Of particular interest is the A7C substitution in NR1 and NR2 that appears to lock the channel in a fully activated conformation upon modification.[135,139,149] A7C can be modified only after receptor activation and accessibility exhibits a linear correlation with agonist efficacy,[135] strongly suggesting a role for M3 in coupling ligand binding to channel gating. Conservation of the A7C phenotype is observed between NR1 and NR2 and across NR2 subtypes, implying functional conservation of M3 among all NMDARs.[139]

Several lines of evidence suggest that the NR1 and NR2 M3 domains may play distinct roles in receptor activation. NR1 and NR2 contribute differently to the M2 loop[150] that forms the inner pore of the channel, and copper coordination of substituted cysteines indicates that their M3 domains may be staggered by a full α helical turn.[151] Consistent with a dimer-of-dimers arrangement, the extracellular vestibule of the homomeric AMPAR exhibits two-fold rotational symmetry, as opposed to the four-fold symmetry of K^+ channels.[152] Recent work evaluating the voltage dependence of cysteine modification suggests that the NR1 M3 domain contributes mainly to the deep portion of the pore, while the NR2 M3 comprises more of the shallow extracellular vestibule.[141]

These subtle structural differences among subunits may be critical for NMDAR activation; for example, kinetic modeling with partial agonists revealed that distinct fast and slow pregating conformational changes are mediated by NR1 and NR2, respectively.[153] Residues displaying state-dependent accessibility were found to cluster together on helical net diagrams, opposite from putative pore-lining residues, suggesting that positions exhibiting *lurcher*-like phenotypes may be located at dynamic interfaces between transmembrane segments.[141] Thus intersubunit interactions at the transmembrane level may be critical for NMDAR activation.

13.3.6 DIMERIZATION OF LIGAND BINDING DOMAINS

One of the most intriguing findings from iGluR structural studies is the tendency of S1S2 LBDs to crystallize as homomeric dimers, initially observed in a set of GluR2 crystal structures.[121] *In vitro*, isolated iGluR LBDs dimerize with a dissociation constant in the millimolar range, most likely due to high protein concentrations. This effect may be mimicked *in vivo* by association of the N terminal domains (NTDs) that occurs at lower protein concentrations.[154] In AMPARs and kainate receptors, the dimer interface is formed exclusively by hydrogen bonds, salt bridges, and hydrophobic interactions between the D1 domains, burying 900 to 1600 Å² of solvent-accessible surface area.[121,155] The D2 domain linking the LBDs to the ion channel portion of the receptor is free to change conformation, suggesting a mechanism by which LBD conformation may be coupled to channel gating.

The functional importance of the LBD dimer interface was initially demonstrated via site-directed mutagenesis, which was used to modulate interactions at the GluR2 interface. Single-point mutations were identified which either attenuated or increased desensitization. Crystallography, ultracentrifugation, and electrophysiological studies established that dimer stability inversely correlates with extent of desensitization.[156] Subsequent studies confirmed and extended this paradigm in AMPARs and kainate receptors.[157,158]

Domain–domain (D1-D1) separation associated with GluR2 desensitization has been estimated to be between 12.4 and 16.2 Å at the top of the interface, based on modification of cysteine interface mutants via bifunctional crosslinkers. Identification of a GluR2 mutant stabilized in the desensitized state (S729C) led to a crystal structure depicting a relaxed, destabilized dimer interface.[158] Crystal structures of GluR2 in complex with several positive allosteric modulators revealed that these molecules exert their potentiating effects via interface stabilization. Cyclothiazide, which blocks receptor desensitization, binds near the edge of the interface, while aniracetam stabilizes the open state by binding near the LBD hinge.[156,159]

The NR1 subunit repeatedly crystallized in monomeric form until the recent cocrystallization of NR1 and NR2A revealed the existence of a heteromeric LBD dimer interface.[125] In contrast to other iGluRs, the LBD dimer interface of the NMDAR contains both D1 and D2 interactions, providing additional opportunities for intersubunit coupling. Furthermore, a critical tyrosine residue in NR1 (Y535) occupies a site homologous to the GluR2 aniracetam binding site and the sizes of substituted amino acids at this position were found to inversely correlate with deactivation rate. Thus, Y535 appears to function as an endogenous positive allosteric modulator. This key difference in interface stability has been proposed to underlie the slow deactivation of NMDARs required for their role in synaptic transmission,[160] suggesting a novel modulatory site for NMDAR-based therapeutics.

The NTDs are clamshell-shaped structures, known to interact with each other and modulate the LBDs, suggesting a possible extension of the interface model.[154] Zn^{2+} binding to the NR2A NTD is allosterically coupled to the glutamate-binding domain. Similar results were obtained for ifenprodil in NR2B.[161,162] An NTD dimer interface may provide the stability for binding of negative modulators to effectively couple to LBDs.

13.3.7 STRUCTURAL BASIS OF PARTIAL AGONISM

Crystal structures of the GluR2 S1S2 domain, in complex with a range of full and partial agonists, established a strong correlation between agonist efficacy and degree of LBD closure. A similar relationship was observed in the GluR6 kainate receptor.[105,122] These results led to a structural model of iGluR activation, in which agonist binding induces LBD closure by rotating D2 toward D1, separating the linker regions and promoting channel opening. According to this hypothesis, partial agonists induce less domain closure and consequently less linker separation, slowing a subunit-specific pregating conformational change.[163]

Engineering steric clashes in the LBD, thus destabilizing the closed cleft state, reduces agonist efficacy and apparent affinity in NR2, GluR2, and GluR6,[164–166] while elimination of agonist–cleft clashes in GluR2 reportedly slows receptor deactivation and increases affinity and efficacy.[167] Stabilization of the NR2 closed cleft state via modulation of an endogenous D1–D2 interaction can also increase open probability, kinetically linking LBD closure and channel gating.[168] Interestingly, crystal structure data obtained from mutagenesis studies has not always followed the cleft closure–agonist efficacy correlation; GluR2 L650T, for example, stabilizes the closed cleft state, increasing kainate-induced efficacy and degree of domain closure, but the AMPA-bound structure revealed both partially and fully closed conformations.[167] A cleft stabilizing GluR6 D1–D2 interaction dramatically increased glutamate sensitivity and slowed deactivation when introduced in GluR2, but did not affect cleft closure. Full agonist-bound GluR6 is almost 6° more closed relative to GluR2.[164]

Crystal structures obtained for the NR1 subunit revealed that full and partial agonists adopt similar degrees of domain closure, indicating that partial agonist action at an NMDAR may follow a different structural paradigm from other iGluRs.[169] Because no partial agonist-bound crystal structure is presently available for the NR2 subunit, it is not yet known which model describes the behavior of the NR2 LBD. Both structural and molecular dynamics studies suggest that the LBD hinge region, particularly the second interdomain β strand, changes conformation according to agonist efficacy.[169,170] Helix F that forms part of the LBD dimer interface has been implicated in coupling agonist efficacy to channel gating in NR2A.[163] NR1-site partial agonism has also been correlated with increased inter-pocket motion, and both glycine and DCS can reportedly move within the pocket without affecting domain closure.[170]

Incomplete and unstable cleft closures have been proposed to affect single channel gating similarly,[166] suggesting that degree of cleft closure may simply be a physical readout of closed cleft stability. Thus, closed cleft stability, and not simply a physical change in conformation between D1 and D2, may be the principal determinant of agonist efficacy at iGluRs.

Understanding the structural basis of partial agonism is of both scientific and clinical interest, since NMDAR hypofunction has been shown to cause cognition and memory defects in animal models.[171,172] Glutamatergic dysfunction has been implicated in the pathophysiology of schizophrenia[173] and receptor augmentation may prove therapeutically valuable. Although both glycine and D-cycloserine (DCS), a well-tolerated partial glycine site agonist, initially exhibited some efficacy in the treatment of schizophrenia,[174–177] a recent multisite double blind randomized trial found both glycine and DCS ineffective for treating negative symptoms and cognitive impairments.[178] DCS has been shown to enhance learning and memory performance in adult and aged rodents[179–182] and monkeys,[183] and is undergoing testing as a cognitive enhancer for treating Alzheimer's disease,[184–186] head trauma,[187] and fear extinction.[188–193] Reported subtype-specific partial agonism at the NMDAR also opens the possibility of individually targeting NR2 subtypes that exhibit regional and developmental variations in expression.[89]

13.3.8 Positive Allosteric Modulators

Few positive modulators, endogenous or otherwise, have been found to act on the extracellular portions of NMDARs. Polyamines such as spermine and spermidine exert both inhibitory and stimulatory effects, depending on concentration and subunit composition (see Chapter 11). Polyamines are generally found within cells and whether they participate in physiological regulation of NMDARs[194] remains unclear. Pregnenolone sulfate (PS), a neurosteroid synthesized *de novo* in the CNS, also exhibits subunit-specific potentiation and inhibition.[195,196] A recent study suggests that PS is released in an activity-dependent manner and may attain synaptic concentrations in the micromolar range, suggesting a possible endogenous modulatory role.[197]

PS potentiates NR2A- and NR2B-containing NMDARs via increases in channel open probability; however, due to a glutamate-induced reduction of receptor affinity for PS, the effect is relatively transient.[198] In contrast, NMDARs containing NR2C or NR2D subunits are inhibited by PS, and all four NR2 subtypes are inhibited by pregnanolone sulfate, a closely related neurosteroid.[196] Planar and bent ring structures appear to favor stimulation and inhibition, respectively, while both necessitate a negatively charged C3 moiety.[199] PS binding has been localized to a steroid modulatory domain, SMD1, comprised of the M4 transmembrane domain and helix J/K located in the S2 segment of the ligand binding domain.[200] Since PS is a charged molecule and potentiation displays no voltage dependence,[201] helix J/K is a more likely candidate for direct binding. This region of S2 has recently been shown to form part of the LBD dimer interface, a region identified as a binding site for numerous positive allosteric modulators of AMPARs.[159]

PS treatment of neuronal cultures exacerbates NMDA-induced excitotoxicity,[199] while intracerebroventricular administration in mice has been shown to cause convulsions, although that effect may be due to inhibitory actions at the GABA$_A$ receptor.[202] Hippocampal PS concentration has been correlated with preserving cognitive functions in aging animals[203] and *in vivo* infusion of PS enhances neurogenesis in mice.[204] PS has also been shown to increase LTP in the hippocampus,[205] possibly providing a mechanism for its role as a cognitive enhancer.

Although the nonselective effects of PS may preclude its clinical use, it may serve as a promising starting point for therapeutic development. Positive allosteric modulators of AMPARS, known as ampakines, improve learning and memory and suppress symptoms of schizophrenia, depression, and ADHD in animal models. Initial studies in humans have also shown positive effects on memory and psychiatric symptoms.[206] Crystal structures of GluR2 in complex with two ampakines, aniracetam and CX614, revealed that both modulators bind within the LBD dimer interface, slowing deactivation and/or desensitization.[159] Ampakines are proposed to enhance cognitive function in a three-fold manner: expansion of cortical networks, facilitation of LTP, and upregulation of BDNF expression.[206] Based on the structural homology among iGluRs, and the role of NMDARs in LTP induction, the development of an NMDAkine is certainly within the realm of possibility.

13.4 CONCLUSIONS

NMDARs play critical roles in both the proper development of the central nervous system and the processes underlying functional and structural plasticity in the adult brain. They are able to perform these tasks because they possess a combination of unique properties: (1) high affinity for the excitatory transmitter L-glutamate, (2) very slow kinetics of (de)activation, (3) pronounced voltage dependence due to external Mg block, (4) high permeability to Ca ions, and (5) large cytoplasmic domains that enable them to become part of and help organize large macromolecular synaptic signaling complexes.

The high affinity for glutamate and relatively slow (de)activation kinetics allow NMDARs to decode synaptic input patterns over prolonged periods (Figure 13.1). Their voltage dependence enables them to act like molecular coincidence detectors, mediating calcium influx only when strong membrane depolarization coincides with synaptic release of glutamate. Ca ions entering through NMDARs act locally at signaling complexes associated with the receptor, to allow long-lasting modification of individual synapses. In addition to this highly localized action, activation of NMDARs in distal dendrites can signal to the nucleus to affect gene transcription by mechanisms that are not fully understood. Finally, the regenerative properties of NMDARs allow them to help generate and propagate dendritic depolarizations, resulting in nonlinear processing of synaptic inputs that may endow neurons with novel computational abilities.

Based on these unique properties, it is not surprising that NMDAR hypofunction or overstimulation can result in many cognitive defects and brain dysfunction, making these receptors prime therapeutic targets. Unfortunately, clinical results with NMDAR drugs have been limited because of a combination of lack of efficacy and unacceptable side effects. However, recent progress in elucidating the molecular mechanisms underlying activation of NMDARs, driven by the availability of high resolution X-ray structures for the ligand binding domains, is likely to revitalize the search for more effective and subtype-specific NMDAR drugs. Particularly promising are the allosteric modulators that either enhance or inhibit NMDAR function to a well-defined extent, without acting as (ant)agonists. Their inherent dose limiting ability indicates that allosteric modulators will be far better tolerated than the compounds investigated to date.

REFERENCES

1. Kemp, J.A. and McKernan, R.M., NMDAR pathways as drug targets, *Nature Neurosci.*, 5, 1039, 2002.
2. Jansen, M. and Dannhardt, G., Antagonists and agonists at the glycine site of the NMDAR for therapeutic interventions, *Eur. J. Med. Chem.*, 38, 661, 2003.
3. Chazot, P.L., The NMDAR NR2B subunit: a valid therapeutic target for multiple CNS pathologies, *Curr. Med. Chem.*, 11, 389, 2004.
4. Farlow, M.R., NMDAR antagonists. A new therapeutic approach for Alzheimer's disease, *Geriatrics*, 59, 22, 2004.
5. Wood, P.L., The NMDAR complex: a long and winding road to therapeutics, *IDrugs*, 8, 229, 2005.
6. Cai, S.X., Glycine/NMDAR antagonists as potential CNS therapeutic agents: ACEA-1021 and related compounds, *Curr. Topics Med. Chem.*, 6, 651, 2006.
7. Missale, C. et al., The NMDA/D1 receptor complex as a new target in drug development, *Curr. Topics Med. Chem.*, 6, 801, 2006.
8. Brown, D.G. and Krupp, J.J., N-methyl-D-aspartate receptor (NMDA) antagonists as potential pain therapeutics, *Curr. Topics Med. Chem.*, 6, 749, 2006.
9. Dingledine, R. et al., Glutamate receptor ion channels, *Pharmacol. Rev.*, 51, 7, 1999.
10. Rothstein, J.D. et al., Localization of neuronal and glial glutamate transporters, *Neuron*, 13, 713, 1994.
11. Huang, Y.H. and Bergles, D.E., Glutamate transporters bring competition to the synapse, *Curr. Opin. Neurobiol.*, 14, 346, 2004.
12. Nowak, L. et al., Magnesium gates glutamate-activated channels in mouse central neurones, *Nature*, 307, 462, 1984.
13. Mayer, M.L., Westbrook, G.L., and Guthrie, P.B., Voltage-dependent block by Mg of NMDA responses in spinal cord neurones, *Nature*, 309, 261, 1984.
14. Seeburg, P.H. et al., The NMDAR channel: molecular design of a coincidence detector, *Recent Prog. Hormone Res.*, 50, 19, 1995.
15. Bliss, T.V.P. and Lomo, T., Long-lasting potentiation of synaptic transmission in the dentate area of the anaesthetized rabbit following stimulation of the perforant path, *J. Physiol.*, 232, 331, 1973.
16. Collingridge, G.L. and Bliss, T.V., Memories of NMDARs and LTP, *Trends Neurosci.*, 18, 54, 1995.

17. Feldman, D.E., Nicoll, R.A., and Malenka, R.C., Synaptic plasticity at thalamocortical synapses in developing rat somatosensory cortex: LTP, LTD, and silent synapses, *J. Neurobiol.*, 41, 92, 1999.
18. Johnston, D., NMDA-receptor independent LTP, *Neurochem. Int.*, 20, 461, 1992.
19. Kapur, A. et al., L-Type calcium channels are required for one form of hippocampal mossy fiber LTP, *J. Neurophysiol.*, 79, 2181, 1998.
20. Dudek, S.M. and Bear, M.F., Homosynaptic long-term depression in area CA1 of hippocampus and effects of N-methyl-D-aspartate receptor blockade, *Proc. Natl. Acad. Sci. USA*, 89, 4363, 1992.
21. Bear, M.F. and Malenka, R.C., Synaptic plasticity: LTP and LTD, *Curr. Opin. Neurobiol.*, 4, 389, 1994.
22. Bortolotto, Z.A., Fitzjohn, S.M., and Collingridge, G.L., Roles of metabotropic glutamate receptors in LTP and LTD in the hippocampus, *Curr. Opin. Neurobiol.*, 9, 299, 1999.
23. Isaac, J., Protein phosphatase 1 and LTD: synapses are the architects of depression, *Neuron*, 32, 963, 2001.
24. Spruston, N. et al., Activity-dependent action potential invasion and calcium influx into hippocampal CA1 dendrites, *Science*, 268, 297, 1995.
25. Stuart, G. et al., Action potential initiation and backpropagation in neurons of the mammalian CNS, *Trends Neurosci.*, 20, 125, 1997.
26. Markram, H. et al., Regulation of synaptic efficacy by coincidence of postsynaptic APs and EPSPs, *Science*, 275, 213, 1997.
27. Debanne, D., Gahwiler, B.H., and Thompson, S.M., Long-term synaptic plasticity between pairs of individual CA3 pyramidal cells in rat hippocampal slice cultures, *J. Physiol.*, 507, 237, 1998.
28. Debanne, D., Gähwiler, B.H., and Thompson, S.M., Asynchronous pre- and postsynaptic activity induces associative long-term depression in area CA1 of the rat hippocampus *in vitro*, *Proc. Natl. Acad. Sci. USA*, 91, 1148, 1994.
29. Abbott, L.F. and Nelson, S.B., Synaptic plasticity: taming the beast, *Nat. Neurosci.*, 3, 1178, 2000.
30. Kepecs, A. et al., Spike-timing-dependent plasticity: common themes and divergent vistas, *Biol. Cybernet.*, 87, 446, 2002.
31. Lisman, J. and Spruston, N., Postsynaptic depolarization requirements for LTP and LTD: a critique of spike timing-dependent plasticity, *Nat. Neurosci.*, 8, 839, 2005.
32. Dan, Y. and Poo, M.M., Spike timing-dependent plasticity: from synapse to perception, *Physiol. Rev.*, 86, 1033, 2006.
33. Drew, P.J. and Abbott, L.F., Extending the effects of spike-timing-dependent plasticity to behavioral timescales, *Proc. Natl. Acad. Sci. USA*, 103, 8876, 2006.
34. Lin, Y.W. et al., Spike-timing-dependent plasticity at resting and conditioned lateral perforant path synapses on granule cells in the dentate gyrus: different roles of N-methyl-D-aspartate and group I metabotropic glutamate receptors, *Eur. J. Neurosci.*, 23, 2362, 2006.
35. Gerkin, R.C. et al., Modular competition driven by NMDAR subtypes in spike-timing-dependent plasticity, *J. Neurophysiol.*, 97, 2851, 2007.
36. Bi, G. et al., Synaptic modification by correlated activity: Hebb's postulate revisited, *Annu. Rev. Neurosci.*, 24, 139, 2001.
37. van Rossum, M.C., Bi, G.Q., and Turrigiano, G.G., Stable Hebbian learning from spike timing-dependent plasticity, *J. Neurosci.*, 20, 8812, 2000.
38. Schiller, J. et al., NMDA spikes in basal dendrites of cortical pyramidal neurons, *Nature*, 404, 285, 2000.
39. Schiller, J. and Schiller, Y., NMDAR-mediated dendritic spikes and coincident signal amplification, *Curr. Opin. Neurobiol.*, 11, 343, 2001.

40. Enoki, R. et al., NMDAR-mediated depolarizing after-potentials in the basal dendrites of CA1 pyramidal neurons, *Neurosci. Res.*, 48, 325, 2004.
41. Gordon, U., Polsky, A., and Schiller, J., Plasticity compartments in basal dendrites of neocortical pyramidal neurons, *J. Neurosci.*, 26, 12717, 2006.
42. Nevian, T. et al., Properties of basal dendrites of layer 5 pyramidal neurons: a direct patch-clamp recording study, *Nat. Neurosci.*, 10, 206, 2007.
43. London, M. and Hausser, M., Dendritic computation, *Annu. Rev. Neurosci.*, 28, 503, 2005.
44. Rhodes, P., The properties and implications of NMDA spikes in neocortical pyramidal cells, *J. Neurosci.*, 26, 6704, 2006.
45. Gasparini, S. and Magee, J.C., State-dependent dendritic computation in hippocampal CA1 pyramidal neurons, *J. Neurosci.*, 26, 2088, 2006.
46. Lisman, J., A mechanism for the Hebb and the anti-Hebb processes underlying learning and memory, *Proc. Natl. Acad. Sci. USA*, 86, 9574, 1989.
47. Brickley, S.G. et al., NR2B and NR2D subunits coassemble in cerebellar Golgi cells to form a distinct NMDAR subtype restricted to extrasynaptic sites, *J. Neurosci.*, 23, 4958, 2003.
48. Thomas, C.G., Miller, A.J., and Westbrook, G.L., Synaptic and extrasynaptic NMDAR NR2 subunits in cultured hippocampal neurons, *J. Neurophysiol.*, 95, 1727, 2006.
49. Hardingham, G.E. and Bading, H., The Yin and Yang of NMDAR signalling, *Trends Neurosci.*, 26, 81, 2003.
50. Vanhoutte, P. and Bading, H., Opposing roles of synaptic and extrasynaptic NMDARs in neuronal calcium signalling and BDNF gene regulation, *Curr. Opin. Neurobiol.*, 13, 366, 2003.
51. Liu, L. et al., Role of NMDAR subtypes in governing the direction of hippocampal synaptic plasticity, *Science*, 304, 1021, 2004.
52. Massey, P.V. et al., Differential roles of NR2A and NR2B-containing NMDARs in cortical long-term potentiation and long-term depression, *J. Neurosci.*, 24, 7821, 2004.
53. Berberich, S. et al., Lack of NMDAR subtype selectivity for hippocampal long-term potentiation, *J. Neurosci.*, 25, 6907, 2005.
54. Morishita, W. et al., Activation of NR2B-containing NMDARs is not required for NMDAR-dependent long-term depression, *Neuropharmacology*, 52, 71, 2007.
55. Husi, H. and Grant, S.G., Proteomics of the nervous system, *Trends Neurosci.*, 24, 259, 2001.
56. Husi, H. et al., Proteomic analysis of NMDAR-adhesion protein signaling complexes, *Nat. Neurosci.*, 3, 661, 2000.
57. Kennedy, M.B., Signal-processing machines at the postsynaptic density, *Science*, 290, 750, 2000.
58. Brambilla, R. et al., A role for the Ras signalling pathway in synaptic transmission and long-term memory, *Nature*, 390, 281, 1997.
59. Gnegy, M.E., Ca2+/calmodulin signaling in NMDA-induced synaptic plasticity, *Crit. Rev. Neurobiol.*, 14, 91, 2000.
60. Sheng, M., The postsynaptic NMDA-receptor-PSD-95 signaling complex in excitatory synapses of the brain, *J. Cell Sci.*, 114, 1251, 2001.
61. Sweatt, J.D., The neuronal MAP kinase cascade: a biochemical signal integration system subserving synaptic plasticity and memory, *J. Neurochem.*, 76, 1, 2001.
62. Riccio, A. and Ginty, D.D., What a privilege to reside at the synapse: NMDAR signaling to CREB, *Nat. Neurosci.*, 5, 389, 2002.
63. Deisseroth, K. et al., Signaling from synapse to nucleus: the logic behind the mechanisms, *Curr. Opin. Neurobiol.*, 13, 354, 2003.

64. Johnson, J.W. and Ascher, P., Glycine potentiates the NMDA response in cultured mouse brain neurons, *Nature*, 325, 529, 1987.
65. Kleckner, N.W. and Dingledine, R., Requirement for glycine in activation of NMDARs expressed in *Xenopus* oocytes, *Science*, 241, 835, 1988.
66. Dingledine, R., Kleckner, N.W., and McBain, C.J., The glycine coagonist site of the NMDAR, *Adv. Exp. Med. Biol.*, 268, 17, 1990.
67. Panatier, A. et al., Glia-derived D-serine controls NMDAR activity and synaptic memory, *Cell*, 125, 775, 2006.
68. Ascher, P., Measuring and controlling the extracellular glycine concentration at the NMDAR level, *Adv. Exp. Med. Biol.*, 268, 13, 1990.
69. Hood, W.F., Compton, R.P., and Monahan, J.B., D-cycloserine: a ligand for the N-methyl-D-aspartate coupled glycine receptor has partial agonist characteristics, *Neurosci. Lett.*, 98, 91, 1989.
70. Henderson, G., Johnson, J.W., and Ascher, P., Competitive antagonists and partial agonists at the glycine modulatory site of the mouse N-methyl-D-aspartate receptor, *J. Physiol.*, 430, 189, 1990.
71. Baran, H., Loscher, W., and Mevissen, M., The glycine/NMDAR partial agonist D-cycloserine blocks kainate-induced seizures in rats: comparison with MK-801 and diazepam, *Brain Res.*, 652, 195, 1994.
72. Rundfeldt, C., Wla, P., and Loscher, W., Anticonvulsant activity of antagonists and partial agonists for the NMDAR-associated glycine site in the kindling model of epilepsy, *Brain Res.*, 653, 125, 1994.
73. Seguin, L. and Millan, M.J., The glycine B receptor partial agonist, (+)-HA966, enhances induction of antinociception by RP 67580 and CP-99,994, *Eur. J. Pharmacol.*, 253, R1, 1994.
74. Tricklebank, M.D. et al., The anticonvulsant and behavioural profile of L-687,414, a partial agonist acting at the glycine modulatory site on the N-methyl-D-aspartate (NMDA) receptor complex, *Brit. J. Pharmacol.*, 113, 729, 1994.
75. Pussinen, R. et al., Enhancement of intermediate-term memory by an alpha-1 agonist or a partial agonist at the glycine site of the NMDAR, *Neurobiol. Learn. Mem.*, 67, 69, 1997.
76. Priestley, T. et al., L-687,414, a low efficacy NMDAR glycine site partial agonist *in vitro*, does not prevent hippocampal LTP *in vivo* at plasma levels known to be neuroprotective, *Brit. J. Pharmacol.*, 124, 1767, 1998.
77. Jones, K.S., VanDongen, H.M., and VanDongen, A.M., The NMDAR M3 segment is a conserved transduction element coupling ligand binding to channel opening, *J. Neurosci.*, 22, 2044, 2002.
78. Ballard, T.M. et al., Severe impairment of NMDAR function in mice carrying targeted point mutations in the glycine binding site results in drug-resistant nonhabituating hyperactivity, *J. Neurosci.*, 22, 6713, 2002.
79. Zylberman, I., Javitt, D.C., and Zukin, S.R., Pharmacological augmentation of NMDAR function for treatment of schizophrenia, *Ann. NY Acad. Sci.*, 757, 487, 1995.
80. Jentsch, J.D. and Roth, R.H., The neuropsychopharmacology of phencyclidine: from NMDAR hypofunction to the dopamine hypothesis of schizophrenia, *Neuropsychopharmacology*, 20, 201, 1999.
81. Coyle, J.T. and Tsai, G., NMDAR function, neuroplasticity, and the pathophysiology of schizophrenia, *Int. Rev. Neurobiol.*, 59, 491, 2004.
82. Pilowsky, L.S. et al., First *in vivo* evidence of an NMDAR deficit in medication-free schizophrenic patients, *Mol. Psych.*, 11, 118, 2006.
83. Neher, E. and Sakmann, B., Single-channel currents recorded from membrane of denervated frog muscle fibres, *Nature*, 260, 799, 1976.

84. Wright, J.M. and Peoples, R.W., NMDAR pharmacology and analysis of patch-clamp recordings, *Meth. Mol. Biol.*, 128, 143, 1999.
85. Rock, D.M. and Macdonald, R.L., The polyamine spermine has multiple actions on N-methyl-D-aspartate receptor single-channel currents in cultured cortical neurons, *Mol. Pharmacol.*, 41, 83, 1992.
86. Wright, J.M., Peoples, R.W., and Weight, F.F., Single-channel and whole-cell analysis of ethanol inhibition of NMDA-activated currents in cultured mouse cortical and hippocampal neurons, *Brain Res.*, 738, 249, 1996.
87. Gibb, A.J. and Colquhoun, D., Glutamate activation of a single NMDAR-channel produces a cluster of channel openings, *Proc. R. Soc. Lond. (Biol.)*, 243, 39, 1991.
88. Popescu, G. et al., Reaction mechanism determines NMDAR response to repetitive stimulation, *Nature*, 430, 790, 2004.
89. Erreger, K. et al., Subunit-specific gating controls rat NR1/NR2A and NR1/NR2B NMDA channel kinetics and synaptic signalling profiles, *J. Physiol.*, 563, 345, 2005.
90. Jahr, C.E. and Stevens, C.F., Glutamate activates multiple single channel conductances in hippocampal neurons, *Nature*, 325, 522, 1987.
91. Stern, P. et al., Single-channel conductances of NMDARs expressed from cloned cDNAs: Comparison with native receptors, *Proc. R. Soc. Lond. (Biol.)*, 250, 271, 1992.
92. Stern, P. et al., Single channel properties of cloned NMDARs in a human cell line—comparison with results from *Xenopus* oocytes, *J. Physiol.*, 476, 391, 1994.
93. Wyllie, D.J. et al., Single-channel currents from recombinant NMDA NR1a/NR2D receptors expressed in *Xenopus* oocytes, *Proc. R. Soc. Lond. (Biol.)*, 263, 1079, 1996.
94. Wood, M.W., VanDongen, H.M.A., and VanDongen, A.M.J., Structural conservation of ion conduction pathways in K channels and glutamate receptors, *Proc. Natl. Acad. Sci. USA*, 92, 4882, 1995.
95. Chapman, M.L., VanDongen, H.M.A., and VanDongen, A.M.J., Activation-dependent subconductance levels in K channels suggest a subunit basis for ion permeation and gating, *Biophys. J.*, 72, 708, 1997.
96. Chapman, M.L. and VanDongen, A.M., K channel subconductance levels result from heteromeric pore conformations, *J. Gen. Physiol.*, 126, 87, 2005.
97. VanDongen, A.M.J., Structure and function of ion channels: a hole in four? *Comm. Theor. Biol.*, 2, 429, 1992.
98. VanDongen, A.M., K channel gating by an affinity-switching selectivity filter, *Proc. Natl. Acad. Sci. USA*, 101, 3248, 2004.
99. Baker, K.A. et al., Conformational dynamics of the KcsA potassium channel governs gating properties, *Nat. Struc. Mol. Biol.*, 14, 1089, 2007.
100. Rosenmund, C., Stern-Bach, Y., and Stevens, C.F., The tetrameric structure of a glutamate receptor channel, *Science*, 280, 1596, 1998.
101. Zheng, J. and Sigworth, F.J., Selectivity changes during activation of mutant Shaker potassium channel, *J. Gen. Physiol.*, 110, 101, 1997.
102. Premkumar, L.S., Qin, F., and Auerbach, A., Subconductance states of a mutant NMDAR channel kinetics, calcium, and voltage dependence, *J. Gen. Physiol.*, 109, 181, 1997.
103. Zheng, J. and Sigworth, F.J., Intermediate conductances during deactivation of heteromultimeric Shaker potassium channels, *J. Gen. Physiol.*, 112, 457, 1998.
104. Schneggenburger, R. and Ascher, P., Coupling of permeation and gating in an NMDA-channel pore mutant, *Neuron*, 18, 167, 1997.
105. Jin, R. et al., Structural basis for partial agonist action at ionotropic glutamate receptors, *Nat. Neurosci.*, 6, 803, 2003.

106. Schorge, S. and Colquhoun, D., Studies of NMDAR function and stoichiometry with truncated and tandem subunits, *J. Neurosci.*, 23, 1151, 2003.
107. Papadakis, M., Hawkins, L.M., and Stephenson, F.A., Appropriate NR1-NR1 disulfide-linked homodimer formation is requisite for efficient expression of functional, cell surface N-methyl-D-aspartate NR1/NR2 receptors, *J. Biol. Chem.*, 279, 14703, 2004.
108. Schuler, T. et al., Formation of NR1/NR2 and NR1/NR3 heterodimers constitutes initial step in N-methyl-D-Aspartate receptor assembly, *J. Biol. Chem.*, in 283, 37, 2008.
109. Hollmann, M. et al., Cloning by functional expression of a member of the glutamate receptor family, *Nature*, 342, 643, 1989.
110. Keinanen, K. et al., A family of AMPA-selective glutamate receptors, *Science*, 249, 556, 1990.
111. Roche, K.W. et al., Transmembrane topology of the glutamate receptor subunit GluR6, *J. Biol. Chem.*, 269, 11679, 1994.
112. Taverna, F.A. et al., A transmembrane model for an ionotropic glutamate receptor predicted on the basis of the location of asparagine-linked oligosaccharides, *J. Biol. Chem.*, 269, 14159, 1994.
113. Wo, Z.G. and Oswald, R.E., Transmembrane topology of two kainate receptor subunits revealed by N-glycosylation, *Proc. Natl. Acad. Sci. USA*, 91, 7154, 1994.
114. Hollmann, M., Maron, C., and Heinemann, S.F., N-glycosylation site tagging suggests a three transmembrane domain topology for the glutamate receptor GluR1, *Neuron*, 13, 1331, 1994.
115. Kuryatov, A. et al., Mutational analysis of the glycine-binding site of the NMDAR: structural similarity with bacterial amino acid-binding proteins, *Neuron*, 12, 1291, 1994.
116. Stern-Bach, Y. et al., Agonist selectivity of glutamate receptors is specified by two domains structurally related to bacterial amino acid–binding proteins, *Neuron*, 13, 1345, 1994.
117. Nakanishi, N., Shneider, N.A., and Axel, R., A family of glutamate receptor genes: evidence for the formation of heteromultimeric receptors with distinct channel properties, *Neuron*, 5, 569, 1990.
118. O'Hara, P.J. et al., The ligand binding domain in metabotropic glutamate receptors is related to bacterial periplasmic binding proteins, *Neuron*, 11, 41, 1993.
119. Kuusinen, A., Arvola, M., and Keinanen, K., Molecular dissection of the agonist binding site of an AMPA receptor, *EMBO J.*, 14, 6327, 1995.
120. Armstrong, N. et al., Structure of a glutamate-receptor ligand-binding core in complex with kainate, *Nature*, 395, 913, 1998.
121. Armstrong, N. and Gouaux, E., Mechanisms for activation and antagonism of an AMPA-sensitive glutamate receptor: crystal structures of the GluR2 ligand binding core, *Neuron*, 26, 165, 2000.
122. Mayer, M.L., Crystal structures of the GluR5 and GluR6 ligand binding cores: molecular mechanisms underlying kainate receptor selectivity, *Neuron*, 45, 539, 2005.
123. Mayer, M.L., Olson, R., and Gouaux, E., Mechanisms for ligand binding to GluR0 ion channels: crystal structures of the glutamate and serine complexes and a closed apo state, *J. Mol. Biol.*, 311, 815, 2001.
124. Furukawa, H. and Gouaux, E., Mechanisms of activation, inhibition and specificity: crystal structures of the NMDAR NR1 ligand-binding core, *EMBO J.*, 22, 2873, 2003.
125. Furukawa, H. et al., Subunit arrangement and function in NMDARs, *Nature*, 438, 185, 2005.
126. Colquhoun, D., Binding, gating, affinity and efficacy: The interpretation of structure-activity relationships for agonists and of the effects of mutating receptors, *Brit. J. Pharmacol.*, 125, 924, 1998.

127. Akabas, M.H. et al., Identification of acetylcholine receptor channel-lining residues in the entire M2 segment of the alpha subunit, *Neuron*, 13, 919, 1994.
128. Akabas, M.H. et al., Acetylcholine receptor channel structure probed in cysteine-substitution mutants, *Science*, 258, 307, 1992.
129. Kalbaugh, T.L., VanDongen, H.M.A., and VanDongen, A.M.J., Ligand-binding residues integrate affinity and efficacy in the NMDAR, *Mol. Pharmacol.*, 66, 209, 2004.
130. Doyle, D.A. et al., The structure of the potassium channel: molecular basis of K conduction and selectivity, *Science*, 280, 69, 1998.
131. Jiang, Y. et al., X-ray structure of a voltage-dependent K channel, *Nature*, 423, 33, 2003.
132. Jiang, Y. et al., Crystal structure and mechanism of a calcium-gated potassium channel, *Nature*, 417, 515, 2002.
133. Hoffmann, J., Gorodetskaia, A., and Hollmann, M., Ion pore properties of ionotropic glutamate receptors are modulated by a transplanted potassium channel selectivity filter, *Mol. Cell. Neurosci.*, 33, 335, 2006.
134. Chen, G.Q. et al., Functional characterization of a potassium-selective prokaryotic glutamate receptor, *Nature*, 402, 817, 1999.
135. Jones, K.S., VanDongen, H.M.A., and VanDongen, A.M.J., The NMDAR M3 segment is a conserved transduction element coupling ligand binding to channel opening, *J. Neurosci.*, 22, 2044, 2002.
136. Cuello, L.G. et al., pH-dependent gating in the Streptomyces lividans K+ channel, *Biochemistry*, 37, 3229, 1998.
137. Yellen, G., The moving parts of voltage-gated ion channels, *Q. Rev. Biophys.*, 31, 239, 1998.
138. Perozo, E., Cortes, D.M., and Cuello, L.G., Structural rearrangements underlying K+-channel activation gating, *Science*, 285, 73, 1999.
139. Yuan, H. et al., Conserved structural and functional control of N-methyl-D-aspartate receptor gating by transmembrane domain M3, *J. Biol. Chem.*, 280, 29708, 2005.
140. Karlin, A. and Akabas, M.H., Substituted-cysteine accessibility method, *Meth. Enz.*, 293, 123, 1998.
141. Sobolevsky, A.I. et al., Subunit-specific contribution of pore-forming domains to NMDAR channel structure and gating, *J. Gen. Physiol.*, 129, 509, 2007.
142. Zhou, J. et al., Neurodegeneration in Lurcher mice caused by mutation in delta2 glutamate receptor, *Nature*, 388, 769, 1997.
143. Kohda, K., Wang, Y., and Yuzaki, M., Mutation of a glutamate receptor motif reveals its role in gating and delta2 receptor channel properties, *Nat. Neurosci.*, 3, 315, 2000.
144. Taverna, F.A. et al., The lurcher mutation of an alpha-amino-3-hydroxy-5-methyl-4-isoxazolepropionic acid receptor subunit enhances potency of glutamate and converts an antagonist into an agonist, *J. Biol. Chem.*, 275, 8475, 2000.
145. Klein, R.M. and Howe, J.R., Effects of the lurcher mutation on GluR1 desensitization and activation kinetics, *J. Neurosci.*, 24, 4941, 2004.
146. Beck, C. et al., NMDAR channel segments forming the extracellular vestibule inferred from the accesibility of substituted cysteines, *Neuron*, 22, 559, 1999.
147. Sobolevsky, A.I., Beck, C., and Wollmuth, L.P., Molecular rearrangements of the extracellular vestibule in NMDAR channels during gating, *Neuron*, 33, 75, 2002.
148. Sobolevsky, A.I., Yelshansky, M.V., and Wollmuth, L.P., Different gating mechanisms in glutamate receptor and K+ channels, *J. Neurosci.*, 23, 7559, 2003.
149. Chen, P.E. et al., Structural features of the glutamate binding site in recombinant NR1/NR2A N-methyl-D-aspartate receptors determined by site-directed mutagenesis and molecular modeling, *Mol. Pharmacol.*, 67, 1470, 2005.

150. Kuner, T. and Schoepfer, R., Multiple structural elements determine subunit specificity of Mg2+ block in NMDAR channels, *J. Neurosci.*, 16, 3549, 1996.

151. Sobolevsky, A.I., Rooney, L., and Wollmuth, L.P., Staggering of subunits in NMDAR channels, *Biophys. J.*, 83, 3304, 2002.

152. Sobolevsky, A.I., Yelshansky, M.V., and Wollmuth, L.P., The outer pore of the glutamate receptor channel has two-fold rotational symmetry, *Neuron*, 41, 367, 2004.

153. Banke, T.G. and Traynelis, S.F., Activation of NR1/NR2B NMDARs, *Nat. Neurosci.*, 6, 144, 2003.

154. Mayer, M.L., Glutamate receptors at atomic resolution, *Nature*, 440, 456, 2006.

155. Nanao, M.H. et al., Structure of the kainate receptor subunit GluR6 agonist-binding domain complexed with domoic acid, *Proc. Natl. Acad. Sci. USA*, 102, 1708, 2005.

156. Sun, Y. et al., Mechanism of glutamate receptor desensitization, *Nature*, 417, 245, 2002.

157. Zhang, Y. et al., Interface interactions modulating desensitization of the kainate-selective ionotropic glutamate receptor subunit GluR6, *J. Neurosci.*, 26, 10033, 2006.

158. Armstrong, N. et al., Measurement of conformational changes accompanying desensitization in an ionotropic glutamate receptor, *Cell*, 127, 85, 2006.

159. Jin, R. et al., Mechanism of positive allosteric modulators acting on AMPA receptors, *J. Neurosci.*, 25, 9027, 2005.

160. Lester, R.A. et al., Channel kinetics determine the time course of NMDAR-mediated synaptic currents, *Nature*, 346, 565, 1990.

161. Zheng, F. et al., Allosteric interaction between the amino terminal domain and the ligand binding domain of NR2A, *Nat. Neurosci.*, 4, 894, 2001.

162. Perin-Dureau, F. et al., Mapping the binding site of the neuroprotectant ifenprodil on NMDARs, *J. Neurosci.*, 22, 5955, 2002.

163. Erreger, K. et al., Mechanism of partial agonism at NMDARs for a conformationally restricted glutamate analog, *J. Neurosci.*, 25, 7858, 2005.

164. Weston, M.C. et al., Interdomain interactions in AMPA and kainate receptors regulate affinity for glutamate, *J. Neurosci.*, 26, 7650, 2006.

165. Hansen, K.B. et al., Tweaking agonist efficacy at N-methyl-D-aspartate receptors by site-directed mutagenesis, *Mol. Pharmacol.*, 68, 1510, 2005.

166. Robert, A. et al., AMPA receptor binding cleft mutations that alter affinity, efficacy, and recovery from desensitization, *J. Neurosci.*, 25, 3752, 2005.

167. Armstrong, N., Mayer, M., and Gouaux, E., Tuning activation of the AMPA-sensitive GluR2 ion channel by genetic adjustment of agonist-induced conformational changes, *Proc. Natl. Acad. Sci. USA*, 100, 5736, 2003.

168. Maier, W. et al., Disruption of interdomain interactions in the glutamate binding pocket affects differentially agonist affinity and efficacy of N-methyl-D-aspartate receptor activation, *J. Biol. Chem.*, 282, 1863, 2007.

169. Inanobe, A., Furukawa, H., and Gouaux, E., Mechanism of partial agonist action at the NR1 subunit of NMDARs, *Neuron*, 47, 71, 2005.

170. Kaye, S.L., Sansom, M.S., and Biggin, P.C., Molecular dynamics simulations of the ligand-binding domain of an N-methyl-D-aspartate receptor, *J. Biol. Chem.*, 281, 12736, 2006.

171. Morris, R.G. et al., Selective impairment of learning and blockade of long-term potentiation by an N-methyl-D-aspartate receptor antagonist, AP5, *Nature*, 319, 774, 1986.

172. McHugh, T.J. et al., Impaired hippocampal representation of space in CA1-specific NMDAR1 knockout mice, *Cell*, 87, 1339, 1996.

173. Mouri, A. et al., Phencyclidine animal models of schizophrenia: approaches from abnormality of glutamatergic neurotransmission and neurodevelopment, *Neurochem. Int.*, 51, 173, 2007.

174. Heresco-Levy, U., Silipo, G., and Javitt, D.C., Glycinergic augmentation of NMDAR-mediated neurotransmission in the treatment of schizophrenia, *Psychopharmacol. Bull.*, 32, 731, 1996.

175. Coyle, J.T. and Tsai, G., The NMDAR glycine modulatory site: a therapeutic target for improving cognition and reducing negative symptoms in schizophrenia, *Psychopharmacology*, 174, 32, 2004.

176. Duncan, E.J. et al., Effects of D-cycloserine on negative symptoms in schizophrenia, *Schizophren. Res.*, 71, 239, 2004.

177. Heresco-Levy, U. and Javitt, D.C., Comparative effects of glycine and D-cycloserine on persistent negative symptoms in schizophrenia: a retrospective analysis, *Schizophren. Res.*, 66, 89, 2004.

178. Buchanan, R.W. et al., Cognitive and Negative Symptoms in Schizophrenia Trial (CONSIST): the efficacy of glutamatergic agents for negative symptoms and cognitive impairments, *Am. J. Psych.*, 164, 1593, 2007.

179. Monahan, J.B. et al., D-cycloserine, a positive modulator of the N-methyl-D-aspartate receptor, enhances performance of learning tasks in rats, *Pharmacol. Biochem. Behav.*, 34, 649, 1989.

180. Billard, J.M. and Rouaud, E., Deficit of NMDAR activation in CA1 hippocampal area of aged rats is rescued by D-cycloserine, *Eur. J. Neurosci.*, 25, 2260, 2007.

181. Gabriele, A. and Packard, M.G., D-Cycloserine enhances memory consolidation of hippocampus-dependent latent extinction, *Learn. Mem.*, 14, 468, 2007.

182. Zlomuzica, A. et al., NMDAR modulation by D-cycloserine promotes episodic-like memory in mice, *Psychopharmacology*, 193, 503, 2007.

183. Matsuoka, N. and Aigner, T.G., D-cycloserine, a partial agonist at the glycine site coupled to N-methyl-D-aspartate receptors, improves visual recognition memory in rhesus monkeys, *J. Pharmacol. Exp. Ther.*, 278, 891, 1996.

184. Randolph, C. et al., D-cycloserine treatment of Alzheimer disease, *Alzheimer's Dis. Assoc. Dis.*, 8, 198, 1994.

185. Schwartz, B.L. et al., d-Cycloserine enhances implicit memory in Alzheimer patients, *Neurology*, 46, 420, 1996.

186. Tsai, G.E. et al., Improved cognition in Alzheimer's disease with short-term D-cycloserine treatment, *Am. J. Psych.*, 156, 467, 1999.

187. Yaka, R. et al., D-cycloserine improves functional recovery and reinstates long-term potentiation (LTP) in a mouse model of closed head injury, *FASEB J.*, 21, 2033, 2007.

188. Bertotto, M.E. et al., Influence of ethanol withdrawal on fear memory: Effect of D-cycloserine, *Neurosci.*, 142, 979, 2006.

189. Guastella, A.J. et al., A randomized controlled trial of the effect of d-cycloserine on exposure therapy for spider fear, *J. Psych. Res.*, 41, 466, 2007.

190. Ledgerwood, L., Richardson, R., and Cranney, J., D-cycloserine facilitates extinction of learned fear: effects on reacquisition and generalized extinction, *Biol. Psych.*, 57, 841, 2005.

191. Ressler, K.J. et al., Cognitive enhancers as adjuncts to psychotherapy: use of D-cycloserine in phobic individuals to facilitate extinction of fear, *Arch. Gen. Psych.*, 61, 1136, 2004.

192. Richardson, R., Ledgerwood, L., and Cranney, J., Facilitation of fear extinction by D-cycloserine: theoretical and clinical implications, *Learn. Mem.*, 11, 510, 2004.

193. Weber, M., Hart, J., and Richardson, R., Effects of D-cycloserine on extinction of learned fear to an olfactory cue, *Neurobiol. Learn. Mem.*, 87, 476, 2007.

194. Munir, M., Subramaniam, S., and McGonigle, P., Polyamines modulate the neurotoxic effects of NMDA *in vivo*, *Brain Res.*, 616, 163, 1993.

195. Wu, F.S., Gibbs, T.T., and Farb, D.H., Pregnenolone sulfate: a positive allosteric modulator at the N-methyl-D-aspartate receptor, *Mol. Pharmacol.*, 40, 333, 1991.

196. Malayev, A., Gibbs, T.T., and Farb, D.H., Inhibition of the NMDA response by pregnenolone sulphate reveals subtype selective modulation of NMDARs by sulphated steroids, *Br. J. Pharmacol.*, 135, 901, 2002.

197. Mameli, M. et al., Neurosteroid-induced plasticity of immature synapses via retrograde modulation of presynaptic NMDARs, *J. Neurosci.*, 25, 2285, 2005.

198. Horak, M. et al., Molecular mechanism of pregnenolone sulfate action at NR1/NR2B receptors, *J. Neurosci.*, 24, 10318, 2004.

199. Weaver, C.E. et al., Geometry and charge determine pharmacological effects of steroids on N-methyl-D-aspartate receptor-induced Ca(2+) accumulation and cell death, *J. Pharmacol. Exp. Ther.*, 293, 747, 2000.

200. Jang, M.K. et al., A steroid modulatory domain on NR2B controls N-methyl-D-aspartate receptor proton sensitivity, *Proc. Natl. Acad. Sci. USA*, 101, 8198, 2004.

201. Park-Chung, M. et al., Distinct sites for inverse modulation of N-methyl-D-aspartate receptors by sulfated steroids, *Mol. Pharmacol.*, 52, 1113, 1997.

202. Kokate, T.G. et al., Convulsant actions of the neurosteroid pregnenolone sulfate in mice, *Brain Res.*, 831, 119, 1999.

203. Vallee, M. et al., Neurosteroids: deficient cognitive performance in aged rats depends on low pregnenolone sulfate levels in the hippocampus, *Proc. Natl. Acad. Sci. USA*, 94, 14865, 1997.

204. Mayo, W. et al., Pregnenolone sulfate enhances neurogenesis and PSA-NCAM in young and aged hippocampus, *Neurobiol. Aging*, 26, 103, 2005.

205. Sliwinski, A. et al., Pregnenolone sulfate enhances long-term potentiation in CA1 in rat hippocampus slices through the modulation of N-methyl-D-aspartate receptors, *J. Neurosci. Res.*, 78, 691, 2004.

206. Lynch, G. and Gall, C.M., Ampakines and the three-fold path to cognitive enhancement, *Trends Neurosci.*, 29, 554, 2006.

14 Presynaptic NMDA Receptors

Ian C. Duguid and Trevor G. Smart

CONTENTS

14.1 INTRODUCTION

Presynaptic receptors, by virtue of their locations, are ideally suited to influence the efficacy of synaptic transmission by affecting neurotransmitter release.[58] In the nervous system, action potential invasion of presynaptic terminals results in a characteristic series of events: initial Ca^{2+} entry, followed by the activation of presynaptic vesicular release machinery, vesicular fusion, and the release of neurotransmitter into the synaptic cleft.[103,105] The efficacy of synaptic transmission is thus governed by the probability of neurotransmitter release, the amount of transmitter released from the presynaptic terminal, the type and number of postsynaptic neurotransmitter receptors, and their response to the released transmitter.

Short- and long-term activity-dependent modulation of the efficacy of a synapse can proceed via a multitude of signaling mechanisms that impact on either the presynaptic release or the receptors that mediate postsynaptic responses.[13,66,102] Such modulatory mechanisms will be crucial for regulating the flow of information throughout the nervous system and have been implicated in many neural processes including learning and memory, vision, motor control, and neuroprotection.

Modulation of transmitter release at a synapse was first demonstrated in the classical studies of Dudel and Kuffler[30] and Eccles[33] who identified that presynaptic GABA receptors inhibited transmitter release from crustacean motor neuron terminals and vertebrate sensory neuron terminals in the spinal cord, respectively. Since then, the modulation of transmitter release by presynaptic receptors is an accepted signaling pathway, and although the focus of attention initially fell on *metabotropic* G-protein coupled receptors,[52,94] it soon became clear that numerous populations of presynaptic *ionotropic* receptors are equally important.[53,58,67]

One receptor that has not featured prominently as a presynaptic regulator of transmitter release is the N-methyl-D-aspartate (NMDA)-sensitive glutamate receptor. It was first proposed to have a presynaptic locus of expression after it was found that exogenously applied NMDA facilitated the release of tritiated neurotransmitter from synaptosomes prepared from noradrenergic terminals in the hippocampus,[86] cerebral cortex,[37] and from dopaminergic terminals in the striatum.[50,56,112] Because of the nature of the preparations, these early studies failed to identify the exact loci of NMDA receptor (NMDAR) subunit expression.

Further evidence for presynaptic NMDARs came from the pioneering work of Liu and colleagues[62] who identified NR1 subunit immunoreactivity in both the dorsal and ventral horns of the rat spinal cord, specifically on axon terminals and very near the active zone, indicating a direct role in the regulation of transmitter release. Similarly, immunoreactivity for NR1 and NR2 was found on presynaptic boutons in rat cerebellar cortex[82,83] and at mossy fiber CA synapses in monkey hippocampus.[98] These early findings provided the necessary impetus to find a more widespread role for presynaptic NMDARs in the regulation of neuronal signaling in the CNS.

In this chapter, we discuss recent advances in our understanding of presynaptic NMDARs as important modulators of synaptic transmission. We consider the potential sources of glutamate for NMDAR activation; the downstream signaling mechanisms that ensue; and the differing forms of synaptic plasticity mediated by presynaptic NMDARs that undoubtedly help sculpt information processing in the brain.

14.2   CRITERIA FOR DEFINING PRESYNAPTIC RECEPTORS

Ideally, before attempting to classify a receptor as having a putative presynaptic location and roles in modulating neurotransmitter release, several criteria should be satisfied:

1. Immunohistochemical or electron microscopic (EM) evidence of a presynaptic location of receptor subunits at a given synapse.
2. Presence of the modulatory transmitter at or adjacent to a synapse.
3. Exogenous application of the transmitter should mimic physiological activation of the receptor.
4. Selective receptor antagonists should block presynaptic receptor activation.
5. Downstream signaling cascades leading to altered transmitter release should be identified in the presynaptic axon terminals.
6. Activation of the presynaptic receptor should affect the frequency of miniature synaptic currents in preference to their amplitudes.
7. The paired-pulse ratio (PPRs) of the amplitudes of two consecutively evoked synaptic currents should be increased or decreased by presynaptic receptor activation. Caution is required when using the PPR as a sole indicator of

presynaptic receptor activation as postsynaptic mechanisms can contribute to a change in PPR.[55]

8. Measurement of a change in the coefficient of variation (CV) of evoked synaptic current amplitudes.

Frequent use is made of comparing the CV with the mean amplitude (m) of evoked synaptic currents to deduce whether variations in synaptic efficacy have their origins at presynaptic or postsynaptic locations. Generally, proportionate changes in CV^{-2} and m indicate presynaptic modulation of transmitter release, whereas changes to m without alteration to CV^{-2} indicate modulation of postsynaptic receptors.

14.3 LOCATIONS OF NMDA RECEPTORS AT SYNAPSES

Although it is widely known that NMDARs are located at the postsynaptic densities of excitatory synapses,[27,78] where they mediate the slow excitatory postsynaptic potential, their locations on presynaptic axon terminals, has been more contentious. Early trafficking studies revealed that NMDARs are potentially mobile by radiolabelling with the antagonist CPP. They were shown to move bidirectionally along the vagus nerve and were restricted by its ligation.[23]

At the light and electron microscopy level, positive immunoreactivity has been observed for NMDAR subunits on a wide variety of asymmetrical synapses throughout the mammalian brain, including the dorsal and ventral horns of the spinal cord,[62] neocortex,[29] hippocampus,[82,98] cerebellar cortex,[82] visual cortex,[4] anterior cingulate cortex,[111] nucleus accumbens,[42] amygdala,[35] and retina.[113] Immunoreactivity for NR2B subunits has also been reported on axonal growth cones isolated from embryonic day 18 (E18) brains.[45] These reports illustrate the potentially widespread distribution of presynaptic NMDARs that may affect excitatory synaptic transmission in the mammalian CNS.

Immunoreactivity for NMDAR subunits can also be found at symmetrical synapses on subpopulations of GABAergic inhibitory boutons in the bed nucleus of the stria terminalis, cerebellar cortex, paraventricular hypothalamic nucleus, neocortex, and arcuate nucleus.[29,32,81] Rich plexuses of NR1-immunoreactive terminal-like varicosities are present in many nuclei in the basal forebrain, midline thalamus, and periventricular hypothalamus.[81] Surprisingly, very few NR1 labeled fibers are apparent in the midbrain, brainstem, and at some cortical levels, suggesting the distribution of presynaptic NMDAR subunits is seemingly confined to brain structures involved in controlling autonomic, neuroendocrine, and limbic functions.

Many earlier observations reported the presence of one type of NMDAR subunit, but this does not constitute proof that functional receptors are present because they are formed only after hetero-oligomerization.[73] Although it is possible to detect immunoreactivity for multiple NMDAR subunits on the same axon terminals (NR1 and NR2A-D are all expressed on axon terminals of interneurons that synapse with Purkinje cells in the cerebellum[32]), this also does not prove that functional NMDAR channels are present. Thus, positive immunoreactivity *per se* does not confirm the presence of functional presynaptic receptors, which can be assessed only via electrophysiological approaches (see criteria).

14.4 PHARMACOLOGY OF PRESYNAPTIC NMDA RECEPTORS

With regard to the NMDAR isoforms present on presynaptic terminals, immunocyto-chemical studies can only suggest potential subunit assemblies, unless, of course, subunit expression is limited to the NR1 subunit and a single NR2 isoform. We know that the minimum requirement for NMDAR cell surface expression is for a single NR1 subunit and at least one NR2 subunit isoform. While the type of NR2 subunit incorporated influences the pharmacological and physiological properties of NMDAR,[26,72,80] very few ligands can definitively distinguish among NMDAR isoforms.

Our selection of certain studies as exemplars highlights the usefulness of selective ligands and the potential problems of interpreting the data. Ifenprodil, an NR2B selective ligand,[116] was used to deduce that NR1 and NR2B receptors are present on axon terminals in the visual cortex[99] and entorhinal cortex.[117] By contrast, in the hippocampus, receptors comprised of NR1 and NR2A were thought to reside on Schaffer collateral terminals, promoting axon excitability and enhancing glutamate release. The presynaptic expression of NR2A was deduced because the increase in axon excitability by applied NMDA was inhibited not by ifenprodil, but by NVP-AAM077[104]—an NR2A, selective, antagonist.[10,63]

Caution is required when using selective NMDAR antagonists to deduce subunit compositions of presynaptic NMDARs. For example, NVP-AAM077 can inhibit receptors comprising NR1 and NR2B subunits.[10] The difference between the K_i values for NVP-AAM077 inhibiting the activation of NR1/NR2A and NR1/NR2B receptors is only 10-fold[80]—the absolute minimum for any ligand to show reasonably useful selectivity between two receptor isoforms. To complicate matters, NVP-AAM077 also has appreciable affinity for receptors containing the NR2C or NR2D subunit; the K_is are only 1.6- and 7-fold higher, respectively, than for NR2A subunit-containing receptors.[36,80] The closeness of the K_is for the current class of competitive antagonists at recombinant NMDARs suggests they are not well suited for distinguishing NMDAR isoforms.

An alternative approach may be utilizing ion channel blockers such as Mg^{2+}, phencyclidine, ketamine, memantine, MK-801, and argiotoxin-636. Moreover, with the exception of argiotoxin-636, these agents are poorly selective among NMDAR isoforms.[80] The most selective is MK-801 that exhibits a 10-fold separation in K_is for NR2A or NR2B over NR2C or NR2D subunit-containing NMDARs.[118] Argiotoxin-636 is of some interest. Although it cannot distinguish NR2A and NR2B subunit-containing receptors, it will select for receptors that contain either of these over NR2C or NR2D subunit-containing receptors for which the K_i is around 50-fold higher.[89]

The final class of compounds in the pharmacological armamentarium for the NMDAR constitutes the allosteric ligands that bind primarily to the N-terminal regions of the NR2 subunits. Ifenprodil, an NR2B selective antagonist, is one of the more useful compounds, exhibiting a greater than 200-fold selectivity over NR2A, 2C, and 2D subunit-containing receptors.[116] This is surpassed by CP101-606[75] and Ro25-6981[38] that showed 750- and 3300-fold greater selectivities, respectively, for the NR2B subunit.

Another N-terminal inhibitor that may prove more useful in distinguishing NMDAR isoforms is the divalent cation Zn^{2+}.[100,101] Zinc ions are at least 100-fold more potent (based on IC_{50}) as inhibitors at NR1/NR2A receptors compared to the next most sensitive NMDAR isoform containing NR1/NR2B subunits[79,107] and approximately 1000- and 500-fold more potent than at NR2C and NR2D subunit-containing receptors.[80,88] Although Zn^{2+} appears ideal for differentially detecting NR2A, and possibly NR2B subunits, it is important to note that the inhibition saturates at less than 100% at high Zn^{2+} concentrations[88]; thus a residual response is always present. Low concentrations of Zn^{2+} can also potentiate glycine receptor function and inhibit some isoforms of $GABA_A$ receptors.[100]

The pharmacology of putative presynaptic NMDARs is therefore complex and our desire to achieve clarity is hampered by a lack of highly selective ligands. A further complication arises from the ability of NMDARs to not just form diheteromers but also triheteromers, possibly incorporating more than one type of NR2 subunit.[22,84] Unfortunately, the pharmacology of these triheteromers is insufficiently distinct for the current crop of ligands to be able to unequivocally distinguish their presence among native NMDARs in neurons.[44] We can also add the possibility that triheteromers may incorporate NR1, NR2, and NR3 subunits.[2,28,93] Although such a subunit combination will exert effects on single channel conductance and permeability to Ca^{2+}, ligands with suitable selectivity are lacking. Whether the NR3A subunit is a subunit partner for presynaptic NMDARs remains to be seen.

14.5 SOURCES OF GLUTAMATE

To better understand the physiological role of presynaptic NMDARs in mammalian brains, we must first consider the sources of glutamate that may activate presynaptic NMDARs: (1) glutamate released from the same terminal on which the receptors are expressed (autoreceptor); (2) direct synaptic excitation of presynaptic boutons (axo-axonic); (3) diffusion of glutamate from an adjacent active terminal (spillover); and (4) following the activity-dependent release of glutamate from the dendrites of a postsynaptic cell (retrograde); and (5) release from surrounding glial cells (paracrine).

14.5.1 PRESYNAPTIC AUTORECEPTORS

Several neurotransmitters have been shown to modulate their own releases through relatively well-defined presynaptic autoreceptor systems.[90] However, the existence and functional role of presynaptic NMDA autoreceptors remains less well-defined and has been complicated by the presence of such receptors on the postsynaptic side of the synapse. The first evidence to suggest that NMDARs could modulate the release of glutamate appeared in the early 1990s when NMDAR antagonists were shown to reduce K^+-evoked glutamate release from CA1 hippocampal neurons.[71] In addition, *in vivo* dialysis of NMDAR agonists resulted in a dose-dependent increase in the extracellular concentration of glutamate in the striatum, indicative of presynaptic NMDARs being located on cortico-striatal nerve endings.[18]

In terms of synaptic transmission, bath application of NMDAR agonists (e.g., NMDA) and antagonists (e.g., D-APV) alters the frequency of spontaneous and miniature EPSCs in the entorhinal cortex,[11,117] spinal cord,[8,92] visual cortex,[99] and cerebellar cortex.[19,41] This type of modulation is indicative of tonic activation of presynaptic NMDARs by, presumably, ambient glutamate. These effects persisted in the background presence of tetrodotoxin (TTX) and when postsynaptic neurons were dialyzed with MK-801, thus excluding the possibility that the effects on synaptic current frequency resulted from a change in NMDAR-dependent network activity or activation of postsynaptic NMDARs. As expected for a coincidence detector, varying the extracellular Mg^{2+} concentration or increasing presynaptic afferent activity enhanced the effects of presynaptic NMDAR activation. Low rates of afferent activation would cause insufficient terminal depolarization, leading to incomplete relief from Mg^{2+} block of NMDARs. Higher activation rates would be more effective by allowing coincident detection of extracellular glutamate.[8,19, 43,92,99,117]

Interestingly, tonic facilitation of transmission in the entorhinal cortex, where NR1 and NR2B subunits are thought to be important,[11,117] decreases during development. However, this is not associated with a switch in subunit expression[87,63] but instead it appears that the loss of autoreceptor function in adult animals may reflect decreases in surface expression of presynaptic NR2B subunits or a redistribution of NR2 subunit-containing receptors to a location relatively inaccessible to ambient glutamate.[120] The loss of function observed in older animals was reversed if the animals suffered chronic epileptic seizures, indicating that this pathophysiological state caused an upregulation of NR2B subunit expression, a redistribution of NMDARs to sites accessible to ambient glutamate, or a change in the composition of the existing nonredistributed NMDARs.[120]

Although the data does not functionally link increased expression of presynaptic NMDA autoreceptors and epileptogenesis, it indicates a possible link between presynaptic autoreceptor function, enhanced basal glutamate release, and elevated network excitability in the entorhinal cortex. It will be of interest to see whether other pathophysiological states are capable of altering presynaptic NMDA autoreceptor activity.

Presynaptic NMDA autoreceptors may also play a role in long-term synaptic plasticity. For example, in neocortical layer 5 pyramidal neurons, a form of long-term depression (LTD) known as timing-dependent LTD (tLTD) relies on postsynaptic action potential firing preceding presynaptic afferent activity. This causes the simultaneous activation of presynaptic endocannabinoid (CB1) receptors and NMDARs.[99] The CB1 receptors detect the extent of postsynaptic activity through the retrograde release of endocannabinoids, whereas the presynaptic NMDA autoreceptors are sensors for presynaptic spiking. The initial spike acts to relieve the Mg^{2+} block, while subsequent spikes lead to glutamate pooling around the terminal and activation of presynaptic NMDA autoreceptors.

In a similar manner, presynaptic NMDA autoreceptors also play a role at layer 4 to layer 2/3 synapses in the somatosensory cortex during the induction of spike timing-dependent plasticity (STDP).[9] Similar to the induction of tLTD in visual cortex, STDP also requires retrograde endocannabinoid signaling and activation of apparently presynaptic classed as (non-postsynaptic) NMDARs.[9]

Presynaptic NMDARs also feature in LTD of parallel fiber inputs in the cerebellum.[19,20] In this case, a presynaptic NMDAR-mediated Ca^{2+} influx activates neuronal nitric oxide synthase (nNOS) to produce NO in the parallel fibers. This is thought to diffuse across the synapse to the Purkinje cell where, by activation of guanylate cyclase, cGMP is produced, resulting in the activation of cGMP-dependent protein kinase (PKG).[59] Ultimately, this signaling pathway could result in inhibition of protein phosphatases (likely to be PP1 and PP2A) in Purkinje cells and the phosphorylation of AMPARs, possibly GluR2 at Ser-880. This will then cause their internalization and ensuing LTD.[48,49] Given the a lack of functional NMDARs on Purkinje cells,[85,91] the logical explanation is that repetitive parallel fiber activation triggers autoreceptor activation of presynaptic NMDARs expressed at parallel fiber axon terminals to initiate a signaling cascade leading to LTD.[19,20]

However, an alternative plausible explanation was suggested by a recent study from Shin and Linden.[97] They proposed that the NMDAR–NO cascade involved in cerebellar LTD is not localized to parallel fibers, but to interneuronal axon terminals.[97] The presynaptic NMDARs are most likely activated by glutamate spillover from the PFs, leading to NO release from interneurons that diffuses to the Purkinje cells to evoke LTD. The difficulty in deducing the correct explanation stems from a lack of detailed EM immunostaining for NMDAR subunits on nerve terminals in the cerebellum.

Although in general the data on presynaptic NMDA autoreceptor function is limited, the widespread distribution of presynaptic NMDARs on asymmetrical synapses, as indicated by electron microscopy studies (see previous section), suggests that autoreceptor regulation of synaptic transmission will be a prevalent feature of information processing throughout the mammalian CNS (Figure 14.1A).

14.5.2 Axo-Axonic Synapses

Axon terminals are normally considered output devices, modulated by presynaptic receptors and ion channels. Extracellular stimulation in the spinal cord results in synaptically evoked excitatory currents in the axons of reticulospinal neurons.[24] As synaptic potentials can be generated in either direction, and are sensitive to block by D-APV,[24,25] they probably originate from axo-axonic synapses (Figure 14.1B).[34]

The excitatory innervation of reticulospinal axons may come from dorsal root ganglion primary afferents,[14] excitatory interneurons,[15,17] and other reticulospinal neurons.[16] These synaptic inputs occur at various locations along axons and the latency of the input is directly associated with the distance of the stimulation electrode from the recording electrode. Tetanic stimulation of inputs to the reticulospinal axon results in a prolonged NMDAR-dependent increase in terminal Ca^{2+} concentration that can last for several seconds.

The depolarization resulting from NMDAR activation is insufficient to activate voltage-activated Ca^{2+} channels, indicating that Ca^{2+} must enter solely via the activated NMDAR ion channels to enhance transmitter release.[25] The presence of axo-axonic excitatory synapses is not unique to the lamprey; they appear to exist also on afferent fibers of frog spinal cord.[109] Although there is no evidence to support a role for excitatory axo-axonic synapses in the mammalian CNS, the diverse expression of presynaptic NMDARs suggests that they could potentially play a role in modulating synaptic transmission.

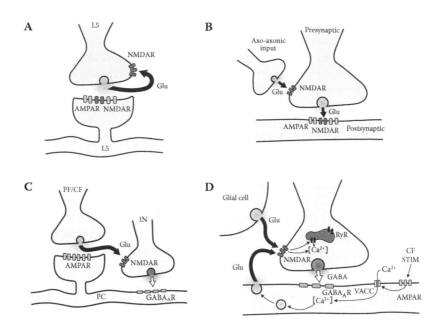

FIGURE 14.1 (See color insert following page 212.) Presynaptic NMDAR activation by released glutamate. (A) Presynaptic autoreceptor activation. High-frequency afferent stimulation (100 Hz) onto a layer 5 (L5) neocortical neuron enables presynaptically released glutamate (Glu) to activate presynaptic NMDA autoreceptors (NMDAR). (B) Axo-axonic NMDAR activation. Direct afferent input onto axon terminals enables released glutamate to activate presynaptic NMDARs, thereby regulating glutamate release onto AMPA receptors (AMPAR) and NMDARs on postsynaptic neurons. (C) Spillover-dependent NMDAR activation. High-frequency stimulation of cerebellar climbing (CF) or parallel fibers (PF) results in glutamate pooling and saturation of juxtaposed excitatory amino acid transporters. Synaptic glutamate can diffuse out of the synapse (spillover) to activate presynaptic NMDARs on inhibitory interneurons. This leads to increased GABA release and postsynaptic GABA$_A$ receptor (GABA$_A$R) activation. (D) Retrograde/paracrine release-dependent NMDAR activation. At the interneuron (IN)–Purkinje cell (PC) synapse, postsynaptic depolarization via CF stimulation and AMPAR activation enables Ca^{2+} influx via voltage-gated Ca^{2+} channels (VGCC) to induce the retrograde release of vesicular glutamate from the PC. Enhanced GABA release results from Ca^{2+} induced Ca^{2+} release from ryanodine-sensitive stores (RyR). Alternatively, activation of the perforant path in the hippocampus leads to the paracrine release of glutamate from adjacent glial cells. Either source of released glutamate activates presynaptic NMDARs to enhance synaptic efficacy.

14.5.3 SPILLOVER

Synaptic transmission during minimal presynaptic afferent activity is largely constrained to the postsynaptic density (PSD). This area is considered to form the boundary of a synapse. The diffusion of neurotransmitter from this boundary is largely controlled, in the case of glutamate, by the density and activity of surrounding excitatory amino acid transporters.[77,106,108] However, during periods of more intense presynaptic activity, the concentration of neurotransmitter in the cleft can rise sufficiently to saturate transporter-based mechanisms, leading to their inability to control the spatial spread of neurotransmitter to peri- and extrasynaptic sites.

The process of transmitter spillover acts to reduce synapse independence, while promoting heterosynaptic activation and synaptic crosstalk (Figure 14.1C).[57,70] The biophysical characteristics and high affinity of NMDARs for glutamate are pivotal for "sensing" transmitter spillover from adjacent sites.[26,119] Such a role for NMDARs has been extensively studied in the cerebellum. Here repetitive stimulation of climbing fiber (CF) inputs to Purkinje cells reduces the amplitude of sIPSCs via NMDAR activation. This was explained by glutamate spillover from CF synapses, resulting in activation of presynaptic NMDARs on adjacent interneuron axon terminals/boutons.[32,41] Inhibition of glutamate uptake by TBOA, a nontransportable neuronal and glial transporter antagonist,[95,96] resulted in a significant enhancement in synaptic crosstalk between excitatory and inhibitory inputs in the cerebellar cortex.[46] Therefore, glial glutamate transporters, and to a lesser extent neuronal transporters, strictly control the spatial spread of glutamate from CF synapses.[3,7]

The spillover of glutamate to neighboring synapses is not restricted to CF synapses in the cerebellum; high-frequency bursts of parallel fiber (PF) activity similar to that observed *in vivo*[21] also produces glutamate pooling and spillover to nearby interneuron axon terminals. The resulting presynaptic NMDAR-dependent increase in inhibitory synaptic efficacy can last several ten of minutes.[64]

Activation of presynaptic NMDARs on GABAergic inputs by released glutamate is not exclusive to the cerebellum. Lien and colleagues[61] showed that light-induced or theta burst stimulation of the optic nerve in the developing *Xenopus* retinotectal system induced LTP of glutamatergic inputs and LTD of GABAergic inputs onto the same tectal neuron. Although both forms of plasticity were blocked by D-APV, only LTP of excitatory afferents was abolished by infusing the tectal cell with MK-801. This suggests that high-frequency stimulation of excitatory afferents resulted in spillover activation of NMDARs on adjacent interneuron terminals, which were depolarized by the same theta burst stimulation. These findings provided the first evidence for the involvement of presynaptic NMDARs in coincidence detection and synaptic plasticity *in vivo*.

Spillover activation of presynaptic NMDARs has also been implicated in a form of associative long-term plasticity of cortical afferents in the amygdala. By coincidently stimulating converging inputs to the amygdala from the thalamus and cortex, LTP resulted, but only at the cortical afferent synapses.[47] Blocking postsynaptic NMDARs with intracellular MK-801 did not prevent LTP induction, indicating that the NMDARs required for associative LTP must have a presynaptic locus of expression in accord with the presence of NR1 subunits on presynaptic terminals in the amygdala.[35] It is conceivable that stimulation of thalamic inputs may directly activate presynaptic NMDARs on adjacent cortical axon terminals. The induction of LTP will then be restricted to those cortical synapses that are active during thalamic afferent stimulation.

14.5.4 RETROGRADE/PARACRINE RELEASE

Retrograde signaling provides an efficient feedback mechanism to enable postsynaptic neurons to communicate with their presynaptic afferents and control transmitter release.

It is thought to operate at a variety of synapses throughout the brain and the classical neurotransmitters, neuropeptides, and endocannabinoids have all been identified as retrograde messengers.[65] However, while endocannabinoids operate almost ubiquitously throughout the brain, retrograde activation of presynaptic NMDARs appears, to date, to be confined to select synapses in the cerebellum and hippocampus.

The exact mechanisms that operate to enable retrograde transmitter release and activation of presynaptic NMDARs remain largely unresolved. Even the identity of the retrograde messenger remains elusive. It is presumed to be glutamate or a glutamate-like molecule and may involve reversed operation of glutamate transporters or exocytosis from postsynaptic vesicles.

Generally, the induction of retrograde 'glutamate' release requires stimulation of the postsynaptic cell, either directly or by activating afferent inputs, and the influx of Ca^{2+}.[60,121] These features are crucial for the expression of a form of inhibitory synaptic plasticity known as depolarization-induced potentiation of inhibition (DPI) at interneuron–Purkinje cell synapses in the cerebellar cortex.[32] Purkinje cell depolarization either directly or by CF activation together with Ca^{2+} influx was sufficient to ensure the activation of APV-sensitive NMDARs (Figure 14.1D).

The insensitivity of Purkinje cells to NMDA indicated that the locus of NMDAR expression must be presynaptic. This was confirmed by revealing immunoreactivity for NR1 and NR2A through D subunits on interneuronal axon terminals.[32] Subsequently following the activation of presynaptic NMDARs, Ca^{2+} influx via NMDA channels, but not via voltage–gated Ca^{2+} channels, induced Ca^{2+} release from ryanodine-sensitive Ca^{2+} stores, resulting in increased release of GABA.[32] The presence of functional presynaptic NMDARs was elegantly confirmed by patch clamp recording of single NMDA ion channel activity on basket/stellate cell terminals.[39]

The retrograde release of glutamate was not inhibited by glutamate transporter blockers,[32] but was substantially reduced by using ligands that disrupt SNARE-dependent vesicular release, e.g., botulinum toxin B, GDP-β-S, or N-ethylmaleimide.[31] Definitive evidence of the source of 'glutamate' release was provided by vibromechanically isolating Purkinje cells with attached GABA-releasing inhibitory axon terminals (nerve–bouton preparation[1,110]). The ability to depolarize Purkinje cells and still activate presynaptic NMDARs proved that the retrograde glutamate was released from individual Purkinje cells in preference to surrounding glia.[31]

During early development, presynaptic NMDARs located on Schaffer collateral axon terminals on hippocampal CA1 neurons are also modulated by the release of a retrograde messenger. However, unlike in the cerebellum, the messenger involved is not glutamate, but a pregnenolone sulfate (PS)-like neurosteroid synthesized *de novo* during afferent stimulation.[54,74] The resulting allosteric modulation of presynaptic NMDARs produces a significant enhancement in mEPSC frequency during early hippocampal development (P3–4), probably due to an increase in the probability of release at excitatory synaptic terminals.[69] This neurosteroid modulation disappears during development (>P6), coinciding with a decline in NR2D subunit expression during the first postnatal week in the murine hippocampus.[76,114,115] This association is unexpected because PS potentiates the function of recombinant NR2A or NR2B subunit-containing NMDARs and inhibits those containing NR2C or NR2D subunits.[40,68] Conceivably, the NMDARs on the Schaffer collaterals may possess different NR1 splice variants or form triheteromers with two distinct types of NR2 subunits.[69] Whether this is sufficient to alter modulation by PS remains to be seen.

An alternative mechanism that may be important for the release of glutamate and activation of presynaptic NMDARs involves paracrine secretion from surrounding glial cells.[12] Studies of cultured hippocampal neurons indicated that astrocyte stimulation during ongoing presynaptic activity enhanced the frequency of miniature postsynaptic currents. This was proposed to occur via paracrine release of glutamate activating extrasynaptic (possibly presynaptic) NMDARs to enhance transmitter release.[5,6]

This hypothesis is supported by a recent landmark study on astrocyte–neuron signaling at perforant path granule cell (PP-GC) synapses in hippocampal slices.[51] Stimulating the PP released ATP that activated P2Y1 receptors on adjacent glial cells. The increased cytosolic Ca^{2+} levels caused the vesicular release of a gliotransmitter and increased the frequency of mEPSCs in granule cells. By using immunogold labeling, glutamate was identified as the gliotransmitter since astrocytic vesicles containing glutamate were found directly opposite NR2B subunit-containing NMDARs on PP presynaptic terminals. By using functional and ultrastructural approaches, this study provided definitive evidence for the physiological control of synaptic transmission via exocytosis of glutamate from astrocytes and the corresponding activation of presynaptic NMDARs.

14.6 CONCLUSIONS

While NMDARs are known for their coincidence detection properties and for underpinning slow EPSPs in central neurons, an increasing number of studies indicate that they have important presynaptic roles in regulating transmitter release. Their location on axon terminals and their ability to initiate a number of Ca^{2+}-dependent signaling mechanisms, from presynaptic transmitter release to phosphorylation of postsynaptic receptors via intermediary messengers, highlights an ever-increasing diversity for presynaptic glutamate receptor signaling at both excitatory and inhibitory synapses.

ACKNOWLEDGMENTS

ICD is supported by a Wellcome Trust Advanced Training Fellowship. TGS is supported by grants from the MRC and BBSRC. We thank Alasdair Gibb and Stuart Cull-Candy for comments on the manuscript and Neil Foubister for providing the illustrations.

REFERENCES

1. Akaike, N. and Moorhouse, A.J., Techniques: applications of the nerve-bouton preparation in neuropharmacology, *Trends Pharmacol. Sci.*, 24, 44, 2003.
2. Al Hallaq, R.A. et al., Association of NR3A with the N-methyl-D-aspartate receptor NR1 and NR2 subunits, *Mol. Pharmacol.*, 62, 1119, 2002.
3. Anderson, C.M. and Swanson, R.A., Astrocyte glutamate transport: review of properties, regulation, and physiological functions, *Glia*, 32, 1, 2000.
4. Aoki, C. et al., Cellular and subcellular localization of NMDA-R1 subunit immunoreactivity in the visual cortex of adult and neonatal rats, *J. Neurosci.*, 14, 5202, 1994.

5. Araque, A., Carmignoto, G., and Haydon, P.G., Dynamic signaling between astrocytes and neurons, *Annu. Rev. Physiol.*, 63, 795, 2001.

6. Araque, A. et al., Calcium elevation in astrocytes causes an NMDAR-dependent increase in the frequency of miniature synaptic currents in cultured hippocampal neurons, *J. Neurosci.*, 18, 6822, 1998.

7. Auger, C. and Attwell, D., Fast removal of synaptic glutamate by postsynaptic transporters, *Neuron*, 28, 547, 2000.

8. Bardoni, R. et al., Presynaptic NMDARs modulate glutamate release from primary sensory neurons in rat spinal cord dorsal horn, *J. Neurosci.*, 24, 2774, 2004.

9. Bender, V.A. et al., Two coincidence detectors for spike timing-dependent plasticity in somatosensory cortex, *J. Neurosci.*, 26, 4166, 2006.

10. Berberich, S. et al., Lack of NMDAR subtype selectivity for hippocampal long-term potentiation, *J. Neurosci.*, 25, 6907, 2005.

11. Berretta, N. and Jones, R.S., Tonic facilitation of glutamate release by presynaptic N-methyl-D-aspartate autoreceptors in the entorhinal cortex, *Neuroscience*, 75, 339, 1996.

12. Bezzi, P. and Volterra, A., A neuron-glia signaling network in the active brain, *Curr. Opin. Neurobiol.*, 11, 387, 2001.

13. Bliss, T.V., Collingridge, G.L., and Morris, R.G., Introduction. Long-term potentiation and structure of the issue, *Philos. Trans. R. Soc. Lond. B. Biol. Sci.*, 358, 607, 2003.

14. Brodin, L., Christenson, J., and Grillner, S., Single sensory neurons activate excitatory amino acid receptors in the lamprey spinal cord, *Neurosci. Lett.*, 75, 75, 1987.

15. Buchanan, J.T., Identification of interneurons with contralateral, caudal axons in the lamprey spinal cord: synaptic interactions and morphology, *J. Neurophysiol.*, 47, 961, 1982.

16. Buchanan, J.T. et al., Reticulospinal neurones activate excitatory amino acid receptors, *Brain Res.*, 408, 321, 1987.

17. Buchanan, J.T. and Grillner, S., Newly identified "glutamate interneurons" and their role in locomotion in the lamprey spinal cord, *Science*, 236, 312, 1987.

18. Bustos, G. et al., Regulation of excitatory amino acid release by N-methyl-D-aspartate receptors in rat striatum: in vivo microdialysis studies, *Brain Res.*, 585, 105, 1992.

19. Casado, M., Dieudonne, S., and Ascher, P., Presynaptic N-methyl-D-aspartate receptors at the parallel fiber-Purkinje cell synapse, *Proc. Natl. Acad. Sci. USA*, 97, 11593, 2000.

20. Casado, M., Isope, P., and Ascher, P., Involvement of presynaptic N-methyl-D-aspartate receptors in cerebellar long-term depression, *Neuron*, 33, 123, 2002.

21. Chadderton, P., Margrie, T.W., and Hausser, M., Integration of quanta in cerebellar granule cells during sensory processing, *Nature*, 428, 856, 2004.

22. Chazot, P.L. et al., Molecular characterization of N-methyl-D-aspartate receptors expressed in mammalian cells yields evidence for the coexistence of three subunit types within a discrete receptor molecule, *J. Biol. Chem.*, 269, 24403, 1994.

23. Cincotta, M. et al., Bidirectional transport of NMDAR and ionophore in the vagus nerve, *Eur. J. Pharmacol.*, 160, 167, 1989.

24. Cochilla, A.J. and Alford, S., Glutamate receptor-mediated synaptic excitation in axons of the lamprey, *J. Physiol.*, 499 (Pt 2), 443, 1997.

25. Cochilla, A.J. and Alford, S., NMDAR-mediated control of presynaptic calcium and neurotransmitter release, *J. Neurosci.*, 19, 193, 1999.

26. Cull-Candy, S., Brickley, S., and Farrant, M., NMDAR subunits: diversity, development and disease, *Curr. Opin. Neurobiol.*, 11, 327, 2001.

27. Danysz, W. and Parsons, A.C., Glycine and N-methyl-D-aspartate receptors: physiological significance and possible therapeutic applications, *Pharmacol. Rev.*, 50, 597, 1998.

28. Das, S. et al., Increased NMDA current and spine density in mice lacking the NMDAR subunit NR3A, *Nature*, 393, 377, 1998.

29. DeBiasi, S. et al., Presynaptic NMDARs in the neocortex are both auto- and hetero-receptors, *Neuroreport*, 7, 2773, 1996.
30. Dudel, J. and Kuffler, S.W., Presynaptic inhibition at the crayfish neuromuscular junction, *J. Physiol.*, 155, 543, 1961.
31. Duguid, I.C. et al., Somatodendritic release of glutamate regulates synaptic inhibition in cerebellar Purkinje cells via autocrine mGluR1 activation, *J. Neurosci.*, 27, 12464, 2007.
32. Duguid, I.C. and Smart, T.G., Retrograde activation of presynaptic NMDARs enhances GABA release at cerebellar interneuron-Purkinje cell synapses, *Nat. Neurosci.*, 7, 525, 2004.
33. Eccles, J.C., Presynaptic inhibition in the spinal cord, *Prog. Brain Res.*, 12, 65, 1964.
34. Eccles, J.C., Schmidt, R., and Willis, W.D., Pharmacological studies on presynaptic inhibition, *J. Physiol.*, 168, 500, 1963.
35. Farb, C.R., Aoki, C., and Ledoux, J.E., Differential localization of NMDA and AMPA receptor subunits in the lateral and basal nuclei of the amygdala: a light and electron microscopic study, *J. Comp. Neurol.*, 362, 86, 1995.
36. Feng, B. et al., Structure-activity analysis of a novel NR2C/NR2D-preferring NMDAR antagonist: 1-(phenanthrene-2-carbonyl) piperazine-2,3-dicarboxylic acid, *Br. J. Pharmacol.*, 141, 508, 2004.
37. Fink, K., Bonisch, H., and Gothert, M., Presynaptic NMDARs stimulate noradrenaline release in the cerebral cortex, *Eur. J. Pharmacol.*, 185, 115, 1990.
38. Fischer, G. et al., Ro 25-6981, a highly potent and selective blocker of N-methyl-D-aspartate receptors containing the NR2B subunit. Characterization *in vitro*, *J. Pharmacol. Exp. Ther.*, 283, 1285, 1997.
39. Fiszman, M.L. et al., NMDARs increase the size of GABAergic terminals and enhance GABA release, *J. Neurosci.*, 25, 2024, 2005.
40. Gibbs, T.T., Russek, S.J., and Farb, D.H., Sulfated steroids as endogenous neuromodulators, *Pharmacol. Biochem. Behav.*, 84, 555, 2006.
41. Glitsch, M. and Marty, A., Presynaptic effects of NMDA in cerebellar Purkinje cells and interneurons, *J. Neurosci.*, 19, 511, 1999.
42. Gracy, K.N. and Pickel, V.M., Ultrastructural immunocytochemical localization of the N-methyl-D-aspartate receptor and tyrosine hydroxylase in the shell of the rat nucleus accumbens, *Brain Res.*, 739, 169, 1996.
43. Hamada, T. et al., NMDA induced glutamate release from the suprachiasmatic nucleus: an *in vitro* study in the rat, *Neurosci. Lett.*, 256, 93, 1998.
44. Hatton, C.J. and Paoletti, P., Modulation of triheteromeric NMDARs by N-terminal domain ligands, *Neuron*, 46, 261, 2005.
45. Herkert, M., Rottger, S., and Becker, C.M., The NMDAR subunit NR2B of neonatal rat brain: complex formation and enrichment in axonal growth cones, *Eur. J. Neurosci.*, 10, 1553, 1998.
46. Huang, H. and Bordey, A., Glial glutamate transporters limit spillover activation of presynaptic NMDARs and influence synaptic inhibition of Purkinje neurons, *J. Neurosci.*, 24, 5659, 2004.
47. Humeau, Y. et al., Presynaptic induction of heterosynaptic associative plasticity in the mammalian brain, *Nature*, 426, 841, 2003.
48. Ito, M., Cerebellar long-term depression: characterization, signal transduction, and functional roles, *Physiol. Rev.*, 81, 1143, 2001.
49. Ito, M., The molecular organization of cerebellar long-term depression, *Nat. Rev. Neurosci.*, 3, 896, 2002.
50. Johnson, K.M. and Jeng, Y.J., Pharmacological evidence for N-methyl-D-aspartate receptors on nigrostriatal dopaminergic nerve terminals, *Can. J. Physiol. Pharmacol.*, 69, 1416, 1991.

51. Jourdain, P. et al., Glutamate exocytosis from astrocytes controls synaptic strength, *Nat. Neurosci.*, 10, 331, 2007.
52. Kandel, E.R. and Schwartz, J.H., Molecular biology of learning: modulation of transmitter release, *Science*, 218, 433, 1982.
53. Khakh, B.S. and Henderson, G., Modulation of fast synaptic transmission by presynaptic ligand-gated cation channels, *J. Auton. Nerv. Syst.*, 81, 110, 2000.
54. Kimoto, T. et al., Neurosteroid synthesis by cytochrome p450-containing systems localized in the rat brain hippocampal neurons: N-methyl-D-aspartate and calcium-dependent synthesis, *Endocrinology*, 142, 3578, 2001.
55. Kirischuk, S., Clements, J.D., and Grantyn, R., Presynaptic and postsynaptic mechanisms underlie paired pulse depression at single GABAergic boutons in rat collicular cultures, *J. Physiol.*, 543, 99, 2002.
56. Krebs, M.O. et al., Glutamatergic control of dopamine release in the rat striatum: evidence for presynaptic N-methyl-D-aspartate receptors on dopaminergic nerve terminals, *J. Neurochem.*, 56, 81, 1991.
57. Kullmann, D.M. and Asztely, F., Extrasynaptic glutamate spillover in the hippocampus: evidence and implications, *Trends Neurosci.*, 21, 8, 1998.
58. Langer, S.Z., Presynaptic autoreceptors regulating transmitter release, *Neurochem. Inc.*, 52, 26, 2008.
59. Lev-Ram, V. et al., Synergies and coincidence requirements between NO, cGMP, and Ca2+ in the induction of cerebellar long-term depression, *Neuron*, 18, 1025, 1997.
60. Levenes, C., Daniel, H., and Crepel, F., Retrograde modulation of transmitter release by postsynaptic subtype 1 metabotropic glutamate receptors in the rat cerebellum, *J. Physiol.*, 537, 125, 2001.
61. Lien, C.C. et al., Visual stimuli-induced LTD of GABAergic synapses mediated by presynaptic NMDARs, *Nat. Neurosci.*, 9, 372, 2006.
62. Liu, H. et al., Evidence for presynaptic N-methyl-D-aspartate autoreceptors in the spinal cord dorsal horn, *Proc. Natl. Acad. Sci. USA*, 91, 8383, 1994.
63. Liu, L. et al., Role of NMDAR subtypes in governing the direction of hippocampal synaptic plasticity, *Science*, 304, 1021, 2004.
64. Liu, S.J. and Lachamp, P., The activation of excitatory glutamate receptors evokes a long-lasting increase in the release of GABA from cerebellar stellate cells, *J. Neurosci.*, 26, 9332, 2006.
65. Ludwig, M. and Pittman, Q.J., Talking back: dendritic neurotransmitter release, *Trends Neurosci.*, 26, 255, 2003.
66. Luscher, C. et al., Synaptic plasticity and dynamic modulation of the postsynaptic membrane, *Nat. Neurosci.*, 3, 545, 2000.
67. MacDermott, A.B., Role, L.W., and Siegelbaum, S.A., Presynaptic ionotropic receptors and the control of transmitter release, *Annu. Rev. Neurosci.*, 22, 443, 1999.
68. Malayev, A., Gibbs, T.T., and Farb, D.H., Inhibition of the NMDA response by pregnenolone sulphate reveals subtype selective modulation of NMDARs by sulphated steroids, *Br. J. Pharmacol.*, 135, 901, 2002.
69. Mameli, M. et al., Neurosteroid-induced plasticity of immature synapses via retrograde modulation of presynaptic NMDARs, *J. Neurosci.*, 25, 2285, 2005.
70. Marcaggi, P. and Attwell, D., Short- and long-term depression of rat cerebellar parallel fibre synaptic transmission mediated by synaptic crosstalk, *J. Physiol.*, 578, 545, 2007.
71. Martin, D. et al., Autoreceptor regulation of glutamate and aspartate release from slices of the hippocampal CA1 area, *J. Neurochem.*, 56, 1647, 1991.
72. Mayer, M.L. and Armstrong, N., Structure and function of glutamate receptor ion channels, *Annu. Rev. Physiol.*, 66, 161, 2004.
73. McBain, C.J. and Mayer, M.L., N-methyl-D-aspartic acid receptor structure and function, *Physiol. Rev.*, 74, 723, 1994.

74. Mellon, S.H. and Vaudry, H., Biosynthesis of neurosteroids and regulation of their synthesis, *Int. Rev. Neurobiol.*, 46, 33, 2001.
75. Mott, D.D. et al., Phenylethanolamines inhibit NMDARs by enhancing proton inhibition, *Nat. Neurosci.*, 1, 659, 1998.
76. Okabe, S. et al., Hippocampal synaptic plasticity in mice overexpressing an embryonic subunit of the NMDAR, *J. Neurosci.*, 18, 4177, 1998.
77. Ozawa, S., Role of glutamate transporters in excitatory synapses in cerebellar Purkinje cells, *Brain Nerve*, 59, 669, 2007.
78. Ozawa, S., Kamiya, H., and Tsuzuki, K., Glutamate receptors in the mammalian central nervous system, *Prog. Neurobiol.*, 54, 581, 1998.
79. Paoletti, P., Ascher, P., and Neyton, J., High-affinity zinc inhibition of NMDA NR1-NR2A receptors, *J. Neurosci.*, 17, 5711, 1997.
80. Paoletti, P. and Neyton, J., NMDAR subunits: function and pharmacology, *Curr. Opin. Pharmacol.*, 7, 39, 2007.
81. Paquet, M. and Smith, Y., Presynaptic NMDAR subunit immunoreactivity in GABAergic terminals in rat brain, *J. Comp. Neurol.*, 423, 330, 2000.
82. Petralia, R.S., Wang, Y.X., and Wenthold, R.J., The NMDAR subunits NR2A and NR2B show histological and ultrastructural localization patterns similar to those of NR1, *J. Neurosci.*, 14, 6102, 1994.
83. Petralia, R.S., Yokotani, N., and Wenthold, R.J., Light and electron microscope distribution of the NMDAR subunit NMDAR1 in the rat nervous system using a selective anti-peptide antibody, *J. Neurosci.*, 14, 667, 1994.
84. Pina-Crespo, J.C. and Gibb, A.J., Subtypes of NMDARs in new-born rat hippocampal granule cells, *J. Physiol.*, 541, 41, 2002.
85. Piochon, C. et al., NMDAR contribution to the climbing fiber response in the adult mouse Purkinje cell, *J. Neurosci.*, 27, 10797, 2007.
86. Pittaluga, A. and Raiteri, M., N-methyl-D-aspartic acid (NMDA) and non-NMDARs regulating hippocampal norepinephrine release. I. Location on axon terminals and pharmacological characterization, *J. Pharmacol. Exp. Ther.*, 260, 232, 1992.
87. Quinlan, E.M. et al., Rapid, experience-dependent expression of synaptic NMDARs in visual cortex in vivo, *Nat. Neurosci.*, 2, 352, 1999.
88. Rachline, J. et al., The micromolar zinc-binding domain on the NMDAR subunit NR2B, *J. Neurosci.*, 25, 308, 2005.
89. Raditsch, M. et al., Subunit-specific block of cloned NMDARs by argiotoxin-636, *FEBS Lett.*, 324, 63, 1993.
90. Raiteri, M., Functional pharmacology in human brain, *Pharmacol. Rev.*, 58, 162, 2006.
91. Renzi, M., Farrant, M., and Cull-Candy, S.G., Climbing-fibre activation of NMDARs in Purkinje cells of adult mice, *J. Physiol.*, 585, 91, 2007.
92. Robert, A., Black, J.A., and Waxman, S.G., Endogenous NMDA-receptor activation regulates glutamate release in cultured spinal neurons, *J. Neurophysiol.*, 80, 196, 1998.
93. Sasaki, Y.F. et al., Characterization and comparison of the NR3A subunit of the NMDAR in recombinant systems and primary cortical neurons, *J. Neurophysiol.*, 87, 2052, 2002.
94. Schlicker, E. and Kathmann, M., Modulation of transmitter release via presynaptic cannabinoid receptors, *Trends Pharmacol. Sci.*, 22, 565, 2001.
95. Shigeri, Y. et al., Effects of threo-beta-hydroxyaspartate derivatives on excitatory amino acid transporters (EAAT4 and EAAT5), *J. Neurochem.*, 79, 297, 2001.
96. Shimamoto, K. et al., DL-threo-beta-benzyloxyaspartate, a potent blocker of excitatory amino acid transporters, *Mol. Pharmacol.*, 53, 195, 1998.
97. Shin, J.H. and Linden, D.J., An NMDAR/nitric oxide cascade is involved in cerebellar LTD but is not localized to the parallel fiber terminal, *J. Neurophysiol.*, 94, 4281, 2005.

98. Siegel, S.J. et al., Regional, cellular, and ultrastructural distribution of N-methyl-D-aspartate receptor subunit 1 in monkey hippocampus, *Proc. Natl. Acad. Sci. USA*, 91, 564, 1994.
99. Sjostrom, P.J., Turrigiano, G.G., and Nelson, S.B., Neocortical LTD via coincident activation of presynaptic NMDA and cannabinoid receptors, *Neuron*, 39, 641, 2003.
100. Smart, T.G., Hosie, A.M., and Miller, P.S., Zn^{2+} ions: modulators of excitatory and inhibitory synaptic activity, *Neuroscientist*, 10, 432, 2004.
101. Smart, T.G., Xie, X., and Krishek, B.J., Modulation of inhibitory and excitatory amino acid receptor ion channels by zinc, *Prog. Neurobiol.*, 42, 393, 1994.
102. Stevens, C.F., Neurotransmitter release at central synapses, *Neuron*, 40, 381, 2003.
103. Stevens, C.F., Presynaptic function, *Curr. Opin. Neurobiol.*, 14, 341, 2004.
104. Suarez, L.M. and Solis, J.M., Taurine potentiates presynaptic NMDARs in hippocampal Schaffer collateral axons, *Eur. J. Neurosci.*, 24, 405, 2006.
105. Sudhof, T.C., The synaptic vesicle cycle, *Annu. Rev. Neurosci.*, 27, 509, 2004.
106. Tovar, K.R. and Westbrook, G.L., The incorporation of NMDARs with a distinct subunit composition at nascent hippocampal synapses in vitro, *J. Neurosci.*, 19, 4180, 1999.
107. Traynelis, S.F. et al., Control of voltage-independent zinc inhibition of NMDARs by the NR1 subunit, *J. Neurosci.*, 18, 6163, 1998.
108. Tzingounis, A.V. and Wadiche, J.I., Glutamate transporters: confining runaway excitation by shaping synaptic transmission, *Nat. Rev. Neurosci.*, 8, 935, 2007.
109. Vesselkin, N.P. et al., Ultrastructural study of glutamate- and GABA-immunoreactive terminals contacting the primary afferent fibers in frog spinal cord. A double postembedding immunocytochemical study, *Brain Res.*, 960, 267, 2003.
110. Vorobjev, V.S., Vibrodissociation of sliced mammalian nervous tissue, *J. Neurosci. Meth.*, 38, 145, 1991.
111. Wang, H. and Pickel, V.M., Presence of NMDA-type glutamate receptors in cingulate corticostriatal terminals and their postsynaptic targets, *Synapse*, 35, 300, 2000.
112. Wang, J.K., Presynaptic glutamate receptors modulate dopamine release from striatal synaptosomes, *J. Neurochem.*, 57, 819, 1991.
113. Wenzel, A. et al., N-methyl-D-aspartate receptors containing the NR2D subunit in the retina are selectively expressed in rod bipolar cells, *Neuroscience*, 78, 1105, 1997.
114. Wenzel, A. et al., Distribution of NMDAR subunit proteins NR2A, 2B, 2C and 2D in rat brain, *NeuroReport*, 7, 45, 1995.
115. Wenzel, A. et al., Developmental and regional expression of NMDAR subtypes containing the NR2D subunit in rat brain, *J. Neurochem.*, 66, 1240, 1996.
116. Williams, K., Ifenprodil discriminates subtypes of N-methyl-D-aspartate receptor: selectivity and mechanisms at recombinant heteromeric receptors, *Mol. Pharmacol.*, 44, 851, 1993.
117. Woodhall, G. et al., NR2B-containing NMDA autoreceptors at synapses on entorhinal cortical neurons, *J. Neurophysiol.*, 86, 1644, 2001.
118. Yamakura, T. et al., Different sensitivities of NMDAR channel subtypes to non-competitive antagonists, *NeuroReport*, 4, 687, 1993.
119. Yamakura, T. and Shimoji, K., Subunit- and site-specific pharmacology of the NMDAR channel, *Prog. Neurobiol.*, 59, 279, 1999.
120. Yang, J., Woodhall, G.L., and Jones, R.S., Tonic facilitation of glutamate release by presynaptic NR2B-containing NMDARs is increased in the entorhinal cortex of chronically epileptic rats, *J. Neurosci.*, 26, 406, 2006.
121. Zilberter, Y., Dendritic release of glutamate suppresses synaptic inhibition of pyramidal neurons in rat neocortex, *J. Physiol.*, 528, 489, 2000.

Index

T - #0362 - 071024 - C4 - 234/156/16 - PB - 9780367386528 - Gloss Lamination